The Antarctic Paleoenvironment: A Perspective on Global Change

Part Two

American Geophysical Union

ANTARCTIC
RESEARCH
SERIES

Physical Sciences

ANTARCTIC OCEANOLOGY
    Joseph L. Reid, *Editor*
ANTARCTIC OCEANOLOGY II: THE AUSTRALIAN-
NEW ZEALAND SECTOR
    Dennis E. Hayes, *Editor*

ANTARCTIC SNOW AND ICE STUDIES
    Malcolm Mellor, *Editor*
ANTARCTIC SNOW AND ICE STUDIES II
    A. P. Crary, *Editor*

ANTARCTIC SOILS AND SOIL FORMING PROCESSES
    J. C. F. Tedrow, *Editor*
DRY VALLEY DRILLING PROJECT
    L. D. McGinnis, *Editor*
GEOLOGICAL INVESTIGATIONS IN NORTHERN
VICTORIA LAND
    Edmund Stump, *Editor*
GEOLOGY AND PALEONTOLOGY OF THE ANTARCTIC
    Jarvis B. Hadley, *Editor*
GEOLOGY OF THE CENTRAL TRANSANTARCTIC
MOUNTAINS
    Mort D. Turner and John F. Splettstoesser,
    Editors
GEOMAGNETISM AND AERONOMY
    A. H. Waynick, *Editor*
METEOROLOGICAL STUDIES AT PLATEAU STATION,
ANTARCTICA
    Joost A. Businger, *Editor*
OCEANOLOGY OF THE ANTARCTIC CONTINENTAL SHELF
    Stanley S. Jacobs, *Editor*
STUDIES IN ANTARCTIC METEOROLOGY
    Morton J. Rubin, *Editor*
UPPER ATMOSPHERE RESEARCH IN ANTARCTICA
    L. J. Lanzerotti and C. G. Park, *Editors*
THE ROSS ICE SHELF: GLACIOLOGY AND GEOPHYSICS
    C. R. Bentley and D. E. Hayes, *Editors*
VOLCANOES OF THE ANTARCTIC PLATE AND SOUTHERN
OCEANS
    W. E. LeMasurier and J. T. Thomson, *Editors*
MINERAL RESOURCES POTENTIAL OF ANTARCTICA
    John F. Splettstoesser and Gisela A. M. Dreschhoff,
    Editors
MARINE GEOLOGICAL AND GEOPHYSICAL ATLAS
OF THE CIRCUM-ANTARCTIC TO 30°S
    Dennis E. Hayes, *Editor*
MOLLUSCAN SYSTEMATICS AND BIOSTRATIGRAPHY
    Jeffrey D. Stilwell and William J. Zinsmeister
THE ANTARCTIC PALEOENVIRONMENT: A PERSPECTIVE
ON GLOBAL CHANGE
    James P. Kennett and Detlef A. Warnke, *Editors*

CONTRIBUTIONS TO ANTARCTIC RESEARCH I
    David H. Elliot, *Editor*
CONTRIBUTIONS TO ANTARCTIC RESEARCH II
    David H. Elliot, *Editor*
CONTRIBUTIONS TO ANTARCTIC RESEARCH III
    David H. Elliot, *Editor*
PHYSICAL AND BIOGEOCHEMICAL PROCESSES
IN ANTARCTIC LAKES
    William J. Green and E. Imre Friedmann, *Editors*

American Geophysical Union

ANTARCTIC RESEARCH SERIES

Volume 60

ANTARCTIC
RESEARCH
SERIES

The Antarctic Paleoenvironment: A Perspective on Global Change

Part Two

James P. Kennett
Detlef A. Warnke

Editors

American Geophysical Union
Washington, D.C.
1993

Volume 60

**ANTARCTIC
RESEARCH
SERIES**

Library of Congress Cataloging-in-Publication Data
(Revised for vol. 2)

The Antarctic paleoenvironment.

(Antarctic research series, 0066-4634 ; v. 56, 60)
Papers from a conference held at the University of California, Santa Barbara, Aug.
28–31, 1991.
Includes bibliographical references.
1. Paleogeography—Antarctic regions—Congresses. 2. Paleocology—Antarctic
regions—Congresses.
I. Kennett, James P.   II. Warnke, Detlef A.
QE501.4.P3A64   1992   560'.45'09989   92-37312
ISBN 0-87590-823-3 (pt. 1)
ISBN 0-87590-838-1 (pt. 2)
ISSN 0066-4634

Published by
American Geophysical Union

Printed in the United States of America.

CONTENTS

The Antarctic Research Series:
STATEMENT OF OBJECTIVES

The Antarctic Research Series provides for the presentation of detailed scientific research results from Antarctica, particularly the results of the United States Antarctic Research Program, including monographs and long manuscripts.

The series is designed to make the results of Antarctic fieldwork available. The Antarctic Research Series encourages the collection of papers on specific geographic areas within Antarctica. In addition, many volumes focus on particular disciplines, including marine biology, oceanology, meteorology, upper atmosphere physics, terrestrial biology, geology, glaciology, human adaptability, engineering, and environmental protection.

Topical volumes in the series normally are devoted to papers in one or two disciplines. Multidisciplinary volumes, initiated in 1990 to enable more rapid publication, are open to papers from any discipline. The series can accommodate long manuscripts and utilize special formats, such as maps.

Priorities for publication are set by the Board of Associate Editors. Preference is given to research manuscripts from projects funded by U.S. agencies. Because the series serves to emphasize the U.S. Antarctic Research Program, it also performs a function similar to expedition reports of many other countries with national Antarctic research programs.

The standards of scientific excellence expected for the series are maintained by the review criteria established for the AGU publications program. Each paper is critically reviewed by two or more expert referees. A member of the Board of Associate Editors may serve as editor of a volume, or another person may be appointed. The Board works with the individual editors of each volume and with the AGU staff to assure that the objectives of the series are met, that the best possible papers are presented, and that publication is timely.

Proposals for volumes or papers offered should be sent to the Board of Associate Editors, Antarctic Research Series, at 2000 Florida Avenue, N.W., Washington, D.C. 20009. Publication of the series is partially supported by a grant from the National Science Foundation.

Board of Associate Editors
Antarctic Research Series

PREFACE

The Antarctic continent and the surrounding Southern Ocean represent one of the major climate engines of the Earth: coupled components critical in the Earth's environmental system. The contributions in this volume help with the understanding of the long-term evolution of Antarctica's environment and biota. The aim of this and the preceding companion volume is to help place the modern system within a historical context.

The environment and biosphere of the Antarctic region have undergone dynamic changes through geologic time. These, in turn, have played a key role in long-term global paleoenvironmental evolution. The development of the Southern Ocean itself, resulting from plate tectonism, created first-order changes in the circulation of the global ocean, in turn affecting meridional heat transport and hence global climates. Biospheric changes responded to the changing oceanic climatic states. Comprehension of the climatic and oceanographic processes that have operated at various times in Antarctica's history is crucial to the understanding of the present-day global environmental system. This knowledge will become increasingly important in parallel with concerns about anthropogenically caused global change. How vulnerable is the Antarctic region, especially its ice sheets, to global warming? The question is not parochial, given the potential of sea level change resulting from any Antarctic cryospheric development. Conversely, how much of a role does the Antarctic region, this giant icebox, play in moderating global, including sea level, change?

This is the second of two volumes in the American Geophysical Union's Antarctic Research Series to present contributions that deal with the paleoenvironmental and biotic evolution of the Antarctic region. The papers are based on work presented at a conference held at the University of California, Santa Barbara, August 28–31, 1991, entitled "The Role of the Southern Ocean and Antarctica in Global Change: An Ocean Drilling Perspective." This conference, jointly sponsored by JOI/USSAC and the Division of Polar Programs, National Science Foundation, was attended by more than 100 scientists from around the world. The primary objectives of the meeting were successful in providing a forum (1) to summarize existing paleoenvironmental data from the Antarctic region; (2) to identify and debate major remaining questions, most of which are thematic in nature; (3) to assist in formulating plans for future Antarctic ocean drilling; and (4) to organize publication of a series of summary/synthesis papers leading to two volumes. Although it has been the intention of the scientific community to produce summary or synthesis volumes of thematic or regional nature related to ocean drilling, few have yet been published. Therefore a major objective of this and the first volume is to help make the results of ocean drilling more widely available to the scientific community.

In addition to these volumes the conference also led to the production of a white paper, compiled by J. Kennett and J. Barron (available from JOI/USSAC, Washington, D.C.), that summarizes major remaining questions related to Southern Ocean paleoenvironmental evolution and outlines further ocean drilling required to assist in answering these questions. Selected material from the white paper was modified and incorporated in the introduction to the first volume.

This volume presents 13 papers of general and synthetic nature on a wide variety of topics related to the environmental and biotic evolution of the Antarctic and southern high-latitude oceans. The contributions incorporate a range of recent concepts that deal with the paleoclimatology, paleoceanography and paleobiogeography of the Antarctic region, especially in relation to the evolution of the continental cryosphere. The volume is organized so that the papers are presented in general order of geologic age, beginning with the Eocene and ending with the last several hundred years. As in the first volume, this arrangement was selected to help emphasize the evolution of the Antarctic environmental and biotic system during the late Phanerozoic. The subject is not without controversy, as shown by a number of the papers included in this volume. The stratigraphic records from the deep sea, continental margins, and land have been examined in these contributions at various stratigraphic resolutions from tens of millions of years to as high as several decades. A wide range of approaches have been employed, either singly or in combination, to decipher the paleoenvironmental record and include oxygen and carbon isotopes, microfossils, plant fossils, sediments, glacial morphology, and seismic stratigraphy.

James P. Kennett and Detlef A. Warnke

ACKNOWLEDGMENTS

A large number of workers have contributed much in providing the necessary reviews of the contributions published in this volume; we heartily thank you all: J. B. Anderson, R. A. Askin, P. J. Barrett, W. A. Berggren, G. W. Brass, L. H. Burckle, P. E. Calkin, P. F. Ciesielski, F. J. Davey, D. J. DeMaster, D. H. Elliot, D. M. Harwood, D. A. Hodell, N. de B. Hornibrook, B. T. Huber, G. Keller, D. E. Kellogg, L. A. Krissek, A. Leventer, H. Y. Ling, S. Locker, P. A. Mayewski, D. C. Mildenhall, T. C. Moore, Jr., C. Nigrini, S. B. O'Connell, M. L. Prentice, L. D. Stott, E. M. Truswell, D. A. Warnke, and J. D. Wright.

Publication of this volume was made possible by JOI/USSAC. We thank Ellen Kappel of JOI/USSAC for her unwavering support of this project, and also H. Zimmerman of the National Science Foundation for his support of the conference leading to this volume.

We also thank Diana M. Kennett, editorial assistant, for her major contributions toward the production of this volume and for her perseverance in keeping publication on schedule.

James P. Kennett and Detlef A. Warnke

THE ANTARCTIC PALEOENVIRONMENT: A PERSPECTIVE ON GLOBAL CHANGE

ANTARCTIC RESEARCH SERIES, VOLUME 60, PAGES 1–25

SOUTHERN OCEAN INFLUENCES ON LATE EOCENE TO MIOCENE DEEPWATER CIRCULATION

JAMES D. WRIGHT

Lamont-Doherty Earth Observatory, Palisades, New York 10964

KENNETH G. MILLER [1]

Rutgers University, Department of Geological Sciences, Piscataway, New Jersey 08903

The Eocene through the Miocene marked the transition from warm polar climates of the early Eocene to the development of near-modern climates and deepwater patterns by the late Miocene. We reconstructed deepwater circulation patterns for the late Eocene through the Miocene using $\delta^{13}C$, $\delta^{18}O$, and sediment distribution. The $\delta^{13}C$ reconstructions and unconformities/hiatuses indicate that the Southern Ocean was the dominant deepwater source for the late Eocene through the Miocene with intervals of increased deepwater production at 40, 36, 30, and 15 Ma. Seismic stratigraphic and carbon isotopic evidence exists for a pulse of Northern Component Water (NCW) production during the earliest Oligocene, indicating that there was bipolar production of deep and bottom waters. Two additional intervals of enhanced NCW production followed in the Miocene, ~20 to 16 Ma and ~12 to 10 Ma. Each of the three intervals of NCW production correlates with an interval of erosion followed by drift development in the deep North Atlantic. Carbon isotope evidence for the production of warm saline deepwater (WSDW) in the mid-latitudes could be found only for the early Miocene, although $\delta^{18}O$ records indicate WSDW production during the late Eocene and early Oligocene.

INTRODUCTION

Warm salty water from the middle latitudes competes with cold, relatively fresh water from the high latitudes to fill the deep ocean basins. In today's ocean, high-latitude sources are dominant because the near-freezing polar conditions produce a denser in situ water mass than the mid-latitude evaporative regions. However, this may not have been true for ancient oceans. *Chamberlin* [1906] and *Brass et al.* [1982] postulated that warmer, saltier water from mid-latitude regions may have been the most dense water mass when polar regions were much warmer (e.g., the Cretaceous and early Eocene), thereby producing a warm deep-ocean circulation in contrast to today's cold deepwater regime.

Faunal, floral, and oxygen isotopic evidence shows that early Eocene high-latitude climates were unusually warm, making the early Eocene the most probable time during the Cenozoic for a dominance of middle latitude sources of warm saline deep water (WSDW) [e.g., *Haq*, 1981]. The warm high-latitude climates were associated with reduced meridional thermal gradients. *Shackleton*

and Boersma [1981] estimated that Eocene sea surface temperature (SST) gradients from the low to high latitudes were approximately one half of the modern meridional gradient. Their planktonic foraminiferal $\delta^{18}O$ reconstruction highlighted two important differences between Eocene and modern surface waters: (1) Eocene equatorial SSTs may have been ~4°C cooler than at present, and (2) Eocene polar SSTs were 10°–15°C warmer than modern polar SSTs (Figure 1a). These estimates have become the cornerstone of modeling efforts that attempt to simulate unusually warm climates at high latitudes [e.g., *Barron*, 1983, 1987; *Manabe and Bryan*, 1985; *Rind*, 1987; *Covey and Barron*, 1988; *Covey and Thompson*, 1989; *Manabe et al.*, 1990; *Rind and Chandler*, 1991].

Paleotemperature estimates from foraminifera require estimates of the $\delta^{18}O_{water}$ in which the foraminifera lived. *Shackleton and Boersma* [1981] estimated Eocene SSTs using a uniform $\delta^{18}O_{water}$ value equal to the estimated global ocean $\delta^{18}O_{water}$ value. While these values are valid for the deepwater estimates, surface waters will have had different values because net evaporation enriches surface waters in ^{18}O in the low to middle latitudes and net precipitation depletes high-latitude surface waters with respect to ^{18}O. Modern surface water $\delta^{18}O_{water}$ values in tropical and subtropi-

---

[1] Also at Lamont-Doherty Earth Observatory, Palisades, New York. 10964.

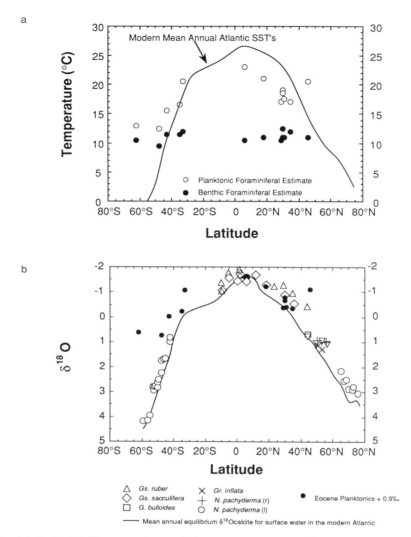

Fig. 1. (a) Estimated Eocene surface and deepwater temperatures from planktonic and benthic foraminiferal $\delta^{18}O$ values [*Shackleton and Boersma*, 1981] relative to the modern mean annual SSTs for the Atlantic Ocean. *Shackleton and Boersma* [1981] used a mean ocean $\delta^{18}O$ of seawater difference of ~0.9‰ between the modern ocean and an ice-free Eocene in their paleotemperature equation. (b) Comparison of the modern planktonic $\delta^{18}O$ profile in the Atlantic (R. G. Fairbanks et al., work in preparation) with the Eocene values from *Shackleton and Boersma* [1981]. We added a constant value of 0.9‰ to the Eocene values [*Shackleton and Kennett*, 1975] to account for the difference between mean ocean $\delta^{18}O$ values during the ice-free Eocene and the present. The Eocene $\delta^{18}O$ gradient is remarkably similar to the modern gradient from 40°N to 40°S. The observation of cooler Eocene tropical temperatures is not supported by this comparison. However, the middle to high latitudes were remarkably different showing a >3‰ difference at 60°S.

cal regions are higher than the mean ocean value by approximately 1‰ [*Craig and Gordon*, 1965]. This difference suggests that Shackleton and Boersma's equatorial estimates may not be correct. A paleotemperature estimate without consideration of $\delta^{18}O_{water}$ variations due to evaporation will produce cooler temperature estimates, and therefore the Eocene SST estimates may have been as much as 4°C too cold. To avoid this problem, we compared *Shackleton and Boersma*'s

[1981] Eocene planktonic foraminiferal $\delta^{18}O$ profile with a modern Atlantic profile (R. G. Fairbanks et al., work in preparation). Both profiles are similar for the low- to mid-latitude regions when ice volume differences between an ice-free Eocene and modern ice volume are taken into account (Figure 1b) (R. G. Fairbanks et al., work in preparation). This comparison suggests that Eocene equatorial SSTs may have been closer to the modern average of 28°C than the 24°C estimated by

Shackleton and Boersma [1981], although more detailed reconstructions show cooler tropical SSTs for certain intervals during the Eocene [*Zachos et al.*, 1993]. With regard to the high-latitudes SSTs, Shackleton and Boersma's comparison (Figure 1) shows that polar surface waters were 10°–15°C warmer during the early Eocene than at present (Figure 1), which is consistent with the conclusions of other studies [e.g., *Shackleton and Kennett*, 1975; *Savin et al.*, 1975; *Shackleton and Boersma*, 1981; *Kennett and Stott*, 1990; *Stott et al.*, 1990; *Barrera and Huber*, 1991; *Zachos et al.*, 1993].

Sea surface temperatures play a direct role in modulating deepwater formation and, hence, govern deepwater temperatures. Polar SST fluctuations are transmitted directly to the deep ocean during intervals when deep water is formed predominantly in polar regions. However, if polar SSTs warm substantially, then the deep waters may acquire properties derived from mid-latitude sources. This may have been the case for the early Eocene [*Kennett and Stott*, 1990; *Pak and Miller*, 1992], when deepwater temperatures were warmest, reaching temperatures of 10°–14°C (*Shackleton and Kennett* [1975], *Savin et al.* [1975], and *Miller et al.* [1987]; Figure 1). From this maximum, both deep and polar surface water temperatures cooled by ~9°C to 2°–4°C by the early Oligocene [*Shackleton and Kennett*, 1975; *Savin et al.*, 1975; *Miller et al.*, 1987; *Miller*, 1992]. This cooling was recorded in benthic and high-latitude planktonic foraminiferal $\delta^{18}O$ values, which increased by ~3.0‰ during this interval with a final increase of ~1.0‰ at the Eocene/Oligocene boundary (Figure 2).

Kennett and Shackleton [1976] proposed that a 1.0‰ $\delta^{18}O$ increase in the benthic foraminifera near the Eocene/Oligocene boundary represented the development of psychrospheric conditions or ventilation of the deep ocean by cold deepwater masses (i.e., thermohaline conditions similar to the present). Another important development was the discovery by *Kennett and Stott* [1990] that late Eocene and early Oligocene benthic foraminiferal $\delta^{18}O$ values on Maud Rise were inverted in relation to modern gradients; high values occurred at intermediate depths (~1500 m), while lower values occurred at greater depths (~2200 m) (Figure 2). Largely on the basis of these $\delta^{18}O$ records, *Kennett and Stott* [1990] proposed that there were three general stages of Cenozoic deepwater circulation:

1. Eocene and possibly Paleocene deepwater circulation was driven by halothermal processes or deepwater convection based largely on salinity-induced density differences. This halothermal circulation was designated the Proteus Ocean, and it consisted of warm saline deepwater overlain by a colder, fresher water mass. WSDW originated in low-latitude regions (Tethyan) where evaporation exceeded precipitation and runoff, creating the densest water mass. A colder, fresher water mass originated in the Southern Ocean, but it penetrated only to intermediate depths because of its lower salinity.

2. Oligocene deepwater circulation (Proto-Oceanus) represented the transition between the halothermal Proteus and thermohaline Oceanus (Neogene). This ocean consisted of three components; two cold, relatively fresh water masses that originated in the high southern latitudes which were separated by a warm saline layer. The deepest water mass was analogous to Antarctic Bottom Water (AABW) and may have been aided by the formation of sea ice around Antarctica.

3. The Neogene deep oceans were ventilated by the thermohaline circulation that resulted from freezing conditions in the high latitudes. There was little deepwater (i.e., below 2 km) ventilation by WSDW production in this psychrospheric ocean designated as Oceanus.

The *Kennett and Stott* [1990] scenario has stimulated much discussion in the paleoceanographic community. In addition to the stable isotopic data from Maud Rise, there is evidence from sediment distributions that broadly supports many of their conclusions. For example, lower and middle Eocene unconformities are rare [*Moore et al.*, 1978], and the conformable nature of seismic reflectors indicates that deepwater circulation was generally sluggish during the early and middle Eocene and possibly in the late Eocene [*Tucholke and Mountain*, 1979, 1986; *Miller and Tucholke*, 1983; *Mountain and Miller*, 1992]. Such a sluggish circulation is consistent with low-temperature gradients with respect to latitude and depth. These lower-temperature gradients would result in low-density gradients and low ocean turnover times relative to the colder intervals which followed. In addition, the calcite compensation depth (CCD) was relatively shallow during the Eocene [*van Andel*, 1975], and the surface to deep $\delta^{13}C$ differences were unusually large [*Boersma et al.*, 1987], which are consistent with slow ventilation of the deep oceans during the Eocene and/or unusually high mean ocean nutrient levels. This period of relative quiescence during the early and middle Eocene ended with an interval of more vigorous deepwater production in the Southern Ocean and the North Atlantic that caused widespread erosion during the late Eocene and earliest Oligocene [*Kennett*, 1977; *Miller and Tucholke*, 1983].

We review the evidence for late Eocene to Miocene deepwater circulation changes in view of the *Kennett and Stott* [1990] hypothesis, focusing on oxygen and carbon isotopic records and sediment distribution. We relate deepwater circulation changes to tectonic and global climate reorganizations that occurred from the late Eocene to the Miocene. This interval is critical to our understanding of climate change because (1) it represents the transition from the warm Eocene climates to the northern hemisphere glacial/interglacial cycles in the Pliocene, (2) it contains the first definitive evidence for Antarctic ice sheet growth, and (3) it

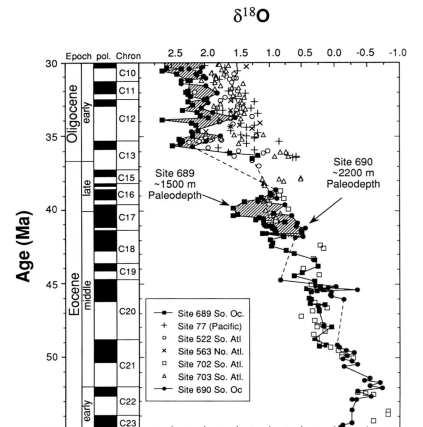

Fig. 2.   Late Paleocene to early Oligocene *Cibicidoides* δ^{18}O records for Ocean Drilling Program (ODP) sites 689 and 690 on Maud Rise [*Kennett and Stott*, 1990] compared to *Cibicidoides* δ^{18}O records from other regions (Site 77 [*Keigwin and Keller*, 1984], Site 522 [*Miller et al.*, 1988], Site 563 [*Miller and Fairbanks*, 1985], Site 702 [*Katz and Miller*, 1992], and Site 703 [*Miller*, 1992]). Late Eocene to early Oligocene paleodepths for sites 689 and 690 are estimated to have been ~1500 m and ~2200 m, respectively. The hachured areas highlight two intervals when the deeper site on Maud Rise (Site 690) recorded lower δ^{18}O values by 0.5 to 1.0‰. Because temperature dominates the δ^{18}O$_{calcite}$ in the modern ocean, deep-ocean *Cibicidoides* δ^{18}O values are expected to increase with depth. However, warmer, but more salty water may become more dense during warm climate intervals. Therefore *Kennett and Stott* [1990] interpreted this "δ^{18}O inversion" as reflecting a warm saline deepwater mass. It is important to note that the deeper site (Site 690) recorded values more similar to mean deepwater conditions than the site at intermediate depths (Site 689).

represents a fundamental change in deepwater circulation from a warm to a cold system.

DEEPWATER CIRCULATION PATTERNS

Carbon Isotope Reconstructions

Carbon isotopes provide a powerful tracer for reconstruction of deepwater circulation patterns. The GEOSECS transects are the foundation for using δ^{13}C distributions to record changes in deepwater circulation. *Kroopnick* [1985] documented that the distribution of δ^{13}C in oceanic dissolved bicarbonate accurately reflects modern deepwater circulation patterns (Figure 3).

Two processes modify deepwater δ^{13}C values once a water mass is isolated from the surface ocean and the atmosphere: (1) mixing with a water mass of different isotopic composition and (2) the accumulation of oxidized organic matter, commonly referred to as "deepwater aging." In the modern Atlantic Ocean, δ^{13}C changes predominantly reflect mixing between North Atlantic Deep Water (NADW) with high δ^{13}C values (1.0‰) and Southern Component Water (SCW) with lower δ^{13}C values (0.4‰) (Figure 3). The residence time of deep water within the Atlantic is short (<350 years), and overlying surface productivity is low, thereby limiting any significant accumulation of oxidized organic

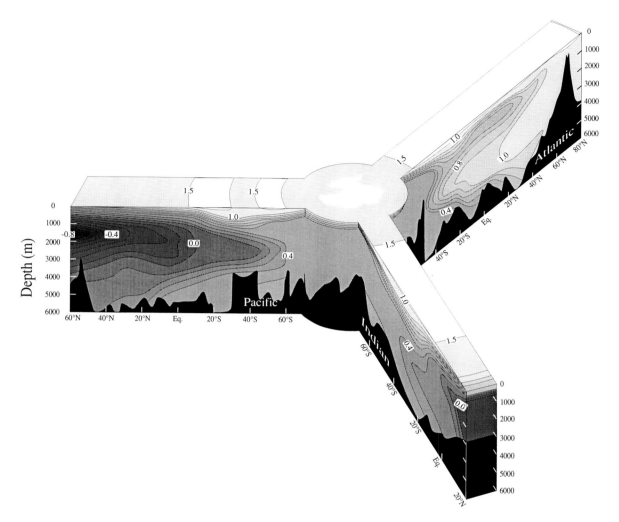

Fig. 3. The distribution of $\delta^{13}C$ in the world's ocean is a reflection of modern deepwater circulation patterns (adapted from *Kroopnick* [1985]). Pacific Ocean $\delta^{13}C$ values approximate the mean ocean water $\delta^{13}C$ value because of its large volume. Deep Atlantic $\delta^{13}C$ values are higher than those in the Pacific because the deep water ventilating the Atlantic, North Atlantic Deep Water (NADW), has a large surface water component. Surface waters are enriched in ^{13}C owing to biological activity which preferentially removes ^{12}C from the surface water. The Southern Ocean $\delta^{13}C$ value is dependent on the deepwater contribution from the Atlantic, Indian, and Pacific oceans [*Oppo and Fairbanks*, 1987]. At present the Southern Ocean is higher than the Pacific because of a high NADW flux. During the last glacial maximum, Pacific Ocean and Southern Ocean $\delta^{13}C$ values were much closer to each other because of the reduced NADW influence [*Oppo and Fairbanks*, 1987; *Charles and Fairbanks*, 1992].

matter within these basins [*Broecker*, 1979]. Conversely, Pacific Ocean and Indian Ocean $\delta^{13}C$ changes result from deepwater aging (Figure 3). Both oceans are ventilated from south to north by one deepwater mass, Southern Component Water. In the Pacific and Indian oceans, longer deepwater residence times (>1000 years for the Pacific [*Broecker et al.*, 1988; *Broecker*, 1989]), relatively high surface water productivity in the equatorial regions, and the lack of a northern deepwater counterpart produce a south to north $\delta^{13}C$ gradient due

to aging (Figure 3). In general, locations proximal to deepwater sources will record higher $\delta^{13}C$ values than more distal sites.

We have a paleorecord of these processes because certain benthic foraminiferal genera (*Cibicidoides and Planulina*) accurately record the $\delta^{13}C$ gradients of the deepwater masses [*Shackleton and Opdyke*, 1973; *Duplessy et al.*, 1970; *Belanger et al.*, 1981; *Graham et al.*, 1981]. Carbon isotope studies have provided detailed reconstructions of Pleistocene deepwater changes using both time series and time slice approaches [e.g., *Shackleton et al.*, 1983; *Boyle and Keigwin*, 1982, 1987; *Curry and Lohmann*, 1982; *Oppo and Fairbanks*, 1987; *Curry et al.*, 1988; *Duplessy et al.*, 1988; *Oppo et al.*, 1990]. We employ similar strategies for Eocene-Miocene reconstructions, using both time series and time slices. The time series approach requires relatively complete records from strategic locations and provides a chronology of deepwater circulation changes [e.g., *Oppo and Fairbanks*, 1987]. It is also necessary to use time slice reconstructions to determine the three-dimensional aspect of circulation changes (e.g., *Duplessy et al.* [1988] for the last glacial maximum and *Woodruff and Savin* [1989] and *Wright et al.* [1992] for the Miocene). The combination of both approaches provides a comprehensive evaluation of deepwater circulation patterns.

Late Eocene to Oligocene $\delta^{13}C$ reconstructions. The largest basin to basin $\delta^{13}C$ differences during the late Eocene to Oligocene were 0.5‰, which is one half of the modern difference (Atlantic-Pacific $\Delta\delta^{13}C$ is currently 1.0‰ (Figure 4)) [*Miller and Fairbanks*, 1985; *Boersma et al.*, 1987]. The low $\delta^{13}C$ differences can reflect (1) lower mean ocean nutrient levels during this interval relative to modern levels [*Boyle*, 1988], (2) a one-component deepwater circulation system, although this requires high ventilation rates in order to minimize aging effects, or (3) multiple sources with similar $\delta^{13}C$ values. While the oligotrophic nature and inferred low nutrients of the Oligocene oceans have been noted [e.g., *Miller and Fairbanks*, 1985; *Boersma et al.*, 1987], better quantification with nutrient proxies (such as Cd/Ca ratios in benthic foraminifera) is needed to substantiate this observation. Lower mean ocean nutrients during the late Eocene through the Oligocene may account for much of the low basin-basin $\delta^{13}C$ difference as also evidenced by low vertical (surface to deep) $\delta^{13}C$ gradients [*Miller and Fairbanks*, 1985; *Boersma et al.*, 1987]. Although the decreased sensitivity (dynamic range) of $\delta^{13}C$ as a tracer results from lower mean ocean nutrients, $\delta^{13}C$ reconstructions still provide information on deepwater circulation changes.

There are few suitable upper Eocene sections available for detailed stable isotopic studies; therefore our interpretation of deepwater circulation patterns based on $\delta^{13}C$ reconstructions is inconclusive for this interval. Available late Eocene $\delta^{13}C$ records show similar values in the major ocean basins, making it difficult to identify a deepwater source region (Figure 4). A three-dimensional representation of the late Eocene $\delta^{13}C$ patterns also shows that values were similar in all the major ocean basins (Figure 5). However, there is an intriguing hint of a deep to intermediate water source originating outside of the Southern Ocean during this interval. Carbon isotope values from Site 549 in the eastern North Atlantic were higher than those in the Southern Ocean from the late middle to early late Eocene (~41 to 38 Ma; Figure 4). (We use the geomag-

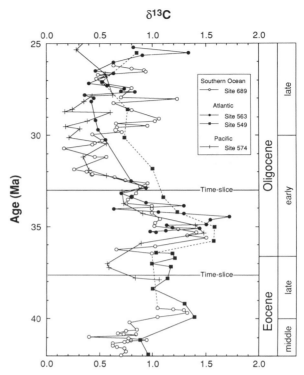

Fig. 4. Late Eocene to Oligocene $\delta^{13}C$ records from the Southern, Atlantic, and equatorial Pacific oceans (Site 689 [*Kennett and Stott*, 1990], Site 574 [*Miller and Thomas*, 1985], Site 563 [*Miller and Fairbanks*, 1985], and Site 549 [*Miller et al.*, 1985]). Basin to basin $\delta^{13}C$ differences were small (≤0.5‰) throughout this interval relative to the modern basin to basin differences (1.0). North Atlantic Site 549 (~2000-m paleodepth) recorded high $\delta^{13}C$ values relative to the Southern Ocean during the late Eocene to Oligocene, indicating its proximity to an intermediate water or deepwater source. This source may have been WSDW flowing out of the western Tethys or an upper limb of NCW. Time slice reconstructions shown in Figures 5 and 7 are marked.

netic polarity time scale (GPTS) of *Berggren et al.* [1985] throughout. Large changes in the Paleogene portions of the GPTS have been proposed [*Cande and Kent*, 1992], with some epoch boundaries changing by more than 2 m.y. Although this change will affect the absolute chronology of the events discussed herein, it will not alter their relative chronologies.) Although more data are needed from Site 549 to establish this difference, it implies that this region was proximal to a deep to intermediate source (late Eocene paleodepth estimates for Site 549 are ~2000 m [*Miller et al.*, 1985]). As a result, we can only infer that the high $\delta^{13}C$ values at Site 549 represent either a source of Northern Component Water (NCW) or western outflow of Tethyan water, but not the depth to which this water penetrated. The present core coverage is inadequate to distinguish between the two.

The Oligocene was dominated by Southern Compo-

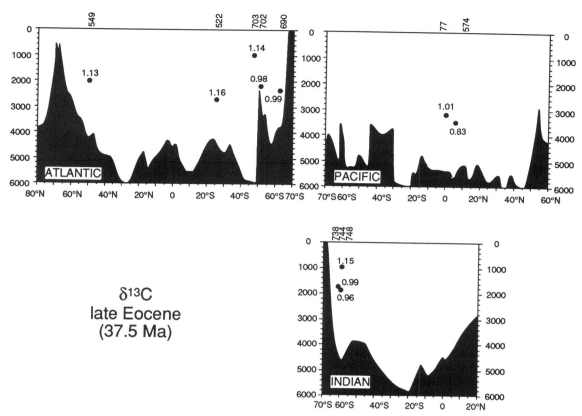

Fig. 5. Late Eocene δ^{13}C cross section of the oceans at 37.5 Ma. There is very little gradient (0.2‰) among the deepwater sites in all three oceans. The highest values were consistently recorded at intermediate depths as well as in the North Atlantic. This is consistent with an intermediate water mass which originated outside of the Southern Ocean. Two probable sources were either WSDW from the Tethys or an upper limb of NCW. Stratigraphic levels and data sources for each site are listed in Table 1*a*.

nent Water (analogous to AABW [*Broecker and Peng*, 1982]). Although the basin to basin δ^{13}C differences during the Oligocene remained low, it is evident that the Southern Ocean δ^{13}C values were equal to or higher than those recorded at other deepwater sites (e.g., sites 563 and 77; Figure 4). *Miller* [1992] noted a brief exception to this during the earliest Oligocene. From ~35.5 to 34.5 Ma, the highest deepwater δ^{13}C values were recorded in the North Atlantic, lowest values were recorded in the Pacific, and intermediate values were recorded in the Southern Ocean (Figure 6). This pattern is similar to the modern distribution of δ^{13}C values that reflect the production of both NCW and SCW. This is the first documented interval of bipolar deepwater production during the Cenozoic [*Miller*, 1992]. It is unlikely that the observed offsets are an artifact of a miscorrelation of the North Atlantic Site 563 record because the age estimates were based on magnetostratigraphic and δ^{18}O correlations. For the remainder of the Oligocene, the δ^{13}C patterns indicate a return to a one-component deepwater system that can be traced to the Southern

Ocean (Figures 4 and 6). Oligocene oxygen isotope comparisons are consistent either with a single source with intermediate depth, Site 689, reflecting the "cold spigot" being diluted by mean ocean water (Figure 2) [*Miller*, 1992; this study], or with two sources, WSDW and Southern Ocean [e.g., *Kennett and Stott*, 1990] (see below for discussion).

The uniformity of early Oligocene δ^{13}C values is illustrated in the 33 Ma time slice (Figure 7). Core coverage is best in the Atlantic Ocean and shows little variation (~0.2‰) at sites with paleodepths greater than 2000 m. With the exception of Site 549, which apparently was influenced by an intermediate to upper deepwater mass of unknown origin, South Atlantic δ^{13}C values were similar to or slightly higher than those in the North Atlantic, which is consistent with a south to north deepwater flow. Two deep Pacific sites recorded values similar to those in the Atlantic and Southern oceans (Figure 7). As was noted previously, near-uniform Oligocene deep-ocean δ^{13}C values probably resulted from relatively low mean ocean nutrients combined with a

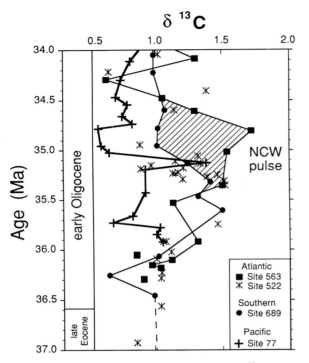

Fig. 6. Earliest Oligocene benthic foraminiferal δ^{13}C records from the North Atlantic (Site 563, solid squares), Southern Ocean (Site 689, solid circles), and Pacific (Site 77, crosses) oceans. From 36.2 to 35.5 Ma, the North Atlantic and Southern oceans recorded similar δ^{13}C values, reflecting SCW filling the North Atlantic. Beginning at 35.5 Ma and continuing through 34.5 Ma, the North Atlantic recorded the highest δ^{13}C values, Southern Ocean values were intermediate values, and the Pacific values were the lowest. This pattern is the same δ^{13}C pattern observed in the modern ocean. This interval from 35.5 to 34.5 Ma is interpreted as the first record of bipolar production of deep water during the Cenozoic [*Miller*, 1992].

dominant Southern Ocean deepwater source. To maintain similar δ^{13}C values with one source requires high ventilation rates; otherwise, aging effects would be more apparent. High ventilation rates for the Oligocene are consistent with a dramatic lowering of the CCD [*van Andel*, 1975] at this time.

Miocene δ^{13}C reconstructions. A consensus has developed concerning the middle and late Miocene deepwater circulation history; however, early Miocene deepwater circulation patterns remain controversial. *Woodruff and Savin* [1989] and *Wright et al.* [1992] compiled two comprehensive syntheses of Miocene deepwater circulation patterns. In general, both studies found that the major deepwater changes followed the subdivisions of this epoch. Early Miocene deepwater circulation patterns consisted of as many as three independent water masses, two of which were relatively warm and salty (Tethyan water [*Woodruff and Savin*, 1989] and NCW [*Wright et al.*, 1992]). Production of these warm deepwater sources diminished during the middle Miocene

before a pattern much like today's developed during the late Miocene.

Almost all studies of Miocene deepwater circulation patterns have concluded that NCW (re)developed during the middle Miocene and reached near-modern fluxes during the late Miocene [*Blanc et al.*, 1980; *Schnitker*, 1980; *Bender and Graham*, 1981; *Woodruff and Savin*, 1989; *Wright et al.*, 1991, 1992]. *Vogt* [1972] was the first to propose that subsidence of the Greenland-Scotland Ridge was responsible for NCW production in an effort to explain evidence for intensified bottom water currents in the North Atlantic between 18 and 10 Ma [*Ruddiman*, 1972]. Two subsequent studies built on this idea. *Blanc et al.* [1980] and *Schnitker* [1980] independently proposed that North Atlantic salinities increased in the middle Miocene as the eastern Tethys closed, shunting warm salty water into the North Atlantic; together with subsidence of the Greenland-Scotland Ridge, this contributed to NCW formation. *Woodruff and Savin* [1989] provided the first comprehensive global synthesis of Miocene benthic foraminiferal isotope data. In regard to global deepwater circulation patterns, *Woodruff and Savin* [1989] concluded that (1) SCW was the dominant water mass ventilating the deep oceans during the Miocene, (2) a warm saline plume originated in the Tethys and was an important component of deepwater circulation and climate during the early Miocene, (3) there is little evidence for NCW prior to 12–10 Ma, and (4) late Miocene deepwater circulation patterns were similar to the modern patterns. Their δ^{13}C results were consistent with those of *Blanc et al.* [1980] and *Schnitker* [1980], indicating that a NCW flux developed during the middle Miocene. The most striking feature of the *Woodruff and Savin* [1989] time slices was a high δ^{13}C water mass in the Indian Ocean during much of the early Miocene. They interpreted this signal as WSDW that originated in the Tethys.

In contrast to these studies, *Miller and Fairbanks* [1983, 1985] suggested that NCW was produced as early as the Oligocene. They noted that North Atlantic δ^{13}C values were higher than those in the Pacific during much of the Miocene, arguing for the presence of NCW during this interval. *Wright et al.* [1992] combined time series and time slice δ^{13}C reconstructions in an attempt to resolve discrepancies concerning early and middle Miocene deepwater circulation patterns. Atlantic-Pacific $\Delta\delta^{13}$C and Indian-Pacific $\Delta\delta^{13}$C records provide good proxies for the large-scale deepwater changes (Figure 8). Both the North Atlantic Ocean (Site 563, 3465 m) and the Indian Ocean (Site 237, 1337 m) recorded similarly high δ^{13}C values from 19 to 16 Ma, indicating that both regions were proximal to a source of high δ^{13}C water (Figure 8). This is even more evident in a time slice reconstruction at 16.2 Ma that shows high δ^{13}C values in both the deep North Atlantic and intermediate Indian oceans (Figure 9). These results reconcile the *Miller and Fairbanks* [1985] and *Woodruff and Savin* [1989] stud-

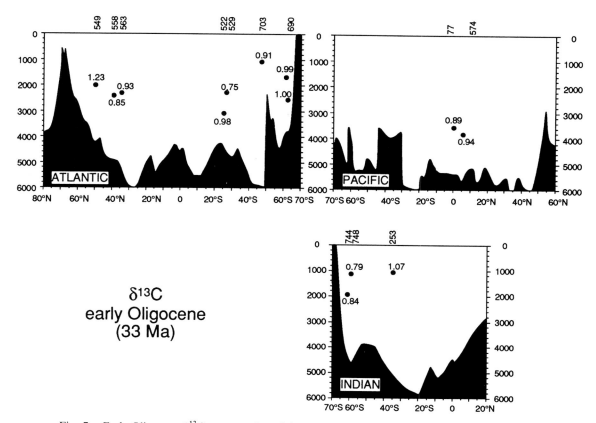

Fig. 7. Early Oligocene $\delta^{13}C$ cross section of the oceans at 33 Ma. As in the late Eocene, there is very little variation in the $\delta^{13}C$ values recorded throughout the oceans. This uniformity is attributed to low mean ocean nutrients and a vigorous deepwater source with a more central location. Again, there is a hint that an intermediate water mass was present in the North Atlantic (Site 549). This could reflect either Tethyan outflow or an intermediate branch of NCW. Stratigraphic levels and data sources for each site are listed in Table 1b.

ies, showing that both NCW and WSDW from the Tethys were produced during the early Miocene.

The convergence of North Atlantic and Pacific $\delta^{13}C$ records after 16 Ma suggests that NCW production ceased at this time (Figure 8). Interbasinal $\delta^{13}C$ differences redeveloped at about 12.5 Ma, indicating a renewal in NCW production. This is consistent with most Miocene deepwater circulation hypotheses [Blanc et al., 1980; Schnitker, 1980; Miller and Fairbanks, 1985; Woodruff and Savin, 1989; Wright et al., 1992]. Furthermore, most Miocene deepwater circulation scenarios suggest that NCW production remained high throughout the late Miocene, representing the evolution of modern deepwater circulation characteristics. Wright et al. [1991] concurred with this general picture; however, a high-resolution Southern Ocean $\delta^{13}C$ record indicates that there were higher-frequency fluctuations in NCW, occurring on the 10^5-year time scales (Figure 8). Wright et al. [1991] further noted that the absolute values and the differences in $\delta^{13}C$ among the North Atlantic, Pacific, and Southern oceans developed between 7 and 6

Ma in response to increased mean ocean nutrients and development of near-modern deepwater circulation patterns. A time slice reconstruction of the late Miocene (6 Ma; Figure 10) shows a pattern that was remarkably similar to the modern deepwater $\delta^{13}C$ distribution (Figure 3), suggesting that near-modern deepwater patterns operated during the late Miocene.

Oxygen Isotope Evidence

The Eocene was an interval of low benthic foraminiferal $\delta^{18}O$ values and high deepwater temperatures. Early Eocene deepwater temperatures were between 10° and 14°C, compared to today's 0°–2.5°C [Shackleton and Kennett, 1975; Savin et al., 1975; Miller et al., 1987]. Following peak warmth during the early Eocene, deep waters cooled by ~10° to 12°C over the next 16 m.y. as recorded by benthic foraminiferal $\delta^{18}O$ values which increased by ~3.0‰ [Shackleton and Kennett, 1975; Savin et al., 1975; Miller et al., 1987]. An initial increase of 1.0‰ occurred in the early middle Eocene

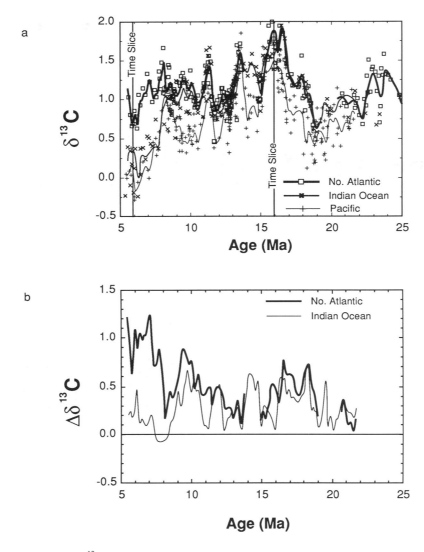

Fig. 8. (a) Miocene δ^{13}C records from the North Atlantic, Indian, and Pacific oceans. The smoothed lines for each record were generated by interpolating to a constant interval of 0.1 m.y. and using a five-point (0.25 m.y.) Gaussian filter. (b) To remove the mean ocean δ^{13}C changes which affected all sites, the smoothed Pacific record was subtracted from the North Atlantic and Indian Ocean records. High δ^{13}C values in the North Atlantic and Indian oceans relative to the Pacific Ocean indicate that the North Atlantic and Indian oceans were sources for deep to intermediate water masses from 20 to 16 Ma. Carbon isotope differences in the North Atlantic, Indian, and Pacific oceans were small between 16 and 12 Ma except for a brief interval in the Indian Ocean around 14 Ma. The small δ^{13}C contrast indicates that SCW was the dominant deepwater source during this interval. Interbasinal differences between the North Atlantic and Pacific oceans developed again by 12.5 Ma, indicating the renewal of NCW production. NCW production remained high through the late Miocene. Modern $\Delta\delta^{13}$C differences between the Atlantic, Pacific, and Indian oceans developed by 7 Ma. Two δ^{13}C cross sections shown in Figures 9 and 10 are marked.

(52–49 Ma), followed by large, relatively rapid increases in the late middle Eocene (~0.75‰ at 43 Ma and ~0.75‰ at 40 Ma) and earliest Oligocene (~1‰ at 36 Ma; Figure 2). The δ^{18}O increases at 43 and 40 Ma preceded the apparent development of large ice sheets near the Eocene/Oligocene boundary, and therefore they

are attributed to deepwater temperature decreases [e.g., *Miller*, 1992]. However, there is some evidence for possible glaciation during the late middle or late Eocene, and part of the increases could be attributed to ice volume changes [*Barron et al.*, 1991; *Robert and Maillott*, 1990; *Ehrmann*, 1991; *Ehrmann and Mackensen*, 1992].

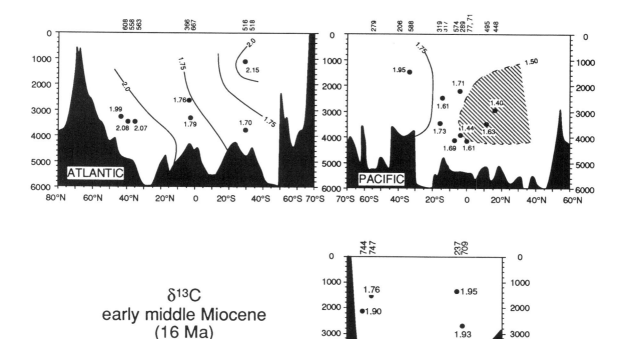

δ^{13}C
early middle Miocene
(16 Ma)

Fig. 9. Early/middle Miocene boundary δ^{13}C reconstruction (~16 Ma) [after *Wright et al.*, 1992]. The highest values were recorded in the deep North Atlantic and intermediate Indian oceans, indicating their proximity to deepwater sources, while the lowest values were still found in the Pacific. Stratigraphic levels and data sources for each site are listed in Table 1c.

Warm Eocene deepwater temperatures are consistent with WSDW production, although it is possible to ascribe these warm deep waters solely to polar warmth. *Shackleton and Boersma*'s [1981] Eocene reconstruction suggests that high-latitude surface waters were equally warm and, therefore, may have been a source of warm deep water (Figure 1). A more detailed reconstruction of Paleogene surface water δ^{18}O gradients compiled by *Zachos et al.* [1993] shows that planktonic and benthic foraminiferal δ^{18}O values were similar to each other in the high southern latitudes for much of the Paleogene. This observation has implications for deepwater sources because they acquire the temperature, salinity, and nutrient characteristics of the surface waters in regions of sinking. Similar δ^{18}O values in high-latitude planktonic and benthic foraminifera indicate that the temperature and salinity in the Southern Ocean were well mixed vertically, which is consistent with convection and a dominant southern ocean source. Alternatively, the similar values can be explained by low-salinity, low-δ^{18}O$_{water}$ high-latitude surface waters

that caused cool surface waters to appear warmer; however, this requires a serendipitous balance resulting in benthic and planktonic foraminifera recording the same δ^{18}O values, a hypothesis that we regard as unlikely. Carbon isotope data support a strong vertical link between the surface and deep waters in high latitudes because δ^{13}C records of planktonic and benthic foraminifera from Maud Rise [*Kennett and Stott*, 1990] and Kerguelen Plateau [*Barrera and Huber*, 1991] paralleled each other for much of the Paleogene.

An alternative explanation for similar planktonic and benthic δ^{18}O and δ^{13}C records in the Southern Ocean is that deep upwelling rather than deep convection caused the vertical mixing. Upwelled deep water can have a strong imprint on surface water characteristics, particularly in regions where there is a strong divergence in the surface waters, such as the modern Southern Ocean. However, given that many of the present-day markers for high Southern Ocean productivity (for example, high biogenic silica and high sedimentation rates) are poorly developed in the Paleogene of this region [*Barker et al.*,

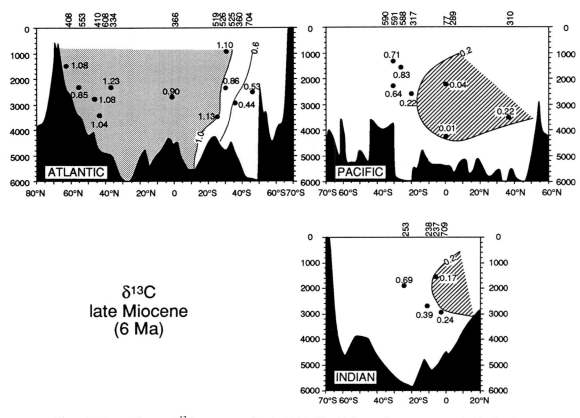

Fig. 10. Late Miocene $\delta^{13}C$ reconstruction (~6 Ma). The highest values were recorded in the deep North Atlantic, intermediate values were found in the Southern Ocean, and the lowest values were still found in the Pacific. The absolute values and distribution of $\delta^{13}C$ throughout the late Miocene ocean is strikingly similar to the modern. Stratigraphic levels and data sources for each site are listed in Table 1*d*.

1990; *Ciesielski et al.*, 1991; *Schlich et al.*, 1992], we regard this hypothesis as less likely.

Other oxygen isotope evidence from the Southern Ocean may indicate WSDW production during the Paleogene [*Kennett and Stott*, 1990]. Inversion of benthic foraminiferal $\delta^{18}O$ values on Maud Rise during the late Eocene through early Oligocene may reflect WSDW below a cooler, fresher intermediate water mass [*Kennett and Stott*, 1990]. Intermediate depth Site 689 (~1500-m paleodepth) recorded higher $\delta^{18}O$ values than the upper deepwater Site 690 (~2200 m) during this interval (Figure 2). The $\delta^{18}O$ difference between these sites averaged 0.5‰ but occasionally reached 1.0‰ (Figure 2) [*Kennett and Stott*, 1990]. The relatively higher $\delta^{18}O$ values at intermediate depths in the Southern Ocean are confirmed on the Kerguelen Plateau [*Barrera and Huber*, 1991]. If this $\delta^{18}O_{calcite}$ difference were based on temperature alone, then it would indicate a temperature inversion of 2°–4°C. *Kennett and Stott* [1990] noted that salinities must have been higher at greater depths to counter the warmer temperatures and maintain a plausible density contrast. The $\delta^{18}O_{water}/$

salinity relationship for the ocean is positive; in other words, higher salinities correspond to higher $\delta^{18}O_{water}$ [*Craig and Gordon*, 1965]. Higher deepwater $\delta^{18}O_{water}$ values make $\delta^{18}O_{calcite}$ values appear to reflect colder temperatures. Therefore the 0.5 to 1.0‰ $\delta^{18}O$ differences must reflect a larger temperature difference than the previous 2°–4°C estimate [*Kennett and Stott*, 1990].

If the *Kennett and Stott* [1990] hypothesis is correct, then a "fingerprint" should exist in mid-latitude planktonic foraminiferal records. One would expect to find similar $\delta^{18}O$ and $\delta^{13}C$ values between these planktonic foraminifera and the benthic foraminifera as were found in the deep and high-latitude surface waters. There are no available records from the key areas (Tethys) to test this hypothesis. However, modeling efforts [*Barron*, 1987; *Rind and Chandler*, 1991] require a low- to mid-latitude flux of heat to the poles; if WSDW was not the source, then there should be evidence for greatly increased heat flux to the Southern Ocean through either surface currents or upwelling of intermediate water masses.

If we consider the Oligocene through Miocene, there is a strong correlation between deepwater temperatures

TABLE 1a. Data Used in the Late Eocene Time Slice Reconstruction (37.5 Ma)

Site	Depth, mbsf	Paleolatitude	Paleodepth	$\delta^{18}O$	$\delta^{13}C$	Source
			Atlantic			
522	151.99	26°S	2700	1.00	1.16	Poore and Matthews [1984]
549	134.76	49°N	2000	0.85	1.13	Miller et al. [1985]
690	93.95	65°S	2450	0.91	0.99	Kennett and Stott [1990]
702	21.33	51°S	2250	0.84	0.98	Katz and Miller [1992]
703	133.00	48°S	1000	0.78	1.14	Miller [1992]
			Pacific			
77	475.30	5°S	3200	0.60	1.01	Keigwin and Keller [1984]
574	518.50	2°S	3500	1.05	0.83	Miller and Thomas [1985]
			Indian			
738	32.84	59°S	1700	1.30	0.99	Barrera and Huber [1991]
744	156.07	58°S	1800	1.22	0.96	Barrera and Huber [1991]
748	123.80	54°S	1000	1.02	1.15	Zachos et al. [1992]

based on benthic foraminiferal $\delta^{18}O$ records and the production of NCW and WSDW that apparently controlled long-term variations (>1 m.y.) in deepwater temperatures [e.g., *Wright et al.*, 1992]. Benthic foraminiferal (*Cibicidoides*) $\delta^{18}O$ values fluctuated about a mean of 1.8‰ through the Oligocene and early Miocene. The constancy of the Oligocene $\delta^{18}O$ values relative to the Eocene and Miocene may be attributed to little change in deepwater sources. The $\delta^{13}C$ evidence indicates that SCW dominated deepwater production during the Oligocene. During the early Miocene, deep waters warmed by ~3°C between 19 and 16 Ma (Figure 11) [*Savin et al.*, 1975; *Shackleton and Kennett*, 1975; *Kennett and Shackleton*, 1976; *Wright et al.*, 1992]. The early Miocene deepwater temperature increase corresponded to increases in the NCW production [*Wright et al.*, 1992] and Tethyan outflow water [*Woodruff and*

Savin, 1989]. Around 16 Ma, deepwater *Cibicidoides* spp. $\delta^{18}O$ values reached 1.0‰ in all three oceans and were the lowest recorded since the late Eocene. Deepwater temperatures are estimated to have been between 6° and 8°C at this time [*Shackleton and Kennett*, 1975; *Savin et al.*, 1975; *Wright et al.*, 1992]. Following peak warmth in the early middle Miocene, deep waters cooled by ~3°C as part of the middle Miocene $\delta^{18}O$ increase (15–12.5 Ma) (actually at least two steps, Mi3 and Mi4; Figure 11). The magnitude of this increase is attributed to a combination of ice growth and deepwater cooling (~3°C), both of which acted to increase the $\delta^{18}O$ value recorded in benthic foraminiferal calcite [*Shackleton and Kennett*, 1975; *Wright et al.*, 1992].

Higher-frequency $\delta^{18}O$ variations with durations of ~1 m.y. were superimposed on the long-term temperature trends. *Miller et al.* [1991] and *Wright and Miller*

TABLE 1b. Data Used in the Early Oligocene Time Slice Reconstruction (33 Ma)

Site	Depth, mbsf	Paleolatitude	Paleodepth	$\delta^{18}O$	$\delta^{13}C$	Source
			Atlantic			
522	108.30	26°S	3100	1.65	0.98	Miller et al. [1988]
529	158.13	28°S	2300	1.49	0.75	Miller et al. [1987]
549	107.43	49°N	2050	1.34	1.23	Miller et al. [1985]
558	391.35	38°N	2400	1.59	0.85	Miller and Fairbanks [1985]
563	346.04	34°N	2300	1.55	0.93	Millet and Fairbanks [1985]
689	110.95	64°S	1650	2.65	0.99	Kennett and Stott [1990]
690	90.20	65°S	2490	2.20	1.00	Kennett and Stott [1990]
703	99.88	48°S	1050	1.44	0.91	Miller [1992]
			Pacific			
77	442.08	4°S	3500	1.59	0.89	Keigwin and Keller [1984]
574	461.52	1°S	3850	1.53	0.94	Miller and Thomas [1985]
			Indian			
253	103.15	35°S	1050	1.36	1.07	Oberhänsli [1986]
744	128.67	58°S	1900	2.23	0.84	Barrera and Huber [1991]
748	105.60	55°S	1100	1.94	0.79	Zachos et al. [1992]

TABLE 1c.   Data Used in the Early Middle Miocene Time Slice Reconstruction (16 Ma) [From *Wright et al.*, 1992]

Site	Depth, mbsf	Paleolatitude	Paleodepth	$\delta^{18}O$	$\delta^{13}C$	Source
			Atlantic			
366	159.23	6°N	2645	1.54	1.76	*Miller et al.* [1989]
516	93.98	30°S	1131	0.86	2.15	*Woodruff and Savin* [1989]
518	65.60	30°S	3746	1.69	1.70	*Woodruff and Savin* [1989]
558	292.31	38°N	3415	1.56	2.08	*Miller and Fairbanks* [1985]
563	248.93	34°N	3465	1.64	2.07	*Wright et al.* [1992]
608	331.40	43°N	3304	1.46	1.99	*Wright et al.* [1992]
667	183.66	5°N	3334	1.59	1.79	*Miller et al.* [1989]
			Pacific			
71	187.50	0°	4112	1.70	1.61	*Woodruff and Savin* [1989]
77	243.02	3.5°S	3910	1.44	1.44	*Savin et al.* [1981]; *Woodruff and Savin* [1989]
289	472.40	4.5°S	2307	1.55	1.71	*Savin et al.* [1981]; *Woodruff and Savin* [1989]
317	165.23	15°S	2519	1.87	1.61	*Woodruff and Savin* [1989]
319	107.40	16°S	3475	1.88	1.73	*Woodruff and Savin* [1989]
448	16.40	16°N	2938	1.46	1.40	*Woodruff and Savin* [1989]
495	236.90	12°N	3478	1.92	1.63	*Barrera et al.* [1985]
475	192.47	8°S	4158	1.98	1.69	*Woodruff and Savin* [1989]
588	314.30	32.5°S	1453	1.30	1.95	*Kennett* [1986]
			Indian			
237	159.20	9°S	1337	1.76	1.95	*Woodruff and Savin* [1989]
709	141.65	6°S	2744	1.78	1.93	*Woodruff et al.* [1990]
744	64.74	61°S	2125	1.90	1.90	*Woodruff and Chambers* [1991]
747	84.90	55°S	1532	1.42	1.76	*Wright and Miller* [1992]

TABLE 1d.   Data Used in the Late Miocene Time Slice Reconstruction (6 Ma)

Site	Depth, mbsf	Paleolatitude	Paleodepth	$\delta^{18}O$	$\delta^{13}C$	Source
			Atlantic			
334	140.00	37°N	2350	2.29	1.23	*Keigwin et al.* [1986]
360	131.59	36°S	2975	2.47	0.44	*Wright et al.* [1991]
366	54.41	6°N	2765	2.89	0.90	*Stein* [1984]
408	100.76	63°N	1466	2.55	1.08	*Keigwin et al.* [1986]
410	175.10	45°N	2735	2.39	1.03	*Keigwin et al.* [1986]
519	111.20	26°S	3510	2.57	1.13	*McKenzie et al.* [1984]
525	22.20	29°S	2380	2.70	0.86	*Shackleton et al.* [1984]
526	49.22	30°S	975	2.41	1.10	*Shackleton et al.* [1984]
553	156.90	56°N	2310	2.25	0.85	*Wright et al.* [1992]
608	142.02	43°N	3470	2.23	1.04	*Wright et al.* [1991]
704	234.84	47°S	2550	2.71	0.53	*Wright et al.* [1991]
			Pacific			
77	92.27	2°S	4050	2.01	0.01	*Savin et al.* [1981]; *Woodruff and Savin* [1989]
289	167.08	2°S	2250	2.39	0.04	*Savin et al.* [1981]
310	60.50	35°N	3480	2.36	0.22	*Woodruff and Savin* [1989]
317	87.60	13°S	2600	2.46	0.51	*Woodruff and Savin* [1989]
588	136.80	28°S	1670	2.03	0.83	*Kennett* [1986]
590	227.60	33°S	1350	1.84	0.71	*Kennett* [1986]
591	240.70	33°S	2200	2.33	0.64	*Kennett* [1986]
			Indian			
237	86.04	8°S	1540	2.48	0.17	*Woodruff and Savin* [1989]
238	137.20	12°S	2740	2.39	0.39	*Vincent et al.* [1980]
253	28.00	27°S	1830	2.45	0.69	*Oberhänsli* [1986]
709	59.25	5°S	2950	2.62	0.24	*Woodruff et al.* [1990]
714	41.40	3°N	1925	2.09	0.55	*Boersma and Mikkelsen* [1990]

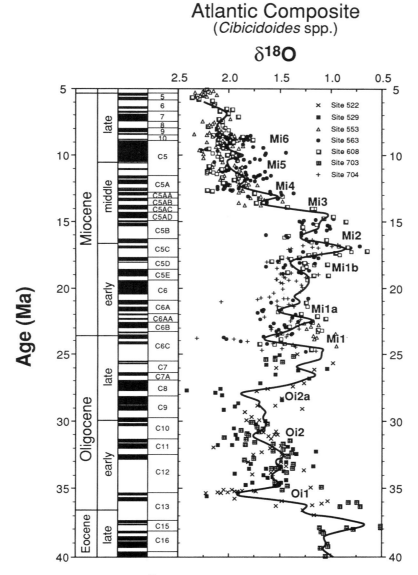

Fig. 11.   Composite *Cibicidoides* $\delta^{18}O$ record for the Atlantic (redrawn from *Miller et al.* [1987]). The smoothed line was generated by interpolating the data to constant intervals of 0.1 m.y. and smoothing with a 21-point Gaussian filter. The Oligocene isotope (''Oi'') and Miocene ''Mi'' zonations of *Miller et al.* [1991] and *Wright and Miller* [1992] are noted as ''Oi'' and ''Mi'' events. These benthic foraminiferal $\delta^{18}O$ increases covary with planktonic foraminiferal $\delta^{18}O$ increases and correlate with intervals of sediment deposition on and around Antarctica consistent with continental ice sheets [*Miller et al.*, 1991].

[1992] suggested that the $\delta^{18}O$ increases in their Oligocene and Miocene oxygen isotope zonations were caused by continental ice growth. They argued that planktonic-benthic foraminiferal $\delta^{18}O$ covariance indicates that these $\delta^{18}O$ increases reflected continental ice growth events followed by ice sheet melting [*Miller et al.*, 1991; *Wright et al.*, 1992]. There were three $\delta^{18}O$ maxima during

the Oligocene that punctuated the long-term trend at ~36, 31, and 28 Ma (equal to bases of isotope zones Oi1, Oi2, and Oi2a) that reflected ice growth and decay on Antarctica (Figure 11). During the Miocene, global increases in $\delta^{18}O$ have been noted at ~24, 22, 20, 18, 16, 13.5, 12.5, 10, and 8 Ma (Figure 11) [*Miller et al.*, 1991]. These increases were interpreted as continental ice growth events as well.

South Atlantic Legs 113 and 114

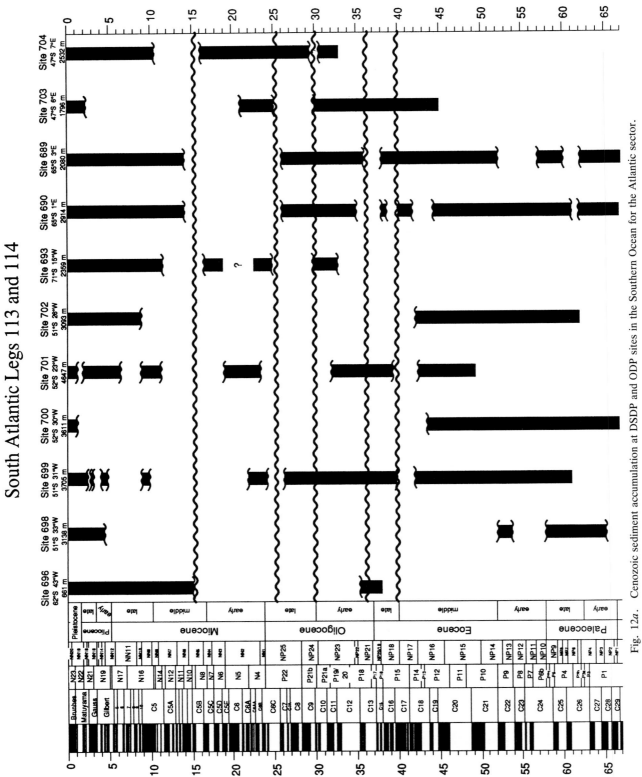

Fig. 12a.   Cenozoic sediment accumulation at DSDP and ODP sites in the Southern Ocean for the Atlantic sector.

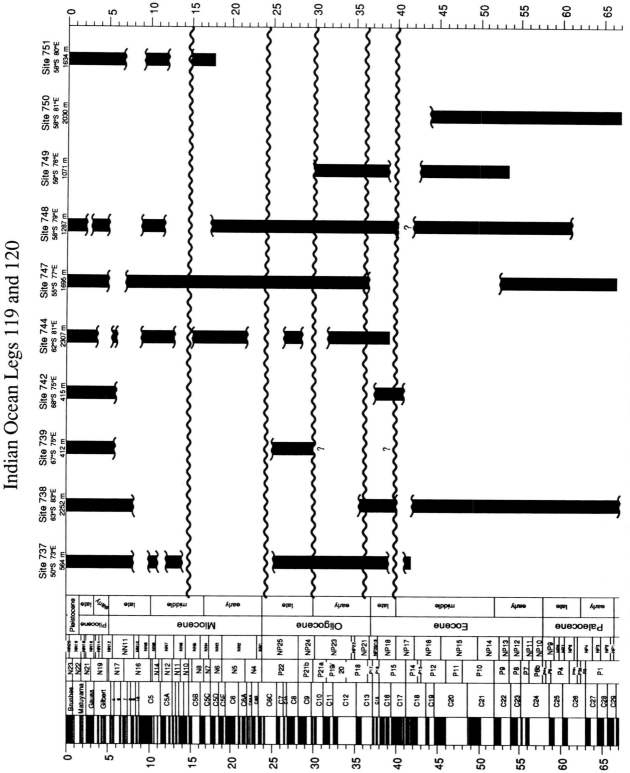

Fig. 12b. Cenozoic sediment accumulation at DSDP and ODP sites in the Southern Ocean for the Indian sector.

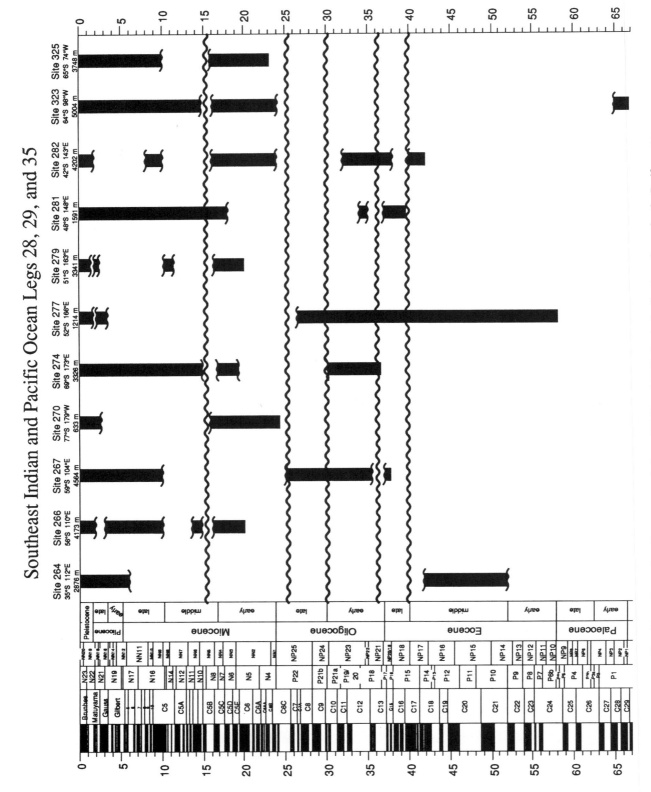

Fig. 12c. Cenozoic sediment accumulation at DSDP and ODP sites in the Southern Ocean for the Pacific sector.

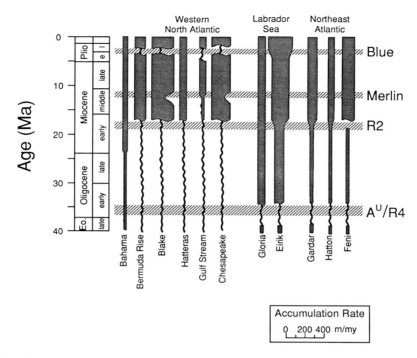

Fig. 13.   The accumulation history of drift deposits in the North Atlantic during the Cenozoic (after G. S. Mountain and K. G. Miller, unpublished data). An extensive erosional unconformity (Reflector R4) in the northern North Atlantic occurred near the Eocene/Oligocene boundary and has been traced to the Greenland-Scotland Ridge [*Roberts*, 1975; *Miller and Tucholke*, 1983]. This unconformity is expressed as Reflector A^u in the western North Atlantic basins [*Miller and Tucholke*, 1983]. Following the initial pulse at 36 Ma, sediments accumulated in chaotic sequences before normal sedimentary features developed by 34 Ma. Two other erosional pulses follow by widespread drift accumulation occurred at 19 to 16 Ma [*Roberts* [1975]; Reflector R2) and 11.5 Ma (Reflector Merlin of *Mountain and Tucholke* [1985]). Each of these three erosional cycles correlate to subsidence on the Greenland-Scotland Ridge, which presumably allowed a dense water mass in the proto-Norwegian and Greenland seas to spill over into the northern North Atlantic.

Sediment Deposition and Erosion

The Southern Ocean.   The late Eocene to early Oligocene climatic cooling led to an increase in the erosional capacity of bottom waters [*Kennett and Shackleton*, 1976]. *Kennett and Shackleton* [1976] noted that hiatuses in the Southern Ocean correspond to the $\delta^{18}O$ increase near the Eocene/Oligocene boundary (36 Ma). Drilling in the Southern Ocean during Ocean Drilling Program legs 113, 114, 119, and 120 [*Barker et al.*, 1990; *Barron et al.*, 1991; *Ciesielski et al.*, 1992; *Schlich et al.*, 1992] provides the foundation to reevaluate sediment distribution and hiatuses in the Southern Ocean.

The sedimentation record varies widely throughout the Cenozoic Southern Ocean (Figure 12). Even sites within close proximity to each other often show dissimilar patterns (for example, Maud Rise sites 689 and 690 [*Thomas et al.*, 1990; *Gersonde et al.*, 1990]). *Kennett and Stott* [1990] argued that differences in sedimentation patterns on Maud Rise resulted from the influence of two different water masses.

We identified five hiatuses that are prevalent, but not ubiquitous, in the Southern Ocean by compiling accumulation records by geographic region despite bathymetric differences in hiatus distributions. Each of these hiatuses corresponds to either an epochal boundary or epochal subdivision. These major erosional events occurred at (1) the middle/late Eocene boundary (~40 Ma), (2) the Eocene/Oligocene boundary (~36 Ma), (3) the early/late Oligocene boundary (~30 Ma), (4) the Oligocene/Miocene boundary (~24 Ma), and (5) near the early/middle Miocene boundary (~15 Ma) (Figure 12). We suggest that these Southern Ocean hiatuses resulted from increases in Southern Ocean deepwater production.

There appears to be a direct correlation between the periods of erosion in the Southern Ocean and global increases in the benthic foraminiferal $\delta^{18}O$ record. The earliest Oligocene and middle Miocene $\delta^{18}O$ increases are well established [*Shackleton and Kennett*, 1975; *Savin et al.*, 1975] and correspond to the erosional unconformities at 36 and 15 Ma [*Kennett and Shackleton*, 1976]. *Miller and Fairbanks* [1985] noted a large

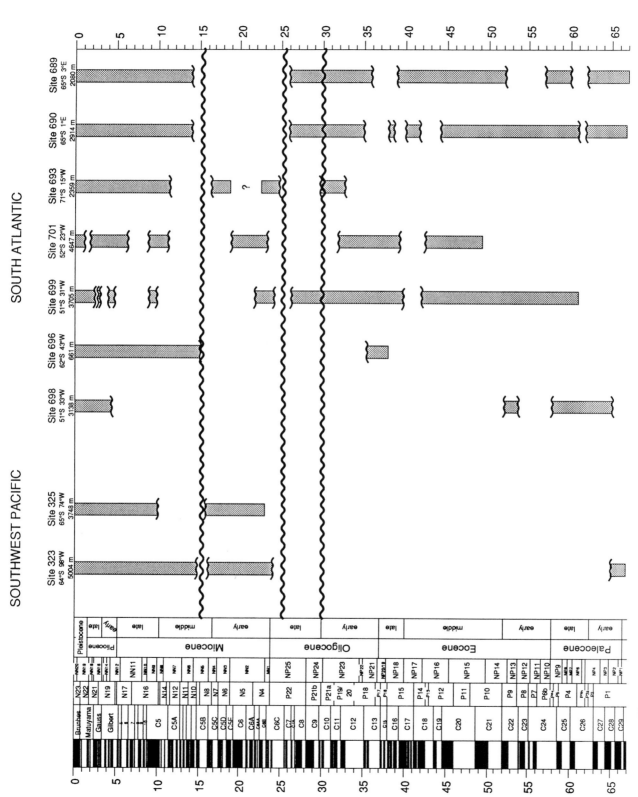

Fig. 14.   Sediment accumulation at DSDP and ODP sites in the southeastern Pacific and southwestern Atlantic during the Cenozoic.

($>1‰$) $\delta^{18}O$ increase at the Oligocene/Miocene boundary that is correlative to the unconformity of similar age (24 Ma). More recently, *Miller et al.* [1991] and *Wright and Miller* [1992] identified additional global $\delta^{18}O$ increases at 32, 28, 22, 18, and 16 Ma as part of their Oligocene and Miocene isotope zonations (Figure 11). Either the 32 or 28 Ma $\delta^{18}O$ increase could be related to the 30 Ma hiatus. The absence of lower Miocene sediments at several Southern Ocean sites may be related to a concatenation of hiatuses associated with the $\delta^{18}O$ increases at 22, 18, and 16 Ma (Figures 11 and 12).

Erosion in the Southern Ocean must reflect an increase in bottom current velocities. In a comparison of model simulations between early Cenozoic and present-day conditions, *Barron et al.* [1991] noted that the strength of SCW appeared to reflect the extent of Antarctic glaciation. Similarly, we suggest that the periodic growth or expansions of the Antarctica ice sheets increased the temperature gradient in the Southern Ocean, leading to more vigorous deepwater circulation originating in the Southern Ocean. This implies that there was a causal relationship between intervals of ice growth noted above, increased SCW production, and erosion.

North Atlantic Erosional and depositional patterns. Depositional patterns in the North Atlantic reflect large-scale deepwater circulation changes beginning in the earliest Oligocene through the middle Miocene. There is extensive evidence for the periodic development of strong deepwater currents with a North Atlantic source from the early Oligocene through middle Miocene [*Jones et al.*, 1970; *Ruddiman*, 1972; *Tucholke and Mountain*, 1979, 1986; *Miller and Tucholke*, 1983; *Mountain and Tucholke*, 1985; *Mountain and Miller*, 1992]. Uppermost Eocene to lowermost Oligocene reflectors R4 in the northern North Atlantic [*Roberts*, 1975; *Miller and Tucholke*, 1983] and A^u along the western edge of the North Atlantic [*Tucholke and Mountain*, 1979, 1986; *Mountain and Tucholke*, 1985; *Mountain and Miller*, 1992] resulted from erosional unconformities (Figure 13). These unconformities can be traced to the Greenland-Scotland Ridge, which is the source of northern deepwater currents. After this initial erosional pulse, large sediment drifts developed in the Labrador Sea and the northeast Atlantic (the Gloria, Eirik, Gardar, Hatton, and Feni drifts) (Figure 13).

The next change in drift deposition occurred in the early Miocene. *Roberts* [1975] identified Reflector R2 in the northern North Atlantic, which is a seismic disconformity with an age estimate of 20–16 Ma [*Ruddiman*, 1972; *Miller and Tucholke*, 1983; *Mountain and Tucholke*, 1985; *Wright et al.*, 1992]. The accumulation rates of North Atlantic drifts increased significantly following this erosional pulse. *Ruddiman* [1972] identified an erosional phase on the Reykjanes Ridge based on the pinchout of seismic reflectors. Following this erosional pulse, sedimentation accumulation increased significantly on the Reykjanes Ridge between 18 and 17

Ma. A third North Atlantic reflector, "Merlin," was identified by *Mountain and Tucholke* [1985] (Figure 13). The age of this event is constrained only to the late middle Miocene ($\sim$12–10 Ma). Since 10 Ma, sediment accumulation on the Reykjanes Ridge has reflected redeposition by nearly continuous deep currents [*Ruddiman*, 1972]. Each of these erosional events and subsequent drift depositions correlate to intervals of NCW production inferred from $\delta^{13}C$ reconstructions. The three major seismic disconformities (A^u/R4, R2, and Merlin) thus correspond to three intervals of large basin-basin $\delta^{13}C$ differences, $\sim$35, 18–16 Ma, and 12–10 Ma. The corroboration of the $\delta^{13}C$ inferences and the seismic stratigraphy indicate that these intervals were of peak NCW flux.

Tectonics and Circumpolar Circulation

There is a direct link between climate, deepwater formation, and tectonics on time scales of a million years or longer [*Barron and Peterson*, 1991]. The gradual change from the quasi-circumequatorial circulation of the Cretaceous and early Paleogene to the development of circumpolar circulation may have been the fundamental cause behind the long-term trend toward colder polar climates during the Cenozoic [*Kennett and Shackleton*, 1976; *Kennett*, 1977; *Barron*, 1987; *Barron and Peterson*, 1991; *Kennett and Stott*, 1990]. *Kennett and Shackleton* [1976] and *Kennett* [1977] suggested that the development of the circumpolar circulation resulted in the thermal isolation of Antarctica from low-latitude heat sources. The removal of two tectonic barriers led to the initiation of circumpolar circulation. *Kennett et al.* [1974] noted that the deep subsidence of the Tasman Rise was a critical tectonic event for circumpolar circulation. They suggested that this occurred during the early to middle Oligocene based on sediment accumulation at DSDP sites from Leg 29, although subsequent work showed that the initial breakup between Antarctica and Australia began around 85 Ma (anomaly 34); normal seafloor spreading began at $\sim$43 Ma (anomaly 19) [*Mutter et al.*, 1985].

The Drake Passage was the second physical barrier to circumpolar circulation. While most tectonic reconstructions of the Drake Passage suggest that it opened near the Oligocene/Miocene boundary [*Barker and Burrell*, 1977, 1982], the complexity of the Drake Passage precludes any model which is based on identifying magnetic anomalies in the Scotia Sea [*Lawver et al.*, 1992]. Shallow connections may have developed between the South Atlantic and Pacific oceans as the tip of South America moved to the east past the Antarctic peninsula near the Eocene/Oligocene boundary, independent of the opening of the Scotia Sea [*Lawver et al.*, 1992]. On the basis of southeast Pacific sedimentation (Deep Sea Drilling Project (DSDP) Leg 35), *Tucholke et al.* [1976] speculated that the initial breach of the Drake

Passage occurred between the late Eocene and early Oligocene, allowing a shallow circumpolar current to develop. The timing of the evolution of deep circumpolar circulation is less certain. *Barker and Burrell* [1982] suggested that the deep connection through the Scotia Sea developed near the Oligocene/Miocene boundary. Sediment accumulation patterns in the South Atlantic and southeast Pacific show a prominent hiatus near 24 Ma (Figure 14). For example, at Site 323 in the southeast Pacific, lower Miocene terrigenous silts and clays that indicate current-controlled deposition were deposited on Danian pelagic sediments. This unconformity was interpreted to reflect a strong deepwater current that developed near the Oligocene/Miocene boundary and eroded the sediment record back to the Cretaceous [*Tucholke et al.*, 1976]. Our compilation (Figures 12 and 14) (1) is consistent with a shallow opening of the Drake Passage by the earliest Oligocene, which triggered further isolation and consequent glaciation, and (2) argues for a distinct erosional event near the Oligocene/Miocene boundary which we attribute to opening of the deep passages.

CONCLUSIONS

The long-term fluctuations in deepwater temperatures responded to the WSDW production during the Cenozoic. Most of the evidence for deepwater circulation patterns presented here indicates that the Southern Ocean has been the dominant deepwater source for the past 40 m.y.:

1. Relatively high benthic foraminiferal $\delta^{13}C$ values were recorded in the Southern Ocean during much of the late Eocene to early Miocene, indicating that this region was proximal to a deepwater source.

2. Relatively high global deepwater $\delta^{18}O$ values are interpreted as cold temperatures, particularly during the Oligocene and the middle Miocene to Recent (see also *Shackleton and Kennett* [1976] and *Kennett and Stott* [1990]).

3. Similar $\delta^{18}O$ values recorded in benthic and high-latitude planktonic foraminiferal records throughout the late Eocene to Recent are consistent with, but are not indisputable proof for deep convection in the Southern Ocean.

4. Late Eocene (~40 Ma), earliest Oligocene (~36 Ma), middle Oligocene (~30 Ma), latest Oligocene (~24 Ma), and middle Miocene (~15 Ma) hiatuses occurred at many Southern Ocean sites, indicating increases in Southern Ocean bottom water currents.

A $\delta^{18}O_{calcite}$ inversion at Maud Rise may reflect the presence of WSDW below a colder, fresher water mass [*Kennett and Stott*, 1990], providing evidence for WSDW in the Paleogene oceans. It remains unclear whether WSDW contributed heat and salt to the deepwater masses by directly filling the deep oceans or through a poleward flux that surfaced in the high south-

ern latitudes and was redistributed by SCW. In either case, a colder and fresher intermediate water mass originated in the Southern Ocean, producing the $\delta^{18}O$ inversion during the late Eocene to Oligocene.

Although SCW generally dominated the Oligocene, there is evidence from carbon isotopes, seismic stratigraphy, and hiatuses that indicate that a pulse of NCW occurred in the earliest Oligocene. Miocene deepwater circulation patterns occurred in two cycles. The first began in the earliest Miocene with a one-component deepwater system dominated by SCW. NCW and Tethyan outflow water production supplemented SCW production between 19 and 16 Ma. The production of NCW and Tethyan outflow water was reduced between 16 and 15 Ma, leaving SCW as the dominant deepwater mass ventilating the deep oceans. At about 12.5 Ma, $\delta^{13}C$ reconstructions indicate that NCW production was renewed and continued through the late Miocene when near-modern deepwater circulation patterns developed.

Acknowledgments.   G. Brass, M. Katz, J. Kennett, G. Mead, and L. Stott provided critical reviews which greatly improved this manuscript. This manuscript also benefited from discussions with C. Charles, R. Fairbanks, J. Lynch-Stieglitz, D. Pak, and L. Burckle. This work was supported by National Science Foundation grant OCE88-11834 to K.G.M. and grant OCE91-17667 to J.D.W. This is Lamont-Doherty Geological Observatory contribution 5027.

REFERENCES

Barker, P. F., and J. Burrell, The opening of Drake Passage, *Mar. Geol.*, *25*, 15–34, 1977.

Barker, P. F., and J. Burrell, The influence upon Southern Ocean circulation, sedimentation, and climate of the opening of Drake Passage, in *Antarctic Geoscience*, edited by C. Craddock, pp. 377–385, University of Wisconsin Press, Madison, 1982.

Barker, P. F., et al., Leg 113, *Proc. Ocean Drill. Program Sci. Results*, *113*, 1033 pp., 1990.

Barrera, E., and B. T. Huber, Paleogene and early Neogene oceanography of the southern Indian Ocean: Leg 119 foraminifer stable isotope results, *Proc. Ocean Drill. Program Sci. Results*, *119*, 693–717, 1991.

Barrera, E., G. Keller, and S. M. Savin, Evolution of the Miocene ocean in the eastern North Pacific as inferred from oxygen and carbon isotopic ratios of foraminifera, The Miocene Ocean: Paleoceanography and Biogeography, *Mem. Geol. Soc. Am.*, *163*, 83–102, 1985.

Barron, E. J., A warm, equable Cretaceous: The nature of the problem, *Earth Sci. Rev.*, *19*, 305–338, 1983.

Barron, E. J., Eocene equator-to-pole surface ocean temperatures: A significant climate problem?, *Paleoceanography*, *2*, 729–739, 1987.

Barron, E. J., and W. H. Peterson, The Cenozoic ocean circulation based on ocean General Circulation Model results, *Palaeogeogr. Palaeoclimatol. Palaeoecol.*, *83*, 1–28, 1991.

Barron, J. A., et al., Leg 119, *Proc. Ocean Drill. Program Sci. Results*, *119*, 1003 pp., 1991.

Belanger, P. E., W. B. Curry, and R. K. Matthews, Core-top evaluation of benthic foraminiferal isotopic ratios for paleoceanographic interpretations, *Palaeogeogr. Palaeoclimatol. Palaeoecol.*, *33*, 205–220, 1981.

Bender, M. L., and D. W. Graham, On late Miocene abyssal hydrography, *Mar. Micropaleontol.*, 6, 451–464, 1981.

Berggren, W. A., D. V. Kent, and J. A. Van Couvering, Neogene geochronology and chronostratigraphy, in *The Chronology of the Geological Record, Mem. 10*, edited by N. J. Snelling, pp. 211–260, Geological Society of London, London, 1985.

Blanc, P.-L., D. Rabussier, C. Vergnaud-Grazzini, and J. C. Duplessy, North Atlantic Deep Water formed by the later middle Miocene, *Nature*, 283, 553–555, 1980.

Boersma, A., and N. Mikkelsen, Miocene-age primary productivity episodes and oxygen minima in the central equatorial Indian Ocean, *Proc. Ocean Drill. Program Sci. Results*, 115, 589–610, 1990.

Boersma, A., I. Premoli Silva, and N. J. Shackleton, Atlantic Eocene planktonic foraminiferal paleohydrographic indicators and stable isotope paleoceanography, *Paleoceanography*, 2, 287–331, 1987.

Boyle, E. A., The role of vertical chemical fractionation in controlling late Quaternary atmospheric carbon dioxide, *J. Geophys. Res.*, 93, 15,701–15,714, 1988.

Boyle, E. A., and L. D. Keigwin, Deep circulation of the North Atlantic over the last 200,000 years, geochemical evidence, *Science*, 218, 784–787, 1982.

Boyle, E. A., and L. D. Keigwin, North Atlantic thermohaline circulation during the past 20,000 years linked to high-latitude surface temperature, *Nature*, 330, 35–40, 1987.

Brass, G. W., J. R. Southam, and W. H. Peterson, Warm saline bottom water in the ancient ocean, *Nature*, 296, 620–623, 1982.

Broecker, W. S., A revised estimate for the radiocarbon age of North Atlantic Deep Water, *J. Geophys. Res.*, 84, 3218–3226, 1979.

Broecker, W. S., Some thoughts about the radiocarbon budget of the glacial Atlantic, *Paleoceanography*, 4, 213–220, 1989.

Broecker, W. S., and T.-H. Peng, *Tracers in the Sea*, 690 pp., Eldigio, Palisades, N. Y., 1982.

Broecker, W. S., M. Andree, G. Bonani, W. Wolfli, H. Oeschger, M. Klas, A. Mix, and W. Curry, Preliminary estimates for the radiocarbon age of deep water in the glacial ocean, *Paleoceanography*, 3, 659–669, 1988.

Cande, S. C., and D. V. Kent, A new geomagnetic polarity time scale for the Late Cretaceous and Cenozoic, *J. Geophys. Res.*, 97, 13,917–13,951, 1992.

Chamberlin, T. C., On a possible reversal of deep-sea circulation and its influence on geologic climates, *J. Geol.*, 14, 363–373, 1906.

Charles, C. D., and R. G. Fairbanks, Evidence from Southern Ocean sediments for the effect of North Atlantic deep-water flux on climate, *Nature*, 355, 416–419, 1992.

Ciesielski, P. F., et al., Leg 114, *Proc. Ocean Drill. Program Sci. Results*, 114, 826 pp., 1991.

Covey, C., and E. Barron, The role of ocean heat transport in climatic change, *Earth Sci. Rev.*, 24, 429–445, 1988.

Covey, C., and S. L. Thompson, Testing the effects of ocean heat transport on climate, *Global Planet. Change*, 1, 331–341, 1989.

Craig, H., and L. Gordon, Deuterium and oxygen-18 variations in the ocean and the marine atmosphere, in *Symposium of Marine Geochemistry, Publ. 3*, 277 pp., Graduate School of Oceanography, University of Rhode Island, Kingston, 1965.

Curry, W. B., and G. P. Lohmann, Carbon isotopic changes in benthic foraminifera from the western South Atlantic: Reconstruction of glacial abyssal circulation patterns, *Quat. Res.*, 18, 218–235, 1982.

Curry, W. B., J.-C. Duplessy, L. D. Labeyrie, and N. J. Shackleton, Quaternary deep-water circulation changes in the distribution of $\delta^{13}C$ of deep water ΣCO_2 between the last

glaciation and the Holocene, *Paleoceanography*, 3, 317–341, 1988.

Duplessy, J.-C., C. Lalou, and A. C. Vinot, Differential isotopic fractionation in benthic foraminifera and paleotemperatures reassessed, *Science*, 168, 250–251, 1970.

Duplessy, J.-C., N. J. Shackleton, R. G. Fairbanks, L. Labeyrie, D. Oppo, and N. Kallel, Deep-water source variations during the last climatic cycle and their impact on the global deepwater circulation, *Paleoceanography*, 3, 343–360, 1988.

Erhmann, W. U., Implications of sediment composition on the southern Kerguelen Plateau for paleoclimate and depositional environment, *Proc. Ocean Drill. Program Sci. Results*, 119, 185–210, 1991.

Erhmann, W. U., and A. Mackensen, Sedimentological evidence for the formation of an East Antarctic ice sheet tin Eocene/Oligocene time, *Palaeogeogr. Palaeoclimatol. Palaeoecol.*, 93, 85–112, 1992.

Gersonde, R., et al., Biostratigraphic synthesis of Neogene siliceous microfossils for the Antarctic Ocean ODP Leg 113 (Weddell Sea), *Proc. Ocean Drill. Program Sci. Results*, 113, 915–936, 1990.

Graham, D. W., B. H. Corliss, M. L. Bender, and L. D. Keigwin, Carbon and oxygen isotopic disequilibria of Recent benthic foraminifera, *Mar. Micropaleontol.*, 6, 483–497, 1981.

Haq, B., Paleogene paleoceanography: Early Cenozoic ocean revisited, *Oceanol. Acta, SP*, 71–82, 1981.

Jones, E. J., M. Ewing, J. I. Ewing, and S. L. Ettreim, Influences of Norwegian Sea overflow water on sedimentation in the northern North Atlantic and Labrador Sea, *J. Geophys. Res.*, 75, 1655–1680, 1970.

Katz, M. E., and K. G. Miller, Early Paleogene benthic foraminiferal assemblages and stable isotopes in the Southern Ocean, *Proc. Ocean Drill. Program Sci. Results*, 120, 481–512, 1992.

Keigwin, L. D., and G. Keller, Middle Oligocene cooling from equatorial Pacific DSDP Site 77B, *Geology*, 12, 16–19, 1984.

Keigwin, L. D., M.-P. Aubry, and D. V. Kent, North Atlantic late Miocene stable-isotope stratigraphy, biostratigraphy, and magnetostratigraphy, *Initial Rep. Deep Sea Drill. Proj.*, 94, 935–963, 1986.

Kennett, J. P., Cenozoic evolution of Antarctic glaciation, the Circum-Antarctic Ocean, and their impact on global paleoceanography, *J. Geophys. Res.*, 82, 3843–3860, 1977.

Kennett, J. P., Miocene to early Pliocene oxygen and carbon isotope stratigraphy in the southwest Pacific, Deep Sea Drilling Project Leg 90, *Initial Rep. Deep Sea Drill. Proj.*, 90, 1383–1411, 1986.

Kennett, J. P., and N. J. Shackleton, Oxygen isotopic evidence for the development of the psychrosphere 38 Myr. ago, *Nature*, 260, 513–515, 1976.

Kennett, J. P., and L. D. Stott, Proteus and Proto-Oceanus: Paleogene oceans as revealed from Antarctic stable isotopic results: ODP Leg 113, *Proc. Ocean Drill. Program Sci. Results*, 113, 865–880, 1990.

Kennett, J. P., et al., Development of the Circum-Antarctic Current, *Nature*, 186, 144–147, 1974.

Kroopnick, P., The distribution of ^{13}C of ΣCO_2 in the world oceans, *Deep Sea Res.*, 32, 57–84, 1985.

Lawver, L. A., L. M. Gahagan, and M. F. Coffin, The development of paleoseaways around Antarctica, in *The Antarctic Paleoenvironment: A perspective on Global Change, Part 1, Antarct. Res. Ser.*, vol. 56, edited by J. P. Kennett and D. A. Warnke, pp. 7–30, AGU, Washington, D. C., 1992.

Manabe, S., and K. Bryan, Jr., CO_2-induced change in a coupled ocean-atmosphere model and its paleoclimatic implications, *J. Geophys. Res.*, 90, 11,689–11,707, 1985.

Manabe, S., K. Bryan, Jr., and M. J. Spelman, Transient

response of a global ocean-atmosphere model to a doubling of atmospheric carbon dioxide, *J. Phys. Oceanogr.*, *20*, 722–749, 1990.

McKenzie, J. A., H. Weissert, R. Z. Poore, R. C. Wright, S. F. Percival, Jr., H. Oberhänsli, and M. Casey, Paleoceanographic implications of stable-isotope data from upper Miocene-lower Pliocene sediments from the southeast Atlantic (Deep Sea Drilling Project Site 519), *Initial Rep. Deep Sea Drill. Proj.*, *73*, 717–735, 1984.

Miller, K. G., Middle Eocene to Oligocene stable isotopes, climate, and deep-water history: The terminal Eocene event?, in *Eocene-Oligocene Climatic and Biotic Evolution*, edited by D. Prothero and W. A. Berggren, pp. 160–177, Princeton University, Princeton, N. J., 1992.

Miller, K. G., and R. G. Fairbanks, Evidence for Oligocene-middle Miocene abyssal circulation changes in the western North Atlantic, *Nature*, *306*, 250–253, 1983.

Miller, K. G., and R. G. Fairbanks, Oligocene to Miocene carbon isotope cycles and abyssal circulation changes, in *The Carbon Cycle and Atmospheric CO_2: Natural Variations Archean to Present, Geophys. Monogr. Ser.*, vol. 32, edited by E. T. Sundquist and W. S. Broecker, pp. 469–486, AGU, Washington, D. C., 1985.

Miller, K. G., and E. Thomas, Late Eocene to Oligocene benthic foraminiferal isotopic record, Site 574, equatorial Pacific, *Initial Rep. Deep Sea Drill. Proj.*, *85*, 771–777, 1985.

Miller, K. G., and B. E. Tucholke, Development of Cenozoic abyssal circulation south of the Greenland-Scotland Ridge, in *Structure and Development of the Greenland-Scotland Ridge*, edited by M. H. P. Bott, S. Saxov, M. Talwani, and J. Thiede, pp. 549–589, Plenum, New York, 1983.

Miller, K. G., W. B. Curry, and D. R. Ostermann, Late Paleogene (Eocene to Oligocene) benthic foraminiferal oceanography of the Goban Spur Region, Leg 80, *Initial Rep. Deep Sea Drill. Proj.*, *80*, 505–538, 1985.

Miller, K. G., R. G. Fairbanks, and G. S. Mountain, Tertiary oxygen isotope synthesis, sea level history, and continental margin erosion, *Paleoceanography*, *2*, 1–19, 1987.

Miller, K. G., M. D. Feigenson, D. V. Kent, and R. K. Olsson, Oligocene stable isotope ($^{87}Sr/^{86}Sr$, $\delta^{18}O$, $\delta^{13}C$) standard section, Deep Sea Drilling Project Site 522, *Paleoceanography*, *3*, 223–233, 1988.

Miller, K. G., J. D. Wright, and R. G. Fairbanks, Unlocking the icehouse: Oligocene-Miocene oxygen isotope, eustasy, and margin erosion, *J. Geophys. Res.*, *96*, 6829–6848, 1991.

Moore, T. C., T. H. van Andel, C. Sancetta, and N. Pisias, Cenozoic hiatuses in pelagic sediments, *Micropaleontology*, *24*, 113–138, 1978.

Mountain, G. S., and K. G. Miller, Seismic and geologic evidence for early Paleogene circulation in the western North Atlantic, *Paleoceanography*, *7*, 423–440, 1992.

Mountain, G. S., and B. E. Tucholke, Mesozoic and Cenozoic geology of the U.S. Atlantic continental slope and rise, in *Geologic Evolution of the United States Atlantic Margin*, edited by C. W. Poag, pp. 293–341, Van Nostrand Reinhold, New York, 1985.

Mutter, J. C., K. A. Hegarty, S. C. Cande, and J. K. Weissel, Breakup between Australia and Antarctica: A brief review in the light of new data, *Tectonophysics*, *114*, 255–279, 1985.

Oberhänsli, H., Latest Cretaceous–early Neogene oxygen and carbon isotopic record at DSDP sites in the Indian Ocean, *Mar. Micropaleontol.*, *10*, 91–115, 1986.

Oppo, D. W., and R. G. Fairbanks, Variability in the deep and intermediate water circulation of the Atlantic Ocean during the past 25,000 years: Northern hemisphere modulation of the Southern Ocean, *Earth Planet. Sci. Lett.*, *86*, 1–15, 1987.

Oppo, D. W., R. G. Fairbanks, and A. L. Gordon, Late Pleistocene Southern Ocean $\delta^{13}C$ variability, *Paleoceanography*, *5*, 43–54, 1990.

Pak, D. K., and K. G. Miller, Paleocene to Eocene benthic foraminiferal isotopes and assemblages: Implications for deepwater circulation, *Paleoceanography*, *7*, 405–422, 1992.

Poore, R. Z., and R. K. Matthews, Late Eocene–Oligocene oxygen and carbon isotope record from South Atlantic Ocean, *Initial Rep. Deep Sea Drill. Proj.*, *73*, 725–735, 1984.

Rind, D., The doubled CO_2 climate: Impact of the sea surface temperature gradient, *Am. J. Atmos. Sci.*, *44*, 3235–3268, 1987.

Rind, D., and M. Chandler, Increased ocean heat transports and warmer climate, *J. Geophys. Res.*, *96*, 7437–7461, 1991.

Robert, C., and H. Maillot, Paleoenvironments in the Weddell Sea area and Antarctic climates, as deduced from clay mineral associations and geochemical data, ODP Leg 113, *Proc. Ocean Drill. Program Sci. Results*, *113*, 51–70, 1990.

Roberts, D. G., Marine geology of the Rockall Plateau and Trough, *Philos. Trans. R. Soc. London, Ser. A*, *278*, 447–590, 1975.

Ruddiman, W. F., Sediment redistribution on the Reykjanes Ridge: Seismic evidence, *Geol. Soc. Am. Bull.*, *83*, 2039–2062, 1972.

Savin, S. M., R. G. Douglas, and F. G. Stehli, Tertiary marine paleotemperatures, *Geol. Soc. Am. Bull.*, *86*, 1499–1510, 1975.

Savin, S. M., G. Keller, R. G. Douglas, J. S. Killingley, L. Shaughnessy, M. A. Sommer, E. Vincent, and F. Woodruff, Miocene benthic foraminiferal isotope records: A synthesis, *Mar. Micropaleontol.*, *6*, 423–450, 1981.

Schlich, R., et al., Leg 120, *Proc. Ocean Drill. Program Sci. Results*, *120*, 1145 pp., 1992.

Schnitker, D., North Atlantic oceanography as possible cause of Antarctic glaciation and eutrophication, *Nature*, *284*, 615–616, 1980.

Shackleton, N. J., and A. Boersma, The climate of the Eocene ocean, *J. Geol. Soc. London*, *138*, 153–157, 1981.

Shackleton, N. J., and J. P. Kennett, Paleotemperature history of the Cenozoic and initiation of Antarctic glaciation: Oxygen and carbon isotopic analyses in DSDP sites 277, 279, and 281, *Initial Rep. Deep Sea Drill. Proj.*, *29*, 743–755, 1975.

Shackleton, N. J., and N. D. Opdyke, Oxygen isotope and paleomagnetic stratigraphy of equatorial Pacific Core V28-238: Oxygen isotope temperatures and ice volumes on a 10^5 year and 10^6 year scale, *Quat. Res.*, *3*, 39–55, 1973.

Shackleton, N. J., J. Imbrie, and M. A. Hall, Oxygen and carbon isotope record of East Pacific Core V19-30: Implications for the formation of deep water in the late Pleistocene, *Earth Planet. Sci. Lett.*, *65*, 233–244, 1983.

Shackleton, N. J., M. A. Hall, and A. Boersma, Oxygen and carbon isotope data from Leg 74 foraminifers, *Initial Rep. Deep Sea Drill. Proj.*, *74*, 599–612, 1984.

Stein, R., Zur neogenen Klimaentwicklung in Nordwest-Afrika und Palaeo-Ozeanographie im Nordost-Atlantik: Ergebnisse von DSDP-sites 141, 266, 297, und 544B, doctoral thesis, Nr. 4, 210 s, Berichte-Reports, Geol.-Palaeontol. Inst. Univ. Kiel, Kiel, Germany, 1984.

Stott, L. D., J. P. Kennett, N. J. Shackleton, and R. M. Corfield, The evolution of Antarctic surface water during the Paleogene: Inferences of the stable isotope composition of planktonic foraminifers on Leg 113, *Proc. Ocean Drill. Program Sci. Results*, *113*, 849–863, 1990.

Thomas, E., et al., Upper Cretaceous–Paleogene stratigraphy of sites 689 and 690, Maud Rise (Antarctica), *Proc. Ocean Drill. Program Sci. Results*, *113*, 901–914, 1990.

Tucholke, B. E., and G. S. Mountain, Seismic stratigraphy, lithostratigraphy, and paleosedimentation patterns in the North American Basin; in *Deep Drilling Results in the Atlantic Ocean: Continental Margins and Paleoenvironments, Maurice Ewing Ser.*, vol. 3, edited by M. Talwani, W.

Hay, and W. B. F. Ryan, pp. 58–86, AGU, Washington, D. C., 1979.

Tucholke, B. E., and G. S. Mountain, Tertiary paleoceanography of the western North Atlantic Ocean, in *The Western North Atlantic Region, Decade of North American Geology*, vol. M, edited by P. R. Vogt and B. E. Tucholke, pp. 631–650, Geological Society of America, Denver, Colo., 1986.

Tucholke, B. E., C. D. Hollister, F. M. Weaver, and W. R. Vennum, Continental rise and abyssal plain sedimentation in the southeast Pacific basin, Leg 35 Deep Sea Drilling Project, *Initial Rep. Deep Sea Drill. Proj.*, 35, 279–294, 1976.

van Andel, T. H., Mesozoic/Cenozoic calcite compensation depth and the global distribution of calcareous sediments, *Earth Planet. Sci. Lett.*, 26, 187–194, 1975.

Vincent, E., J. S. Killingley, and W. H. Berger, The Magnetic Epoch—6 carbon shift: A change in the ocean's $^{13}C/^{12}C$ ratio 6.2 million years ago, *Mar. Micropaleontol.*, 5, 185–203, 1980.

Vogt, P. R., The Faeroe-Iceland-Greenland Aseismic Ridge and the Western Boundary Undercurrent, *Nature*, 239, 79–81, 1972.

Woodruff, F., and S. R. Chambers, Mid-Miocene benthic foraminiferal oxygen and carbon isotopes and stratigraphy, Southern Ocean ODP Site 744, *Proc. Ocean Drill. Program Sci. Results*, 115, 935–940, 1991.

Woodruff, F., and S. M. Savin, Miocene deepwater oceanography, *Paleoceanography*, 4, 87–140, 1989.

Woodruff, F., S. M. Savin, and L. Abel, Miocene Indian Ocean benthic foraminiferal oxygen and carbon isotopes: ODP Site 709, *Proc. Ocean Drill. Program Sci. Results*, 115, 519–528, 1990.

Wright, J. D., and K. G. Miller, Miocene stable isotope stratigraphy, Site 747, Kerguelen Plateau, *Proc. Ocean Drill. Program Sci. Results*, 120, 855–866, 1992.

Wright, J. D., K. G. Miller, and R. G. Fairbanks, Evolution of deep-water circulation: Evidence from the late Miocene Southern Ocean, *Paleoceanography*, 6, 275–290, 1991.

Wright, J. D., K. G. Miller, and R. G. Fairbanks, Miocene stable isotopes: Implications for deepwater circulation and climate, *Paleoceanography*, 7, 357–389, 1992.

Zachos, J. C., W. A. Berggren, M.-P. Aubrey, and A. Mackensen, Isotope and trace element geochemistry of Eocene and Oligocene foraminifers from Site 748, Kerguelen Plateau, *Proc. Ocean Drill. Program Sci. Results*, 120, 839–854, 1992.

Zachos, J. C., L. D. Stott, and K. C. Lohmann, Evolution of early Cenozoic marine temperatures, *Paleoceanography*, in press, 1993.

(Received August 18, 1992;
accepted January 5, 1993.)

THE ANTARCTIC PALEOENVIRONMENT: A PERSPECTIVE ON GLOBAL CHANGE
ANTARCTIC RESEARCH SERIES, VOLUME 60, PAGES 27–48

LATE EOCENE TO OLIGOCENE VERTICAL OXYGEN ISOTOPIC GRADIENTS IN THE SOUTH ATLANTIC: IMPLICATIONS FOR WARM SALINE DEEP WATER

GREGORY A. MEAD, DAVID A. HODELL, AND PAUL F. CIESIELSKI

Department of Geology, University of Florida, Gainesville, Florida 32611

Oxygen isotopic data from Ocean Drilling Program (ODP) sites 699 and 703 confirm the isotopic inversion found in the Southern Ocean by Kennett and Stott (1990) and interpreted by them as indicating the presence of Warm Saline Deep Water (WSDW) during the Eocene and Oligocene. We have used relationships between seawater temperature, salinity, density, and oxygen isotope ratios to further estimate salinities and temperatures of the bottom water in these regions, using a minimally stable density stratification. Our estimates of temperatures and salinities of the water overlying Site 699 lead us to suggest that the deep water was not WSDW sensu strictu (i.e., produced directly in semienclosed basins in low latitudes by excess evaporation). Instead, we suggest that the deep water was Warm Saline Deep Water produced by mixing and/or cooling of warm saline water at higher northern latitudes or by mixing of warm saline water produced in restricted basins with colder water during downward advection. A uniform 1‰ increase in $\delta^{18}O$ occurred during the earliest Oligocene in ODP sites 689, 690, and 699, ranging from ~1400-m to ~3400-m paleodepth. Assuming a 0.3–0.4‰ ice volume effect as suggested by other studies, the data indicate cooling in both the cooler intermediate water and the warmer deep water by ~2°–2.5°C. Since the shallower and deeper water masses had very different origins, we interpret the data to indicate that a concomitant cooling occurred in the two source areas. Depending on the exact location of the source areas, this interpretation suggests either an increase in the planetary thermal gradient by ~2°–2.5°C at this time or a worldwide cooling of the same magnitude. Between 35.8 and 35.2 Ma, a carbon isotopic enrichment of 0.5–1.0‰ occurred at all depths. This increase, along with evidence for increased occurrence of unconformities, suggests increased rates of overturn of the oceans during the earliest Oligocene beginning at the $\delta^{18}O$ increase.

INTRODUCTION

The history of climate change during the Tertiary has been one of stepwise cooling since the early Eocene [*Shackleton and Kennett*, 1975]. A great deal of evidence suggests high (10°–15°C [*McKenna*, 1980; *Wolfe*, 1980]) oceanic temperatures near the poles in the early Eocene, including faunal, floral [*Estes and Hutchinson*, 1980; *McKenna*, 1980; *Wolfe*, 1980], and oxygen isotopic evidence [*Shackleton and Kennett*, 1975; *Douglas and Savin*, 1975; *Miller et al.*, 1987]. The oxygen isotopic evidence has been somewhat equivocal, however, because the oxygen isotopic ratio of biogenic calcite is affected both by temperature and by the oxygen isotopic composition of seawater (and hence ice volume and salinity). *Shackleton and Kennett* [1975] and *Savin et al.* [1975] suggested that a 1‰ increase in $\delta^{18}O$ values during the earliest Oligocene was due to a decrease in high-latitude temperature and the production of cold bottom water, with another increase in the middle Miocene due to the first permanent accumulation

of ice on East Antarctica. In contrast, *Matthews and Poore* [1980] reinterpreted the oxygen isotopic signal to suggest that the earliest Oligocene oxygen isotopic increase was due to ice accumulation on East Antarctica and that a cooling of bottom waters caused the middle Miocene $\delta^{18}O$ increase. Further work has failed to resolve this question, although a consensus [*Wise et al.*, 1991] is emerging of at least limited Oligocene glaciation, particularly in light of recent discoveries of earliest Oligocene ice-rafted debris [*Zachos et al.*, 1992b; *Barrera and Huber*, this volume]. Neither interpretation has considered in detail the implications of the theories on deep-ocean circulation.

Production of deep and bottom waters today results from evaporation, cooling, and sea ice formation at high latitudes. As a result, the deep water in the oceans is cold and moderately saline. *Chamberlin* [1906] and *Brass et al.* [1982] pointed out that during times when the planetary temperature gradient was small and polar regions were warmer than today, cold dense water

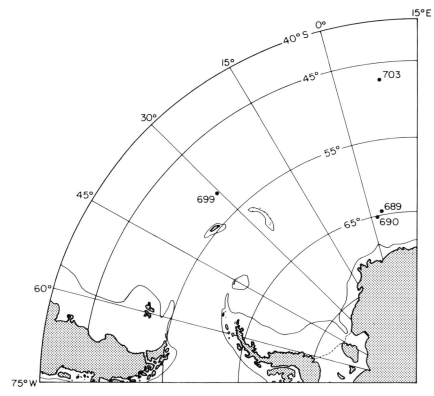

Fig. 1.   Location map of ODP sites 689, 690, 699, and 703.

could not have been produced at high latitudes. Instead, *Brass et al.* [1982] suggested, on the basis of modeling, that production of warm, highly saline water in low-latitude marginal seas may have produced Warm Saline Deep Water (WSDW) which would have been the densest water in the oceans in the absence of cold dense water.

A recent study [*Kennett and Stott*, 1990], which analyzed benthic foraminiferal oxygen isotopes from Ocean Drilling Program (ODP) Site 689 (paleodepth of ~1400 m) and Site 690 (paleodepth of ~2200 m) (Figure 1), provided the first evidence for a Paleogene deep circulation quite different from that of the late Neogene. Neogene to Recent deep circulation, with cold bottom water, results in a pattern of decreasing temperature with increasing depth in the ocean. Because $^{18}O/^{16}O$ ratios in biogenic calcite increase with decreasing temperature, analysis of Recent benthic foraminiferal oxygen isotopic ratios shows a pattern of increased $\delta^{18}O$ with depth, resulting in a positive ($\delta^{18}O$ versus depth) gradient. Instead of this configuration, *Kennett and Stott* [1990] found that at several times in the late Eocene and during much of the Oligocene, the gradient of benthic foraminiferal $\delta^{18}O$ versus depth reversed, indicating increased temperatures with increased depth. They suggested that this temperature inversion could be

explained by the presence of Warm Saline Deep Water produced in lower latitudes. Their ''Proteus'' scenario began with WSDW sporadically bathing Site 690 in the Eocene. *Kennett and Stott* [1990] suggested that in the Oligocene (''Proto-Oceanus''), the production of Proto-Antarctic Bottom Water (Proto-AABW) raised the WSDW up to the level of Site 690, creating a near-permanent oxygen isotopic inversion. In the late Oligocene, the establishment of modern (''Oceanus'') circulation occurred, with the dominance of AABW in the oceans.

Kennett and Stott [1990] speculated that the onset of AABW formation began after the earliest Oligocene oxygen isotopic enrichment, but in the absence of direct data from a deepwater site, this hypothesis has gone untested. To test this hypothesis, first we need to determine whether the reversed isotope gradient is present at other sites with different depths, in order to better outline the bathymetric, geographic, and temporal extent of WSDW. Second, we need to test whether or not Proto-AABW may have been present below the WSDW in the Oligocene. If in fact the production of Proto-AABW below WSDW begins at the earliest Oligocene oxygen isotope shift, it should be signaled by the establishment of a trend of increased $\delta^{18}O$ (decreased temperatures) with increasing depth below the WSDW-AABW boundary.

Leg 114 of the Ocean Drilling Program drilled two holes that help address these questions. Hole 703A was drilled in relatively shallow (1796 m) water on the Meteor Rise, sampling midwater paleodepths, and Hole 699A was drilled in deeper water (3705 m), sampling abyssal water masses (Figure 1). We analyzed the isotopic composition of benthic foraminiferal calcite from these two sites to delineate, in conjunction with sites 689 and 690, the vertical water mass structure of the South Atlantic during the late Eocene to Oligocene, from ~42 to 26 Ma.

METHODS

Age Model

The age models used by *Kennett and Stott* [1990] for sites 689 and 690 are based on magnetostratigraphy [*Spiess*, 1990; *Hamilton*, 1990] and planktonic foraminiferal biostratigraphy [*Stott and Kennett*, 1990] and were calibrated to the *Berggren et al.* [1985] time scale. This time scale was used rather than the newer *Cande and Kent* [1992] geomagnetic polarity time scale because of the lack of calibration (as yet) of biostratigraphic datums to the newer time scale. An alternative to the *Kennett and Stott* [1990] age model has been developed for Site 690 by *Thomas et al.* [1990] that is significantly different in some respects. In most cases, these differences have little effect on oxygen isotopic conclusions, with 1 m.y. averages differing by 0.22‰ or less. The only major difference occurs with respect to an unconformity near 91–94 meters below seafloor (mbsf). *Kennett and Stott*'s [1990] isotopic data suggest that the position of the unconformity should be nearer to 93.20 mbsf as opposed to the 91.58 mbsf level suggested by *Thomas et al.* [1990]. If the *Thomas et al.* [1990] position is used, the earliest Oligocene isotopic shift occurs below the unconformity, at an age greater than 38.5 Ma. Because this isotopic shift is known to occur very near 36 Ma [*Oberhänsli et al.*, 1984; *Miller et al.*, 1988; *Zachos et al.*, 1992a], we accept the position of the unconformity at 93.20 mbsf. Unfortunately, biostratigraphic and magnetostratigraphic studies [*Spiess*, 1990; *Hamilton*, 1990; *Stott and Kennett*, 1990; *Thomas et al.*, 1990; *Wei*, 1991] do not delineate the boundary ages well enough to determine which of the ages of the unconformity is correct. We provisionally accept the remainder of the *Kennett and Stott* [1990] age model. This site is key for Antarctic paleoceanography, and more work is needed to define better the age of sediments at this site.

A combination of paleomagnetic datums [*Hailwood and Clement*, 1991a, b] and calcareous nannofossil biostratigraphic markers were used to construct the age models for sites 699 and 703 (Table 1). The ages were derived from the *Berggren et al.* [1985] time scale. The

datum levels below 35.9 Ma in Site 699 and the lowest two datum levels in Site 703 are biostratigraphic rather than magnetostratigraphic, which might lead to doubts about the age resolution, especially in the high sedimentation rate zone between the first appearance datums (FAD) of the calcareous nannofossils *Reticulofenestra oamaruensis* and *Ismolithus recurvus* (Site 699). However, *Wei* [1991, 1992] has shown these datums to be reliable markers for the latitudes from 47°S to 65°S. Similarly, *Wei* [1991, 1992] found the 41.0 Ma biostratigraphic datum (FAD of *Chiasmolithus oamaruensis*) and the nearly correlative datum just below (last appearance datum (LAD) of *C. solitus*, 41.4 Ma) also to be reliable biostratigraphic markers.

One unconformity in Site 699 and four unconformities in Site 703 were identified within the time represented by the isotopic records. In Site 699, an unconformity was identified between 33.9 and 33.0 Ma, marked by the joint LADs of *C. oamaruensis* and *I. recurvus* between 254.33 and 252.94 mbsf. There is a 1.4-m uncertainty in the position of this unconformity, below which the sedimentation rate was 16.0 m/m.y., but even when this uncertainty and the uncertainty in the age of the base of the unconformity (±0.14 m.y.) are taken into account, the range of possible sedimentation rates is only 14.6 to 17.7 m/m.y. Age assignments of samples will not vary by more than 0.28 m.y. (the total uncertainty in the age of the base of the unconformity). In Site 703 a hiatus is present across the Eocene/Oligocene boundary between 37.1 and 35.0 Ma. The lower age of this unconformity is based on a combination of biostratigraphic and magnetostratigraphic factors, combined with an age extrapolated from sedimentation rates. The younger age is extrapolated from the sedimentation rate defined by the LAD of *I. recurvus* and the C12N/C12R paleomagnetic boundary. Another unconformity occurs between 30.4 and 29.5 Ma, for which the lower boundary is defined by another extrapolation between the C12N/C11R and C11N/C10R boundaries. The third unconformity, between 25.8 and 24.0 Ma, is defined at the base by an abundance of *Rocella vigilans* and at the top by a combination of biostratigraphic and magnetostratigraphic datums. The last unconformity, between 23.2 and 21.4 Ma, is defined by extrapolation of sedimentation rates based on paleomagnetic datums.

The age models of sites 699 and 703 were based on a different set of biostratigraphic datums than sites 689 and 690. The planktonic foraminiferal faunas of the Leg 113 sites are different from those of the Leg 114 sites, resulting in different biostratigraphic zonations [*Stott and Kennett*, 1990]. Calcareous nannofossil biostratigraphy [*Wei and Wise*, 1990] was primarily used to guide the magnetostratigraphy of the Leg 113 sites [*Spiess*, 1990]. Several of the reliable markers used by *Wei* [1991, 1992] are present in both sets of sites, suggesting to us that miscorrelations, if any, between these sites are small.

TABLE 1.   Age Models for ODP Holes 699A and 703A

Depth, mbsf	Age, Ma	Datum
	Hole 699A	
127.69 ± 0.15	25.97	C7N/C7R
137.09 ± 0.05	26.38	C7R/C7AN
140.38	26.56	C7AN/C7AR
143.84 ± 0.10	26.86	C7AR/C8N
145.73 ± 0.08	26.93	C8N.1
146.21 ± 0.04	27.01	C8N.2
156.14 ± 0.10	27.74	C8N/C8R
166.95 ± 0.75	28.15	C8R/C9N
182.69 ± 0.35	29.21	C9N/C9R
189.34 ± 0.10	29.73	C9R/C10N
192.55 ± 0.10	30.03	C10N.5
193.05 ± 0.10	30.09	C10N.6
198.35 ± 1.55	30.33	C10N/C10R
208.09 ± 0.25	31.23	C10R/C11N
216.05 ± 0.75	32.06	C11N/C11R
249.70	32.90	C12N/C12R
253.64 ± 0.70	33.00 ± 0.02	
unconformity		
253.64 ± 0.70	33.94 ± 0.14	
284.55 ± 0.86	35.87	C13N2/C13R2
321.28	38.16	FAD *Reticulofenestra oamaruensis*
330.70	38.40	FAD *Ismolithus recurvus*
350.56	41.00	LAD *Chiasmolithus oamaruensis*
	Hole 703A	
22.72 ± 0.37	19.35	C5ER/C6N
31.35 ± 0.10	20.45	C6N/C6R
32.59 ± 0.05	20.88	C6R/C6AN
36.54 ± 0.09	21.16	C6AN.3
39.14 ± 0.10	21.38	C6AN.6
39.44 ± 0.30	21.40	
unconformity		
39.44 ± 0.30	23.22	
40.00 ± 0.10	23.27	C6BN/C6CN
41.89 ± 0.05	23.44	C6CN.2
44.54 ± 0.40	23.55	C6CN.3
45.99 ± 0.25	23.79	C6CN.6
52.95 ± 0.55	24.03	
unconformity		
52.95 ± 0.55	25.79 ± 0.10	
56.00 ± 0.04	25.97	C7N/C7R
57.25 ± 0.05	26.38	C7R/C7AN
57.40	26.56	C7AN/C7AR
58.00 ± 0.05	26.86	C7AR/C8N
63.40 ± 0.10	27.74	C8N/C8R
64.00	28.15	C8R/C9N
65.89 ± 0.05	29.21	C9N/C9R
67.81 ± 1.49	29.47	
unconformity		
67.81 ± 1.49	30.21 ± 0.31	
78.34 ± 0.10	31.23	C10R/C11N
79.35 ± 0.10	31.58	C11N.4
80.30 ± 0.15	31.64	C11N.5
86.02 ± 1.02	32.06	11
90.99 ± 1.35	32.46	top C12
95.74 ± 0.40	32.90	C11R/C12N
107.52	34.80	LAD *Ismolithus recurvus*
109.70	35.15 ± 0.01	

TABLE 1.   (continued)

Depth, mbsf	Age, Ma	Datum
unconformity		
109.70	37.10 ± 0.05	
130.70 ± 0.05	38.10	16
146.50 ± 0.55	38.79	C16N.5
154.95 ± 0.10	39.24	C16N/C16R
173.71	41.00	FAD *Chiasmolithus oamaruensis*

Subsidence Models

Paleodepths of the sites were calculated using the thermal subsidence model of *Parsons and Sclater* [1977]. Paleodepths were calculated with isostatic corrections made using sediment densities presented in the site reports [*Barker et al.*, 1988*a*; *Ciesielski et al.*, 1988]. *Hayes* [1988] recently recalculated subsidence constants for the South Atlantic and found that different segments and sides of the South Atlantic mid-ocean ridge subsided at different rates. Subsidence constants were taken from *Hayes* [1988] if applicable; otherwise an average value of 300 was used (*Hayes* [1988]; Table 2). A computer program was written to isostatically correct for sediment loading, and oceanic crustal subsidence was calculated with a crossover age of 30 Ma separating "young" and "old" crust (as defined by the differing styles of subsidence).

Average wet bulk density between each successive isotopic measurement was multiplied by the depth difference between each measurement and summed to get the isostatic loading, and the isostatic corrections were made assuming a mantle density of 3.30 g/cm^3 and seawater density of 1.03 g/cm^3. If drilling did not reach basement, the depth to basement was taken from seismic data in the site reports: 332 mbsf for Site 689 [*Barker et al.*, 1988*a, b*] and ~740 mbsf for Site 699 [*Ciesielski et al.*, 1988]. Densities of these deeper sediments were assumed to be equal to the average of the lowermost few density measurements. All sites were assumed to have subsided according to the *Parsons and Sclater* [1977] model. The only difference was that the starting or maximum depths were different based on the present depth, and calculations were based on the parameters presented in Table 2. Subsidence curves are plotted in Figure 2.

Isotopic Methods

Well-preserved specimens of *Cibicidoides* were picked from the washed >150-μm fraction of samples from holes 699A and 703A. Organic carbon was removed from the picked specimens by either crushing and roasting at 375°C for 1 hour (part of Hole 703A samples) or by reaction in 12.5% H_2O_2 for 1/2 to 1 hour (part of Hole 703A and all Hole 699A samples). No

TABLE 2. Parameter List for Subsidence Calculations

Site	Crustal Age, Ma	Present Depth, m	Total Sediment Thickness, mbsf	Subsidence Constant, m/m.y.$^{0.5}$	Depth at Zero Age, m
689	84.0[a]	2080.0	320	300[f]	−495
690	84.0[a]	2914.0	332[d]	300[f]	325
699	100.0[b]	3705.5	740[e]	380[g]	775
703	60.0[c]	1796.1	364	290[h]	−310

Unless otherwise noted, all water depths and sediment thicknesses are from site reports in the Ocean Drilling Program Initial Reports [*Barker et al.*, 1988*a*; *Ciesielski et al.*, 1988].

[a]*Kennett and Stott* [1990] and *Barker et al.* [1988*b*] note that the oldest sediments correlate with the Campanian/Maestrichtian boundary (74.5 Ma). However, *Shipboard Scientific Party* [1988*a*] suggests a Cenomanian age for basement, while *Shipboard Scientific Party* [1988*b*] suggests 10 Ma of subaerial exposure after formation at Site 689. We therefore use an age of 84 Ma for the formation of basement at this site.

[b]Assumed to be the same age as nearby Site 700, dated by *Kristoffersen and LaBrecque* [1991].

[c]From *Raymond et al.* [1991].

[d]Hole 690 (297.3 mbsf maximum penetration) was abandoned an estimated 25–45 m above basement [*Shipboard Scientific Party*, 1988*b*]. We use 35 m as an average value, resulting in a total sediment load of 332 m.

[e]An estimated 700 m of seismically defined sediment overlies an approximately 80-m-thick basal unit [*Shipboard Scientific Party*, 1988*c*]. Basement is not well defined and is assumed to be midway within the basal unit.

[f]Revised average value [*Hayes*, 1988] for the entire ocean was used for these sites because revised values for the Southern Ocean were not available.

[g]Western basin of the southern South Atlantic (Zone C of *Hayes* [1988]).

[h]Eastern basin of the southern South Atlantic (Zone C of *Hayes* [1988]).

significant difference was found in results from these two methods of pretreatment. Specimens were ultrasonically cleaned in methanol for 10–20 s to remove adhering debris and were examined microscopically to ensure the effectiveness of the treatment.

Specimens were reacted in 100% orthophosphoric acid at either 70° or 90°C to release CO_2 in an automated preparation system. The released CO_2 was analyzed in a VG Prism triple collector mass spectrometer. Isotopic standards were Carrara Marble and NBS-19, corrected to Pee Dee Belemnite [*Hodell et al.*, 1989]. Precision (1 standard deviation) was ±0.13‰ and ±0.05‰ for oxy-

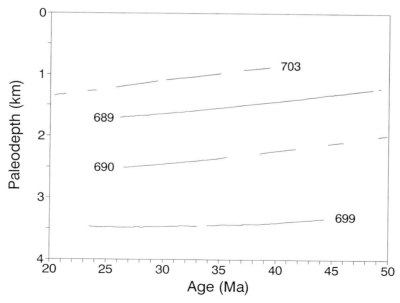

Fig. 2. Subsidence curves for the sites used in this study, calculated by the method of *Parsons and Sclater* [1977].

TABLE 3.   Comparison of Isotopic Analyses of NBS-19
and NBS-20

	$\delta^{13}C$	$\delta^{18}O$
NBS-19		
Coplen et al. [1983] standard value	1.92	−2.19
Hodell et al. [1989]	1.90	−2.19
Kennett and Stott [1990]	1.92	−2.19
NBS-20		
Coplen et al. [1983] standard value	−1.06	−4.14
Hodell et al. [1989]	−1.10	−4.19
Kennett and Stott [1990]	−1.06	−4.14

Values from *Hodell et al.* [1989] are those used for this study,
and the values for *Kennett and Stott* [1990] were provided by J.
P. Kennett (personal communication, 1990).

gen and carbon isotopes, respectively, for the Carrara
Marble standard and ±0.09‰ and ±0.04‰, respectively, for the NBS-19 standard. A total of 14% of all
sample analyses were replicated. The mean difference
between replicate pairs (or standard deviation when
more than two replicates were made) was 0.16‰ for
$\delta^{18}O$ and 0.12‰ for $\delta^{13}C$.

This study includes isotopic results from ODP sites
689 and 690, produced by *Kennett and Stott* [1990], and
sites 699 and 703 from this study. Interlaboratory calibration, especially with reference to possible offsets, is
therefore a concern. To reduce this possibility, both
data sets have been calibrated to PDB by using the same
intermediate standards (NBS-19 and NBS-20) or have
been intercalibrated with that standard via Carrara
Marble. Repeated analyses of NBS-19 and NBS-20 in
our laboratory are not significantly different from the
results of *Kennett and Stott* [1990] or the standard
values of *Coplen et al.* [1983] (Table 3).

RESULTS

Subsidence Curves

Table 4 presents a synopsis of paleodepths at 26, 30,
35, and 40 Ma for the five sites. The four ODP sites
show nonoverlapping paleodepths (Figure 2). Note that
the minimum separation between any two of these sites
is 495 m (sites 689 and 703 at 26 Ma), so that even if a
total error of as much as 495 m is present in any pair of
the paleodepth calculations, the relative depths of sites
689, 690, 699, and 703 will remain the same.

The Maud Rise, on which sites 689 and 690 are
located, is an aseismic ridge, making assumptions of
normal oceanic subsidence [*Parsons and Sclater*, 1977]
an oversimplification [*Thomas*, 1989]. However, as
noted above, even errors in depth estimates of >0.5 km
will not change the ordination of the ODP sites. In
addition, our estimates are comparable to estimates
derived by other methods [*Thomas*, 1989, 1990, 1992,
written communication, 1991].

Oxygen Isotopes

Site 703. The isotopic record of 162 samples from
Hole 703A extends from 39.6 to 20.4 Ma (159.13–30.72
mbsf) (appendix Table A1 and Figure 3). The average
sampling interval is 0.79 m (80 kyr, excluding unconformities). The upper Eocene portion of the $\delta^{18}O$ record
averages ~1‰. The earliest Oligocene oxygen isotope
shift is ~0.5‰, smaller than the usual 1.0‰ seen in most
areas of the deep sea. After the earliest Oligocene
oxygen isotope increase, which occurs within the unconformity present between 37.10 and 35.2 Ma (109.30–
109.85 mbsf), $\delta^{18}O$ values decreased from near 2‰ to
about 1.6‰ until 33.0 Ma (96 mbsf), an increase of 0.6‰
over Eocene values.

Near 32 Ma, $\delta^{18}O$ values began to increase, reaching
peak values of over 2.8‰ from 30.8 to 30.9 Ma. A
decrease to 2.2–2.5‰ follows to the unconformity at
67.81 mbsf (30.2 Ma). When sedimentation resumed (at
29.3 Ma), $\delta^{18}O$ values generally decreased from a maximum of 2.4‰ near 28 Ma to 1.7–1.8‰ between 27.0 and
25.8 Ma (59.0–52.95 mbsf), after which another unconformity occurred. During the short period of sedimentation between this unconformity and the next at 39.44
mbsf, sediments are present with ages ranging from 24.0
to 23.2 Ma. In this interval, $\delta^{18}O$ ratios increased from
initial values of 1.7‰ to near 2.0‰ from 23.8 to 23.5 Ma
and then decreased back to 1.6–1.8‰. Above the unconformity at 39.44 mbsf, this record extends from 21.4
Ma, when sedimentation resumed, to 20.4 Ma (30.72
mbsf). Oxygen isotope ratios average about 1.9‰ in this
interval, including one value of 2.44‰ at 21 Ma.

Site 699. The Hole 699A record (appendix Table A2
and Figure 3) begins at 41.6 Ma (356.35 mbsf) and
continues to 26.3 Ma (134.71 mbsf), with a total of 130
samples and an average sampling interval of 111 kyr
(1.72 m). The record ends at 134.71 mbsf because not
enough carbonate is present above this level for isotopic
analysis. Initial $\delta^{18}O$ values are lower than those of
equivalent age in Hole 703A despite the greater paleodepth of Hole 699A. Beginning at 41.6 Ma, $\delta^{18}O$
values average 0.2–0.3‰, increasing to 0.7–0.8‰ between 39 and 36 Ma (288 mbsf). The earliest Oligocene
oxygen isotope shift occurs between 35.87 and 35.77 Ma
(284.60 and 283.02 mbsf). Additional unpublished oxygen isotopic data from analysis of *Stilostomella aculeata* reveal that the shift is further constrained be-

TABLE 4.   Paleodepths of Sites 689, 690, 699, and 703 at
Selected Times, Based on Subsidence Curves Calculated
Using Parameters in Tables 1 and 2

Site	26 Ma	30 Ma	35 Ma	40 Ma
703	1206	1093	981	865
689	1701	1638	1543	1438
690	2518	2455	2356	2240
699	3480	3466	3431	3401

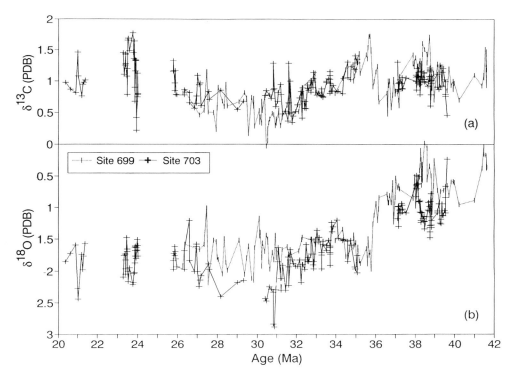

Fig. 3.   Oxygen and carbon isotopic records from ODP sites 699 and 703, plotted versus age. Gaps in records indicate positions of nonconformities.

tween 35.87 and 35.82 Ma (284.60 and 283.79 mbsf), suggesting that the change occurs on a time scale of 50 kyr or less. The rapidity of the shift is in agreement with data from other sites [*Kennett and Shackleton*, 1976; *Oberhänsli et al.*, 1984; *Zachos et al.*, 1992*a*; *Barrera and Huber*, this volume]. At Site 699 the oxygen isotopic shift is larger than that at Site 703, with a magnitude of about 1‰, comparable to most benthic oxygen isotopic signals found in other regions (North Atlantic Ocean [*Miller and Fairbanks*, 1983], South Atlantic Ocean [*Poore and Matthews*, 1984; *Corliss et al.*, 1984; *Oberhänsli and Toumarkine*, 1985; *Keigwin and Corliss*, 1986; *Miller et al.*, 1987, 1988], Pacific Ocean [*Keigwin*, 1980; *Corliss et al.*, 1984; *Miller and Thomas*, 1985], and Indian Ocean [*Corliss et al.*, 1984; *Zachos et al.*, 1992*a*, *b*; *Barrera and Huber*, this volume]).

Initial postshift $\delta^{18}O$ values average 1.76‰, thereafter averaging mostly between 1.55 and 1.81‰ to 26.3 Ma. Greater values of $\delta^{18}O$ are generally found at 36–35 Ma and 31–28 Ma. Values commonly exceed 1.8‰, suggesting glacial conditions [*Miller and Fairbanks*, 1983; *Miller et al.*, 1985*b*, 1987]. Oxygen isotopic ratios of 2.0–2.15‰ are found especially between 29.85 and 29.7 Ma. High-frequency variability is seen in the Site 699 oxygen isotopic record, with values varying by ±0.5‰. However, owing to the generally low sampling interval, it is difficult to interpret this variability. The

oxygen isotopic values in Site 699 are generally comparable to or lower than those of Site 703, despite the considerably greater paleodepth of Site 699. In addition, the average values from Site 699 are lower than those found in either Site 689 or Site 690.

Carbon Isotopes

Site 703.   Initial $\delta^{13}C$ values average ~1‰ from the base of the record to the upper Eocene–lower Oligocene unconformity at 109.30–109.85 mbsf. Above the unconformity, $\delta^{13}C$ values increase to as much as 1.41‰ for a period of about 680 kyr (4.2 m). After this interval, $\delta^{13}C$ values decrease first to between 0.8 and 0.9‰ by ~34 Ma, decrease further to a minimum of ~0.5‰ at 31.4 Ma, and then increase again to ~0.8‰ at the unconformity at 30.2 Ma.

During the unconformity-bounded interval between 29.5 and 25.8 Ma (67.81–52.95 mbsf), $\delta^{13}C$ values remained at about 0.7–0.8‰ until increasing to over 1.1‰ just before the upper unconformity. After sedimentation resumed at 24.0 Ma, carbon isotopic ratios decreased rapidly from values near 0.9‰ to as low as 0.22‰ by 24.0 Ma before increasing again to maximum values of between 1.33 and 1.77‰ from 23.9 to 23.5 Ma. Carbon isotopic ratios decreased again to between 0.7 and 1.45‰ just before the unconformity. The record above

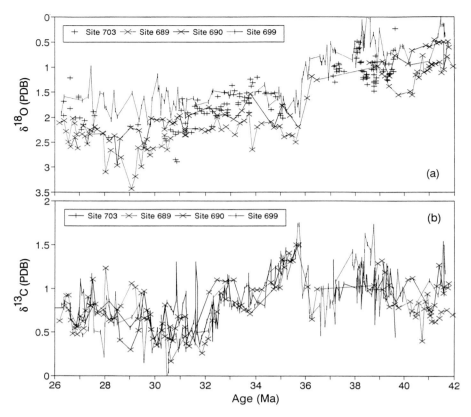

Fig. 4.   Isotopic records from ODP sites 689, 690, 699, and 703 for the time period 26–42 Ma: (a) $\delta^{18}O$ and (b) $\delta^{13}C$.

the unconformity from 21.4 to 20.4 Ma averages ~1‰, with one short interval of high $\delta^{13}C$ associated with high $\delta^{18}O$ values at 21 Ma.

Site 699.   Results for Site 699 are generally similar to those of Site 703. From 41.6 to 39.2 Ma (356.35–337.1 mbsf), $\delta^{13}C$ averaged ~1‰. A sudden increase, not seen in Site 703, began at this point and continued until 38.7 Ma (333 mbsf) to values as high as 1.7‰. Afterward a decrease to preceding values occurred, reaching values of ~0.8–1.0‰ by 37.2 Ma (306.1 mbsf). Coeval with the earliest Oligocene $\delta^{18}O$ increase beginning at about 35.8 Ma (284 mbsf), another short-term increase, similar to that seen in Site 703, occurred with $\delta^{13}C$ increasing to between 1.49‰ and 1.77‰ until 35.5 Ma (279 mbsf). Carbon isotopic ratios average 1.0–1.2‰ until the hiatus at 33.9–33.0 Ma (253.64 mbsf). With the onset of renewed sedimentation at 33 Ma (253.64 mbsf), $\delta^{13}C$ values began to decrease from about 1‰ to a brief minimum of 0‰ at about 30.5 Ma. Values averaged near 0.5–1.0‰ thereafter, with high-frequency variability of about 0.5‰.

DISCUSSION

Comparison of the two oxygen isotopic records (Figure 3) reveals that during most of the time from 26 to 42

Ma (the time during which overlap occurred between the records for sites 689, 690, 699, and 703), an inversion in the $\delta^{18}O$ gradient was present, with lighter $\delta^{18}O$ values deeper in the water column. This configuration is similar to that reported by *Kennett and Stott* [1990]. Oxygen (Figure 4a) and carbon (Figure 4b) isotopic records of sites 689, 690, 699, and 703 plotted together reveal that strong $\delta^{18}O$ gradients were present between the sites during most of this interval, but there was little difference between the $\delta^{13}C$ signals. We have contoured the four $\delta^{18}O$ (Figure 5a) and $\delta^{13}C$ (Figure 5b) data sets (uncorrected *Cibicidoides*) as a function of paleodepth and time, to better observe the changes taking place during this interval. We used weighted 1 m.y. averages of each of the four isotopic data sets to decrease effects of uncertain dating, high-frequency variability, and noise in the isotopic signal. In calculating weighted averages, replicate analyses were first averaged, then data in 100-kyr intervals were averaged, and finally the 100-kyr values were combined into 1 m.y. averages. The only part of the data not treated in this way were the data on either side of the earliest Oligocene oxygen isotopic shift in Site 699. Data on opposite sides of the shift were averaged separately. The 1 m.y. averages were weighted in this way in order to eliminate bias

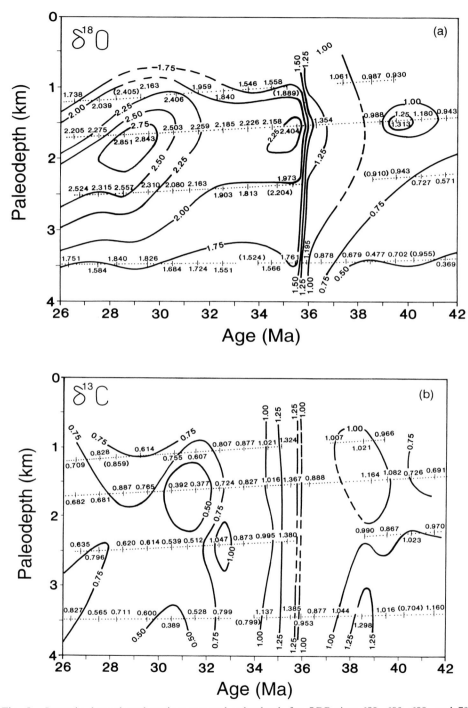

Fig. 5. Isotopic data plotted against age and paleodepth for ODP sites 689, 690, 699, and 703, contoured on the basis of the ODP sites. (a) Data for $\delta^{18}O$. Note the near-permanent $\delta^{18}O$ minimum near the paleodepth of Site 689, indicating a temperature minimum. Oxygen isotopic ratios decrease downward to maximum paleodepths, indicating the presence of a warmer, more saline deepwater mass. (b) Data for $\delta^{13}C$. Note the relative lack of vertical gradients at all times and the enrichment in $\delta^{13}C$ over the 36–35 Ma interval.

based on unequal sample distributions within each 1 m.y. interval. This also has the unfortunate effect of reducing the amplitude of some of the maxima and minima when the duration of the extreme values is substantially less than 1 m.y. For example, peak $\delta^{13}C$ values of over 1.7‰ at ~35.6 Ma in Site 699 are not apparent in the contoured data.

Beginning at 42 Ma (before the Site 703 record begins), an oxygen isotopic gradient of about 0.6‰ existed between Site 689 (1.4-km paleodepth, 0.94‰) and Site 699 (3.4-km paleodepth, 0.37‰). Except for one measurement from Site 699A at 40.24 Ma, this gradient is maintained at 0.5–0.6‰ from 42 to 31 Ma. From 31 to 28 Ma, the gradient between Sites 689 and 699 strengthened to between 0.8 and 1.0‰. At this time the core of the high $\delta^{18}O$ values, initially always present at Site 689 (1.3- to 1.6-km paleodepth), seems to have increased in depth, to about 2.5 km, with the core of greater $\delta^{18}O$ (cooler temperatures) near the depth of Site 690. By 26 Ma the magnitude of the gradient also weakened slightly to 0.7–0.8‰. The gradient between high $\delta^{18}O$ values at Site 689 and lower values at the overlying Site 703 was 0.4‰ at the beginning of this overlapping sequence, decreasing to 0‰ from 40 to 38 Ma. Data are not present for either site from 38 to 35 Ma, but by 35 Ma a gradient was well established, with magnitudes of 0.6–0.7‰ for 2 m.y. and decreasing to 0.1‰ by 30 Ma. From 30 to 29 Ma, the gradient increased to 0.7‰. Following only a single data point for the 29–28 Ma interval in Site 703 and the deepening of the colder core, the magnitude of the gradient between Site 703 and 690 increased from 0.3‰ to 0.8‰ again.

This pattern extends *Kennett and Stott*'s [1990] identification of inverted oxygen isotope gradients to greater depths, suggesting the presence of a warmer water mass down from paleodepths of ~2100–2400 m to nearly 3500 m. However, alternative explanations for these gradients must first be addressed. We have identified three alternative possibilities.

First, the gradients could be artifacts of interlaboratory calibration errors, since our results have generally (for a given age) lower $\delta^{18}O$ values than the results of *Kennett and Stott* [1990]. However, because of the intercalibration already discussed and because the gradients are generally larger than the uncertainty, we feel this cannot be the cause of the gradients.

Second, errors in dating could be a source of these gradients as well. One of the reasons for averaging 1 m.y. increments was to help eliminate small errors in dating. Examination of the gradients shows that except near the earliest Oligocene oxygen isotope shift, even errors in dating of 1 m.y. would not change the character of the gradients. Near the earliest Oligocene, dating is well constrained by the position of the oxygen isotope shift, which has been shown to be synchronous worldwide [*Corliss et al.*, 1984; *Poore and Matthews*, 1984;

Miller et al., 1985*b*, 1988; *Oberhänsli and Toumarkine*, 1985; *Corliss and Keigwin*, 1986].

Third, another possible reason for these isotopic patterns is a latitudinal mixing effect. Water masses mix with overlying and underlying water masses during northward and southward transport in the Atlantic, changing composition and temperature. It is possible that the reason for the lower $\delta^{18}O$ ratios found in sites 699 and 703 is that they contain a larger northerly component than sites 689 and 690 simply because they are located 15° to the north. To evaluate this, we constructed equilibrium oxygen isotopic profiles using modern hydrographic data from near the site locations [*Bainbridge*, 1981*a*, *b*; *Ostlund et al.*, 1987]. We calculated equilibrium $\delta^{18}O$ values using the temperatures and oceanic $\delta^{18}O$ values at the modern location and paleodepths of the four sites at four different times (39.5, 35.5, 30.5, and 26.5 Ma), corrected to the *Cibicidoides* offset of −1.02‰ [*Graham et al.*, 1981]. We use a correction factor of −1.02‰ rather than the more commonly used −0.64‰ because, except for *Shackleton and Opdyke* [1973], most investigations into taxa-specific fractionation of oxygen isotopes have found an offset of about −1‰ between the $\delta^{18}O$ of *Cibicidoides* and of equilibrium calcite [*Duplessy et al.*, 1970; *Belanger et al.*, 1981; *Graham et al.*, 1981]. Figure 6 shows that indeed a latitudinal effect may be present. Equilibrium $\delta^{18}O$ values of Site 699, corrected to *Cibicidoides*, are seen to be up to about 0.25‰ lower than those found at the position and paleodepths of Site 689. This difference is primarily due to the warmer in situ temperatures of Site 699 (1.0°C) than those at Site 689 (−0.3°C). However, isotopic gradients between these sites during the late Eocene and early Oligocene are nearly always at least twice the 0.25‰ seen in the present-day situation. In addition, while the calculated $\delta^{18}O$ values for the present day show lower $\delta^{18}O$ at Site 703 than at Site 699 (as would be expected for a shallower site), the oxygen isotopic composition is lighter at Site 699 than at Site 703 during most of the 14 m.y. time period when the two records overlap. Another consideration is that low latitudinal thermal gradients in surface waters during the Eocene and Oligocene relative to today would reduce the total range of temperatures seen in the ocean [*Keigwin and Corliss*, 1986; *Zachos et al.*, 1992*a*]. These factors lead us to believe that the gradients seen between the Maud Rise and sites 703 and 699 are in fact real, although perhaps up to 0.25‰ smaller than indicated in Figure 6. We conclude that our evidence supports a warmer, more highly saline water mass underlying a cooler, less saline water mass in the South Atlantic during most of the period from 42 to 26 Ma.

Circulation Patterns

Kennett and Stott [1990] suggested that the reason for the strengthened inversion present after the Eocene/

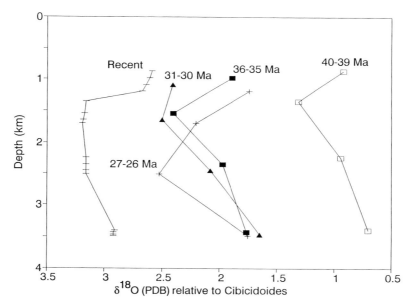

Fig. 6. Vertical oxygen isotopic gradients from *Cibicidoides* for the present (calculated on the basis of hydrographic data and an offset from equilibrium calcite of $-1.02‰$ by *Cibicidoides*) and for the time periods 40–39 Ma, 36–35 Ma, 31–30 Ma, and 27–26 Ma. Gradients are larger in the Paleogene sections than in Recent sections, indicating that any latitudinal mixing effect is not solely responsible for the gradients seen in the Paleogene sections.

Oligocene boundary is the onset of Antarctic Bottom Water formation, and a "Proto-Oceanus" configuration. If this was the case, we would expect to see an eventual cooling of deep water and an increase in $\delta^{18}O$ somewhere deeper than Site 690. Our data do not confirm or deny this. We find instead a continuous and permanent (at least to 26 Ma) $\sigma^{18}O$ inversion from ~1500-m to −3400-m paleodepth. We interpret the configuration to show, from 42 to 28 Ma, a core of cooler water centered near 1500 m (the paleodepth of Site 689) with temperatures increasing upward to Site 703 and downward from Site 689 through Site 690 and at least to the paleodepth of Site 699 (3400–3500 m). A mixing zone between the cooler water at Site 689 and the warmer water at Site 699 would have been present, but if a discrete boundary layer was present, it is not identifiable within the resolution of this study. At 28 Ma there is a suggestion that this core of cooler water may have begun to deepen somewhat. If proto-AABW was present in the Oligocene, it must have been present at greater paleodepths than we find at Site 699 (3400–3500 m).

Examination of the carbon isotopic composition of foraminiferal calcite at these four sites may be useful in understanding deep circulation during this period. Today, substantial vertical and horizontal gradients exist in $\sigma^{13}C$ in the Atlantic Ocean [*Kroopnick*, 1980]. This is a result of mixing of younger North Atlantic Deep Water (NADW) with Antarctic Bottom Water. There is little effect of aging on NADW because the residence time of

water in the Atlantic basin is short [*Broecker*, 1979]; thus the relatively high $\delta^{13}C$ of NADW is a useful tracer throughout the Atlantic [*Kroopnick*, 1985; *Oppo and Fairbanks*, 1987; *Charles and Fairbanks*, 1992]. A vertical section of carbon isotopes (Figure 5b), constructed in the same way as Figure 5a, shows that almost no vertical gradient existed during most of the 16 m.y. time interval examined in this paper. Two possible reasons for this are (1) that the deep Southern Ocean could have been well mixed during this time or (2) that there could indeed have been a singular source of deepwater production. However, the presence of the oxygen isotope gradients renders both of these explanations unlikely. Previous studies have shown that during most of the late Eocene and Oligocene, there was little difference between carbon isotope records from different oceans, as well as small surface-to-deep carbon isotopic gradients [*Miller and Fairbanks*, 1985; *Miller*, 1992; *Wright and Miller*, this volume]. Although *Miller* [1992] and *Wright and Miller* [this volume] suggested that this may be evidence that the Antarctic was the sole source of bottom water during this time, they also noted that the Oligocene ocean was oligotrophic, which may have led to low input of organic carbon from the photic zone. As a result, there would be little carbon isotopic evidence for aging of bottom water during this time period. There may also have been multiple sources of bottom water during this time [*Wright and Miller*, this volume; J. C. Zachos, personal communication, 1991]. The existence of multiple sources would not allow significant aging of

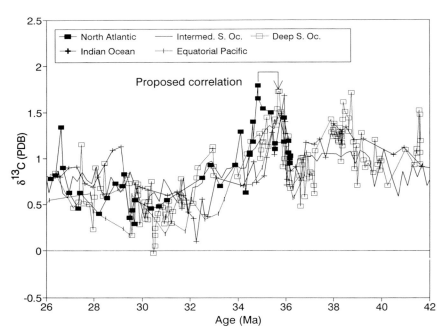

Fig. 7.   Carbon isotopic ratios for DSDP and ODP sites from the North Atlantic (Site 563 [*Miller and Fairbanks*, 1985]), intermediate Southern Ocean (sites 689, *Kennett and Stott*, 1990; this paper]), deep Southern Ocean (Site 699 (this paper)), Indian Ocean (Site 748 [*Zachos et al.*, 1992a]), and equatorial Pacific (Site 574 [*Miller and Thomas*, 1985]).

any water mass and could be another reason for the lack of global δ^{13}C gradients in deep water.

Miller [1992] and *Wright and Miller* [this volume] compared Oligocene to Miocene benthic foraminiferal δ^{13}C records from the North Atlantic and South Atlantic oceans, the Southern Ocean, and the equatorial Pacific Ocean and noted an increase in North Atlantic δ^{13}C values from 35.5 to 34.5 Ma, above values at that time in other sites. Based on this δ^{13}C maximum and on seismic stratigraphic data [*Miller and Fairbanks*, 1983, 1985; *Miller et al.*, 1985b; *Miller*, 1992], they inferred that a pulse of Northern Component Water (NCW), similar to North Atlantic Deep Water in origin, occurred near 35.5–34.5 Ma.

We suggest that it is possible that the δ^{13}C maximum at 35.5–34.5 Ma [*Miller et al.*, 1985a; *Miller*, 1992] may instead be correlated more correctly to the ubiquitous 36–35 Ma δ^{13}C maximum which occurs just after the δ^{18}O shift at 35.8 Ma in records from the Atlantic, Pacific, Indian, and Southern oceans (Figure 7). Our logic stems in part from a comparison of the two reasonably complete Oligocene North Atlantic isotopic records (Deep Sea Drilling Project (DSDP) sites 558 and 563 [*Miller and Fairbanks*, 1985]). Only one, Site 563, shows the early Oligocene carbon isotopic maximum. Although *Miller et al.* [1985a] suggest that Site 558 may be diagenetically altered within the lower Oligocene, diagenesis in fact tends to affect carbon isotopic records

to a small degree owing to the low concentration of dissolved non-carbonate-derived carbon in pore water. Yet the δ^{13}C record from Site 558 does not exhibit the maximum seen at Site 563 [*Miller and Fairbanks*, 1985], despite the fact that Site 558 has a longer record and despite the close proximity and similar depths [*Bougault et al.*, 1985], suggesting that there may be a problem with dating of these records. It should be noted that, with the exception of the carbon isotope correlation we are suggesting, the biostratigraphy and magnetostratigraphy used by *Miller et al.* [1985] and *Miller* [1992] is internally consistent [*Wright and Miller*, this volume]. If our suggestions are correct, revisions of the biostratigraphy and magnetostratigraphy of Site 563 will be necessary.

Our alternate interpretation results in the following observations. First, the δ^{18}O values from the Site 563 record are relatively low and generally comparable to values from Site 699. Second, the δ^{13}C maximum seen in Site 563 is larger than any other cited in this paper, with the single exception again of Site 699, indicating that Site 563 is closest to the source of deep water. This raises the possibility that the pulse of NCW suggested by *Miller* [1992] and *Wright and Miller* [this volume] is the source of WSDW suggested by this paper.

There is abundant stratigraphic evidence that bottom current speeds increased near the Eocene/Oligocene boundary. Unconformities near this time are common

TABLE 5. Oxygen Isotopic Temperatures of Bottom Water at Sites 689 and 699, Based on $\delta^{18}O$ of Average Ocean Water Being $-1.2‰$ (Ice-Free Conditions) in the Eocene and $-0.85‰$ (Limited Glaciation Conditions) in the Oligocene

Age, Ma	Site 689		Site 699		$T_{699} - T_{689}$, °C
	Equilibrium $\delta^{18}O$	Temperature, °C	Equilibrium $\delta^{18}O$	Temperature, °C	
26.5	3.225	0.71	2.771	2.35	1.64
27.5	3.295	0.46	2.604	2.96	2.50
28.5	3.871	−1.55	2.797	2.26	3.81
29.5	3.863	−1.52	2.826	2.15	3.67
30.5	3.523	−0.34	2.704	2.60	2.94
31.5	3.279	0.52	2.744	2.45	1.93
32.5	3.205	0.78	2.571	3.09	2.30
34.5	3.178	0.88	2.586	3.03	2.15
35.5	3.424	0.01	2.781	2.31	2.31
36.5	2.374	2.52	1.898	4.29	1.77
38.5	2.008	3.88	1.497	5.81	1.94
39.5	2.333	2.67	1.722	4.96	2.28
41.5	1.963	4.05	1.389	6.23	2.18

Temperatures calculated using *Shackleton* [1974] paleotemperature equation. Because of an unconformity at Site 699 at 34–33 Ma, a lack of data at Site 689 at 38–37 Ma, and a lack of data at Site 699 at 41–42 Ma, no calculations are shown for these time periods.

worldwide [*Kennett et al.*, 1972; *Moore and Heath*, 1977; *Moore et al.*, 1978; *Miller et al.*, 1985a, b, 1987]. There are also unconformities at this time in two of the sections discussed in this study, sites 690 and 703. The general increase in $\delta^{13}C$ at all depths in this area is also seen in other ocean basins [*Miller and Fairbanks*, 1985; *Keigwin and Corliss*, 1986; *Miller and Thomas*, 1985; *Miller et al.*, 1988; *Zachos et al.*, 1992a, b] but is not found in surface waters [*Zachos et al.* 1992a, b]. This suggests that there was a generalized increase in the rate of overturn in the deep sea, including bottom, deep, and intermediate water masses as well.

Magnetic anomaly patterns indicate that a connection between the Arctic Ocean and North Atlantic may have been opened at the time of the $\delta^{13}C$ maximum [*Rowley and Lottes*, 1988]. This connection, combined with extensive North Atlantic unconformities of this age [*Miller et al.*, 1985a, b] indicates that the opening of the Arctic may have contributed to the increased rates of overturn of deep water.

Salinity and Temperature in WSDW

The pattern seen in the contoured $\delta^{18}O$ data can assist in estimating possible temperatures of the WSDW and in the colder core of seawater overlying it. We can use the paleotemperature equation of *Shackleton* [1974] ($T = 16.9 - 4.38(\delta_c - \delta_o) + 0.10(\delta_c - \delta_o)^2$) to calculate the temperatures of these water masses during this time period, provided that the $\delta^{18}O$ of deep water (δ_o) is determined.

The values of δ_o we need will be different from the present average value ($-0.28‰$) for two reasons. First, the amount of continental ice present during the Eocene

and Oligocene was less than that of today. Second, the oxygen isotopic composition throughout the water column must have changed, because of the greater salinity of the deeper, warmer water. The determination of the proper values of δ_o used is critical to the calculation of correct bottom water temperatures.

To determine the average oceanic $\delta^{18}O$ value, we assume that the Eocene ocean was ice free (with an average $\delta^{18}O$ of $-1.2‰$ [*Shackleton and Kennett*, 1975]). *Keigwin and Corliss* [1986], *Miller et al.* [1987], and *Zachos et al.* [1992a] found that the shift in lower-latitude planktonic foraminiferal $\delta^{18}O$ in the earliest Oligocene was ~0.3 to 0.4‰. Assuming constancy of tropical oceanic temperatures [*Matthews and Poore*, 1980; *Adams et al.*, 1990], this increase was due to an increase in Antarctic ice volume. *Zachos et al.* [1992a, b] found a similar oxygen isotopic shift in benthic (1.2‰) and planktonic (0.5‰) foraminifera at high latitudes (58°S) on the Kerguelen Plateau at ODP Site 748 coincident with the first appearance of ice-rafted debris. They concluded, from comparison with lower-latitude sites, that 0.3–0.4‰ of the isotopic signal was due to ice volume effect. Assuming that the Eocene polar regions were ice free, we conclude that the average oceanic $\delta^{18}O$ after the earliest Oligocene would be $-0.85‰$ (= $-1.2‰ + 0.35‰$). Thus 60–70% of the 1‰ shift in benthic foraminifera was due to a decrease in temperatures of about 2.0°–2.5°C in most of the water column, and the remaining 30–40% was due to buildup of ice on Antarctica. Table 5 presents our temperature estimates based on using $\delta_o = -1.2‰$ during the Eocene and $\delta_o = -0.85‰$ after the earliest Oligocene.

Several features are illustrated by these data. First,

TABLE 6.   Estimated Hydrographic Properties of Site 689 and 699 Based on $\delta^{18}O$ and Salinity of Ocean Water at Site 689 Being 1.2‰ and 34.0 ppt in the Eocene and −0.85‰ and 34.266 ppt in the Oligocene

Age, Ma	Site 689		Site 699					
	Equilibrium $\delta^{18}O$	Temperature, °C	Equilibrium $\delta^{18}O$	Temperature, °C	Salinity, ppt	$\delta^{18}O_w$, ‰	$T_{699} - T_{689}$	$S_{699} - S_{689}$
26.5	3.225	0.71	2.771	2.58	34.44	−0.788	1.87	0.18
27.5	3.295	0.46	2.604	3.33	34.55	−0.750	2.87	0.29
28.5	3.871	−1.55	2.797	2.66	34.58	−0.740	4.21	0.31
29.5	3.863	−1.52	2.826	2.53	34.57	−0.745	4.06	0.30
30.5	3.523	−0.34	2.704	2.98	34.56	−0.747	3.32	0.29
31.5	3.279	0.52	2.744	2.72	34.47	−0.777	2.20	0.21
32.5	3.205	0.78	2.571	3.44	34.54	−0.754	2.66	0.27
34.5	3.178	0.88	2.586	3.36	34.52	−0.760	2.48	0.26
35.5	3.424	0.01	2.781	2.61	34.50	−0.769	2.60	0.23
36.5	2.374	2.52	1.898	4.65	34.27	−1.105	2.13	0.27
38.5	2.008	3.88	1.497	6.32	34.37	−1.069	2.44	0.37
39.5	2.333	2.67	1.722	5.46	34.38	−1.067	2.79	0.38
41.5	1.963	4.05	1.389	6.83	34.44	−1.045	2.79	0.44

Temperature at Site 689 calculated using *Shackleton* [1974] paleotemperature equation. Temperature and salinity at Site 699 estimated assuming a minimally stable water column (i.e., approximately the same density as at Site 689), following *Railsback et al.* [1989]; $\delta^{18}O$ of ocean water at Site 699 calculated from equilibrium $\delta^{18}O$ of calcite and estimated temperature from paleotemperature equation.

bottom water temperatures at Site 689 averaged 3.28°C in the late Eocene, while Site 699 temperatures averaged 5.32°C (2.0°C warmer). During the first 6 m.y. of the Oligocene, temperature differences are similar, with Site 689 temperatures averaging 0.51°C and Site 699 temperatures averaging 2.70°C, a difference of 2.2°C. However, between 30 and 28 Ma, isotopic temperatures calculated for Site 689 dropped radically, to an unrealistically low temperature of −1.5°C. This argues for additional ice volume effects at this time, as suggested by *Miller et al.* [1985*b*], *Keigwin and Keller* [1984], *Miller and Thomas* [1985], *Keigwin and Corliss* [1986], and subsequent studies.

Because the deeper water at Site 699 most probably originated at least in part in low latitudes through evaporation in enclosed basins [*Brass et al.*, 1982], it would have had a higher $\delta^{18}O$ than the overlying cooler water due to enrichment during evaporation. The correct temperature offset between the WSDW and the colder overlying water will therefore be greater than that calculated by simply assuming an average oceanic oxygen isotopic composition.

By offsetting today's deep salinity versus $\delta^{18}O$ relationship using different values of average δ_o and examining the relationship between temperature, salinity, and density, we can explore the range of possible temperatures and salinities necessary to produce a stable water column. We used the technique of *Railsback et al.* [1989] to generate curves of constant density and constant calcite $\delta^{18}O$ in temperature-salinity space in order to constrain these temperatures and salinities at Site 699, assuming that average oceanic values of δ_o and salinity are present at Site 689. In these calculations, we assume that the modern, empirically determined $\delta^{18}O/$

salinity ratio of 0.35‰/parts per thousand (ppt) [*Railsback et al.*, 1989] still holds (ppt will be used when referring to salinity in this discussion instead of the commonly used per mil, in order to avoid confusion with the per mil terminology of stable isotopes). We have assumed that the δ_o values at Site 689 are representative of average ocean water, calculated the temperature at that point, and then calculated the minimum salinity and maximum temperature at Site 699 needed to produce a stable water column. We used *Railsback et al.*'s [1989] technique to determine the salinity and temperature required for a minimally stable water column assuming (1) an average salinity of 34.0 ppt and an average oceanic $\delta^{18}O$ of −1.2‰ (42–36 Ma) and (2) average values of 34.266 ppt and −0.85‰ (36–26 Ma) at Site 689 to approximate the properties of the water mass bathing Site 699 between 42 and 26 Ma (Table 6). Results from these calculations show that greater thermal gradients are indeed found; gradients range from 1.9° to 4.2°C and average 2.8°C, 0.4°C higher than if salinity changes are not considered. Salinities at Site 699 average 0.29 ppt greater than those at Site 689.

One final modification to this model needs to be made. During the Messinian salinity crisis in the late Miocene, about 6% of the ocean's salt was removed [*Ryan*, 1973], leading to a freshening of the oceans. To calculate oceanic salinity before this event, the average salinity values (34.0 ppt for the Eocene and 34.266 ppt for the Oligocene) were divided by 0.94. The *Railsback et al.* [1989] technique was again applied to calculate temperatures, salinities, and $\delta^{18}O_w$ at Site 699 necessary for a stable water column (Table 7). With the exception of higher absolute salinities at Site 699 as well, the model is robust: temperatures increase by less than 0.07°C,

TABLE 7. Estimated Hydrographic Properties of Sites 689 and 699 After Adding 6% Additional Salt Extracted During the Messinian Salinity Crisis

Age, Ma	Site 689		Site 699					
	Equilibrium $\delta^{18}O$	Temperature, °C	Equilibrium $\delta^{18}O$	Temperature, °C	Salinity, ppt	$\delta^{18}O_w$, ‰	$T_{699} - T_{689}$	$S_{699} - S_{689}$
26.5	3.225	0.71	2.771	2.60	36.65	−0.781	1.89	0.20
27.5	3.295	0.46	2.604	3.37	36.77	−0.740	2.91	0.32
28.5	3.871	−1.55	2.797	2.72	36.81	−0.725	4.26	0.36
29.5	3.863	−1.52	2.826	2.59	36.79	−0.731	4.11	0.34
30.5	3.523	−0.34	2.704	3.02	36.78	−0.735	3.36	0.33
31.5	3.279	0.52	2.744	2.74	36.68	−0.770	2.22	0.23
32.5	3.205	0.78	2.571	3.48	36.76	−0.744	2.70	0.30
34.5	3.178	0.88	2.586	3.40	36.74	−0.751	2.52	0.28
35.5	3.424	0.01	2.781	2.65	36.71	−0.760	2.64	0.26
36.5	2.374	2.52	1.898	4.68	36.46	−1.097	2.16	0.29
38.5	2.008	3.88	1.497	6.36	36.57	−1.059	2.48	0.40
39.5	2.333	2.67	1.722	5.50	36.58	−1.056	2.83	0.41
41.5	1.963	4.05	1.389	6.87	36.64	−1.034	2.83	0.47

Salinity of ocean water at Site 689 would have been 36.170 ppt during the Eocene under ice-free conditions and 36.453 ppt in the Oligocene.

$\delta^{18}O_w$ decreases by less than 0.02‰, and salinity gradients increase by less than 0.06 ppt, to a maximum of 0.47 ppt.

The temperatures (Eocene, 4°–7°C; Oligocene, 2°–4°C) and salinities <0.5 ppt greater than mean ocean values) calculated at Site 699 from the isotopic data are too low to characterize a Warm Saline Deep Water sensu strictu. Seawater produced today from excess evaporation, such as Mediterranean Outflow Water (MOW), has more extreme properties than our calculations suggest, with temperatures and salinities of the order of 9°C and 36 ppt (1.3 ppt greater than average oceanic salinity) at the Straits of Gibraltar, to 13°–15°C and 37–39 ppt within the Mediterranean [Reid, 1979; Pickard and Emery, 1982].

Therefore it is evident that the source of this WSDW cannot be direct production from excess evaporation in semienclosed basins in low to middle latitudes. MOW today is entrained into the North Atlantic gyre and flows northward into the Norwegian Greenland Sea. North Atlantic Deep Water is produced by the cooling of saline water derived in part from MOW [Reid, 1979]. One hypothesis for the production of WSDW is that during the late Eocene and early Oligocene a similar process occurred, with either a larger flux of warm saline intermediate water or less efficient cooling, producing Warm Saline Deep Water in high northern latitudes instead of low to middle latitudes. Bipolar cooling at high latitudes during the earliest Oligocene could have caused cooling of the entire deepwater column without affecting low-latitude surface temperatures. However, this hypothesis may not be justified, especially in the Eocene, in light of suggested warm Arctic temperatures [Estes and Hutchinson, 1980; McKenna, 1980; Wolfe, 1980] during the Paleogene.

This hypothesis also seems to be contradicted by the carbon isotopic data, which fail to show lower $\delta^{13}C$ values at the warmer Site 699 relative to cooler Site 689, an expected trend if WSDW had been produced in the more distant northern hemisphere. The $\delta^{13}C$ gradient, in fact, is opposite. Another hypothesis, however, allows for the derivation of cooler WSDW from more proximate sources. Perhaps warm saline water does in fact sink near the point of origin and, during its passage through cooler, less saline waters, entrains and mixes with that water to decrease the temperature and salinity to the values measured at Site 699. This hypothesis would adequately explain the observed data but would not explain why temperatures decreased equally in intermediate and deep waters in the earliest Oligocene unless they decreased equally at polar and lower latitudes.

Evidence from benthic foraminiferal assemblages is difficult to reconcile with the calculated salinity and temperature gradients. The benthic foraminiferal faunas of sites 689 and 699 [Thomas, 1989, 1990, 1992] show essentially the same patterns over the interval studied here rather than showing the effect of differing water masses. No benthic foraminiferal faunal analyses have been made to date over this age interval at Site 699, but our observations are that the faunas are broadly similar to those described from sites 689 and 690. It is possible that benthic foraminiferal faunas do not respond to the relatively low-temperature gradients identified here but rather to other factors such as rain rates of organic particles or to concentration differences in CO_2. Given the relatively low productivity [Miller, 1992] during the late Eocene-Oligocene and the low CO_2 gradients implied by the low total dissolved $\delta^{13}C$ gradients, differences in bottom water characteristics may not have been large enough to induce different benthic foramin-iferal faunas at the different sites.

Keigwin and Corliss [1986] examined thermal gradi-ents in surface water in the Eocene, Oligocene, and Recent based on planktonic isotopic gradients and found low thermal gradients in the late Eocene and early Oligocene. However, little data were present from high latitudes. Additional data from the Kerguelen Plateau and Maud Rise has been interpreted [*Zachos et al.*, 1992*a*] to mean that the surface water latitudinal ther-mal gradient was higher at high southern latitudes. Our interpretation of the isotopic data from sites 689–699 indicates that either an additional increase in the plan-etary thermal gradient took place across the earliest Oligocene oxygen isotope shift and that a bipolar cool-ing of about 2.5°C occurred or that an overall worldwide temperature decrease of about 2.5°C occurred.

CONCLUSIONS

Analysis of new benthic foraminiferal oxygen isotopic data from high southern latitude ODP Site 699 (Eocene/Oligocene paleodepth of ~3400 m) and site 703 (Eocene/Oligocene paleodepth of ~950 m) has led to three advances in the interpretation of Paleogene paleocean-ography. The first supports the hypothesis [*Kennett and Stott*, 1990] that a warm saline water mass was pro-duced during much of the Paleogene. These two sites, together with Kennett and Stott's previously published isotopic data from ODP Site 689 (paleodepth of ~1500 m) and Site 690 (paleodepth of ~2300 m), define a maximum in calcite $\delta^{18}O$ in Site 689 (~1500 m). The decreases above and below indicate a temperature min-imum near 1400 m, with water temperature increasing steadily in the three sites from 1500 to 3400 m. A second finding of this study is a near-constant oxygen isotopic gradient before and after the earliest Oligocene oxygen isotope enrichment event. When considered in the light of suggestions of an approximately 0.3–0.4‰ ice volume effect, we suggest that there was a cooling at the time of the oxygen isotope shift of about 2.5°C. Finally, we suggest that the immediate source of the ''WSDW'' was not directly from subtropical, semienclosed basins with high evaporation rates, as has been suggested by *Brass et al.* [1982]. After initial production in that environ-ment, warm saline waters were cooled and/or mixed with cooler water at high northern latitudes to form warm saline deep waters, similar in genesis to the North Atlantic Deep Water today, or during downward advec-tion of the warm saline water. Two factors support this. First, the same level of cooling occurred in both the cooler intermediate waters and the deeper, warmer waters during the earliest Oligocene. Second, the water properties defined by oxygen isotope ratios in Paleogene sediments at Site 699 are neither warm nor saline enough to have been produced by direct evaporation and sinking at low- to mid-latitude source regions.

Two contrasting conclusions may be drawn from the different possible generation mechanisms: either an increase in the planetary thermal gradient occurred, with bipolar cooling of about 3°C, or an overall plane-tary decrease in temperatures of the same amount occurred. Warm Arctic temperatures during the Paleo-gene support the latter interpretation.

Benthic foraminiferal studies [*Corliss*, 1981; *Thomas*, 1989, 1990, 1992] find no significant change in faunas across the Eocene/Oligocene oxygen isotopic shift or between sites, suggesting little geographic or temporal temperature change. It is possible that benthic faunas may not respond to the low-temperature gradients (mostly 2–3°C) calculated here. However, low inter-ocean $\delta^{13}C$ gradients suggest little difference in age of bottom water or rain rate of organic carbon, and benthic foraminiferal faunas may be more sensitive to those factors than to small changes in temperature or salinity.

A maximum in $\delta^{13}C$ occurs near 36–35 Ma in virtually all benthic deep-sea records from the world ocean. This fact, in combination with the development of widespread unconformities at this time, leads us to suggest an increase in deep-sea circulation rates, possibly related to a pulse of warm saline water from the North Atlantic.

TABLE A1.   Oxygen and Carbon Isotope Analyses of *Cibicidoides* spp. for ODP Hole 703A

c-s	int	Depth (mbsf)	Age (Ma)	$\delta^{13}C$	$\delta^{18}O$
4-5,	82	30.72	20.370	0.982	1.854
4-6,	41	31.81	20.610	0.868	1.724
4-6,	132	32.72	20.889	0.821	1.592
5-1,	82	34.22	20.996	1.467	2.275
5-1,	127	34.67	21.027	1.078	2.441
5-3,	82	37.22	21.218	0.762	1.741
5-3,	127	37.67	21.256	0.952	1.979
5-4,	40	38.30	21.309	0.979	1.765
5-4,	43	38.33	21.311	1.585	0.897*
5-4,	127	39.17	21.382	1.017	1.568
5-5,	82	40.22	23.290	1.457	2.092
5-5,	127	40.67	23.330	1.109	1.757
5-6,	43	41.33	23.390	1.253	1.736
5-6,	82	41.72	23.425	1.424	1.827
				1.441	1.964
				1.277	2.045
5-6,	127	42.17	23.452	1.083	1.482
5-7,	43	42.83	23.479	1.206	1.696
6-1,	41	43.31	23.499	0.784	1.657
6-1,	99	43.89	23.523	1.655	2.003

TABLE A1. (continued)

c-s	int	Depth (mbsf)	Age (Ma)	δ¹³C	δ¹⁸O
6-1,	141	44.31	23.540	1.691	2.164
6-2,	41	44.81	23.595	1.490	1.950
6-2,	141	45.81	23.760	1.769	2.212
6-3,	41	46.31	23.801	1.458	2.181
6-3,	99	46.89	23.821	1.365	2.031
6-4,	41	47.81	23.853	1.638	2.024
6-4,	99	48.39	23.873	0.680	1.626
				0.900	1.870
6-4,	141	48.81	23.887	1.162	1.611
6-5,	41	49.31	23.904	1.338	1.764
				1.329	1.682
6-5,	99	49.89	23.924	1.336	1.802
6-6,	41	50.81	23.956	0.223	1.682
6-6,	45	50.85	23.958	0.424	1.634
6-7,	6	51.96	23.996	0.750	1.509
6-7,	31	52.21	24.004	1.126	1.673
6-7,	45	52.35	24.009	0.788	1.604
6-7,	66	52.56	24.017	0.802	1.762
7-1,	61	53.01	25.794	1.156	1.970
				1.324	1.831
7-1,	111	53.51	25.823	1.171	.728
7-2,	31	54.21	25.864	0.9721	1.755
7-2,	61	54.51	25.882	0.848	1.618
7-2,	111	55.01	25.912	1.175	1.697
7-3,	3	55.43	25.936	0.784	1.811
7-3,	61	56.01	25.973	0.776	1.930
7-4,	31	57.21	26.367	0.776	1.969
				0.863	1.681
7-4,	61	57.51	26.615	0.819	1.211
				0.658	1.829
7-4,	111	58.01	26.862	0.573	2.009
7-5,	31	58.71	26.976	0.763	1.581
				0.647	1.615
7-5,	61	59.01	27.025	1.094	2.009
7-5,	97	59.37	27.083	0.941	2.242
7-5,	140	59.80	27.153	0.971	2.142
7-CC,	10	60.12	27.205	0.612	2.069
8-1,	42	62.32	27.564	0.833	1.887
8-1,	64	62.54	27.600	0.707	1.972
8-2,	64	64.04	28.172	0.859	2.405
8-3,	64	65.54	29.014	0.550	2.178
8-4,	42	66.82	29.336	0.677	2.148
8-6,	42	69.82	30.405	0.774	2.431
8-6,	101	70.41	30.462	0.845	2.475
9-1,	75	72.15	30.630	0.777	2.248
9-2,	25	73.15	30.727	0.722	2.290
9-2,	125	74.15	30.824	0.861	2.337
				1.284	2.844
9-3,	25	74.65	30.873	0.877	2.892
9-3,	92	75.32	30.937	0.593	2.296
9-4,	58	76.48	31.050	0.677	1.858
9-4,	91	76.81	31.082	0.817	1.631

TABLE A1. (continued)

c-s	int	Depth (mbsf)	Age (Ma)	δ¹³C	δ¹⁸O
9-5,	75	78.15	31.212	0.947	2.118
9-5,	92	78.32	31.228	0.790	2.300
9-5,	115	78.55	31.303	0.479	1.911
9-6,	10	79.00	31.459	0.528	2.301
9-6,	58	79.48	31.588	0.372	1.781
9-6,	75	79.65	31.599	0.652	1.931
9-6,	92	79.82	31.610	1.284	2.222
9-7,	58	80.98	31.690	1.003	1.661
10-1,	8	80.98	31.690	0.490	1.826
10-1,	125	82.15	31.776	0.349	1.828
10-2,	75	83.15	31.849	0.444	1.878
10-3,	73	84.63	31.958	0.520	1.925
10-4,	60	86.00	32.059	0.626	1.922
10-5,	60	87.50	32.179	0.733	1.827
10-6,	1	88.41	32.252	0.833	2.180
10-6,	48	88.88	32.290	0.414	1.892
10-6,	60	89.00	32.300	0.288	0.400*
10-6,	71	89.11	32.309	0.894	1.892
10-7,	60	90.50	32.421	0.568	1.968
11-1,	30	90.70	32.437	0.679	1.823
11-1,	91	91.31	32.490	0.560	1.589
11-2,	30	92.20	32.572	0.942	1.848
11-2,	68	92.58	32.607	0.886	1.787
11-2,	91	92.81	32.629	0.853	1.859
11-3,	68	94.08	32.746	0.989	1.734
				0.823	1.505
11-3,	91	94.31	32.768	1.008	1.502
11-4,	30	95.20	32.850	0.906	1.734
				0.984	1.904
11-4,	68	95.58	32.885	1.041	1.848
				1.142	1.961
11-5,	30	96.70	33.055	0.780	1.363
11-5,	68	97.08	33.116	0.820	1.549
11-5,	91	97.31	33.153	0.817	1.471
11-6,	29	98.19	33.295	0.767	1.713
				0.773	1.965
				0.877	1.745
11-6,	68	98.58	33.358	0.741	1.534
				0.752	1.588
11-7,	30	99.70	33.539	0.982	1.675
11-7,	37	99.77	33.550	1.001	1.603
12-1,	41	100.31	33.637	1.042	1.514
12-1,	85	100.75	33.708	1.032	1.613
				0.972	1.634
12-1,	91	100.81	33.718	0.811	1.908
12-1,	94	100.84	33.723	1.158	1.709
				1.094	1.731
12-1,	130	101.20	33.781	0.830	1.242
12-2,	3	101.43	33.818	0.846	1.385
12-2,	40	101.80	33.877	0.935	1.312
12-3,	3	102.93	34.060	0.836	1.201
12-3,	40	103.30	34.119	0.843	1.486

TABLE A1. (continued)

c-s	int	Depth (mbsf)	Age (Ma)	δ¹³C	δ¹⁸O
12-4,	40	104.80	34.361	0.801	1.528
12-4,	72	105.12	34.413	1.052	1.503
12-5,	40	106.30	34.603	1.309	1.532
				1.062	1.848
12-5,	130	107.20	34.748	1.025	1.515
12-6,	3	107.43	34.785	1.237	1.593
12-6,	118	108.58	34.970	1.193	1.930
12-6,	130	108.70	34.989	1.409	1.964
12-7,	40	109.30	35.086	1.293	2.024
				1.355	1.753
13-1,	45	109.85	37.107	1.308	1.178
13-1,	90	110.30	37.129	0.875	0.989
13-2,	45	111.35	37.179	0.971	1.300
				0.847	0.977
13-2,	145	112.35	37.226	0.980	1.040
13-3,	45	112.85	37.250	1.024	1.097
13-4,	61	114.51	37.329	0.973	0.929
13-4,	87	114.77	37.341	0.886	0.959
				1.067	1.075
13-4,	133	115.23	37.363	0.876	1.023
13-CC,	36	118.86	37.536	1.054	1.085
15-1,	136	129.76	38.055	1.006	0.637
				1.073	0.702
15-1,	146	129.86	38.060	1.104	0.825
15-2,	41	130.31	38.081	0.860	0.486
15-2,	82	130.72	38.101	1.209	0.703
15-2,	132	131.22	38.123	1.080	0.631
15-3,	41	131.81	38.148	0.986	0.508
15-3,	82	132.22	38.166	0.968	0.616
15-3,	132	132.72	38.188	1.147	0.904
15-4,	82	133.72	38.232	0.949	0.910
15-4,	118	134.08	38.248	1.022	1.070
				0.785	0.894
15-5,	82	135.22	38.297	1.023	1.075
15-5,	132	135.72	38.319	1.164	1.241
				1.086	1.207
15-6,	82	136.72	38.363	1.070	1.086
16-1,	15	138.05	38.421	0.999	1.175
16-1,	82	138.72	38.450	1.142	1.150
				1.144	1.224
16-1,	122	139.12	38.468	1.135	1.143
16-CC,	23	139.58	38.488	1.034	1.217
				0.986	1.343
17-1,	14	143.54	38.661	0.927	1.043
17-1,	100	144.40	38.698	0.994	0.913
17-2,	32	145.22	38.734	1.026	1.051
				1.079	0.974
17-2,	100	145.90	38.764	1.187	1.149
17-2,	106	145.96	38.766	1.180	1.313
17-2,	110	146.00	38.768	1.053	1.287
				1.217	1.478
17-2,	110	146.00	38.768	1.157	1.383

TABLE A1. (continued)

c-s	int	Depth (mbsf)	Age (Ma)	δ¹³C	δ¹⁸O
17-3,	32	146.72	38.802	0.610	0.897
17-3,	61	147.01	38.817	0.787	0.776
17-3,	100	147.40	38.838	1.086	1.275
				1.018	1.315
17-CC,	10	147.60	38.849	0.974	1.210
18-1,	10	153.00	39.136	1.117	1.087
18-1,	38	153.28	39.151	0.907	1.054
18-2,	38	154.78	39.231	0.875	0.939
18-2,	95	155.35	39.278	1.265	1.246
18-2,	140	155.80	39.320	1.237	1.265
				1.236	1.183
18-3,	38	156.28	39.365	0.922	0.936
18-3,	71	156.61	39.396	1.152	1.127
18-3,	95	156.85	39.418	1.207	1.130
18-4,	38	157.78	39.506	0.583	0.707*
				1.028	1.046
18-4,	95	158.35	39.559	0.997	1.088
18-4,	99	158.39	39.563	0.795	0.661
				0.767	0.839
18-5,	23	159.13	39.632	0.466	0.232

Column headings: ''c-s'' indicates core-section; ''int'' indicates interval.

*Analyses considered invalid and not plotted or used in calculating million-year averages.

TABLE A2. Oxygen and Carbon Isotope Analyses of *Cibicidoides* spp. for ODP Hole 699A

c-s	int	Depth (mbsf)	Age (Ma)	δ¹³C	δ¹⁸O
15-2,	111	134.71	26.276	0.812	.919
15-4,	45	137.05	26.378	0.841	1.582
16-5,	10	147.70	27.120	0.464	1.667
16-CC,	20	150.80	27.347	0.530	1.583
17-1,	145	152.55	27.476	1.153	1.007
17-2,	110	153.70	27.561	0.509	2.185
17-5,	103	158.13	27.815	0.501	1.519
18-1,	140	162.00	27.962	0.230	1.540
18-2,	115	163.25	28.010	0.582	1.789
18-3,	80	164.40	28.053	0.899	1.821
18-5,	92	167.52	28.188	0.093	1.459*
18-6,	56	168.66	28.265	0.676	2.049
19-1,	8	170.18	28.368	0.591	1.496
19-1,	100	171.10	28.429	0.946	1.863
19-2,	57	172.17	28.502	0.601	1.987
19-3,	130	174.40	28.652	-0.184	2.045*
20-1,	133	180.93	29.091	0.673	1.515
20-2,	100	182.10	29.170	0.143	1.687*
20-4,	57	184.67	29.365	0.818	2.054
20-5,	35	185.95	29.465	0.729	1.793
20-6,	28	187.38	29.577	0.170	1.610
21-1,	12	189.22	29.721	0.737	2.115

TABLE A2. (continued)

c-s	int	Depth (mbsf)	Age (Ma)	δ¹³C	δ¹⁸O
21-1,	122	190.32	29.822	0.602	2.150
				0.656	2.081
				0.724	2.063
21-2,	37	190.97	29.882	0.283	1.637
21-3,	117	193.27	30.100	0.531	1.168
21-4,	27	193.87	30.127	0.346	1.452
21-4,	107	194.67	30.163	0.431	1.396
21-5,	120	196.30	30.237	0.335	1.779
				0.676	1.927
21-6,	42	197.02	30.270	0.535	1.564
22-1,	30	198.90	30.381	0.751	1.694
22-1,	119	199.79	30.463	-0.029	1.440
22-2,	43	200.53	30.531	0.054	1.518
				0.246	1.551
				0.169	1.502
22-2,	122	201.32	30.604	0.340	1.853
22-3,	10	201.70	30.640	0.402	1.974
22-3,	140	203.00	30.760	0.485	1.706
22-4,	75	203.85	30.838	0.472	1.826
22-5,	15	204.75	30.921	0.292	1.439
23-1,	12	205.22	30.965	0.398	1.798
				0.333	1.674
23-2,	5	206.65	31.097	0.538	1.724
				0.453	1.677
23-2,	130	207.90	31.212	0.296	1.995
23-3,	36	208.46	31.269	0.427	1.660
23-4,	137	210.97	31.530	0.615	1.619
23-5,	65	211.75	31.612	0.783	1.736
23-5,	98	212.08	31.646	0.494	1.759
24-3,	77	218.37	32.118	0.515	1.689
25-1,	32	224.42	32.269	0.793	1.506
25-2,	31	225.91	32.306	0.900	1.471
27-5,	119	250.29	32.915	1.085	1.523
27-6,	82	251.42	32.944	1.130	1.588
28-1,	15	252.75	32.977	0.916	1.540
28-1,	30	252.90	32.981	0.814	1.499
28-1,	134	253.94	33.959	0.799	1.524
29-1,	125	260.35	34.359	1.049	1.350
29-3,	66	262.76	34.509	1.185	1.772
30-1,	25	268.85	34.890	1.270	1.628
30-1,	66	269.26	34.915	1.002	1.431
30-1,	130	269.90	34.955	1.089	1.593
30-2,	61	270.71	35.006	1.170	1.559
30-2,	120	271.30	35.043	1.179	1.895
30-3,	80	272.40	35.111	1.294	1.624
30-4,	68	273.78	35.198	1.447	1.843
30-4,	138	274.48	35.241	1.126	1.876
31-1,	105	279.15	35.533	1.770	1.744*
31-1,	112	279.22	35.537	1.428	1.647
31-2,	40	280.00	35.586	1.360	1.886
31-2,	107	280.67	35.628	1.538	1.493
31-3,	30	281.40	35.673	1.721	1.759

TABLE A2. (continued)

c-s	int	Depth (mbsf)	Age (Ma)	δ¹³C	δ¹⁸O
31-3,	112	282.22	35.725	1.639	2.003*
				1.729	1.699
31-4,	42	283.02	35.774	1.490	1.971
				1.507	1.967
31-5,	50	284.60	35.873	0.909	1.262
31-5,	129	285.39	35.922	0.996	1.127
31-6,	130	286.90	36.017	1.152	0.932
32-1,	65	288.25	36.101	1.242	1.211
32-1,	110	288.70	36.129	0.902	0.304*
32-1,	140	289.00	36.147	0.738	1.027
32-2,	25	289.35	36.169	0.673	0.839
32-6,	76	295.86	36.575	0.810	0.788
32-6,	130	296.40	36.609	0.482	0.853
32-7,	26	296.86	36.637	1.007	0.925
33-1,	20	297.30	36.665	0.930	0.924
33-1,	97	298.07	36.713	1.009	0.806
33-2,	59	299.19	36.783	0.582	0.923
33-2,	81	299.41	36.796	0.724	0.639*
33-3,	20	300.30	36.852	0.777	0.603
33-3,	111	301.21	36.909	1.036	1.231
33-4,	81	302.41	36.984	0.796	0.837
33-4,	111	302.71	37.002	0.852	0.529
33-5,	81	303.91	37.077	1.170	0.956*
33-5,	111	304.21	37.096	0.840	0.679
33-6,	81	305.41	37.171	0.704	0.654
33-6,	111	305.71	37.189	0.613	0.692
33-CC		306.11	37.214	1.083	0.708
35-1,	86	316.96	37.891	1.428	0.858
35-1,	90	317.00	37.893	1.283	0.787
35-2,	30	317.90	37.949	1.233	0.664
35-2,	86	318.46	37.984	1.316	0.509
35-2,	145	319.05	38.021	1.378	0.976
35-3,	86	319.96	38.078	1.528	1.379*
35-4,	10	320.70	38.124	1.236	0.643
35-4,	86	321.46	38.165	1.259	0.572
35-5,	10	322.20	38.183	1.189	0.696
35-5,	86	322.96	38.203	1.178	0.515
35-6,	10	323.70	38.222	1.284	0.512
35-6,	81	324.41	38.240	1.264	0.416
35-7,	10	325.20	38.260	1.335	0.342
36-1,	21	325.81	38.275	1.254	0.094
36-1,	126	326.86	38.302	0.968	0.229
36-2,	73	327.83	38.327	1.234	0.430
				1.171	0.537
36-2,	115	328.25	38.338	1.291	0.644
36-3,	21	328.81	38.352	1.131	0.317
36-3,	116	329.76	38.376	1.425	0.572
36-4,	21	330.31	38.390	1.613	0.388
36-4,	116	331.26	38.473	1.511	-0.023
36-5,	73	332.33	38.613	1.292	0.564
				1.438	0.056
36-6,	21	333.31	38.742	1.713	0.339

TABLE A2. (continued)

c-s	int	Depth (mbsf)	Age (Ma)	δ¹³C	δ¹⁸O
36-6,	75	333.85	38.812	1.177	0.492
36-6,	116	334.26	38.866	1.051	1.075
				0.885	0.768
36-7,	21	334.81	38.938	1.136	0.652
36-7,	50	335.10	38.976	0.704	0.251
37-1,	128	336.38	39.144	1.282	0.726
37-2,	54	337.14	39.243	0.980	0.920
37-2,	96	337.56	39.298	0.899	0.737
37-3,	54	338.64	39.439	0.949	0.721
37-3,	86	338.96	39.481	0.667	0.941*
37-4,	54	340.14	39.636	0.848	0.645
37-4,	118	340.78	39.720	1.203	0.808
37-5,	54	341.64	39.832	0.933	0.593
				0.871	0.590
37-5,	122	342.32	39.921	1.002	0.584
37-6,	15	342.75	39.978	0.973	0.597
37-CC,	22	343.14	40.029	0.765	0.128*
38-1,	16	344.76	40.241	0.704	0.955
38-CC		350.60	41.004	1.096	0.887
39-1,	26	354.36	41.367	0.929	0.379
39-1,	110	355.20	41.472	1.103	0.027
				1.053	0.161
39-2,	26	355.86	41.567	1.521	0.168
39-2,	51	356.11	41.603	0.936	0.176
				1.060	0.376
				1.476	0.267
39-2,	75	356.35	41.638	1.191	0.360

Column headings: "c-s" indicates core-section; "int" indicates interval.
*Analyses considered invalid and not plotted or used in calculating million-year averages.

Acknowledgments. We would like to express our appreciation to Jim Kennett for sponsoring our participation at The Role of the Southern Ocean and Antarctica in Global Change: An Ocean Drilling Perspective meeting at the University of California, Santa Barbara, in August 1991, where this paper was presented and where we received many useful comments. Help in the laboratory by Rich Cooke is greatly appreciated. Ray Thomas is thanked for technical help with the mass spectrometer. Comments by E. J. Barron, J. P. Kennett, K. G. Miller, and E. Thomas on early versions of this manuscript were extremely useful, as were later reviews by G. Brass and J. Wright. This work was supported by National Science Foundation grants OCE-8858012 and DPP-8717854.

REFERENCES

Adams, C. G., D. E. Lee, and B. R. Rosen, Conflicting isotopic and biotic evidence for tropical sea-surface temperatures during the Tertiary, *Palaeogeogr. Palaeoclimatol. Palaeoecol.*, 77, 289–313, 1990.

Bainbridge, A., *GEOSECS Atlantic Expedition*, vol. 1, *Hydrographic Data 1972–1973*, 121 pp., National Science Foundation, Washington, D. C., 1981a.

Bainbridge, A., *GEOSECS Atlantic Expedition*, vol. 2, *Sections and Profiles*, 198 pp., National Science Foundation, Washington, D. C., 1981b.

Barker, P. F., et al., Leg 113, *Proc. Ocean Drill. Program Initial Rep.*, 113, 1033 pp., 1988a.

Barker, P. F., et al., Weddell Sea palaeoceanography: Preliminary results of ODP Leg 113, *Palaeogeogr. Palaeoclimatol. Palaeoecol.*, 67, 75–102, 1988b.

Barrera, E., and B. T. Huber, Eocene to Oligocene oceanography and temperatures in the Antarctic Indian Ocean, this volume.

Belanger, P. E., W. B. Curry, and R. K. Matthews, Core-top evaluation of benthic foraminiferal isotopic ratios for paleoceanographic interpretations, *Palaeogeogr. Palaeoclimatol. Palaeoecol.*, 33, 205–220, 1981.

Berggren, W. A., D. V. Kent, J. J. Flynn, and J. A. Van Couvering, Cenozoic geochronology, *Geol. Soc. Am. Bull.*, 96, 1499–1510, 1985.

Bougault, H., et al., Leg 82, *Initial Rep. Deep Sea Drill. Proj.*, 82, 667 pp., 1985.

Brass, G. W., J. R. Southam, and W. H. Peterson, Warm Saline Bottom Water in the ancient ocean, *Nature*, 296, 620–623, 1982.

Broecker, W. S., A revised estimate for the radiocarbon age of North Atlantic Deep Water, *J. Geophys. Res.*, 84, 3218–3226, 1979.

Cande, S. C. and D. V. Kent, A new geomagnetic polarity time scale for the Late Cretaceous and cenozoic, *J. Geophys. Res.*, 97, 13,917–13,951, 1992.

Chamberlin, T. C., On a possible reversal of deep-sea circulation and its influence on geological climates, *J. Geol.*, 14, 363–373, 1906.

Charles, C. D., and R. G. Fairbanks, Evidence from Southern Ocean sediments for the effect of North Atlantic deep-water flux on climate, *Nature*, 355, 416–419, 1992.

Ciesielski, P. F., et al., Leg 114, *Proc. Ocean Drill. Program Initial Rep.*, 114, 815 pp., 1988.

Coplen, T. B., C. Kendall, and J. Hopple, Comparison of stable isotope reference samples, *Nature*, 302, 236–238, 1983.

Corliss, B. H., Deep-sea benthonic foraminiferal faunal turnover near the Eocene/Oligocene boundary, *Mar. Micropaleontol.*, 6, 367–384, 1981.

Corliss, B. H., and L. D. Keigwin, Jr., Eocene-Oligocene paleoceanography, in *Mesozoic and Cenozoic Oceans, Geodynam. Ser.*, vol. 15, edited by K. J. Hsü, pp. 101–118, AGU, Washington, D. C., 1986.

Corliss, B. H., M.-P. Aubry, W. A. Berggren, J. M. Fenner, L. D. Keigwin, and B. Keller, The Eocene/Oligocene boundary event in the deep sea, *Science*, 226, 806–810, 1984.

Douglas, R. G., and S. M. Savin, Oxygen and carbon isotope analyses of Tertiary and Cretaceous microfossils from Shatsky Rise and other sites in the North Pacific Ocean, *Initial Rep. Deep Sea Drill. Proj.*, 32, 509–520, 1975.

Duplessy, J. C., C. Labou, and A. C. Vinot, Differential isotopic fractionation in benthic foraminifera and paleotemperatures reassessed, *Science*, 250–251, 1970.

Estes, R., and J. H. Hutchinson, Eocene lower vertebrates from Ellesmere Island, Canadian Arctic Archipelago, *Palaeogeogr. Palaeoclimatol. Palaeoecol.*, 30, 325–347, 1980.

Graham, D. W., B. H. Corliss, M. L. Bender, and L. D. Keigwin, Jr., Carbon and oxygen isotopic disequilibria of Recent deep-sea benthic foraminifera, *Mar. Micropaleontol.*, 6, 483–497, 1981.

Hailwood, E. A., and B. M. Clement, Magnetostratigraphy of sites 699 and 700, east Georgia Basin, *Proc. Ocean Drill. Program Sci. Results*, 114, 337–357, 1991a.

Hailwood, E. A., and B. M. Clement, Magnetostratigraphy of sites 703 and 704, southeastern South Atlantic, *Proc. Ocean Drill. Program Sci. Results*, 114, 367–385, 1991b.

Hamilton, N., Mesozoic magnetostratigraphy of Maud Rise, Antarctica, *Proc. Ocean Drill. Program Sci. Results*, 113, 255–260, 1990.

Hayes, D. E., Age-depth relationships and depth anomalies in the southeast Indian Ocean and South Atlantic Ocean, *J. Geophys. Res.*, 93, 2937–2954, 1988.

Hodell, D. A., R. H. Benson, J. P. Kennett, and K. Rakic-El Bied, Stable isotope stratigraphy of latest Miocene sequences in northwest Morocco: The Bou Regreg section, *Paleoceanography*, 4, 467–482, 1989.

Keigwin, L. D., Paleoceanographic change in the Pacific at the Eocene-Oligocene boundary, *Nature*, 287, 722–725, 1980.

Keigwin, L. D. and B. H. Corliss, Stable isotopes in late middle Eocene to Oligocene foraminifera, *Geol. Soc. Am. Bull.*, 97, 335–345, 1986.

Keigwin, L. D., and G. Keller, Middle Oligocene cooling from equatorial Pacific DSDP Site 77B, *Geology*, 12, 16–19, 1984.

Kennett, J. P. and N. J. Shackleton, Oxygen isotopic evidence for the development of the psychrosphere 38 Myr ago, *Nature*, 260, 513–515, 1976.

Kennett, J. P. and L. D. Stott, Proteus and Proto-Oceanus: Ancestral Paleogene oceans as revealed from Antarctic stable isotopic results, ODP Leg 113, *Proc. Ocean Drill. Program Sci. Results*, 113, 865–880, 1990.

Kennett, J. P., R. E. Burns, J. E. Andrews, M. Churkin, Jr., T. A. Davies, P. Dumitrica, A. R. Edwards, J. S. Galehouse, G. H. Packham, and G. J. van der Lingen, Australian-Antarctic continental drift, palaeocirculation changes and Oligocene deep-sea erosion, *Nature Phys. Sci.*, 239, 51–55, 1972.

Kristoffersen, Y. and J. LaBrecque, On the tectonic history and origin of the northeast Georgia Rise, *Proc. Ocean Drill. Program Sci. Results*, 114, 23–38, 1991.

Kroopnick, P., The distribution of ^{13}C in the Atlantic Ocean, *Earth Planet. Sci. Lett.*, 49, 469–484, 1980.

Kroopnick, P., The distribution of ^{13}C of ΣCO_2 in the world oceans, *Deep Sea Res.*, 32, 57–84, 1985.

Matthews, R. K., and R. Z. Poore, Tertiary del ^{18}O record and glacio-eustatic sea-level fluctuations, *Geology*, 8, 501–504, 1980.

McKenna, M. C., Eocene paleolatitude, climate, and mammals of Ellesmere Island, *Palaeogeogr. Palaeoclimatol. Palaeoecol.*, 30, 349–362, 1980.

Miller, K. G., Middle Eocene to Oligocene stable isotopes, climate, and deep-water history: The terminal Eocene event?, in *Eocene-Oligocene Climatic and Biotic Evolution*, edited by D. R. Prothero and W. A. Berggren, pp. 160–177, Princeton University Press, Princeton, N. J., 1992.

Miller, K. G., and R. G. Fairbanks, Evidence for Oligocene-middle Miocene abyssal circulation changes in the western North Atlantic, *Nature*, 306, 250–253, 1983.

Miller, K. G., and R. G. Fairbanks, Oligocene-Miocene global carbon and abyssal circulation changes, in *The Carbon Cycle and Atmospheric CO$_2$: Natural Variations Archean to Present, Geophys. Monogr. Ser.*, vol. 32, edited by E. T. Sundquist and W. S. Broecker, pp. 469–486, AGU, Washington, D. C., 1985.

Miller, K. G., and E. Thomas, Late Eocene to Oligocene benthic foraminiferal isotopic record, Site 574, equatorial Pacific, *Initial Rep. Deep Sea Drill. Proj.*, 85, 771–777, 1985.

Miller, K. G., M.-P. Aubry, M. J. Khan, A. J. Melillo, D. V. Kent, and W. A. Berggren, Oligocene-Miocene biostratigra-phy, magnetostratigraphy, and isotopic stratigraphy of the western North Atlantic, *Geology*, 13, 257–261, 1985a.

Miller, K. G., G. S. Mountain, and B. E. Tucholke, Oligocene glacio-eustasy and erosion on the margins of the North Atlantic, *Geology*, 13, 10–13, 1985b.

Miller, K. G., R. G. Fairbank, and G. S. Mountain, Tertiary oxygen isotope synthesis, sea level history, and continental margin erosion, *Paleoceanography*, 2, 1–19, 1987.

Miller, K. G., M. D. Feigenson, D. V. Kent, and R. K. Olsson, Upper Eocene to Oligocene isotope ($^{87}Sr/^{86}Sr$, $\delta^{18}O$, $\delta^{13}C$) standard section, Deep Sea Drilling Project Site 522, *Paleoceanography*, 3, 223–233, 1988.

Moore, T. C., and G. R. Heath, Survival of deep-sea sedimentary sections, *Earth Planet. Sci. Lett.*, 37, 71–80, 1977.

Moore, T. C., T. H. van Andel, C. Sancetta, and N. Pisias, Cenozoic hiatuses in pelagic sediments, *Micropaleontology*, 24, 113–138, 1978.

Oberhänsli, H., and M. Toumarkine, The Paleogene oxygen and carbon isotope history of sites 522, 523, and 524 from the central South Atlantic, in *South Atlantic Paleoceanography*, edited by K. J. Hsü and H. J. Weissert, pp. 125–147, Cambridge University Press, New York, 1985.

Oberhänsli H., J. McKenzie, M. Toumarkine, and H. Weissert, A paleoclimatic and paleoceanographic record of the Paleogene in the central South Atlantic (Leg 73, sites 522, 523, and 524), *Initial Rep. Deep Sea Drill. Proj.*, 73, 737–747, 1984.

Oppo, D. W., and R. G. Fairbanks, Variability in the deep and intermediate water circulation of the Atlantic Ocean during the past 25,000 years: Northern hemisphere modulation of the Southern Ocean, *Earth Planet. Sci. Lett.*, 86, 1–15, 1987.

Ostlund, H. G., Craig, H., Broecker, W. S., and Spencer, D., *GEOSECS Atlantic, Pacific, and Indian Ocean Expeditions, Vol. 7, Shorebased data and graphics*, IDOE, NSF, 198 pp., 1987.

Parsons, B., and J. G. Sclater, An analysis of the variation of ocean floor bathymetry and heat flow with age, *J. Geophys. Res.*, 82, 803–827, 1977.

Pickard, G. L., and W. J. Emery, *Descriptive Physical Oceanography: An Introduction*, 249 pp., Pergamon, New York, 1982.

Poore, R. Z., and R. K. Matthews, Late Eocene-Oligocene oxygen- and carbon-isotope record from South Atlantic Ocean, Deep Sea Drilling Project Site 522, *Initial Rep. Deep Sea Drill. Proj.*, 73, 725–735, 1984.

Railsback, L. B., T. F. Anderson, S. C. Ackerly, and J. L. Cisne, Paleoceanographic modeling of temperature-salinity profiles from stable isotopic data, *Paleoceanography*, 4, 585–591, 1989.

Raymond, C. A., J. L. LaBrecque, and Y. Kristofferson, Islas Orcadas Rise and Meteor Rise: The tectonic and depositional history of two aseismic plateaus from ODP sites 702, 703, and 704, *Proc. Ocean Drill. Program Sci. Results*, 114, 5–22, 1991.

Reid, J. L., On the contribution of the Mediterranean Sea outflow to the Norwegian-Greenland Sea, *Deep Sea Res.*, 26, 1199–1223, 1979.

Rowley, D. B., and A. L. Lottes, Plate-kinematic reconstructions of the North Atlantic and Arctic: Late Jurassic to Present, *Tectonophysics*, 155, 73–120, 1988.

Ryan, W. B. F., Geodynamic implications of the Messinian crisis of salinity, in *Messinian Events in the Mediterranean*, edited by C. W. Drooger, pp. 26–38, North-Holland, New York, 1973.

Savin, S. M., R. G. Douglas, and F. G. Stehli, Tertiary marine paleotemperatures, *Geol. Soc. Am. Bull.*, 86, 1499–1510, 1975.

Shackleton, N. J., Attainment of isotopic equilibrium between ocean water and the benthonic foraminifera genus Uvigerina: Isotopic changes in the ocean during the last glacial, *Colloq. Int. C.N.R.S.*, *219*, 203–209, 1974.

Shackleton, N. J., and J. P. Kennett, Paleotemperature history of the Cenozoic and the initiation of Antarctic glaciation: Oxygen and carbon isotope analysis in DSDP sites 277, 279, and 281, *Initial Rep. Deep Sea Drill. Proj.*, *29*, 743–755, 1975.

Shackleton, N. J., and N. D. Opdyke, Oxygen isotope and palaeomagnetic stratigraphy of equatorial Pacific Core V28-238: Oxygen isotope temperatures and ice volumes on a 10^5 year and 10^6 year scale, *Quat. Res.*, *3*, 39–55, 1973.

Shipboard Scientific Party, Site 689, *Proc. Ocean Drill. Program Initial Rep.*, *113*, 89–181, 1988*a*.

Shipboard Scientific Party, Site 690, *Proc. Ocean Drill. Program Initial Rep.*, *113*, 183–292, 1988*b*.

Shipboard Scientific Party, Site 699, *Proc. Ocean Drill Program Initial Rep.*, *114*, 621–795, 1988*c*.

Spiess, V., Cenozoic magnetostratigraphy of Leg 113 drill sites, Maud Rise, Weddell Sea, Antarctica, *Proc. Ocean Drill. Program Sci. Results*, *113*, 261–315, 1990.

Stott, L. D., and J. P. Kennett, Antarctic Paleogene planktonic foraminifer biostratigraphy: ODP Leg 113, sites 689 and 690, *Proc. Ocean Drill. Program Sci. Results*, *113*, 549–569, 1990.

Thomas, E., Development of Cenozoic deep-sea benthic foraminiferal faunas in Antarctic waters, Origins and Evolution of the Antarctic Biota, *Geol. Soc. Spec. Publ.*, *47*, 283–296, 1989.

Thomas, E., Late Cretaceous through Neogene deep-sea benthic foraminifers (Maud Rise, Weddell Sea, Antarctica), *Proc. Ocean Drill. Program Sci. Results*, *113*, 571–594, 1990.

Thomas, E., Middle Eocene late Oligocene bathyal benthic foraminifera (Weddell Sea): Faunal changes and implications for ocean circulation, in *Eocene-Oligocene Climatic and Biotic Evolution*, edited by D. A. Prothero and W. A. Berggren, pp. 245–271, Princeton University Press, Princeton, N. J., 1992.

Thomas, E., E. Barrera, N. Hamilton, B. T. Huber, J. P. Kennett, S. B. O'Connell, J. J. Pospichal, V. Spiess, L. D. Stott, W. Wei, and S. W. Wise, Jr., Upper Cretaceous–

Paleogene stratigraphy of sites 689 and 690, Maud Rise, Antarctica, *Proc. Ocean Drill. Program Sci. Results*, *113*, 901–914, 1990.

Wei, W., Middle Eocene–lower Oligocene calcareous nannofossil magnetobiochronology of ODP holes 699A and 703A in the Subantarctic South Atlantic, *Mar. Micropaleontol.*, *18*, 143–165, 1991.

Wei, W., Paleogene chronology of Southern Ocean drill holes: An update, in *The Antarctic Paleoenvironment: A Perspective on Global Change, Part One, Antarct. Res. Ser.*, vol. 56, edited by J. P. Kennett and D. A. Warnke, pp. 75–96, AGU, Washington, D. C., 1992.

Wei, W., and S. W. Wise, Jr., Middle Eocene to Pleistocene calcareous nannofossils recovered by (Ocean Drilling Program) Leg 113 in the Weddell Sea, *Proc. Ocean Drill. Program Sci. Results*, *113*, 639–666, 1990.

Wise, S. W., Jr., J. R. Breza, D. M. Harwood, and W. Wei, Paleogene glacial history of Antarctica, in *Controversies in Modern Geology*, edited by D. W. Muller, J. A. McKenzie, and H. Weissert, pp. 133–171, Academic, San Diego, Calif., 1991.

Wolfe, J. A., Tertiary climates and floristic relationships at high latitudes in the northern hemisphere, *Palaeogeogr. Palaeoclimatol. Palaeoecol.*, *30*, 313–323, 1980.

Wright, J. D., and K. G. Miller, Southern Ocean influences on late Eocene to Miocene deepwater circulation, this volume.

Zachos, J. C., W. A. Berggren, M.-P. Aubry, and A. Mackensen, Isotope and trace element geochemistry of Eocene and Oligocene foraminifers from Site 748, Kerguelen Plateau, *Proc. Ocean Drill. Program Sci. Results*, *120*, 839–854, 1992*a*.

Zachos, J. C., J. R. Breza, and S. W. Wise, Early Oligocene ice-sheet expansion on Antarctica: Stable isotope and sedimentological evidence from Kerguelen Plateau, southern Indian Ocean, *Geology*, *20*, 569–573, 1992*b*.

(Received February 5, 1993;
accepted March 1, 1993.)

THE ANTARCTIC PALEOENVIRONMENT: A PERSPECTIVE ON GLOBAL CHANGE
ANTARCTIC RESEARCH SERIES, VOLUME 60, PAGES 49–65

EOCENE TO OLIGOCENE OCEANOGRAPHY AND TEMPERATURES IN THE ANTARCTIC INDIAN OCEAN

ENRIQUETA BARRERA

Department of Geological Sciences, University of Michigan, Ann Arbor, Michigan 48109

BRIAN T. HUBER

Department of Paleobiology, Smithsonian Institution, Washington, D. C. 20560

Oxygen and carbon isotopic analyses of benthic and planktonic foraminiferal species from the lower Eocene to Oligocene section at Ocean Drilling Program sites 738 and 744 in the Indian Ocean provide insights into the response of polar surface and deep waters during a major climatic cooling and development of continental glaciation on Antarctica. Based on isotopic ranking and $\delta^{13}C$ values of planktonic foraminiferal species, there is no evidence for major changes in upwelling or productivity as a consequence of cooling of the high latitudes from early Eocene to early Oligocene time. The planktonic foraminiferal data indicate that the thermal structure of the water column changed little not only from early to late middle Eocene but also from late Eocene to early Oligocene time. An ~1.0‰ decrease in $\delta^{18}O$ values of both planktonic and benthic foraminifera occurred at about 42 Ma. Results of high-resolution analyses across the Eocene-Oligocene transition indicate the following: (1) a 1.4‰ positive shift in foraminiferal $\delta^{18}O$ values occurred in sediments at the base of Chron 13N deposited at ~35.85 Ma; (2) the highest Oligocene *Cibicidoides* $\delta^{18}O$ values (~2.5‰) occurred from ~35.85 to 35.24 Ma; (3) increased continental ice accumulation in the early Oligocene increased the average oceanic $\delta^{18}O$ composition by at least 0.6‰.

INTRODUCTION

The inaccessibility of the Antarctic sedimentary record due to the present thick ice sheet has caused the climatic and glacial history of the continent to be primarily reconstructed from oxygen isotopic studies of benthic foraminifera from deep-sea sediments. This reconstruction is based on the idea that oxygen isotopic compositions of deep-sea benthic foraminifera reflect temperatures at the high latitudes where bottom waters form [*Emiliani*, 1956; *Savin et al.*, 1975]. Isotopic studies on Antarctic pelagic sediments of Paleogene age recovered during Deep Sea Drilling Project (DSDP) Leg 29 in the southwestern Pacific Ocean [*Shackleton and Kennett*, 1975], and more recently, Weddell Sea Ocean Drilling Program (ODP) Leg 113 [*Kennett and Stott*, 1990, 1991; *Stott et al.*, 1990] and southern Indian Ocean ODP Leg 119 [*Barrera and Huber*, 1991; *Barrera et al.*, 1990, 1991] and Leg 120 [*Zachos et al.*, 1992a] have confirmed previous inferences from lower-latitude sequences and provided more reliable and detailed indirect evidence of the continent's climatic evolution during this time. In this paper, we discuss foraminiferal stable oxygen and carbon isotopic results from pelagic sediments spanning the lower Eocene through Oligocene at sites 738 and 744, which were drilled during Leg 119 in the southern Kerguelen Plateau (Figure 1). The high-latitude location and the proximity to the margins of East Antarctica make these sites ideal monitoring stations of climatic events and associated oceanographic changes within surface to intermediate water depths in the surrounding Antarctic Ocean.

Previous isotopic studies have documented the maximum high-latitude warming of the Cenozoic in the early Eocene [*Savin et al.*, 1975; *Stott et al.*, 1990; *Barrera and Huber*, 1991] when latitudinal temperature gradients were very low [*Shackleton and Boersma*, 1981] and high-latitude bottom water production appears to have been reduced [*Kennett and Stott*, 1991; *Barrera and Huber*, 1991; *Barrera and Keller*, 1991]. Subsequent to the early Eocene $\delta^{18}O$ minimum, benthic foraminifera and high- and mid-latitude planktonic $\delta^{18}O$ values increased by about 2‰ during the middle and late Eocene, reflecting gradual cooling of the high latitudes during this time [*Shackleton and Kennett*, 1975; *Oberhänsli et al.*, 1984; *Shackleton et al.*, 1984; *Kennett and Stott*, 1990; *Barrera and Huber*, 1991]. The largest Paleogene increase of about 1 to 1.5‰ in benthic foraminiferal

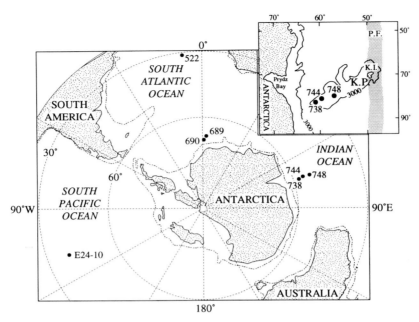

Fig. 1. Geographic reconstruction at 43 Ma in the middle Eocene, according to *Scotese and Denham* [1988] showing the estimated location of deep-sea sites and cores discussed in this study. The dotted line surrounding the continents represents the edge of the continental shelf. The area drawn in the upper right-hand corner shows the present position of the Kerguelen Plateau (K.P.) relative to the continental margin of East Antarctica and Prydz Bay. The present polar front (P.F.) position is indicated by the stippled pattern at about the latitude of Kerguelen Island (K.I.), and contours are drawn at the 3000-m isobath.

$\delta^{18}O$ values was recorded just above the Eocene/ Oligocene boundary at the base of Chron 13N, with an estimated age of 35.9 Ma [*Oberhänsli et al.*, 1984; *Miller et al.*, 1988; *Barrera and Huber*, 1991; *Barrera et al.*, 1991]. Unlike earlier increases in benthic foraminifer $\delta^{18}O$ values, which have been attributed to declines in temperatures, the early Oligocene ^{18}O increase has been interpreted as a combination of both high-latitude cooling and increased Antarctic ice [*Kennett*, 1977; *Matthews and Poore*, 1980; *Keigwin and Keller*, 1984; *Keigwin and Corliss*, 1986; *Miller et al.*, 1987].

The onset of Cenozoic glaciation in Antarctica has been the subject of much debate. One school of thought [*Matthews and Poore*, 1980; *Prentice and Matthews*, 1988] has suggested that East Antarctic ice growth to nearly modern volumes occurred in steps during the middle Eocene at about 49, 46, and 44 Ma. An ice sheet slightly larger than the present was established by 42 Ma, and larger ice volumes dominated throughout the rest of the Neogene. Their argument is based on the assumption that temperatures of tropical surfaces waters have remained constant and thus variations in $\delta^{18}O$ values of planktonic foraminifera from the low latitudes reflect ice volume. Deep-sea benthic foraminifera $\delta^{18}O$ records both high-latitude temperature and ice volume changes. A different approach to the interpretation of the deep-sea oxygen isotopic record in terms of Antarc-

tic glaciation has been followed by others based on comparison of modern and ancient temperatures for deep waters [*Keigwin and Keller*, 1984; *Miller and Thomas*, 1985; *Miller et al.*, 1987]. Accordingly, calculated low isotopic temperatures of deep waters are similar to those of the present at about 36, 30, and 23 Ma, indicating that Antarctica had a large accumulation of continental ice at these times [*Miller et al.*, 1987].

Sedimentologic evidence from the continent corroborates the scenario of extensive Antarctic ice accumulation at the beginning of the Oligocene. The first evidence came from recovery of a 700-m-thick lower to upper Oligocene sequence of glaciomarine sediments and glacial diamictites from the Ross Sea CIROS 1 hole [*Barrett*, 1989]. However, the significance of these deposits was compromised by their proximity to the Transantarctic Mountains and thus the possibility that the glacial extent was local. Glacial sediments drilled on the continental margin of East Antarctica during ODP Leg 119 indicate that ice extended along the shelf at Prydz Bay during the early Oligocene [*Barron et al.*, 1991a]. Prydz Bay is the discharge area of the Lambert Glacier that today drains about 20% of East Antarctica. It is likely that this area, located several hundreds of kilometers from the pelagic sites discussed in this study, played a similar key role during the late Paleogene.

In an earlier publication [*Barrera and Huber*, 1991],

we discussed the thermal and carbon isotope evolution of surface and deep waters at sites 738 and 744 during the Paleocene to early Miocene. Briefly, we indicated that the cooling of the southern high latitudes appeared to have begun first at the surface and later at the near surface and at depth during the late early Eocene, suggesting a low-latitude source of deep waters at this location. In the transition from the late Eocene to early Oligocene, the $\delta^{18}O$ shift previously associated with increased cryospheric development was found to correspond with a thick layer of ice-rafted material. These sediments were most likely deposited from floating icebergs calved from an ice sheet in East Antarctica. In this contribution, we investigate temperature and productivity changes in surface waters and the water column during the cooling of the southern high latitudes from the early Eocene to the early Oligocene. For this, we have determined the isotopic ranking of planktonic foraminiferal species in Antarctic waters. Next, we focus on the Eocene-Oligocene transition and discuss implications and estimates for the extent of Antarctic glacial conditions in the Oligocene. The Eocene isotopic data generated in this study are used to estimate surface temperatures and to evaluate whether the ice-rafted debris reported in middle Eocene sediments at this site is in situ and may represent evidence for continental glaciation.

METHODS

All sediment samples were processed according to standard micropaleontological techniques described in the work of *Barrera and Huber* [1991]. Sediment plugs from Core 119-738B-10H and Core 119-738B-11H at Site 738 were searched for ice-rafted grains. To prevent contamination with extraneous material, we avoided sediment close to the core liner, on the surface of the core split, and very near to the top in the core. Except for the specific sizes of samples listed in Table 1, planktonic foraminiferal specimens between 150 and 250 μm were analyzed isotopically. Specimens within this size range are adults because, in general, high-latitude foraminiferal faunas are composed of smaller specimens than those from the low latitudes. Benthic foraminiferal specimens were larger than 150 μm. To remove adhering fine particles, individual specimens were ultrasonically agitated in distilled water.

Prior to isotopic analyses, foraminiferal samples of about 0.05 to 0.1 mg were roasted under vacuum at 380°C for 1 hour to remove volatile organic contaminants. For isotopic analysis, samples were reacted with 100% H_3PO_4 at 75°C in an automated Kiel carbonate extraction system coupled directly to a Finnegan MAT-251 mass spectrometer for analysis of the evolved CO_2 gas, at the stable isotope laboratory of the University of Michigan. Isotopic measurements were made in relation to a laboratory standard CO_2 gas, which is calibrated to

international standards through analysis of NBS-18, NBS-19 and NBS-20 powdered standards. Isotopic values are reported in the delta (δ) notation as per mil deviations from the Peedee Belemnite (PDB) standard. Standard replicates have a precision of ±0.05‰ for both $\delta^{18}O$ and $\delta^{13}C$.

We use the equation of *O'Neil et al.* [1969] as recast by *Shackleton* [1974] to calculate seawater paleotemperatures. Modern seawater $\delta^{18}O$ values in the polar Indian Ocean were calculated using the measured salinity and temperature data reported for GEOSEC Station 430 [*Ostlund*, 1987] and the relationship between salinity and $\delta^{18}O$ from the data of *Craig and Gordon* [1965] for seawater in the southwestern Pacific Ocean. We assume that in an ice-free world the average $\delta^{18}O$ of seawater would have been 0.9‰ lower than at present [*Shackleton and Kennett*, 1975]. Where noted, *Cibicidoides* $\delta^{18}O$ values have been increased by 0.5‰ to compensate for "vital" or disequilibrium effects [*Shackleton et al.*, 1984]. This estimate is smaller in comparison with the value of 0.64‰ calculated for Miocene *Cibicidoides* data [*Savin et al.*, 1981].

All of the stable isotopic results from sites 738 and 744 used and discussed in this paper are either in Tables 1 and 2 or in the work of *Barrera and Huber* [1991].

STRATIGRAPHY AND FORAMINIFERAL PRESERVATION

Age determination of sediments at Site 744 is based on the integration of calcareous and siliceous plankton biostratigraphic data with magnetic polarity data and strontium isotope stratigraphy [*Barron et al.*, 1991b; *Barrera et al.*, 1991]. Sediment ages at Site 738 are based on calcareous microfossil datums [*Huber*, 1991; *Barron et al.*, 1991b] tied to the chronology of *Berggren et al.* [1985]. Specifically, we follow the planktonic foraminiferal zonation of *Stott and Kennett* [1990] as applied in the work of *Huber* [1991]. Results discussed in this paper are from the lower Eocene to upper Eocene section at Site 738 and the upper Eocene to lower Oligocene section at Site 744 where sedimentation during this interval appears to have been nearly 100% complete. Sediments are calcareous nannofossil ooze.

Preservation of the original calcite in foraminiferal species isotopically analyzed was evaluated at selected levels based on scanning electron microscope observation of broken surfaces and interior test surfaces. Both replacement by secondary calcite and secondary calcite overgrowth are minor and probably not volumetrically significant enough to affect $^{18}O/^{16}O$ ratios. In general, $\delta^{13}C$ values of foraminifera in deep-sea sediments should be less affected by dissolution-reprecipitation processes because the carbon reservoir in pore waters is significantly smaller than in the carbonate reservoir. Solution features are common in almost all of the tests examined, and it is possible that dissolution has biased

TABLE 1. Oxygen and Carbon Isotope Ratios of Selected Planktonic Foraminiferal Species

Species	Sample 744-16H-1 (89–94) $\delta^{18}O$	$\delta^{13}C$	Sample 744-20H-5 (95–100) $\delta^{18}O$	$\delta^{13}C$	Sample 738-10H-3 (23–25) $\delta^{18}O$	$\delta^{13}C$	Sample 738-11H-1 (45–47) $\delta^{18}O$	$\delta^{13}C$	Sample 738–5R-1 (78–80) $\delta^{18}O$	$\delta^{13}C$	Sample 738-7R-2 (104–106) $\delta^{18}O$	$\delta^{13}C$	Sample 738-11R-1 (15–17) $\delta^{18}O$	$\delta^{13}C$
Morozovella aequa													−0.92 (250–210)	2.76
Morozovella gracilis													−0.96 (180–150)	2.63
Morozovella cf. quetra					0.35 (250–150)	2.76	0.53 (250–150)	2.69						
Acarinina acarinata													−1.07 (210–180)	2.73
Acarinina interposita													−0.91 (250–210)	2.54
Acarinina nitida													−1.19 (250–210)	2.99
Acarinina triplex													−1.15 (250–210)	2.80
Acarinina bullbrooki									−0.59 (250–210)	2.34	−0.79 (250–210)	2.25		
Acarinina broedermanni									−0.79 (250–210)	2.15				
Acarinina collactea									−0.54 (250–210)	2.00	−0.72 (250–210)	1.90		
Acarinina pentacamerata									−0.97 (250–210)	2.61	−0.72 (250–210)	2.12		
Acarinina primitiva					0.26 (250–150)	2.86	0.08 (250–150)	3.11	−1.02 (250–150)	2.71	−1.26 (250–210)	2.81		
Acarinina pseudotopilensis											−0.76 (250–210)	2.17		
Chiloguembelina cubensis	1.66 (150–106)	1.99	0.59 (150–106)	2.27	0.82 (125–63)	1.96	0.88 (250–150)	1.95						
Chiloguembelina wilcoxensis													−0.04 (210–150)	1.10
Tenuitella gemma	1.83 (150–106)	1.39	1.19 (150–106)	1.05										
Globigerinatheka index			0.78 (250–180)	1.58	1.31 (250–150)	2.20	0.38 (250–150)	2.85						
Globanomalina australiformis									−0.19 (250–210)	1.20	−0.68 (250–210)	1.16		
Globanomalina reissi													−0.35 (210–150)	1.64
Globanomalina planoconica									−0.17 (250–210)	1.23			−0.39 (210–150)	1.51
Subbotina angiporoides	2.22 (250–180)	1.28	1.22 (250–180)	1.57	1.17 (250–150)	1.40	0.92 (250–150)	1.37						
Subbotina corpulenta	2.32 (250–180)	1.31												
Subbotina eocaena	2.31 (250–180)	1.39	1.27 (250–180)	1.67	1.07 (250–150)	1.42			−0.03 (250–210)	1.36				
Subbotina hornibrooki									−0.03 (250–210)	1.45	−0.31 (250–210)	1.25		
Subbotina linaperta	2.08 (250–180)	1.44	1.27 (250–180)	1.65	0.95 (250–150)	1.42	1.18 (250–150)	1.6	−0.19 (250–210)	1.35			−0.22 (250–210)	1.34
Subbotina linaperta					0.79 (>250)	1.8					−0.24 (250–210)	1.29		
Subbotina pseudoeocaena									−0.28 (300–250)	1.07	−0.35 (300–250)	1.22		
Globorotaloides suteri	2.13 (150–106)	1.05	1.38 (150–106)	1.26	1.31 (250–150)	1.51	1.2 (250–150)	1.46						
Catapsydrax unicava					0.9 (250–150)	1.48	1.11 (250–150)	1.38						
Nuttallides spp.	2.08	0.92	1.01	0.89	0.73	0.93	1.22	1.06	−0.50	0.76	−0.82	0.58	−0.41	0.74

Numbers in parentheses are the size range of specimens in micrometers.

TABLE 2. Oxygen and Carbon Isotope Ratios of Selected Foraminiferal Taxa

Core-Section, Interval, cm	Depth, m	Age, m.y.	Antarctic Foraminiferal Biozone	Cibicidoides sp. δ¹⁸O	Cibicidoides sp. δ¹³C	Nuttallides spp. δ¹⁸O	Nuttallides spp. δ¹³C	Nuttallides truempyi δ¹⁸O	Nuttallides truempyi δ¹³C	Chiloguembelina spp. δ¹⁸O	Chiloguembelina spp. δ¹³C	Globorotaloides suteri δ¹⁸O	Globorotaloides suteri δ¹³C	Subbotina linaperta δ¹⁸O	Subbotina linaperta δ¹³C	Acarinina primitiva δ¹⁸O	Acarinina primitiva δ¹³C
Hole 744A																	
16-1, 89-94	138.11	35.07	AP13	2.19	1.31	2.08	0.92			1.42	1.49						
16-3, 64-66	140.85	35.42	AP13	2.53	1.13					1.82	2.09						
16-3, 145-147	141.66	35.48	AP13	2.50	1.23					1.85	2.11						
16-4, 5-7	141.76	35.49	AP13	2.47	1.41					1.87	1.99						
16-4, 56-58	142.27	35.53	AP13	2.54	1.50					1.81	2.00						
16-4, 95-100	142.67	35.56	AP13	2.28	1.27					1.92	2.07						
16-5, 5-7	142.96	35.58	AP13	2.37	1.15					2.01	1.98						
16-5, 56-58	143.47	35.62	AP13	2.37	1.32					2.09	2.27						
16-5, 145-147	144.35	35.69	AP13	2.24	1.44					1.80	2.26						
16-6, 4-6	144.45	35.70	AP13	2.35	1.33					1.72	2.39						
16-6, 56-58	144.96	35.74	AP13	2.32	1.67					1.56	2.05	2.39	1.32				
16-6, 95-100	145.37	35.77	AP13														
16-6, 145-147	145.86	35.81	AP13	2.51	1.46							2.51	1.15				
16-7, 12-14	146.03	35.83	AP13	2.58	1.21							2.51	1.29				
16-7, 55-57	146.46	35.86	AP13	1.97	1.23							1.75	1.14				
16CC	146.65	35.87	AP13														
17-1, 17-19	146.87	35.89	AP13	1.63	1.08							1.98	1.34				
17CC, 17-19	147.34	36.15	AP13	1.77	1.08					1.26	1.95	1.62	1.18				
18-1, 15-17	147.75	36.20	AP13	1.52	1.22					1.23	2.20						
18-1, 95-100	148.57	36.30	AP13	1.50	0.78												
18-5, 95-100	154.57	37.12	AP13	1.25	1.05												
18CC	157.10	37.48	AP13	1.08	1.00												
20-5, 95-100	173.57	39.36	AP13			1.01	0.89										
Hole 738B																	
4H-4, 90-95	28.42	36.78	AP13	1.14	0.93					0.86	1.79	1.61	0.87				
4H-5, 90-95	29.92	37.04	AP13	1.01	1.07					0.96	1.49	1.42	0.85				
4H-6, 90-95	31.42	37.31	AP13	1.14	0.91					1.04	1.51	1.42	0.61				
7H-1, 90-95	52.42	39.96	AP12	1.41	0.93					1.13	1.80	1.43	0.85				
7H-3, 90-95	55.42	40.08	AP12	1.59	0.85							1.49	1.15				
7H-4, 90-95	56.92	40.31	AP12	1.35	0.66							1.52	0.81				
7H-6, 90-95	59.92	40.46	AP12	1.44	0.86							1.36	1.27				
8H-1, 90-95	61.92	40.57	AP11	1.55	1.02							1.31	0.79				
8H-2, 90-95	63.42	40.68	AP11	1.31	0.77							1.50	0.95				
8H-3, 90-95	64.92	40.80	AP11	1.40	1.13							1.46	1.27				
8H-4, 90-95	66.42	41.02	AP11	1.31	0.73					0.99	1.48	1.57	0.88				
8H-6, 90-95	69.42	41.08	AP11	1.52	0.84					0.69	1.39						
8H-7, 20-25	70.22	41.14	AP11	1.25	1.00					1.20	1.51						
9H-1, 90-95	71.42	41.15	AP11	0.69	1.51												
9H-1, 119-121	71.70	41.19	AP11	0.87	0.68												
9H-2, 90-95	72.92	41.20	AP11							0.43	1.09	1.26	0.66				
9H-2, 119-121	73.20	41.25	AP11	0.83	0.57												
9H-3, 90-95	74.42		AP11	0.58	1.01							1.45	1.05			0.30	2.60

TABLE 2. (continued)

Hole 738B (continued)

Core-Section, Interval, cm	Depth, m	Age, m.y.	Antarctic Foraminiferal Biozone	Cibicidoides sp. $\delta^{18}O$	Cibicidoides sp. $\delta^{13}C$	Nuttallides spp. $\delta^{18}O$	Nuttallides spp. $\delta^{13}C$	Nuttallides truempyi $\delta^{18}O$	Nuttallides truempyi $\delta^{13}C$	Chiloguembelina spp. $\delta^{18}O$	Chiloguembelina spp. $\delta^{13}C$	Globorotaloides suteri $\delta^{18}O$	Globorotaloides suteri $\delta^{13}C$	Subbotina linaperta $\delta^{18}O$	Subbotina linaperta $\delta^{13}C$	Acarinina primitiva $\delta^{18}O$	Acarinina primitiva $\delta^{13}C$
9H-3, 90–95	74.42	41.25	AP11	0.58	1.01					0.43	1.09	1.45	1.05			0.30	2.60
9H-4, 90–95	75.92	41.30	AP11	1.09	1.01												
9H-5, 40–42	76.91	41.34	AP11	1.12	0.75												
9H, CC (5 cm)	80.00	41.45	AP10	1.11	0.94												
10H-1, 90–95	80.93	41.49	AP10	0.82	0.87			0.36	1.07	0.53	1.68					0.30	2.22
10H-1, 105–107	81.06	41.49	AP10	0.94	0.75												
10H-2, 90–95	81.92	41.52	AP10							0.77	1.92						
10H-2, 100–102	82.01	41.53	AP10	0.96	1.13											0.64	2.49
10H-3, 23–25	83.24	41.57	AP10					0.73	0.93								
10H-3, 25–27	83.26	41.57	AP10	0.91	1.15			0.71	0.87			0.85	2.04			0.31	3.22
10H-3, 90–95	84.01	41.60	AP10	0.71	0.97					0.87	1.78					0.25	2.87
10H-3, 110–112	84.11	41.61	AP10	0.86	1.16			0.82	1.01			0.56	2.34			0.34	2.87
10H, CC (10 cm)	85.00	41.64	AP10	0.94	1.2			0.86	0.89			1.15	1.67			-0.04	3.37
11H-1, 20–22	85.20	41.64	AP10	1.07	1.31			1.01	0.98			1.00	1.79				
11H-1, 40–42	85.41	41.65	AP10	1.05	1.38												
11H-1, 45–47	85.46	41.65	AP10					1.22	1.06			0.93	2.05			0.47	2.56
11H-1, 85–87	85.86	41.67	AP10	0.88	1.32			0.85	1.07	0.68	2.00						
11H-1, 90–95	85.92	41.67	AP10					0.91	0.85								
11H-2, 30–32	86.81	41.71	AP10	0.84	1.33			0.63	0.70			0.84	1.74				
11H-2, 90–95	87.42	41.73	AP10	1.03	1.54					0.82	1.94					0.27	2.95
11H-2, 118–120	87.68	41.74	AP10					0.87	1.13			1.21	1.55				
11H-3, 30–32	88.31	41.76	AP10	0.70	1.43			0.71	0.94			0.98	1.45				
11H-3, 90–95	88.92	41.78	AP10							0.53	1.91						
11H-4, 30–32	89.81	41.82	AP10	0.42	1.21			0.38	0.94								
11H-4, 90–95	90.42	41.84	AP10							0.64	2.08						
11H-4, 130–132	90.81	41.85	AP10	0.87	1.23												
11H-5, 30–32	91.33	41.87	AP10	0.51	1.13							0.87	1.36			0.19	2.74
11H-5, 90–95	91.93	41.90	AP10	0.47	1.15			0.48	0.82	-0.39	2.63						
11H-6, 30–32	92.81	41.93	AP10	-0.16	0.53											0.19	2.03
11H-6, 90–95	93.42	41.95	AP10	0.11	0.96			-0.15	0.38			0.47	1.28				
11H-7, 10–12	94.11	41.98	AP10	0.11	1.01			-0.07	0.66							0.39	1.93
11H, CC	94.48	41.99	AP10													-0.23	2.14

Sample	Depth	Age	Zone										
12H-1, 87–92	95.40	42.02	AP10	0.58	0.99			0.78	0.49			1.50	1.03
12H-2, 87–92	96.90	42.08	AP10	0.42	0.82			0.47	0.40			1.34	0.91
12H-3, 87–92	98.40	42.21	AP10	0.55	0.83			0.42	0.40			1.66	0.90
12H-4, 86–91	99.90	42.38	AP10	0.58	0.68			0.42	0.57			1.77	0.87
12H-5, 87–92	101.40	42.56	AP10	0.72	0.81			0.38	0.75			1.80	0.87
12H-6, 87–92	102.90	42.73	AP10	1.04	0.55							1.58	0.86
12H, CC	103.98	42.85	AP10	0.99	0.70							1.73	0.73
13H-1, 95–100	104.97	42.97	AP10	0.97	0.79			0.45	0.86			2.11	0.14
13H-2, 95–100	106.47	43.14	AP10	0.89	0.64			0.30	0.97			2.06	0.23
13H-3, 30–35	107.26	43.23	AP10	1.34	0.97			0.60	0.78			2.16	0.30
13H, CC	108.15	43.33	AP10	0.85	0.59							2.13	0.40
14X-1, 90–95	109.12	43.44	AP10	0.62	0.50								
14X-3, 90–95	112.12	43.79	AP10	0.78	0.41								
14X, CC	117.75	44.43	AP10	0.64	0.42								
15X-3, 90–95	121.72	44.89	AP10	0.64	0.49			0.11	0.67			1.85	0.15
15X-4, 90–95	123.22	45.06	AP10	0.35	0.81								
15X-5, 90–92	124.72	45.10	AP10	0.57	0.98								
16X-1, 29–34	127.82	45.18	AP10	0.58	0.69			0.16	0.42			2.02	0.24
16X-3, 39–41	130.90	45.38	AP10										
16X, CC	137.18	45.81	AP10										
17X-1, 90–95	138.12	45.87	AP10	0.44								2.29	0.37
17X-3, 90–95	141.12	46.07	AP10									1.76	0.43
21X-1, 90–95	176.72	47.21	AP8									2.70	0.32
23X-1, 90–95	195.92	48.98	AP7					0.29	−0.36				
23X, CC	204.60	50.18	AP7					0.14	0.02				
24X, CC	214.30	51.23	AP7					0.36	−0.71				
Hole 738C													
4R-1, 14–15	216.04		AP7					−0.40	0.77	−0.47	1.19	−0.46	2.08
4R-1, 50–52	216.41		AP7					−0.11	1.32	−0.11	1.32	−1.17	2.85
4R-1, 119–121	217.10		AP7					−0.41	0.83	−0.01	1.40	−0.44	2.45
4R-2, 15–17	217.56		AP7					−0.60	0.59	−1.12	0.77	−0.83	2.77
4R-2, 50–52	217.91		AP7					−0.58	0.49	−0.10	1.13	−1.20	2.46
4R-2, 100–102	218.41		AP7					−0.39	0.67	−0.06	1.29	−0.83	2.63
4R-3, 10–12	219.01		AP7					−0.38	0.75	−0.12	1.28	−0.75	2.40
4R-2, 51–53	219.42		AP7					−0.38	0.49	−0.12	1.28	−0.91	2.40
4R-3, 87–89	219.78		AP7					−0.38	0.49				
5R-1, 78–80	226.32		AP7	−0.50	0.76							−1.19	2.99
7R-1, 145–147	246.22		AP7	−0.84	0.59			−0.01	1.24				
7R-2, 104–106	247.22		AP7	−0.82	0.58			−0.31	1.25			−1.26	2.82
11R-1, 15–17	283.52		AP5	−0.41	0.74	−0.04	1.11						
11R-1, 60–62	283.97		AP5	−0.41	0.63	−0.14	1.01					−0.94	3.23

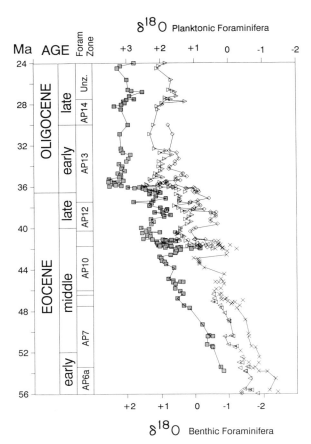

Fig. 2. Eocene to Oligocene oxygen isotopic records of planktonic and benthic foraminifera versus sediment age at sites 738 and 744. Isotopic values are listed in Table 2 and in the work of *Barrera and Huber* [1991]. Note that different scales have been used for oxygen isotopic compositions of planktonic and benthic foraminifera. No adjustment for disequilibrium has been applied to the benthic foraminiferal $\delta^{18}O$ data. Symbols are as follows: *Cibicidoides* spp., squares; *Globorotaloides suteri*, triangles pointing right; *Subbotina* spp., triangles pointing left; *Chiloguembelina* spp., diamonds; *Acarinina primitiva*, crosses.

isotopic records with the selective preservation of ^{18}O-enriched tests.

PALEOGENE FORAMINIFERAL $\delta^{18}O$ TIME SERIES AT SITES 738 AND 744

The $\delta^{18}O$ time series for sites 738 and 744 in Figure 2 are based on the isotopic composition of the benthic foraminifera *Cibicidoides* and the planktonic foraminiferal species that best represent surface and near-surface water conditions in early Eocene to early Oligocene time. The selection of these planktonic taxa was based on a study of their isotopic ranking which is discussed in the next section. In general, $\delta^{18}O$ curves for surface, near-surface, and intermediate waters fluctuate sympathetically, with greater resemblance between the sea-

floor and near-surface variations. A marked, but gradual, $\delta^{18}O$ increase of about 1.5‰ in surface and deep waters is observed in lowermost middle Eocene sediments from about 54 to 42 Ma similar to that recorded at many other deep-sea sites [*Shackleton et al.*, 1984; *Shackleton*, 1986; *Oberhänsli*, 1986; *Miller et al.*, 1987; *Katz and Miller*, 1991; *Oberhänsli et al.*, 1984; *Oberhänsli and Toumarkine*, 1985]. However, in the southern Indian Ocean the middle Eocene record of cooling is interrupted by an event of considerably lower $\delta^{18}O$ values at about 42 Ma. This event, which appears to reflect a sudden warming of the southern high latitudes lasting ~1 m.y., is not observed in the $\delta^{18}O$ record of Weddell Sea Site 689 or lower-latitude sites. It may be restricted to the Indian Ocean because it was recorded by benthic foraminifera in the mid-latitude Indian Ocean Site 219 [*Keigwin and Corliss*, 1986; E. Barrera, unpublished data, 1991]. The nature of this event is not yet known, although it had a slightly greater effect on waters at the seafloor than at the surface in the high latitudes. It occurs in an interval where gravel and sand-sized grains of glacial origin were found at ~43.5 and 42 Ma at Site 738 [*Ehrmann*, 1991]. The isotopic record of this interval and the implications for the glacial history of Antarctica will be discussed below.

The highest $\delta^{18}O$ values of the Eocene are recorded in the transition from the middle to late Eocene at ~41 Ma at Site 738. Afterward late Eocene temperatures at both the surface and seafloor were on the average higher and variable. The largest and most rapid $\delta^{18}O$ change in the composite section at sites 738 and 744 occurred in the transition just above the Eocene/Oligocene boundary where values increased by over 1‰ [*Barrera and Huber*, 1991]. This shift corresponds to both an increase in the amount of glacial ice accumulating on Antarctica and cooling of the high latitudes. Both of these topics are discussed in detail in a later section. It is apparent that after this time, the high latitudes remained cold during the Oligocene.

ISOTOPIC SYSTEMATICS AND STRATIFICATION OF PLANKTONIC FORAMINIFERA

Background

Stable isotope studies of planktonic foraminiferal species collected in plankton tows in tropical and subtropical areas indicate that specimens can be widely distributed in the photic zone (about the uppermost 100 m of the water column), but most of the test calcification occurs within a narrow depth range above or within the thermocline [*Fairbanks and Wiebe*, 1980; *Fairbanks et al.*, 1980, 1982; *Bouvier-Soumagnac and Duplessy*, 1985; *Ravelo et al.*, 1990]. The $\delta^{18}O$ values of most planktonic species are either slightly lower (as in the case of spinose species) or close to equilibrium values [*Fairbanks et al.*, 1982]. In the sediments, $\delta^{18}O$ compositions of the dominant species approximate the annual range of calculated $\delta^{18}O$ for calcite precipitated in equilibrium with surface waters, considering their annual variability in temperature and salinity [*Ravelo and Fairbanks*, 1992]. The $\delta^{13}C$ of Modern planktonic foraminifera in the water column is close to the isotopic value of total dissolved CO_2 at the depth where calcification is inferred to have occurred [*Williams et al.*, 1977; *Fairbanks et al.*, 1982]. Total dissolved CO_2 $\delta^{13}C$ values decrease with depth owing to the oxidation at depth of surface-produced organic matter [*Deuser and Hunt*, 1969; *Kroopnick*, 1974]. Therefore it is possible to use oxygen and carbon isotopic data of monospecific planktonic specimens in the sediments to determine changes in the general physico-chemical structure of the upper level of the water column [*Emiliani*, 1954; *Douglas and Savin*, 1978; *Shackleton et al.*, 1985; *Oppo and Fairbanks*, 1989; *Ravelo et al.*, 1990].

Tertiary and Cretaceous planktonic foraminiferal species from low- and mid-latitude regions have also been found to be stratified in the water column according to their $\delta^{18}O$ and $\delta^{13}C$ compositions [*Douglas and Savin*, 1978; *Boersma and Shackleton*, 1981; *Shackleton et al.*, 1985; *Boersma et al.*, 1987]. There is a direct relation between $\delta^{13}C$ values and test size, particularly for the shallow-dwelling species [*Douglas and Savin*, 1978; *Shackleton et al.*, 1985; E. Barrera, unpublished data], which is similar to that measured in modern forms [*Bouvier-Soumagnac and Duplessy*, 1985; *Oppo and Fairbanks*, 1989]. Unlike the modern nonspinose species, the isotopic composition of acarininids and morozovellids indicates a habitat in the photic zone above the spinose subbotinids [*Shackleton et al.*, 1985]. Among the acarininids, the largest tests seem to reflect conditions closest to the surface [*Shackleton et al.*, 1985]. In this paper, we have used $\delta^{18}O$ and $\delta^{13}C$ compositions of Eocene and Oligocene planktonic foraminiferal species from sites 738 and 744 (1) to determine which species best represent surface and near-surface conditions in Antarctic waters, (2) to compare isotopic rankings to rankings of tropical species, and (3) to relate changes in the $\delta^{18}O$-$\delta^{13}C$ structure and temperature of Antarctic waters with south polar climatic variations.

Isotopic Stratification of Paleogene Species in Antarctic Waters

Isotopic results are presented in Table 1 and shown in Figure 3. The relationship between the $^{18}O/^{16}O$ and $^{13}C/^{12}C$ data, which is similar to that in the Modern water column, indicates the reliability of the foraminiferal data for inferring the depth stratification. In the early Eocene (Zone AP5 and Zone AP7), species of acarininids and morozovellids had, on the average, high $\delta^{13}C$ values and low $\delta^{18}O$ values relative to subbotinids, turborotalids, globanomalinids, and chiloguembelinids, suggesting calcification close to the sea surface. The warm water morozovellids species were present only

briefly in the southern high-latitude oceans during the early Eocene maximum warming [*Stott et al.*, 1990; *Huber*, 1991; *Barrera and Keller*, 1991]. Isotopic ranking somewhat similar to that in the early Eocene is observed in middle Eocene samples (Zone AP10). At this time, *Morozovella* cf. *quetra* (in the sense of *Huber* [1991]) appear to have calcified in shallow surface waters near *Acarinina primitiva*, whereas the chiloguembelinids moved closer to the surface and had intermediate isotopic values between the surface water species and the deepwater subbotinids, catapsydracids, and *Subbotina angiporoides*. Just before the extinction of the acarininids in the transition from the middle to the late Eocene (Zone AP11 to Zone AP12), $\delta^{18}O$ values of chiloguembelinids and *A. primitiva* from the same sample are very close, although there are large differences in their $\delta^{13}C$ values (Tables 1 and 2 and Figure 3). It appears that the chiloguembelinids came to inhabit the shallow surface layer in the high latitudes during the late Eocene (Foraminiferal Zone AP12) and Oligocene (Foraminiferal Zone AP13) as previously seen in middle- and low-latitude regions [*Keigwin and Corliss*, 1986; *Boersma et al.*, 1987]. During the late Eocene and early Oligocene, subbotinids and globorotaloidids continued to grow in deeper waters. The isotopic ranking of these early Eocene to Oligocene taxa from Antarctic waters is analogous to that observed in lower-latitude Pacific and Atlantic sites during this time interval [*Douglas and Savin*, 1978; *Shackleton et al.*, 1985; *Keigwin and Corliss*, 1986; *Boersma et al.*, 1987].

In the polar Indian Ocean the vertical $\delta^{13}C$ range in surface waters was ~1.5‰ in the early Eocene with shallow-dwelling species measuring ~3‰ and the deep-dwelling taxa near 1 to 1.5‰ (Figure 3). The $\delta^{18}O$ stratification in the surface layer was near 1‰ at this time. The late middle Eocene samples exhibit a similar range of isotopic compositions and absolute $\delta^{13}C$ values but overall significantly higher $\delta^{18}O$ values. Thus even with major cooling of the high latitudes inferred by ~1.5‰ increase in benthic and planktonic foraminiferal $\delta^{18}O$ values, the isotopic data suggest no significant changes in the thermal structure in the upper portion of the water column (Figures 2 and 3).

There is no indication of any significant change in surface water productivity from early to late middle Eocene. Productivity changes can be inferred from variations in the $\delta^{13}C$ difference between surface and deep water since phytoplankton preferentially removes ^{12}C from surface waters during photosynthesis [*Deuser and Hunt*, 1969]. By the late Eocene the major shallow-dwelling taxa of the middle Eocene had either become extinct or decreased in abundance at both high- and low-latitude sites [*Boersma et al.*, 1987; *Huber*, 1991]. At this time the range of $\delta^{13}C$ and $\delta^{18}O$ values in Antarctic waters decreased as the inferred shallow-dwelling species had lower $\delta^{13}C$ and slightly higher $\delta^{18}O$ values than those before. The deeper-dwelling

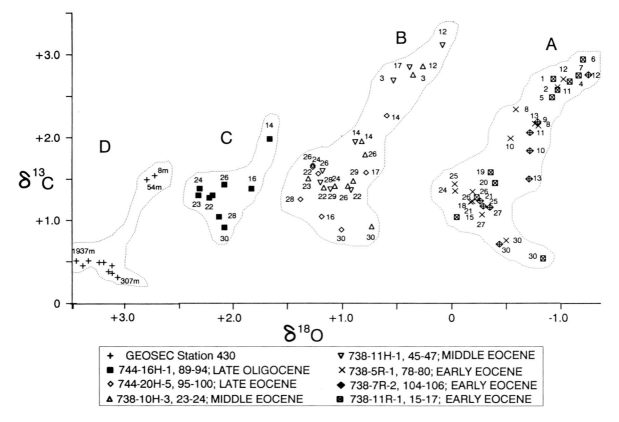

Fig. 3. Oxygen versus carbon isotope values of planktonic foraminiferal species in Indian Ocean polar waters in the (*a*) early Eocene, (*b*) late middle and late Eocene, and (*c*) early Oligocene. (*d*) Carbon isotopic values for Modern waters and calculated $\delta^{18}O$ values for calcite precipitated in equilibrium with those water masses are from data at GEOSEC Station 430 in the work of *Ostlund* [1987], and the numbers next to the data points represent water depth in meters. This station (59°59'S, 60°58'E) is south of the polar front and west of the southern Kerguelen Plateau. Foraminiferal isotopic values are from Table 1. Numbers represent foraminiferal species as follows: (1) *Morozovella aequa*, (2) *M. gracilis*, (3) *M.* cf. *quetra*, (4) *Acarinina acarinata*, (5) *A. interposita*, (6) *A. nitida*, (7) *A. triplex*, (8) *A. bullbrooki*, (9) *A. broedermanni*, (10) *A. collactea*, (11) *A. pentacamerata*, (12) *A. primitiva*, (13) *A. pseudotopilensis*, (14) *Chiloguembelina cubensis*, (15) *C. wilcoxensis*, (16) *Tenuitella gemma*, (17) *Globigerinatheka index*, (18) *Globanomalina australiformis*, (19) *G. reissi*, (20) *G. planoconica*, (21) *G.* cf. *pseudomenardii*, (22) *Subbotina angiporoides*, (23) *S. corpulenta*, (24) *S. eocaena*, (25) *S. hornibrooki*, (26) *S. linaperta*, (27) *S. pseudoeocaena*, (28) *Globorotaloides suteri*, (29) *Catapsydrax unicava*, and (30) benthic taxon *Nuttallides* spp.

taxa continued to live in waters as warm as those in the late middle Eocene. During the early Oligocene the structure of the upper water column resembled that of the late Eocene, except for markedly higher $\delta^{18}O$ values (by ~1‰) reflecting a combination of ice volume and temperature change, as discussed below. This is further indicated by the planktonic foraminiferal assemblages that exhibit no change in diversity or relative abundance of species from the late Eocene to the early Oligocene, except for the disappearance of *Globigerinatheka index*, which is a global event just below the Eocene/Oligocene boundary [*Berggren et al.*, 1985].

The change to lower surface $\delta^{13}C$ values by the late Eocene may be species dependent and not necessarily indicative of decreased productivity. The disappearance of many surface-dwelling species globally and the permanent increase in silicious microfossils in the southern Indian Ocean in late Eocene time indicate a change in surface water conditions and assemblage composition [*Huber*, 1991; *Barron et al.*, 1991*b*]. At the same time, there is no apparent change in $\delta^{13}C$ values of either the chiloguembelinids or the deep-dwelling taxa that remained within about a 0.5‰ range of the early Eocene to early Oligocene values. This feature suggests relatively stable conditions at the depth inhabited by these species and no change in upwelling conditions in spite of significant changes in water temperature.

Calculated $\delta^{18}O$ values of calcite precipitated in equi-

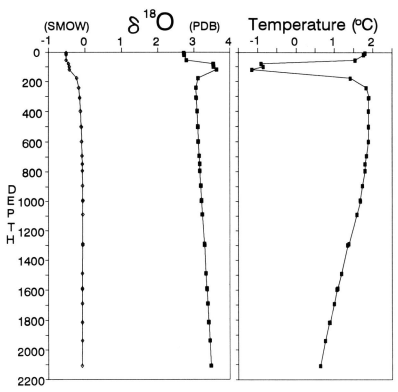

Fig. 4. Temperature, calculated $\delta^{18}O$ values of seawater (relative to standard mean ocean water (SMOW)) and calcite (relative to PDB) precipitated in equilibrium with the upper 2000 m of the water column at GEOSEC Station 430 near Site 738 in the Indian Ocean. Seawater $\delta^{18}O$ values were calculated using the salinity-temperature relation of *Craig and Gordon* [1965] for the southwestern Pacific Ocean. Diamonds, water $\delta^{18}O$; squares, calcite $\delta^{18}O$.

librium and measured $\delta^{13}C$ values of total dissolved CO_2 of surface to deep waters at GEOSEC Station 430 permit comparison of the Modern thermal and $\delta^{13}C$ structures of the water column to those in the early Oligocene (Figure 4). Modern $\delta^{13}C$ values of surface and deep waters in the polar Indian Ocean are lower by ~0.5‰ to those of the early Oligocene as it has been previously documented at other locations [*Shackleton et al.*, 1984; *Shackleton*, 1986]. The Modern surface water to deepwater $\delta^{13}C$ difference of ~1‰ is similar, whereas the $\delta^{18}O$ difference of ~0.7‰ is slightly smaller than that of the early Oligocene. The latter assumes that both measured early Oligocene *Nuttallides* and *Cibicidoides* $\delta^{18}O$ were depleted by 0.5‰ in relation to calcite precipitated in equilibrium (Tables 1 and 2; *Shackleton et al.* [1984]). Because foraminiferal $\delta^{18}O$ values reflect both temperatures and seawater $\delta^{18}O$ composition, which is affected by the volume of continental ice accumulation, it is not possible to estimate how different early Oligocene seawater temperatures were from Modern ones. However, the larger early Oligocene surface

water to deepwater $\delta^{18}O$ difference suggests that seasonal temperature/salinity variation was greater and the water column was not as cold throughout the year as today.

Do isotopic values and relative abundance of planktonic foraminiferal species provide insight to the depth of the thermocline in the southern Kerguelen Plateau in Eocene-Oligocene time? In areas of the modern ocean where the thermocline is shallow and well developed, many niches are available leading to great species diversity and greater $\delta^{18}O$ differences between species [*Thunell and Honjo*, 1987; *Ravelo et al.*, 1990]. If, at any given time, temperature and salinity variations within the photic zone were small, as in the high latitudes today, the assemblage would be dominated by very few species and the $\delta^{18}O$ differences between species would be small [*Berger*, 1969, 1971; *Bé*, 1977]. In general, the decrease in the number of species, particularly the shallow-dwelling forms, and the $\delta^{18}O$ differences between species corresponds to the cooling and reduction in the vertical thermal gradient of high-latitude surface waters from the Eocene to the Oligocene.

MIDDLE EOCENE TO OLIGOCENE TEMPERATURES AND OCEANOGRAPHIC CHANGES

The $\sim 1.5\%_0$ increase in $\delta^{18}O$ values of both planktonic and benthic foraminifera from early Eocene to late middle Eocene occurred gradually over ~ 10 m.y. In contrast, the $\sim 1\%_0$ increase in $\delta^{18}O$ values from the late Eocene to the early Oligocene occurred more suddenly (Figures 2 and 3). This change in benthic and planktonic foraminiferal $\delta^{18}O$ values has been interpreted to reflect a combination of high-latitude cooling and increased continental ice accumulation on Antarctica [*Keigwin and Keller*, 1984; *Miller and Thomas*, 1985; *Keigwin and Corliss*, 1986; *Miller et al.*, 1987]. The evidence for the latter is the lower than present isotopic temperatures calculated for earliest Oligocene deep waters based on benthic foraminiferal samples from low- and mid-latitude sites and the assumption that continental ice volume did not significantly affect average seawater $\delta^{18}O$ values [*Keigwin and Keller*, 1984; *Miller and Thomas*, 1985; *Miller et al.*, 1987]. Evidence for continental ice accumulation and ice sheet development in the early Oligocene and the linking of foraminiferal $\delta^{18}O$ data to this event were described by *Barrera and Huber* [1991] and *Barron et al.* [1991a] from sediments drilled during ODP Leg 119. At Prydz Bay, glacial diamictites were recovered from an ~ 150-m sequence dated between 34.8 and 36 Ma on the basis of its fossil content. An ~ 200-m sequence of glacial diamictites of uncertain age, possibly lower Oligocene to upper middle Eocene, underlies this well-dated sequence [*Hambrey et al.*, 1991; *Barron et al.*, 1991a]. These sediments were deposited near the continental shelf break and beyond the limit of the present ice shelf, suggesting more extensive glaciation in this area during the early Oligocene than today.

A thin layer of ice-rafted quartz grains coincides with an increase of $\sim 1.4\%_0$ in benthic and planktonic foraminiferal $\delta^{18}O$ values in pelagic sediments of the southern Kerguelen Plateau dated from magnetobiostratigraphy at ~ 35.85 Ma [*Barrera and Huber*, 1991; *Ehrmann*, 1991]. A similar layer of ice-rafted sediments correlative with the $\delta^{18}O$ shift was subsequently reported for lower Oligocene sediments from Site 748 (ODP Leg 120) by *Zachos et al.* [1992a, b] and *Breza and Wise* [1992]. These data suggest that icebergs drifted at least 1000 km from the continent in the earliest Oligocene. Icebergs large enough to travel this distance in waters warmer than at present would have been derived from an ice shelf instead of alpine or valley glaciers [*Mercer*, 1978]. Furthermore, the record of glacial sediments in widely separated parts of the continent, including those at the base of the CIROS 1 drillhole in the Ross Sea dated at

~ 34 to 36 Ma [*Barrera*, 1989; *Barrett*, 1989], suggests the presence of an ice sheet of continental proportions for at least a brief time during the early Oligocene.

Middle Eocene Temperatures

Foraminiferal $\delta^{18}O$ data and estimated temperatures for seafloor and shallow surface waters at sites 738 and 744 provide information on the magnitude of the earliest Oligocene ice advance and also the possibility of a related Eocene event (Figure 5). Gravel and angular to subangular sands in middle Eocene sediments near the top of Core 119-738B-11H and Core 119-738B-13H and within Section 119-738B-14X-1 and Section 119-738B-16X-3 at Site 738 have been interpreted by *Ehrmann* [1991] as ice-rafted sediments. The association of these grains with clays typical of the Eocene sequence (chlorite and kaolinite), instead of clays typical of lower Oligocene and younger sediments (illite), along with abundant Eocene radiolarians, has been cited as evidence that the grains are in situ [*Ehrmann and Mackensen*, 1992]. We found feldspar and quartz grains in samples taken from the center of cores 10H-3 and 11H-1. These grains are smaller (200–300 μm) and less angular but are otherwise similar to lower Oligocene and Modern quartz grains affected by glacial processes [*Krinsley and Donahue*, 1968; *Margolis and Kennett*, 1971].

Estimated surface water temperatures were between $\sim 5^\circ$ and 8°C in the late middle Eocene interval in which the glacial sediments were found (Figure 5). These isotopic temperatures were calculated assuming no continental ice was present at the time and average seawater $\delta^{18}O$ was $0.9\%_0$ lower than at present [*Shackleton and Kennett*, 1975]. Deepwater temperatures were between 4° and 7°C. Both surface water and deepwater temperatures would have been even higher if there was continental ice accumulation. Although model experiments using coastal sea surface temperatures 7°C higher than present suggest maintenance of an Antarctic ice sheet [*Oglesby*, 1989, 1991], there is presently no geologic evidence to support this scenario. While ice may have accumulated in the continental interior, particularly at high elevations in the middle and late middle Eocene, without direct geological evidence from the continent, the significance of grains within Site 738 middle Eocene sediments remains a mystery.

Early Oligocene Ice Volume and Temperatures

At Site 744, detailed foraminiferal isotopic data permit inferences not only concerning the magnitude of early Oligocene ice volume but also temperature stratification of the water column. The $\delta^{18}O$ values of the benthic foraminifera *Cibicidoides* and the near-surface dweller *Globorotaloides suteri* increased $\sim 1.4\%_0$ from

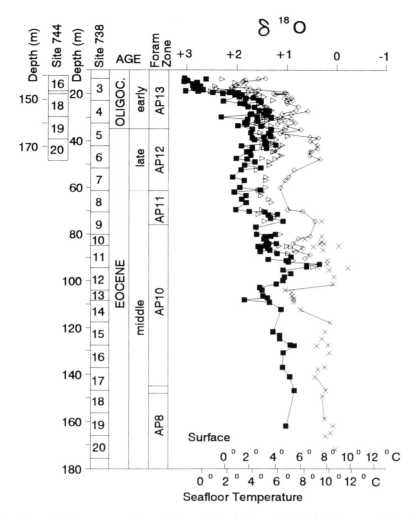

Fig. 5. Oxygen isotope compositions of foraminifera in the middle to late Eocene at sites 738 and 744. Symbols are as in Figure 2. The $\delta^{18}O$ values of *Cibicidoides* have been adjusted by 0.5‰ as discussed in the method section. Surface and seafloor paleotemperatures were calculated, assuming no ice volume effect.

~36.5 Ma (Section 119-744A-18H-2, 90–95 cm, 150.07 mbsf) to ~35.83 Ma (Section 119-744A-16H-7, 12–14 cm, 146.03 mbsf) (Figure 6). The sedimentary record across the prominent $\delta^{18}O$ shift is nearly complete. Although Core 17H is short, it does exhibit 83% recovery. Core 16H has 100% recovery [*Barron et al.*, 1989]. The highest Oligocene values from the benthic taxon, the intermediate dwelling planktonic taxon, and the surface dwelling *C. cubensis* occurred within Chron 13N deposited from ~35.83 Ma to 35.24 Ma at 139.67 mbsf. In this interval of ~0.5 m.y., the $\delta^{18}O$ composition of *Cibicidoides*, a taxon that precipitated calcite out of equilibrium by either −0.5 or −0.64‰, according to *Shackleton et al.* [1984] and *Savin et al.* [1981], respectively, was as high as 2.5‰. Calcite precipitated in equilibrium with today's waters at 0.8°C and 1800-m

depth (the paleodepth of Site 744) at GEOSEC Station 430 would have $\delta^{18}O$ values close to 3.4‰, which is only ~0.3–0.4‰ higher than the estimated values for calcite precipitated in equilibrium with early Oligocene waters (Figures 4 and 6). Therefore if continental ice accumulation at the time of the maximum $\delta^{18}O$ shift was sufficient to affect the $\delta^{18}O$ value of average ocean waters by ~0.5–0.6‰, deepwater temperatures at Site 744 would have been similar to today's temperatures.

This scenario of continental ice accumulation in the earliest Oligocene is different from previous ones. *Keigwin and Corliss* [1986] proposed that the increase in ice accumulation contributed ~0.4‰ to the ~1.0‰ shift in benthic foraminiferal $\delta^{18}O$, with the remainder attributed to a decrease in deepwater temperatures. Their estimate was based on the magnitude of the isotopic

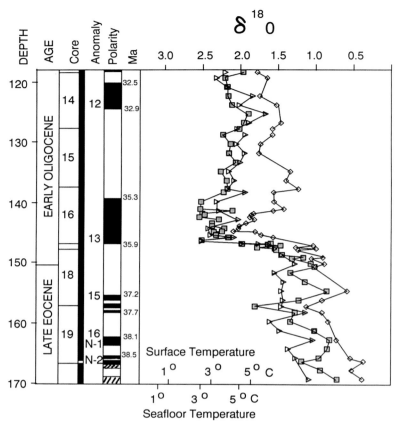

Fig. 6.   Oxygen isotope compositions of foraminifera across the Eocene-Oligocene transition at Site 744. Symbols are as in Figure 2. Surface and seafloor temperatures were calculated for the early Oligocene $\delta^{18}O$ shift using estimated seawater $\delta^{18}O$ at GEOSEC Station 430 and assuming average seawater $\delta^{18}O$ was enriched by 0.6‰ owing to increased ice volume. Lower values for ice volume would yield low temperatures.

increase in planktonic foraminifera and the assumption that the temperatures of tropical surface waters remained constant. A similar estimate was subsequently obtained with the addition of planktonic foraminiferal data from the high latitudes [Zachos et al., 1992a]. Alternatively, on the basis of calcareous nannofossil abundances, Wei [1991] suggested that the $\delta^{18}O$ positive shift reflects only 0.2‰ change in ice volume and more than 3°C decrease in southern high-latitude surface water. One consequence of our interpretation of the Cibicidoides $\delta^{18}O$ data is that adoption of an ice volume effect smaller than 0.5‰ would result in calculated temperatures that would be below freezing unless some volume of continental ice before the $\delta^{18}O$ shift is assumed. In contrast, Zachos et al. [1992b] proposed that the effect of ice volume on ocean $\delta^{18}O$ in the earliest Oligocene was similar to that of today.

Comparison of calculated temperatures for deep waters just before and after the $\delta^{18}O$ shift indicates that most of the change in $\delta^{18}O$ values was caused by an increase in global ice volume and that cooling was

minor. For example, just prior to the $\delta^{18}O$ shift estimated surface and near-surface water temperatures over the southern Kerguelen Plateau were ~6°–7°C and ~4°–5.5°C, respectively, assuming that the world was ice free (Figure 5). Deepwater temperatures were ~5°–7°C and perhaps slightly higher than those of near-surface waters. At the time of the maximum ^{18}O enrichment, calculated surface water temperatures were ~4°–5°C and near-surface temperatures were ~2.5°–4°C, if the total effect of increased ice volume on average ocean $\delta^{18}O$ was 0.6‰. Deepwater temperatures would have been at ~0.8°C or similar to today's temperatures. These calculated temperatures indicate that the cooling mainly occurred in the deep waters. This conclusion is consistent with our previous inference (presented in the section on the depth stratification of Paleogene species) of a very small change in the thermal structure of the upper water column from late Eocene to early Oligocene time (Figure 3).

The pattern of foraminiferal $\delta^{18}O$ values at Site 744 shows that surface and deep waters responded somewhat differently across the early Oligocene isotopic shift. After the isotopic shift, $\delta^{18}O$ values from *Cibicidoides* and *G. suteri* remained relatively high but variable for ~0.5 m.y. (Figure 6). These high but variable $\delta^{18}O$ values for this interval are in contrast to the published results for Site 748 obtained by *Zachos et al.* [1992b] and suggest either instability in the extent of ice sheets/continental ice on Antarctica or highly variable deepwater temperatures. After ~35.24 Ma (139.67 mbsf) and for the remainder of the early Oligocene, values from *Cibicidoides* became lower by 0.3–0.5‰. In contrast, the highest $\delta^{18}O$ values (~2.0‰) of the surface dweller *C. cubensis* occurred earlier than ~35.58 Ma (142.96 mbsf) within the base of Chron 13N. The length of time or the magnitude of the ^{18}O enrichment of this taxon could not be determined precisely because specimens were too scarce in Core 16H-7 for isotopic analysis. After this time, $\delta^{18}O$ values of *C. cubensis* decreased by 0.5‰ until ~35.44 Ma (141.17 mbsf) and remained relatively low within the upper part of Chron 13N and above (Figure 6). The change in *Chiloguembelina* $\delta^{18}O$ values during the earliest Oligocene may be interpreted to reflect increased ice volume and decreasing temperatures, followed by increasing surface water temperatures, lower ice volume, or both. Because $\delta^{18}O$ values of the surface-dwelling taxon decreased earlier than those of the benthic and near-surface-dwelling taxa, either the surface waters warmed significantly or deeper waters continued to cool after 35.58 Ma. Comparison of these high-latitude isotopic records with those from lower-latitude sites can aid in determining the most likely scenario.

At lower-latitude deep-sea sites such as Site 522, benthic foraminiferal $\delta^{18}O$ records exhibit high $\delta^{18}O$ values near the base of Chron 13N followed by lower values near the top [*Miller et al.*, 1988]. This may indicate that the amount of ice volume decreased, deepwater temperatures increased, or both. Because this $\delta^{18}O$ pattern is similar to that observed in surface waters at Site 744 (Figure 6), it suggests that they are both controlled to some extent by the same factor. It is possible that this factor represents a decrease in the amount of continental ice on Antarctica. If this is the case, deep waters at Site 744 must have cooled substantially from ~35.85 to ~35.24 Ma, and the effect of continental ice and ice sheets on ocean $\delta^{18}O$ values at the early Oligocene isotopic shift must have been larger than we estimated above. After 35.2 Ma, $\delta^{18}O$ values of the surface-dwelling taxon generally increase, opposite to the trend in *Cibicidoides* $\delta^{18}O$ values. Thus the absence of covariance in these records suggests that they are likely not caused by changes in ice volume. To resolve the relative influence of ice volume and temperature changes on the Oligocene isotopic records, we need to compare these results with other high-resolution records. Such records are not yet available. Nevertheless, we have previously noted [*Barrera et al.*, 1990]

that early and late Oligocene benthic foraminiferal $\delta^{18}O$ values at Site 744 are generally higher than 2‰, which in an ice-free world would yield deepwater temperatures close to the present ones, implying that a variable amount of continental ice persisted during this time.

CONCLUSIONS

Stable isotopic analyses of planktonic and benthic foraminiferal species from the lower Eocene to Oligocene section at Site 738 and Site 744 in the Indian Ocean have provided information on changes in the structure of the Antarctic water column during major climatic cooling and glaciation of the southern polar regions. The major conclusions of this study are as follows:

1. The $\delta^{18}O$ and $\delta^{13}C$ values of planktonic foraminiferal species indicate that their relative stratification in the Antarctic water column was similar to that in lower latitudes from the early Eocene to the early Oligocene.

2. The $\delta^{18}O$ data indicate no significant changes in the thermal structure in the upper portion of the water column during cooling of the southern high latitudes from the early to the late middle Eocene. During the same interval, there is no evidence from $\delta^{13}C$ data for major changes in productivity or upwelling because there is no apparent change in the $\delta^{13}C$ difference between surface and deep waters.

3. There is little change in the thermal structure of the upper water column from late Eocene to early Oligocene time. This conclusion is consistent with the suggestion that the early Oligocene $\delta^{18}O$ increase is largely the result of an increase in the volume of continental ice accumulation rather than cooling.

4. A climatic episode characterized by a decrease in both benthic and planktonic $\delta^{18}O$ values of ~1.0‰ occurred at Site 738 at about 42 Ma. This event is also apparent in the benthic foraminiferal $\delta^{18}O$ record of Indian Ocean Site 219.

5. A $\delta^{18}O$ shift of 1.4‰ occurs at Site 744 from ~36.5 to ~35.83 Ma during the early Oligocene.

6. Comparison of *Cibicidoides* $\delta^{18}O$ values at the earliest Oligocene maximum ^{18}O enrichment with calculated values of calcite precipitated in equilibrium with Modern waters at GEOSEC Station 430 in the Antarctic Indian Ocean indicates that the total effect of continental ice accumulation at this time on the average composition of the oceans was at least 0.6‰. Lower estimates yield deepwater temperatures at Site 744 which are lower than present temperatures.

7. The highest Oligocene $\delta^{18}O$ values of the benthic foraminifera *Cibicidoides*, the intermediate dwelling taxon *G. suteri*, and the surface dweller *C. cubensis* were recorded in sediments within Chron 13N deposited from ~35.83 to 35.24 Ma. During this interval, $\delta^{18}O$ values of *C. cubensis* decreased by 0.5‰ after ~35.58 Ma, whereas $\delta^{18}O$ values of the other taxa remained

high until ~35.24 Ma. This trend suggests that the change in surface water values may reflect decreasing ice volume while deep waters cooled.

8. After the early Oligocene maximum ^{18}O enrichment, Oligocene *Cibicidoides* $\delta^{18}O$ values at Site 744 were generally higher than 2‰, which in an ice-free world yields temperatures close to the present ones. This situation implies that a variable amount of continental ice persisted during this time.

Acknowledgments. The authors are grateful to K. C. Lohmann and Jim O'Neil for the stable isotope analyses from their laboratory facilities at the University of Michigan and to Christopher Hamilton for drafting Figure 1. Lowell Stott, Jim Kennett, Mike Tevesz, and an anonymous reviewer provided useful comments and suggestions to improve the manuscript. E.B. thanks Gerta Keller for help with picking planktonic foraminiferal specimens. This research was supported entirely by NSF grant DPP-9096290 to E.B.

REFERENCES

Barrera, E., Strontium isotope ages, *DSIR Bull. N. Z.*, *245*, 151–152, 1989.

Barrera, E., and B. T. Huber, Paleogene and early Neogene oceanography of the southern Indian Ocean: Leg 119 foraminifer stable isotope results, *Proc. Ocean Drill. Program Sci. Results*, *119*, 693–717, 1991.

Barrera, E., and G. Keller, Late Paleocene to early Eocene climatic and oceanographic events in the Antarctic Indian Ocean, *Geol. Soc. Am. Abstr. Programs*, *22*(6), A179, 1991.

Barrera, E., et al., Stable isotope and sedimentologic evidence for late middle Eocene to early Oligocene glaciation in East Antarctica: Results from ODP Leg 119 in the southern Indian Ocean (abstract), *Eos Trans. AGU*, *71*, 1398, 1990.

Barrera, E., J. Barron, and A. Halliday, Strontium isotope stratigraphy of the Oligocene–lower Miocene section at Site 744, southern Indian Ocean, *Proc. Ocean Drill. Program Sci. Results, 119*, 731–738, 1991.

Barrett, P. J. (Ed.), Antarctic Cenozoic history from the CIROS-1 drillhole, McMurdo Sound, *DSIR Bull. N. Z.*, *245*, 254 pp., 1989.

Barron, J. A., et al., Leg 119, *Proc. Ocean Drill. Program Initial Rep.*, *119*, 942 pp., 1989.

Barron, J. A., B. Larsen, and J. G. Baldauf, Evidence for late Eocene to early Oligocene Antarctic glaciation and observations on late Neogene glacial history of Antarctica: Results from Leg 119, *Proc. Ocean Drill. Program Sci. Results*, *119*, 869–891, 1991*a*.

Barron, J. A., J. G. Baldauf, E. Barrera, J. P. Caulet, B. T. Huber, B. H. Keating, D. Lazarus, H. Sakai, H. R. Thierstein, and W. Wei, Biochronologic and magnetochronologic synthesis of ODP Leg 119 sediments from the Kerguelen Plateau and Prydz Bay, Antarctica, *Proc. Ocean Drill. Program Sci. Results*, *119*, 813–847, 1991*b*.

Bé, A. W. H., An ecological, zoogeographic and taxonomic review of recent planktonic foraminifera, in *Oceanic Micropaleontology*, vol. 1, edited by A. T. S. Ramsay, pp. 1–100, Academic, San Diego, Calif., 1977.

Berger, W. H., Planktonic foraminifera: Basic morphology and ecologic implications, *J. Paleontol.*, *43*, 1369–1383, 1969.

Berger, W. H., Sedimentation of planktonic foraminifera, *Mar. Geol.*, *11*, 325–358, 1971.

Berggren, W. A., D. V. Kent, and J. A. Van Couvering, Paleogene geochronology and chronostratigraphy, in *The Chronology of the Geologic Record*, edited by N. J. Snelling, pp. 141–195, Geological Society of London, London, 1985.

Boersma, A., and N. J. Shackleton, Oxygen and carbon isotope variations and planktonic foraminiferal depth habitats: Late Cretaceous to Paleocene, Central Pacific, DSDP sites 463 and 465, Leg 65, *Initial Rep. Deep Sea Drill. Proj.*, *65*, 513–526, 1981.

Boersma, A., I. Premoli Silva, and N. J. Shackleton, Atlantic Eocene planktonic foraminiferal paleohydrographic indicators and stable isotope paleoceanography, *Paleoceanography*, *2*, 287–331, 1987.

Bouvier-Soumagnac, Y., and J.-C. Duplessy, Carbon and oxygen isotopic composition of planktonic foraminifera from laboratory culture, plankton tows and recent sediment: Implications for the reconstruction of paleoclimatic conditions and of the global carbon cycle, *J. Foraminiferal Res.*, *15*, 302–320, 1985.

Breza, J., and S. W. Wise, Jr., Lower Oligocene ice-rafted debris on the Kerguelen Plateau: Evidence for East Antarctic continental glaciation, *Proc. Ocean Drill. Program Sci. Results*, *120*, 161–178, 1992.

Craig, H., and L. Gordon, Deuterium and oxygen-18 variation in the ocean and marine atmosphere, *Univ. Rhode Island Occas. Publ.*, *3*, 277–374, 1965.

Deuser, W. G., and J. M. Hunt, Stable isotope ratios of dissolved inorganic carbon in the Atlantic, *Deep Sea Res.*, *16*, 221–225; 1969.

Douglas, R. G., and S. M. Savin, Oxygen isotopic evidence for the depth stratification of Tertiary and Cretaceous planktonic foraminifera, *Mar. Micropaleontol.*, *3*, 175–196, 1978.

Ehrmann, W. U., Implications of sediment composition of the southern Kerguelen Plateau for paleoclimate and depositional environment, *Proc. Ocean Drill. Program Sci. Results*, *119*, 185–210, 1991.

Ehrmann, W. U., and A. Mackensen, Sedimentological evidence for the formation of an East Antarctic ice sheet in Eocene/Oligocene time, *Palaeogeogr. Palaeoclimatol. Palaeoecol.*, *93*, 85–112, 1992.

Emiliani, C., Depth habitats of some species of pelagic foraminifera as indicated by oxygen isotope ratios, *Am. J. Sci.*, *252*, 149–158, 1954.

Emiliani, C., On paleotemperatures of Pacific bottom waters, *Science*, *123*, 460–461, 1956.

Fairbanks, R. G., and P. H. Wiebe, Foraminifera and chlorophyll maximum: Vertical distribution, seasonal succession, and paleoceanographic significance, *Science*, *209*, 1524–1525, 1980.

Fairbanks, R. G., P. H. Wiebe, and A. W. H. Bé, Vertical distribution and isotopic composition of living planktonic foraminifera in the western North Atlantic, *Science*, *207*, 61–63, 1980.

Fairbanks, R. G., R. F. Sverdlove, P. H. Wiebe, and A. W. H. Bé, Vertical distribution and isotopic fractionation of living planktonic foraminifera from the Panama Basin, *Nature*, *298*, 841–844, 1982.

Hambrey, M. J., W. U. Ehrmann, and B. Larsen, Cenozoic glacial record of the Prydz Bay continental shelf, East Antarctica, *Proc. Ocean Drill. Program Sci. Results*, *119*, 77–132, 1991.

Huber, B. T., Paleogene and early Neogene planktonic foraminifer biostratigraphy of ODP Leg 119 sites 738 and 744, Kerguelen Plateau (southern Indian Ocean), *Proc. Ocean Drill. Program Sci. Results*, *119*, 427–449, 1991.

Katz, M. E., and K. G. Miller, Early Paleogene benthic foraminiferal assemblage and stable isotope composition in the Southern Ocean, Ocean Drilling Program, Leg 114, *Proc. Ocean Drill. Program Sci. Results*, *114*, 481–513, 1991.

Keigwin, L. D., and B. H. Corliss, Stable isotopes in late middle Eocene to Oligocene forams, *Geol. Soc. Am. Bull.*, *97*, 335–345, 1986.

Keigwin, L. D., and G. Keller, Middle Oligocene cooling from equatorial Pacific DSDP Site 77B, *Geology*, *12*, 16–19, 1984.

Kennett, J. P., Cenozoic evolution of Antarctic glaciation, the Circum-Antarctic Current, and their impact on global paleoceanography, *J. Geophys. Res.*, *82*, 3843–3860, 1977.

Kennett, J. P., and L. D. Stott, Proteus and Proto-Oceanus: Paleocene oceans as revealed from Antarctic stable isotope results, *Proc. Ocean Drill. Program Sci. Results*, *113*, 865–880, 1990.

Kennett, J. P., and L. D. Stott, Abrupt deep-sea warming, palaeoceanographic changes and benthic extinctions at the end of the Palaeocene, *Nature*, *353*, 225–229, 1991.

Krinsley, D., and J. Donahue, Environmental interpretation of sand grain surface features by electron microscopy, *Geol. Soc. Am. Bull.*, *79*, 743–748, 1968.

Kroopnik, P., The dissolved O_2-CO_2-^{13}C system in the eastern equatorial Pacific, *Deep Sea Res.*, *21*, 211–277, 1974.

Margolis, S. V., and J. P. Kennett, Cenozoic paleoglacial history of Antarctica recorded in Subantarctic deep-sea cores, *Am. J. Sci.*, *271*, 1–36, 1971.

Matthews, R. K., and R. Z. Poore, Tertiary $\delta^{18}O$ record and glacio-eustatic sea-level fluctuations, *Geology*, *8*, 501–504, 1980.

Mercer, J. H., Glacial development and temperature trends in the Antarctic and in South America, in *Antarctic Glacial History and World Palaeoenvironments*, edited by E. M. van Zinderen Bakker, pp. 73–93, Balkema, Rotterdam, Netherlands, 1978.

Miller, K. G., and E. Thomas, Late Eocene to Oligocene benthic foraminiferal isotopic record, Site 574, equatorial Pacific, *Initial Rep. Deep Sea Drill. Proj.*, *85*, 771–777, 1985.

Miller, K. G., R. G. Fairbanks, and G. S. Mountain, Tertiary oxygen isotope synthesis, sea level history, and continental margin erosion, *Paleoceanography*, *2*, 1–19, 1987.

Miller, K. G., M. D. Feigson, and R. K. Olsson, Upper Eocene to Oligocene isotope (^{87}Sr/^{86}Sr, $\delta^{18}O$, $\delta^{13}C$) standard section, Deep Sea Drilling Project Site 522, *Paleoceanography*, *3*, 223–233, 1988.

Oberhänsli, H., Latest Cretaceous–early Neogene oxygen and carbon isotopic record at DSDP sites in the Indian Ocean, *Mar. Micropaleontol.*, *10*, 91–115, 1986.

Oberhänsli, H., and M. Toumarkine, The Paleogene oxygen and carbon isotope history of sites 522, 523, and 524 from the central South Atlantic, in *South Atlantic Paleoceanography*, edited by K. Hsü and A. Weissert, pp. 124–147, Cambridge University Press, New York, 1985.

Oberhänsli, H., J. McKenzie, M. Toumarkine, and H. Weissart, A paleoclimatic and paleoceanographic record of the Paleogene in the Central South Atlantic (Leg 73, sites 522, 523, 524), *Initial Rep. Deep Sea Drill. Proj.*, *73*, 737–748, 1984.

Oglesby, R. J., A GCM study of Antarctic glaciation, *Clim. Dyn.*, *3*, 135–156, 1989.

Oglesby, R. J., Joining Australia to Antarctica: GCM implications for the Cenozoic record of Antarctic glaciation, *Clim. Dyn.*, *6*, 13–22, 1991.

O'Neil, J. R., R. N. Clayton, and T. K. Mayeda, Oxygen isotope fractionation in divalent metal carbonates, *J. Chem. Phys.*, *51*, 5547–5558, 1969.

Oppo, D. W., and R. G. Fairbanks, Carbon isotope composition of tropical surface water during the past 22,000 years, *Paleoceanography*, *4*, 333–351, 1989.

Ostlund, H. G., *GEOSECS Atlantic, Pacific and Indian Ocean Expeditions*, vol. 7, 226 pp., National Science Foundation, Washington, D. C., 1987.

Prentice, J. L., and R. K. Matthews, Cenozoic ice-volume history: Development of a composite oxygen isotope record, *Geology*, *17*, 963–966, 1988.

Ravelo, A. C., and R. G. Fairbanks, Oxygen isotopic composition of multiple species of planktonic foraminifera: Recorders of the Modern Photic Zone temperature gradient, *Paleoceanography*, *7*, 815–831, 1992.

Ravelo, A. C., R. G. Fairbanks, and S. G. H. Philander, Reconstructing tropical Atlantic hydrography using planktonic foraminifera and an ocean model, *Paleoceanography*, *5*, 409–431, 1990.

Savin, S. M., R. G. Douglas, and F. G. Stehli, Tertiary marine paleotemperatures, *Geol. Soc. Am. Bull.*, *86*, 1499–1510, 1975.

Savin, S. M., R. G. Douglas, G. Keller, J. S. Killingley, L. Shaughenssy, M. A. Sommer, E. Vincent, and F. Woodruff, Miocene benthic foraminiferal isotope records: A synthesis, *Mar. Micropaleontol.*, *6*, 423–450, 1981.

Scotese, C. R., and C. R. Denham, Terra Mobilis: Plate tectonics for the Macintosh, Earth in Motion Technologies, Austin, Tex., 1988.

Shackleton, N. J., Attainment of isotopic equilibrium between ocean water and benthonic foraminifera genus *Uvigerina*: Isotopic changes in the ocean during the last glacial, Les Methodes Quantitative d'Etude des Variations au Cours du Pleistocene, *Colloq. Int. C.N.R.S.*, *219*, 203–209, 1974.

Shackleton, N. J., Paleogene stable isotope events, *Palaeogeogr. Palaeoclimatol. Palaeoecol.*, *57*, 91–102, 1986.

Shackleton, N. J., and A. Boersma, The climate of the Eocene ocean, *J. Geol. Soc. London*, *138*, 153–157, 1981.

Shackleton, N. J., and J. P. Kennett, Paleotemperature history of the Cenozoic and the initiation of Antarctic glaciation: Oxygen and carbon isotope analyses in DSDP sites 277, 279, and 281, *Initial Rep. Deep Sea Drill. Proj.*, *29*, 743–755, 1975.

Shackleton, N. J., M. A. Hall, and A. Boersma, Oxygen and carbon isotope data from Leg 74 foraminifers, *Initial Rep. Deep Sea Drill. Proj.*, *74*, 599–612, 1984.

Shackleton, N. J., R. M. Corfield, and M. A. Hall, Stable isotope data and the ontogeny of Paleocene planktonic foraminifera, *J. Foraminiferal Res.*, *15*, 321–336, 1985.

Stott, L. D., and J. P. Kennett, Antarctic Paleogene planktonic foraminiferal biostratigraphy: ODP Leg 113, sites 689 and 690, *Proc. Ocean Drill. Program Sci. Results*, *113*, 549–570, 1990.

Stott, L. D., J. P. Kennett, N. J. Shackleton, and R. M. Corfield, The Evolution of Antarctic surface waters during the Paleogene: Inferences from the stable isotopic composition of planktonic foraminifers, ODP Leg 113, *Proc. Ocean Drill. Program Sci. Results*, *113*, 849–863, 1990.

Thunell, R. C., and S. Honjo, Seasonal and interannual changes in planktonic foraminiferal production in the North Pacific, *Nature*, *328*, 335–337, 1987.

Wei, W., Evidence for an earliest Oligocene abrupt cooling in the surface waters of the Southern Ocean, *Geology*, *19*, 780–783, 1991.

Williams, D. F., M. A. Sommer, and M. L. Bender, Carbon isotopic composition of recent planktonic foraminifer of the Indian Ocean, *Earth Planet. Sci. Lett.*, *36*, 391–403, 1977.

Zachos, J. C., W. A. Berggren, M.-P. Aubry, and A. Mackensen, Isotope and trace element geochemistry of Eocene and Oligocene foraminifers from Site 748, Kerguelen Plateau, *Proc. Ocean Drill. Program Sci. Results*, *120*, 839–854, 1992*a*.

Zachos, J. C., J. R. Breza, and S. W. Wise, Early Oligocene ice-sheet expansion on Antarctica: Stable isotope and sedimentological evidence from Kerguelen Plateau, southern Indian Ocean, *Geology*, *20*, 569–573, 1992*b*.

(Received May 3, 1992;
accepted April 6, 1993.)

THE ANTARCTIC PALEOENVIRONMENT: A PERSPECTIVE ON GLOBAL CHANGE

ANTARCTIC RESEARCH SERIES, VOLUME 60, PAGES 67–73

NOTHOFAGUS FOSSILS IN THE SIRIUS GROUP, TRANSANTARCTIC MOUNTAINS: LEAVES AND POLLEN AND THEIR CLIMATIC IMPLICATIONS

ROBERT S. HILL

Department of Plant Sciences, University of Tasmania, Hobart, Tasmania 7001, Australia

ELIZABETH M. TRUSWELL

Australian Geological Survey Organisation, Canberra, Australian Capital Territory 2601, Australia
Antarctic Co-operative Research Centre, Hobart, Tasmania, Australia

Glacially deposited sediments of the Sirius Group in the Transantarctic Mountains have yielded wood, leaf, and pollen fossils, all of which can be shown, on the basis of the present evaluation, to represent a single species of the southern beech, *Nothofagus*. This paper presents new data on the morphology and affinities of both leaves and pollen and discusses the implications of this information for climatic reconstruction. The leaves show considerable similarity to those of the extant Tasmanian subalpine to alpine species *Nothofagus gunnii*, although there are differences. Pollen of *Nothofagus* recovered from the unit all belongs to a single fossil species and most closely resembles the species *Nothofagidites lachlaniae*, first described from the Pleistocene of New Zealand. Although problems of interpretation remain, it can be argued that *N. lachlaniae* belongs to the same pollen group as *N. gunnii*, showing a consistent affinity between leaf and pollen remains. There is nothing to suggest that any of these fossils are recycled. All organs preserved reflect a stunted, impoverished plant community dominated by a single species of *Nothofagus*, existing probably under extreme conditions, but nevertheless requiring temperatures as high as 5°C for at least 3 months of the year, as well as supplies of liquid water during the summer.

INTRODUCTION

The complex history of the Antarctic ice sheet, its origins and subsequent fluctuations, has been the subject of recent reviews [*Webb*, 1990; *Barron et al.*, 1991]. The waxing and waning of the terrestrial vegetation of Antarctica, i.e., that of vascular plants, in response to the repeated growth and decay of ice sheets and the associated climatic pressures during the Cenozoic, remain poorly understood. This is largely a result of poor exposure, scarce fossils, and the difficulties of preservation in the glacial domain, where the recycling of organic remains is a major problem.

There are few Antarctic Cenozoic plant macrofossil or microfossil assemblages outside the Antarctic Peninsula region that can unequivocally be shown to be in situ. The highest concentration of available information comes from the Ross Sea region, where most of the existing data are palynological. There, pollen suites from Deep Sea Drilling Project (DSDP) drill sites, notably from Site 270 [*Kemp and Barrett*, 1975], suggest that a temperate forest vegetation with *Nothofagus* and

podocarpaceous conifers as dominants was present in coastal regions in the early Tertiary. *Mildenhall* [1989, p. 124], on the basis of pollen assemblages from the CIROS 1 drillhole in McMurdo Sound, described this vegetation as "essentially a beech forest, with podocarps, proteas and other shrubby angiosperms" and indicated that it probably persisted in the coastal regions of the Ross Sea until at least the late Oligocene. A high diversity of *Nothofagus* is indicated by the pollen spectrum; it is notable that this diversity includes a number of taxa not known from the Tertiary of Australia or New Zealand. Aspects of the pollen assemblages at that site, such as the occurrence of pollen in adhering clumps, suggest that the record is in situ: this interpretation is confirmed by the presence of a *Nothofagus* leaf [*Hill*, 1989] at a level between glacial beds in the late Oligocene.

The post-Oligocene history of this vegetation remains enigmatic, and there are no unequivocal Miocene pollen assemblages known. Pollen assemblages from the J9 drill site, beneath the Ross Ice Shelf, were identified as middle Miocene by *Brady and Martin* [1979], but there is other evidence that suggests that these assemblages may not be in place either [*Kellogg and Kellogg*, 1981].

The difficulties of obtaining a coherent record of Antarctic vegetation history during the Cenozoic make the record preserved in the Sirius Group of the Transantarctic Mountains one of outstanding value. Outcrops identified as belonging to the Sirius Group have been recorded from some 40 localities along the flanks of the Transantarctic Mountains [*McKelvey et al.*, 1991]. Sediments consist of lithified or semilithified diamictites and associated fluvial and glacial lacustrine deposits. Age control is provided by recycled marine microfossils, mainly diatoms [*Harwood*, 1986] which are believed to originate from material deposited in the Wilkes and Pensacola basins. These microfossils suggest that the Sirius Group may be of middle to late Pliocene in age. A different view was presented by *Burckle and Pokras* [1991], who speculated that these sediments might be no younger than Oligocene.

Interest in the climatic significance of the Sirius Group was heightened by the discovery in 1986 of excellently preserved and apparently in situ woody twigs and stems in the Beardmore Glacier region [*Webb and Harwood*, 1987]. The fossil wood was described by *Carlquist* [1987], who reported that it was referable to the southern beech *Nothofagus* and that it showed anatomical similarities to wood of the Tasmanian *N. gunnii* and also to the South American *N. betuloides*. Leaf remains within the same deposit were described in a preliminary way by *Hill* [1991] and *Hill et al.* [1991]. The present paper presents data bearing on the affinities of the leaves and discusses the climatic constraints implied by the vegetation they reflect. Further information is presented on the associated pollen assemblages; this is in accord with the macrofossil data in terms of reconstructing the nature and affinities of the parent flora.

THE LEAF FOSSILS

Morphology and Affinity

The morphology of the leaf fossils recovered from the Sirius Formation at Oliver Bluffs (85°07'S, 166°49'E) was described by *Hill* [1991]. All the morphological evidence allies the leaves most closely with *Nothofagus gunnii*, which grows today in subalpine to alpine habitats in Tasmania. The leaves are, however, morphologically further removed from *N. gunnii* than is the species from the Oligocene section in the CIROS 1 drill hole (R. S. Hill, D. M. Harwood, and P.-N. Webb, work in progress). There is considerable similarity in leaf form to the South American endemic *N. pumilio*. It should be noted however that the two extant species are not closely related and that they belong to different subgenera and produce different pollen types (see below).

Deciduousness Reflected in the Fossils

Three separate lines of evidence indicate clearly that the *Nothofagus* species of the Sirius Formation was deciduous in habit. These are reflected in the following:

Comparison with living species. The living species morphologically closest to the fossil leaves, the Tasmanian *N. gunnii* and the South American *N. pumilio*, are both winter deciduous.

Leaf vernation type. Vernation is the way the leaves are arranged in the bud prior to expansion, and in *Nothofagus* there are four types (Figure 1). In three of these types the leaf blade forms as a flat surface and stays that way after expansion. The fourth type, known as plicate vernation, which is unique to the deciduous species, has the leaves folded like a fan in the bud, and in many cases evidence for this can be seen in the fully mature leaves by the presence of a ridge between the secondary veins. This is clearly apparent in the fossil leaves and is compelling evidence for their deciduous nature.

Presence of a leaf bed. The fossil leaves occur in dense but very thin mats, which look very much like a short-term event, and probably reflect an autumn leaf fall. Such a dense mat is unlikely to have resulted had the leaves been transported far, for instance by wind. The probable low growth habit, suggested by the wood, also argues against mass transport of leaves by wind.

A Local Source for the Fossils

There are several lines of evidence which confirm a local source for the leaf fossils, and some of these overlap with the points noted above.

The leaves occur in a thin but dense mat. They have organic preservation but are very thin (as are all the extant deciduous species). There is no possibility that they could have been reworked from older sediments. There is no evidence of damage to the very delicate laminas (either mechanical or due to decay), suggesting that the leaves were incorporated into the sediment soon after they were shed.

THE POLLEN DATA

Samples from the leaf and wood-bearing beds at Oliver Bluffs, Dominion Range, have yielded poorly diversified pollen assemblages on maceration, in addition to a great deal of parenchymatous material, and finely divided wood. In the course of the present study, pollen was recovered from three of six samples macerated.

The information derived from these palyniferous sediments confirms and builds on data published by *Askin and Markgraf* [1986]. Palynomorphs were observed to be sparse and predominantly of angiospermous origin, but a number of thick-walled fungal resting spores were

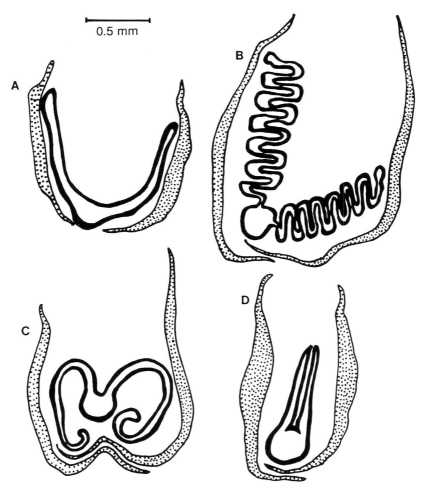

Fig. 1. Transverse section of lamina and stipules to show vernation types. Thick black lines show the leaf blades; the stipules are stippled. (*a*) Plane vernation (evergreen leaves). (*b*) Plicate vernation (deciduous leaves). (*c*) Revolute vernation (evergreen leaves). (*d*) Conduplicate vernation (evergreen leaves). In Figure 1*b* the midvein is shown as the large circular center of the leaf blade. The lateral (secondary) veins occur at the base of each of the folds going out on either side of the midvein. That is, all the lateral veins are in the same plane. Between the lateral veins the leaf blade is folded upward like a fan. When the leaf expands, it often does not completely flatten, and the lamina between lateral veins is folded upward, often with a distinct crease. This can be clearly observed in the fossil leaves. (After *Philipson and Philipson* [1978].)

evident, as were spherical bodies of presumed algal origin. The pollen suite is dominated by *Nothofagidites*, consistent with a *Nothofagus*-dominated vegetation. Other types present, but rare, include a distinctive tricolpate angiosperm with a coarsely reticulate exine, a type which *Askin and Markgraf* [1986] suggested might represent Lamiaceae or Polygonaceae. *Mildenhall* [1989, Plate 1, Figure 13] reported the species from the CIROS 1 borehole, and *Truswell* [1983] noted the same form as a recycled element in recent Ross Sea muds. Its precise affinity remains to be determined. Also reported was a single trilete spore reminiscent of Lycopodiaceae.

Nothofagus Pollen Morphology

All of the observed pollen grains of *Nothofagidites* represent a single morphotype (Figure 2) and conform with the species figured by *Askin and Markgraf* [1986, p. 34, Figure 3]. Most are thin walled; this may result from corrosion, but it may also be in part a primary condition. There is a remarkable consistency in the morphology of the grains examined. Most are rounded (peritreme) in outline. The grain surface bears a cover of uniformly distributed coni or short spines.

There are (usually) seven short, gaping colpi with thickened rims. The structure of the thickened colpus

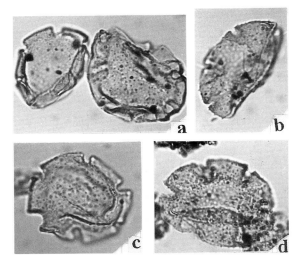

Fig. 2. Pollen grains referable to *Nothofagidites* species (compare *N. lachlanae*), from Sirius Group sediments, all ×1000. (*a*) From Oliver Bluffs section 8A, above and below leaf horizon. (*b, c, d*) Specimens from Oliver Bluffs section 5.

margins is not wholly clear from the material observed. Exine stratification is not apparent with light microscopy, so that it is not clear whether the observed thickening is wholly internal or whether there is an outwardly thickened component to the endexine. Some grains do, however, show a weakly developed annulus. The ratio of colpus length to equatorial diameter, a feature which *Praglowski* [1982] considered to be of value in discriminating between pollen of extant species, is low (<0.2) in the Sirius species.

Identification and Affinities

The fossil pollen species which the Sirius species most closely resembles is *Nothofagidites lachlaniae* (Couper) *Pocknall and Mildenhall*, 1984, which was originally described from an early Pleistocene sample in Toetoes Bay, South Island, New Zealand. *N. lachlaniae* has been described [*Couper*, 1953; *Pocknall and Mildenhall*, 1984; *Dettmann et al.*, 1990] as having an exine in which stratification is not apparent and which has gaping colpi with partially thickened and ''collared'' rims. The colpus thickenings indicate relationship with the *fusca* group of extant taxa which includes Tasmanian and South American species.

Pollen referable to *Nothofagidites lachlaniae* has been widely reported from Antarctica. It was first identified by *Truswell* [1983], who reported that it was present as a recycled element in most modern sediment samples from the Ross Sea, and has since been reported from Eocene (possibly middle Eocene) sediments in the Weddell Sea [*Mohr*, 1990] and from the Oligocene sequence in the CIROS 1 borehole in McMurdo Sound [*Mildenhall*, 1989].

In their comprehensive review of the fossil species *Nothofagidites*, *Dettmann et al.* [1990] subdivided *fusca* pollen types into two categories, defined according to the nature of the exine thickening adjacent to the colpi. In *fusca* type a pollen, produced by species within subgenus *Fuscospora* [see *Hill*, 1992], the thickenings are less conspicuous and only directed inward; in type b, produced by species within *Nothofagus* subgenus *Nothofagus*, the margins have an outward expression of thickening and an annulate appearance. Dettmann et al. assigned *N. lachlaniae* to type b, by which action they suggested that the affinities of the fossil species lay with South American groups, including *N. pumilio*.

In the case of *N. lachlaniae*, however, reference to these subdivisions is not clear cut, for two reasons. First, there is ambiguity in the nature of the pollen grain wall, as indicated in the description of *Couper*'s [1953] holotype given by *Pocknall and Mildenhall* [1984]. Second, there is some confusion which arises as a direct result of the poor condition of Couper's type material. Because of deterioration in the holotype, *Dettmann et al.* [1990] selected a new reference or topotype specimen. This specimen appears to possess colpi that are more distinctly collared and generally more robustly thickened than the original holotype and may represent a move away from the original morphological concept. For these reasons, the Sirius Group species is compared with *N. lachlaniae* rather than referred formally to that species. Certainly on the basis of information currently available on the morphology of the Sirius Group species, they are considered to be closer to the *fusca* a group, which includes the extant *N. gunnii*, than to the *fusca* b group.

Direct comparison of the fossil pollen with the pollen of *N. gunnii* reveals a basic similarity of structure, but does show some differences, with *N. gunnii* pollen having a more distinctly stratified pollen wall and more pronounced colpal thickenings. Differences between the fossil form and *N. pumilio* are more pronounced, especially in the rather abruptly thickened colpal margins possessed by *N. pumilio* (and by *N. antarctica*, which is in the same subsection as *N. pumilio*).

Pollen Transport and a Local Source

The fact that there is clearly only one morphological type of *Nothofagus* pollen in the Sirius Formation sediments examined to date suggests that the pollen is in situ rather than recycled as *Askin and Markgraf* [1986] surmised. Information presently available shows that early Tertiary sediments in Antarctica contain a high diversity of *Nothofagus* species. Recycling would thus be expected to produce more of a pollen mix. Further, in some samples *Nothofagidites* occurs in clumps suggestive of shedding from a single anther, and it is unlikely that such would survive intensive recycling. The pollen evidence, therefore, supports the leaf evi-

TABLE 1. Frost Resistance of *Nothofagus* Species From High Latitudes (in Degrees Centrigrade)

Species	Leaf	Bud	Cortex	Xylem	Collection Site
		Evergreen			
N. cunninghamii	−16.5	⋯	⋯	⋯	Tasmania, 980 m asl
N. moorei	−9.0	⋯	⋯	⋯	New South Wales, 1500 m
N. fusca	−8	−10	−10	−17	New Zealand, 230 m
N. solandri var. cliffortioides	−13	−15	−15	−15	New Zealand, 1370 m
		Deciduous			
N. antarctica	⋯ *	−22	−22	−22	Chile (planted in New Zealand)
N. gunnii	⋯ *	−17	−17	−17	Tasmania, 1000 m

From *Sakai et al.* [1981] and *Read and Hill* [1988, 1989]. The abbreviation asl means above sea level.
*Leaf resistance not relevant in deciduous species.

dence in indicating a local source for the plant fossil material in the Sirius Formation.

SUMMARY OF MORPHOLOGICAL DATA

Available evidence from the fossils suggests that wood, leaves, and pollen all derive from the same species. The leaf data suggest affinity with the lineage represented today by the Tasmanian *Nothofagus gunnii*, although the fossil leaves are distinct from those of the living species; the pollen evidence is not at odds with such a postulated affinity, but it also shows minor differences between fossil and living taxa. Preservation of the three different organs in the sediment is confirmation of a local source for the parent plants, which may have been growing right at the site of deposition. The available data suggest low stands of almost monospecific *Nothofagus* communities.

The pollen data suggest the presence of one or two other angiosperm species, possibly a conifer and a species of *Lycopodium*. Macrofossil data (R. S. Hill, unpublished data) indicate the presence of an unidentified moss. While data are too sparse to be definitive, the general impression is of a vegetation type similar to, although less diverse than, alpine communities growing now in Tasmania, with *N. gunnii*, a variety of conifers, and other taxa [*Macphail*, 1979].

CLIMATIC CONSTRAINTS

The vast literature on living and fossil species of *Nothofagus* suggests that they evolve slowly, both morphologically and physiologically. The similarity of the fossil leaves, both from the Sirius Group and from the Oligocene of the Ross Sea, suggests that this species was also conservative, at least in terms of morphology. This conservatism means that the physiological responses of living species can be used to infer the physiology of the fossil species. Two aspects are of particular interest: frost resistance and the temperature and water regime.

Frost Resistance

In winter a deciduous species loses its leaves, but its wood and dormant buds must still survive the low winter temperatures. The available data on frost resistance in *Nothofagus* suggest that minimum temperatures below about −20°C will place plants in jeopardy (Table 1). Although the data are limited, it is interesting that some of the evergreen species appear to be as frost resistant as some of the deciduous species, and the reason that a deciduous species has survived rather than an evergreen may be due to photoperiod rather than frost resistance. Snow cover may have been of some assistance in surviving winter extremes, but this is difficult to judge.

Temperature and Water Regime

During summer the plants must reach a temperature substantially above 0°C in order to grow and, importantly, to reproduce. Temperatures in excess of 5°C were probably required for a prolonged period of time, at least for 3 months of the year [*Hill et al.*, 1988]. It is also notable that the fossil leaves are substantially larger than some of the extant deciduous species, which again may be the effect of the photoperiod, but suggests milder conditions than those experienced, for example, in alpine Tasmania. In order to survive, there must have been a plentiful supply of liquid water during summer (also suggesting temperatures above 0°C). The open Ross Sea may have served as an important source of moisture and may have had a controlling effect on the prevailing microclimate.

AGE RELATIONSHIPS

A middle to late Pliocene age for the Sirius Group sediments has been determined on the basis of known age ranges of marine diatoms recycled into the unit [*Harwood*, 1986; personal communication, 1992]. Difficulties of accommodating the magnitude of inferred climate change in the Pliocene led *Burckle and Pokras* [1991] to suggest that the Sirius Group sediments and *Nothofagus* flora may be much older, possibly not younger than late Oligocene.

The fossils described in this paper add little to the resolution of the age debate. The leaf fossil record is too sparse for age determination. The palynomorph species are long ranging; the dominant form, *Nothofagidites lachlaniae*, is recorded from the Eocene in Antarctica. The age ranges of the other, minor, components of the pollen assemblage are unknown. It should be noted, however, that the Sirius assemblage differs greatly in aspect from assemblages known to be of Oligocene age in the Ross Sea region. Those described by *Mildenhall* [1989] from CIROS 1 contain a much greater diversity of palynomorphs, including a greater diversity of *Nothofagus* species. This diversity may include both recycled and in situ components. The difference between them and the Sirius assemblages may be due to different environmental settings; alternatively, it could reflect significant age differences, with the Sirius being markedly younger.

SUMMARY AND CONCLUSIONS

Wood, leaves, and pollen recovered from the Sirius Group all point to the plant fossil remains in this sedimentary unit being essentially in situ. They are dominantly the remains of a single species of *Nothofagus*, with a suggestion of other, minor, floral elements. They appear to reflect a stunted vegetation in which the dominant form was a species of *Nothofagus* with affinities with the living *N. gunnii* of subalpine and alpine Tasmania, although all organs are morphologically distinct from those of modern *N. gunnii*.

The presence of this plant community indicates that temperature (and moisture) conditions very different from those of the present cold glacial regime prevailed during the middle to late Pliocene. Temperatures as high as 5°C for part of the year would have been required for these communities to survive and reproduce. Moist conditions must also have prevailed during summer months. The evidence thus suggests that pre-Pleistocene glaciations in Antarctica may have involved warmer and wetter ice sheets.

It is generally recognized that living species of *Nothofagus* have weak powers of long-distance dispersal, and it would have been extremely difficult to recolonize Antarctica once communities were lost. The available evidence, therefore, suggests that elements of this vegetation must have been continuously present since the early Tertiary or Late Cretaceous. Examination of palynological data from the Ross Sea suggests that the pattern may have been one of progressive depauperization from an earlier vegetation in which *Nothofagus* was very diverse. The evidence does not suggest a progression through a tundra phase as this is now understood,

although there are some indications that taxa such as the Asteraceae and grasses, which are prominent in modern tundra types of vegetation, may at least have been represented, if not strongly, in the Ross Sea region in the Oligocene. Considerably more data are required to document the response of the Antarctic Tertiary vegetation to the growth and major fluctuations in the ice sheet during this interval.

Acknowledgments. We are indebted to David Harwood for provision of the material on which this study is based. E.M.T. publishes with permission of the Executive Director, Bureau of Mineral Resources.

REFERENCES

Askin, R. A., and V. Markgraf, Palynomorphs from the Sirius Formation, Dominion Range, Antarctica, *Antarct. J. U. S.*, *21*(5), 34–35, 1986.

Barron, J., B. Larson, and J. G. Baldauf, Evidence for late Eocene to early Oligocene Antarctic glaciation and observations on late Neogene glacial history of Antarctica: Results from Leg 119, *Proc. Ocean Drill. Program Sci. Results*, *119*, 869–891, 1991.

Brady, H. T., and H. A. Martin, The Ross Sea region in the middle Miocene—A glimpse into the past, *Science*, *203*, 437–438, 1979.

Burckle, L. H., and E. M. Pokras, Implications of a Pliocene stand of *Nothofagus* (southern beech) within 500 kilometers of the south pole, *Antarct. Sci.*, *3*, 389–403, 1991.

Carlquist, S., Upper Pliocene–lower Pleistocene *Nothofagus* wood from the Transantarctic Mountains, *Aliso*, *11*, 571–583, 1987.

Couper, R. A., Upper Mesozoic and Cainozoic spores and pollen grains from New Zealand, *N. Z. Geol. Surv. Paleontol. Bull.*, *22*, 77 pp., 1953.

Dettmann, M. E., D. T. Pocknall, E. J. Romero, and M. de C. Zamaloa, *Nothofagidites* Erdtman ex Potonié 1960: A catalogue of species with notes on the palaeogeographic distribution of *Nothofagus* Bl. (southern beech), *N. Z. Geol. Surv. Paleontol. Bull.*, *60*, 79 pp., 1990.

Harwood, D. M., Recycled siliceous microfossils from the Sirius Formation, *Antarct. J. U. S.*, *21*(5), 101–103, 1986.

Hill, R. S., Fossil leaf, Antarctic Cenozoic History From the CIROS-1 Drillhole, McMurdo Sound, *DSIR Bull. N. Z.*, *245*, 143–144, 1989.

Hill, R. S., Tertiary *Nothofagus* (Fagaceae) macrofossils from Tasmania and Antarctica and their bearing on the evolution of the genus, *Bot. J. Linn. Soc.*, *105*, 73–112, 1991.

Hill, R. S., *Nothofagus*: Evolution from a southern perspective, *Trends Ecol. Evol.*, *7*, 190–194, 1992.

Hill, R. S., J. Read, and J. Busby, The temperature dependence of photosynthesis of some Australian temperate rainforest trees and its biogeographical significance, *J. Biogeogr.*, *15*, 431–449, 1988.

Hill, R. S., D. M. Harwood, and P.-N. Webb, Last remnant of Antarctica's Cenozoic flora: Pliocene *Nothofagus* of the Sirius Group, Transantarctic Mountains, in *Eighth Gondwana Subcommission Symposium, Abstracts*, Hobart, Tasmania, 1991.

Kellogg, T. B., and D. E. Kellogg, Pleistocene sediments beneath the Ross Ice Shelf, *Nature*, *293*, 130–133, 1981.

Kemp, E. M., and P. J. Barrett, Antarctic glaciation and early Tertiary vegetation, *Nature*, *258*, 507–508, 1975.

Macphail, M. K., Vegetation and climates in southern Tasmania since the last glaciation, *Quat. Res.*, *11*, 306–341, 1979.

McKelvey, B. C., P.-N. Webb, D. M. Harwood, and M. C. G. Mabin, The Dominian Range Sirius Group—A record of the late Pliocene–early Pleistocene Beardmore Glacier, in *Geological Evolution of Antarctica*, edited by M. R. A. Thomson, J. A. Crame and J. W. Thomson, pp. 675–683, Cambridge University Press, New York, 1991.

Mildenhall, D. C., Terrestrial palynology, Antarctic Cenozoic History From the CIROS-1 Drillhole, McMurdo Sound, *DSIR Bull. N. Z.*, *245*, 119–127, 1989.

Mohr, B. A. R., Eocene and Oligocene sporomorphs and dinoflagellate cysts from Leg 113 drill sites, Weddell Sea, Antarctica, *Proc. Ocean Drill. Program Sci. Results*, *113*, 595–612, 1990.

Philipson, W. R., and M. N. Philipson, Leaf vernation in *Nothofagus*, *N. Z. J. Bot.*, *17*, 417–421, 1978.

Pocknall, D. T., and D. C. Mildenhall, Late Oligocene–early Miocene spores and pollen from Southland, New Zealand, *N. Z. Geol. Surv. Paleontol. Bull.*, *51*, 66 pp., 1984.

Praglowski, J., Fagaceae L. Fagoidae, *World Pollen Spore Flora*, *11*, 1–28, 1982.

Read, J., and R. S. Hill, Comparative responses to temperature of the major canopy species of Tasmanian cool temperate rainforest and their ecological significance, I, Foliar frost resistance, *Aust. J. Bot.*, *36*, 131–143, 1988.

Read, J., and R. S. Hill, The response of some Australian temperate rainforest tree species to freezing temperatures and its biogeographical significance, *J. Biogeogr.*, *16*, 21–27, 1989.

Sakai, A., D. M. Paton, and P. Wardle, Freezing resistance of trees of the south temperate zone, especially subalpine species of Australia, *Ecology*, *62*, 563–570, 1981.

Truswell, E. M., Recycled Cretaceous and Teritary pollen and spores in Antarctic marine sediments: A catalogue, *Palaeontographica*, *186B*, 121–174, 1983.

Webb, P.-N., The Cenozoic history of Antarctica and its global impact, *Antarct. Sci.*, *2*, 3–21, 1990.

Webb, P.-N., and D. M. Harwood, Late Neogene terrestrial flora of Antarctica: Its significance in interpreting late Cenozoic glacial history, *Antarct. J. U. S.*, *22*(2), 7–11, 1987.

(Received April 13, 1992; accepted October 15, 1992.)

THE ANTARCTIC PALEOENVIRONMENT: A PERSPECTIVE ON GLOBAL CHANGE
ANTARCTIC RESEARCH SERIES, VOLUME 60, PAGES 75–89

CENOZOIC GLACIAL SEQUENCES OF THE ANTARCTIC CONTINENTAL MARGIN AS RECORDERS OF ANTARCTIC ICE SHEET FLUCTUATIONS

ALAN K. COOPER AND STEPHEN EITTREIM

U.S. Geological Survey, Menlo Park, California 94025

URI TEN BRINK

U.S. Geological Survey, Woods Hole, Massachusetts 02543

IGOR ZAYATZ

Marine Arctic Geological Expedition, Murmansk, 183012 Russia

Seismic reflection profiles across the Antarctic continental margin show thick Cenozoic prograding sequences (CPS) of likely glacial and interglacial origin unconformably overlying preglacial Paleozoic, Mesozoic, and Cenozoic sedimentary and basement rocks. Many parts of the continental shelf, underlain by CPS, may have been permanently overdeepened (>150-m water) like today, at glacial onset and initial advance(s) of marine-based grounded ice sheets across the continental shelf. Subsequently, glaciomarine strata were deposited as topset and foreset strata on the overdeepened shelf by nearby grounded ice sheets. Later grounded ice sheets eroded, reworked, and redistributed strata to give the present bathymetric profile that includes glacially shaped banks and troughs. The banks are underlain by varied geometries, but principally topset strata that are thicker under shallower banks. The CPS have similar geometries on different parts of the continental margin and contain unconformities that can sometimes be traced across the margin. The similarity and continuity suggest that the CPS are good sites to core and drill along transects to resolve longstanding debates about circum-Antarctic Cenozoic paleoclimatic (ice volume), paleoceanographic (sea level), and paleogeographic events in the Southern Ocean and Antarctic regions. Such drilling would also link the proximal shelf record (mostly unsampled) with the distal ocean record (highly sampled). A general transect drilling strategy is suggested.

INTRODUCTION

Since the 1960s, several hundred thousand kilometers of acoustic reflection data, including over 130,000 km of multichannel seismic reflection data, have been recorded across the Antarctic continental margin [e.g., *National Oceanic and Atmospheric Administration*, 1984; *Behrendt*, 1990; *Hayes*, 1991; *Cooper et al.*, 1991a, 1992]. The data show an overdeepened continental shelf (i.e., water depths >150 m), diverse bathymetric features, and thick sedimentary deposits along at least five broad segments of the margin (Figure 1): Wilkes Land [e.g., *Wannesson*, 1990], Ross Sea [e.g., *Hinz and Block*, 1984], Antarctic Peninsula [e.g., *Larter and Barker*, 1991], Weddell Sea [e.g., *Kuvaas and Kristoffersen*, 1991], and Prydz Bay [e.g., *Stagg*, 1985; *Kuvaas and Leitchenkov*, 1993].

A generalized section across the Antarctic margin, based on multichannel seismic reflection profiles from the five areas and on limited drilling information, is shown in Figure 2 (see reviews of drilling in the work of *Hayes and Frakes* [1975], *Barker et al.* [1988], *Barron et al.* [1989], and *Barrett* [1989] and of acoustic and geologic data in the work of *Davey* [1985], *Cooper et al.* [1991a], and *Anderson* [1991]).

The margin is covered by an upper unit of likely Cenozoic glacial and interglacial strata that prograde and aggrade the shelf and drape across the slope and rise. Beneath the middle and outer shelf, seaward dipping foreset beds and nearly flat-lying topset strata commonly have a cumulative thickness of 1–2 km, with a maximum thickness of 3.5–5.0 km in the Weddell and Ross seas [*Kuvaas and Kristoffersen*, 1991; *Hinz and Block*, 1984].

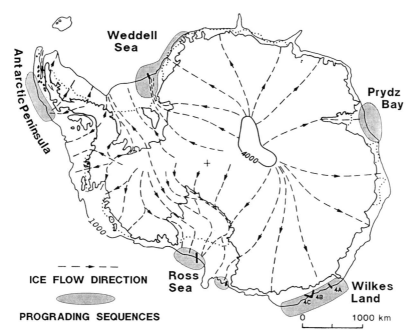

Fig. 1. Index map of Antarctica showing locations of prograding sedimentary sequences, ice flow directions during the last glacial maximum, and locations of seismic profiles (modified from *Cooper et al.* [1991*a*]).

The relative thickness of topset and foreset units varies along the margin and depends largely on position relative to axes of glacial banks and troughs (see the discussion below). The oldest ages of glacial strata in the upper unit are probably diachronous around Antarctica and extend back to at least early Oligocene in Prydz Bay, Ross Sea, and Weddell Sea and to at least late Miocene in the Antarctic Peninsula (see the above drilling references).

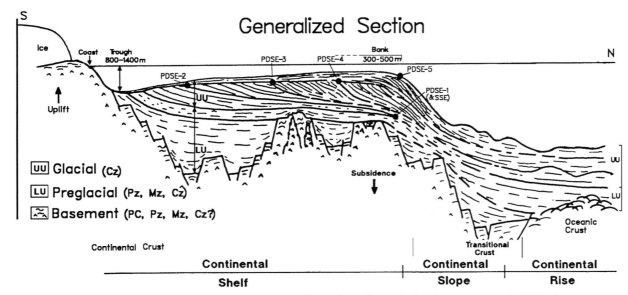

Fig. 2. Generalized section across the Antarctic continental margin showing an upper unit (UU) of likely Cenozoic glacial sequences unconformably overlying a lower unit (LU) of Paleogene and older preglacial rocks filling rift basins and covering oceanic crust. PDSE is paleodepositional shelf edge and SSE is structural shelf edge. The large landward shift, followed by seaward progression, of the PDSE (i.e., PDSE 1 to PDSE 5) is evidence of likely permanent overdeepening of the continental shelf since early glacial times when grounded ice sheets first advanced onto the shelf. See also Figure 5.

A major unconformity is inferred between the upper unit and a lower unit of Mesozoic and possibly nonglacial Paleogene strata that fill the deep extensional basins on the shelf and slope. The lower unit also covers and onlaps oceanic crust beneath the continental rise. The structural settings and depositional styles for both units vary in response to different histories of rifting and tectonism, margin subsidence and flexure, sediment supply, eustacy, and depositional processes around Antarctica during nonglacial and glacial periods [*Bleil and Thiede*, 1990; *Cooper et al.*, 1991*a*; *Anderson*, 1991].

In this paper, we concentrate on the upper unit of Cenozoic strata that cover the continental margin. We show models and seismic reflection data to address two aspects of the evolution of the Cenozoic glacial sequences and their relationship to the waxing and waning of grounded ice sheets. These aspects are the overdeepening and prograding of the continental shelf and the evolution of some glacial banks and troughs on the continental shelf. Also, we show seismic data to help elucidate the effect that Antarctic ice sheet variations have had on sedimentation on the adjacent continental slope and rise in the eastern Ross Sea and Wilkes Land sectors.

Scientific drilling and coring are needed to test hypotheses about Cenozoic glacial sequences and underlying Mesozoic rift-related deposits. In our conclusions, we suggest general areas where such drilling might be done.

CENOZOIC SEQUENCES

The Cenozoic sequences of the Antarctic continental margin have been widely studied [e.g., *Haugland et al.*, 1985; *Hinz and Kristoffersen*, 1987; *Anderson and Molnia*, 1989; *Larter and Barker*, 1989, 1991; *Cooper et al.*, 1991*a*; *Bartek et al.*, 1991; *Larter and Cunningham*, 1993] and are currently the topic of many international studies (e.g., see *Cooper and Webb* [1990]). The following is summarized from these studies. The Cenozoic sequences include seafloor and subsurface features that range from centimeters to hundreds of kilometers in size [*Anderson et al.*, 1983; *Cooper et al.*, 1991*a*; *Anderson*, 1991]. Seafloor coring, drilling, and seismic reflection data indicate that the features are or may be of glacial and interglacial origin. Herein, we refer to the documented and inferred glacial and interglacial sequences beneath broad regions of the middle and outer parts of the continental shelf and upper slope as Cenozoic prograding sequences (CPS). These sequences also aggrade the margin, but to a lesser degree.

In general, the CPS have complex sigmoidal and oblique geometries similar to those that occur on low-latitude (i.e., nonpolar) continental margins. But some geometries within the Antarctic glacial sequences differ from those of low-latitude margins and include the overdeepened, commonly landward dipping, and highly eroded bathymetric profile of the continental shelf and the glacially formed features, such as trough mouth fan [*Kuvaas and Kristoffersen*, 1991], till tongue and liftoff moraine [*King et al.*, 1991], till delta [*Alley et al.*, 1989], subglacial delta, and shelf margin delta fan [*Anderson and Bartek*, 1992] (Figure 3). Although these geometries are affected by processes that act on all continental margins (for example, subsidence, eustacy, sediment transport, and sediment compaction), they are also strongly controlled by processes found only on ice-covered margins (for example, crustal flexure from ice loading, glacial erosion, and subglacial sedimentation) (Figure 3).

The prograding sequences are as old as late middle Eocene age in Prydz Bay (40 Ma [*Barron et al.*, 1991]) and as young as Holocene from coring all around Antarctica [*Anderson*, 1991]. In the eastern Ross Sea the Cenozoic prograding sequences consist principally of glaciomarine strata [*Hayes and Frakes*, 1975] that have been deposited near or directly in front of marine-based ice sheets [*Anderson and Bartek*, 1992]. These ice sheets have been grounded episodically on the inner continental shelf of the Ross Sea since at least middle to late Oligocene time [*Bartek et al.*, 1991] and probably out to the paleodepositional continental shelf edge by at least early Miocene time [*Cooper et al.*, 1991*a*]. However, earlier studies of the sequences in the Ross Sea and Weddell Sea proposed that these sequences are fluviomarine delta lobes deposited by Antarctic coastal currents and/or dense Antarctic bottom waters [*Hinz and Block*, 1984; *Hinz and Kristoffersen*, 1987].

OVERDEEPENING OF THE CONTINENTAL SHELF

Most areas of the Antarctic continental shelf are overdeepened with water depths of >300 m. The shelf commonly, but not everywhere, has a reverse profile with water depths of about 400 m at the shelf edge (Figure 4*a*) and as great as 1400 m in troughs near the coast (Figure 2). Many factors are responsible for the overdeepening and reverse profile of the shelf [*ten Brink and Cooper*, 1990, 1992; *Cooper et al.*, 1991*a*; *Anderson*, 1991], but the principal factor is Paleogene and younger glacial erosion of the shelf that redistributed sediments from the inner shelf to the outer shelf and slope areas, resulting in differential sediment loading and flexure of the continental margin [*ten Brink and Cooper*, 1990, 1992].

The overdeepening of the continental shelf and the landward dips of the seafloor have influenced the depositional processes and subsurface stratal geometries beneath the continental margin. For example, sea level changes of up to 150 m [*Haq et al.*, 1987] will not expose the overdeepened shelf to subaerial erosion, and the reverse bathymetric profile will favor filling of inner

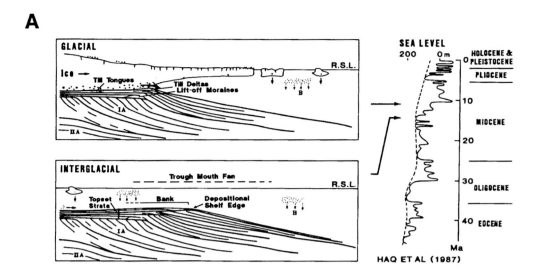

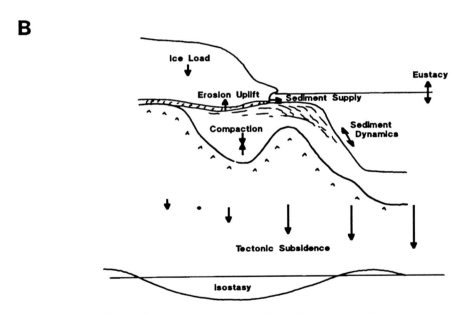

Fig. 3. (*a*) Model for subglacial and marine deposition of Cenozoic glacial sequences by eustatically related fluctuations of grounded ice sheets across the continental shelf. R.S.L. is relative sea level. IA and IIA are different types of acoustic sequences. (Modified from *Cooper et al.* [1991*a*].) (*b*) Schematic drawing illustrating the numerous factors that affect the deposition of sedimentary sequences on the Antarctic margin, where, unlike low-latitude margins, the effects of massive grounded ice sheets episodically moving across the shelf must also be considered.

shelf troughs and transporting of sediment along, rather than seaward across, the shelf [*Anderson et al.*, 1983]. When open marine conditions with no grounded ice exist, a large volume is available for accommodating sediments. Stratal geometries for prograding sequences should be like high-sea-level deposits (for example, high-stand deposits [*Posamentier and Vail*, 1988]).

However, when marine-based ice sheets are grounded on the continental shelf, no accommodation space exists beneath the ice sheet, and different stratal geometries of the subice deposits (for example, tills, liftoff moraines, and till tongues) occur. Strata that are deposited into the water column directly in front of grounded ice sheets (for example, till delta, trough mouth fan, and shelf

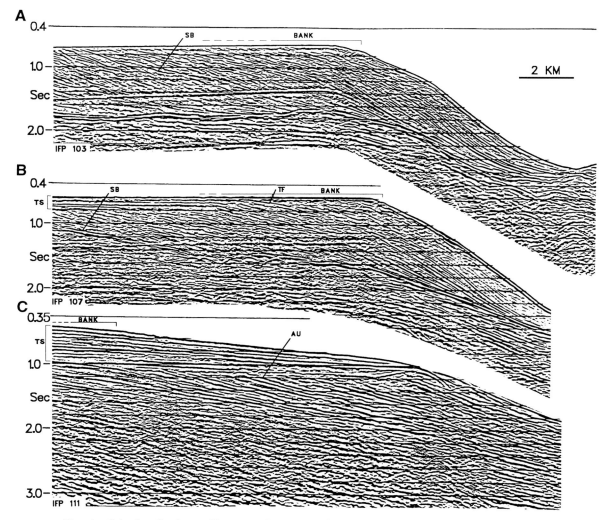

Fig. 4. Seismic reflection profiles across the outer continental shelf of Wilkes Land showing the large variations in reflector geometry and continuity that occur in topset strata beneath shelf "banks" and in foreset beds and unconformities across paleocontinental slopes. AU is angular unconformity, SB is sequence boundary, TF is topset-to-foreset bed, and TS is topset strata.

margin delta fan) may, however, have geometries like low-latitude margins.

Drill cores [e.g., *Leckie and Webb*, 1983; *Hambrey et al.*, 1992] and seismic reflection profiles [*Cooper et al.*, 1991*a*, *b*; *Anderson and Bartek*, 1992; *Eittreim et al.*, 1993] across the Antarctic continental margin provide evidence that broad areas of the continental shelf were probably overdeepened permanently during the initial advances of grounded ice onto the continental shelf before most of the Cenozoic prograding sequences were deposited. By permanent overdeepening, we mean that the seafloor remained at least 150 m below sea level at all times following the initial overdeepening of the shelf and was not subject to subaerial erosion.

In our discussion, we refer to the structural shelf edge (SSE) (Figure 2) as the point at which basement and

overlying strata deepen rapidly into the adjacent ocean basin. The SSE was the shelf edge prior to overdeepening of the shelf and in many areas still is the shelf edge. We refer to a paleodepositional shelf edge (PDSE) (Figure 2) as the large break in depositional slope at which water depths and accommodation volume increased abruptly seaward. In many Antarctic areas the location of the PDSE lies far landward of the SSE. A similar geometric pattern occurs on low-latitude margins where *van Wagoner et al.* [1988] indicate that the "depositional-shoreline break" may be 160 km or more landward of the "shelf break."

Three general observations provide acoustic evidence that broad areas of the continental shelf were permanently overdeepened during the initial advances of marine-based grounded ice sheets onto the continental shelf. First, a buried regional unconformity is commonly observed to deepen from the coast (or from a midshelf structural high) toward the structural shelf edge (Figure 2) and can be traced (or inferred to continue) beneath the continental slope and rise. Geometries and/or the character of acoustic units differ across the unconformity. The Cenozoic prograding sequences downlap onto this and overlying unconformities. Second, in several areas around Antarctica, there is a lateral large shift of the paleodepositional shelf edge from its initial position at the structural shelf edge to a point up to 85 km landward. The PDSEs then unidirectionally prograde and aggrade to the position of the present depositional shelf edge (i.e., from PDSE 1 to PDSE 2 to PDSE 5 in Figure 2). Third, stratal geometries indicative of low-sea-level deposits, such as incised fluvial valleys on paleoshelves and low-stand fan deposits on paleocontinental slopes of normal depth margins, are missing in most, but not all, parts of the Antarctic margin.

Some geometries of the prograding sequences, especially on the paleo-outer-shelves, suggest that eustatic fluctuations have influenced depositional processes, even though the shelf may have been permanently overdeepened. As on low-latitude margins, the geometries include angular unconformities along the top of the paleo-outer-shelves, downlap surfaces, and complex sigmoidal and oblique topset-to-foreset beds that can be traced across the paleo-outer-shelf onto the paleo-upper-slope. For Antarctica, we and others [*Jeffers and Anderson*, 1990; *Larter and Barker*, 1989, 1991; *Cooper et al.*, 1991a; *Bartek et al.*, 1991; *Larter and Cunningham*, 1993] attribute these geometries to erosion and deposition of sediments beneath and in front of marine-based ice sheets that are episodically grounded out to the seaward edge of the continental shelf. These geometries are not attributed to the cyclic subaerial exposure and immersion of the continental shelf, as occurs on low-latitude shelves with normal water depth [e.g., *Posamentier and Vail*, 1988].

Figure 5 illustrates how the Antarctic continental margin could have acquired an overdeepened and reverse depth profile during Cenozoic advances of the ice sheet to the continental shelf edge. Our model is simplified from that of *ten Brink and Cooper* [1992], who use a statistical analysis of random ice sheet movements to illustrate how a bathymetric profile along the transect of drill sites on the Prydz Bay shelf may have evolved. The morphology of the seafloor is shown for a 230-km-wide normal depth continental shelf (Figure 5a) and for a similar width but eroded and overdeepened shelf (Figures 5b and 5c). The variable shapes result from progressive ice loading by an advancing, grounded ice sheet. For simplicity, the seafloor shape for each ice front position (Figures 5a and 5c) is the laterally shifted, flexural ice-loading curve for a variable strength elastic lithosphere [*ten Brink and Cooper*, 1992].

We assume that prior to the major buildup of ice on Antarctica, the continental shelf had a normal water depth profile (Figure 5a, profile 1). As the massive Antarctic ice sheet crosses the original coastline and moves out onto the continental shelf, the lithosphere flexes and the shape of the seafloor changes, depending upon the location of the front of the grounded ice.

Once the ice sheet retreats and leaves the shelf, the shape of the seafloor depends upon the differential erosion across the shelf. If, for example, sediments are not eroded or deposited, then the shelf would return to a normal profile (Figure 5a, profile 1) in about 10,000 years [*Cathles*, 1975]. If, however, sediments are differentially eroded from the shelf by the ice sheet, as commonly occurs in broad glacial troughs, then the shape of the seafloor would be permanently altered. For illustration, we assume that the ice sheet erodes a sediment thickness that varies uniformly from 300 m at the coast to 30 m at the paleo-shelf-edge (Figure 5b). This is about the minimum amount of erosion that, after compensation for erosion uplift and flexural rebound, is required to permanently overdeepen the shelf by at least 150 m, thereby leaving the shelf below sea level for all eustatic changes of up to 150 m (Figure 5c, profile 1D). *Ten Brink and Cooper* [1992] give statistical evidence that an erosion distribution generally similar to Figure 5b can be caused by a randomly waxing and waning ice sheet. As the ice sheet again advances onto the continental shelf (Figure 5c, profiles 2D to 8D), the seafloor remains overdeepened.

In the simple models (Figures 5a and 5c), if sediments are deposited on the continental shelf, they could have landward or seaward stratal dips depending upon where the front of the ice sheet is located. For example, when the ice sheet is less than halfway across the shelf, sediments draped across the outer shelf would dip gently seaward to the shelf edge and more steeply down the continental slope like normal topset and foreset beds (Figure 5, profiles 2–5, 2D, and 3D). In contrast, when the ice sheet lies between the middle and the edge of the shelf, the sediment would be trapped between the front of the ice sheet and the shelf edge and would dip landward or be flat in front of the ice (Figure 5, profiles 6–8 and 4D–8D). While the ice is at the shelf edge, sediments would be deposited directly onto the continental slope (Figure 5, profiles 8 and 8D).

Seismic data suggest a more complex history for some highly eroded parts of the shelf, as indicated by the shifted locations of the paleo-shelf-edges. The conceptual sketch in Figure 5d shows a greatly overdeepened part of the shelf that is being buried by prograding sequences. Initially, the shelf may have been overdeepened by multiple glacial erosion events [e.g., *ten Brink and Cooper*, 1990, 1992] and other factors (for example, thermal subsidence). Thereafter, the paleo-shelf-edge shifts landward from the structural shelf edge (SSE in Figure 5d) to the new paleodepositional shelf edge

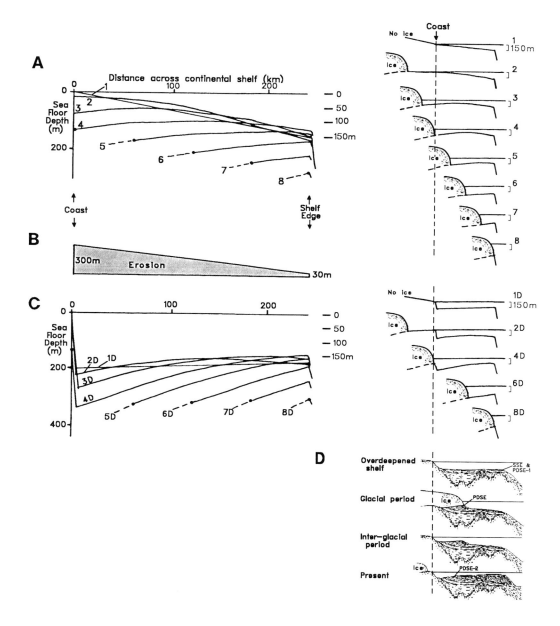

Fig. 5. Simplified models and conceptual sketch. Models show seafloor shapes due to flexural loading by grounded ice on the continental shelf. The sketch illustrates a possible mode of glacial/interglacial sedimentation on an overdeepened shelf. (*a*) Depth profiles and ice sheet models for a continental shelf with an initial normal depth seafloor. (*b*) Minimum variable amounts of erosion of the normal depth seafloor (profile 1) needed to permanently overdeepen the continental shelf (profile 1D) after compensation for erosion uplift and ice loading. (*c*) Depth profiles and ice sheet models for a continental shelf with an initial overdeepened and landward dipping seafloor (i.e., due to variable erosion in Figure 5*b*). Only relatively small amounts of erosion (30–300 m) are needed to overdeepen the shelf. The seafloor beneath the outer shelf can dip either seaward or landward depending on the position of the ice front. Numbers (for example, 1, 2, · · · , 1D, 2D, · · ·) refer to seafloor profiles for various ice front positions. Profiles are dashed where seafloor lies under grounded ice. (*d*) Sketch illustrating how Cenozoic glacial sequences could be deposited on a permanently overdeepened part of the continental shelf and how paleodepositional shelf edges (PDSE) shift landward at the initial overdeepening of the shelf and then move unidirectionally seaward to the present shelf edge location. Geometries are based on Figure 2.

(PDSE 2 in Figure 5*d*). Grounded ice sheets deposit sediments onto the overdeepened shelf during glacial periods of advanced ice. During interglacial periods like today, little sediment is delivered to the overdeepened shelf. However, the rates and types of sediments delivered to the shelf during earlier interglacial periods could be quite large and variable as seen in the Ross Sea [*Anderson and Bartek*, 1992]. The prograding sequences would be deposited during several glacial and interglacial episodes that caused the paleodepositional shelf edge to move seaward to its present location.

Deposition, erosion, and sediment supply processes during glacial and interglacial times are more variable and complex than we imply in our models and sketch (Figure 5). Even so, Figure 5 is helpful in conceptually explaining the geometries of strata and unconformities evident in seismic profiles collected across the prograding sequences that lie beneath many parts of the Antarctic margin. The models and sketch give several important insights:

1. Only a relatively small amount of erosion (30–300 m, Figure 5*b*) is required to permanently overdeepen part or all of the continental shelf. Although the number of grounded ice sheet advances is unknown, any erosion downward below previous seafloor levels leads to progressively greater seafloor depths, from profile 1 to profile 1D to the present (Figures 5 and 2).

2. The shape of the seafloor changes, owing to flexural loading, as the grounded ice sheet moves onto and across the shelf. Marine sediments deposited in front of the ice sheet and present on the continental shelf could dip either landward or seaward depending on the location of the ice sheet and the amount of prior erosion of the shelf.

3. The initial advance(s) of the ice sheet to the structural shelf edge could erode strata and produce the basal regional unconformity over which later paleodepositional shelf edges unidirectionally prograde and onto which strata downlap, getting younger seaward. Erosion resulting from subaerial exposure of the shelf is not required (Figure 5*d*).

We acknowledge that thermal subsidence, crustal thinning during rifting, and crustal flexure from sediment loading cause subsidence and could, like glacial erosion, result in overdeepening of the shelf, especially in times of relatively low sediment accumulation. Indeed, rapid thermal subsidence has been suggested for the Antarctic Peninsula [*Larter and Barker*, 1989, 1991] and eastern Ross Sea [*Cooper et al.*, 1991*c*]. The interrelationship of these factors in causing overdeepening is not yet clearly understood, but we suspect that glacial erosion may be the dominant factor on the basis of existing seismic and drilling data.

STRATIGRAPHY OF GLACIAL BANKS AND TROUGHS

Most areas of the Antarctic continental shelf have seafloor banks and troughs that range in size from a few kilometers to hundreds of kilometers in width, have reliefs of up to 1 km, and have edges that are commonly parallel to the coast and cross the shelf [*Anderson*, 1991]. The banks (i.e., water depths shallower than 500 m) also are underlain by diverse subsurface acoustic stratigraphies that commonly cannot be predicted from bathymetric profiles and contour maps. Nevertheless, the banks are an integral part of the Cenozoic glacial and interglacial sequences beneath the outer half of the continental shelf.

The shallowest bank areas are on the middle and outer shelf, where multichannel seismic reflection profiles show a variety of depositional structures. The structures include steeply seaward dipping strata (foreset beds) that are truncated within a few meters of the seafloor (Figure 4*a*), or nearly flat-lying strata (topset beds) that can be continuously traced to paleodepositional shelf edges and then downdip (foreset beds) onto the paleocontinental slopes (Figures 4*b*, 6*b*, and 6*c*), or nearly flat-lying strata that truncate underlying seaward dipping foreset beds and are themselves eroded at their seaward ends (Figures 4*c* and 6*a*). Although the examples are from only the Ross Sea, Weddell Sea, and Wilkes Land segments of the margin, similar geometries are observed under other parts of the outer continental shelf of the Ross Sea [*Hinz and Block*, 1984; *Cooper et al.*, 1991*a*; *Anderson and Bartek*, 1992], the Antarctic Peninsula [*Anderson et al.*, 1990; *Larter and Barker*, 1991], the Weddell Sea [*Haugland et al.*, 1985; *Kuvaas and Kristoffersen*, 1991], and Prydz Bay [*Stagg*, 1985; *Cooper et al.*, 1991*b*; *Kuvaas and Leitchenkov*, 1993]. In this paper, we collectively refer to topset beds and other nearly flat-lying strata on the continental shelf as topset strata. By definition, topset beds connect with foreset beds. However, some Antarctic topset strata do not connect with foreset beds.

The thickness and character of the topset strata vary within the 15- to 40-m resolution of the multichannel seismic reflection data (Figures 4 and 6). In single-channel seismic data [e.g., *Anderson and Bartek*, 1992], thinner beds and greater stratigraphic detail are visible. The total thickness of topset strata ranges from a few meters to several hundred meters and may either increase seaward (Figure 6*b*) or landward (Figures 4*b* and 4*c*) across the continental shelf edge. Topset strata are thin or absent along the axis of major seafloor glacial troughs that cross the continental shelf, and the axial topset strata thicken toward the edge of the trough (for example, Prydz Bay [*Cooper et al.*, 1991*b*], eastern Ross Sea [*Alonso et al.*, 1992], and Wilkes Land [*Eittreim et al.*, 1993]). Individual beds within the topset strata commonly are highly reflective, thicken seaward

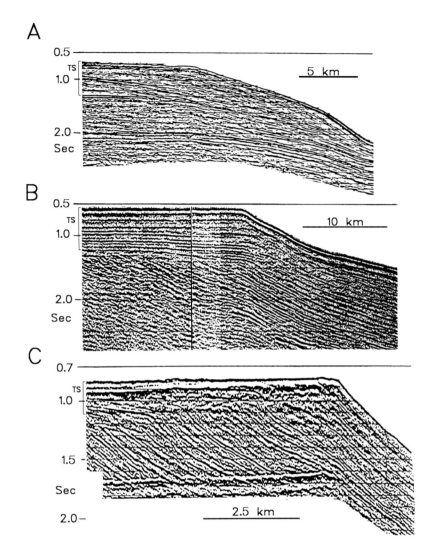

Fig. 6. Seismic reflection profiles across the Antarctic outer continental shelf. (*a*) Western Ross Sea [from *Berger et al.*, 1989]. (*b*) Eastern Ross Sea (Figure 7). (*c*) Weddell Sea [from *Kuvaas and Kristoffersen*, 1991]. Topset strata (TS) exhibit large variability in thickness, lateral extent, and dips (i.e., both landward and seaward) at, and deeply buried beneath, the seafloor.

near the paleo-shelf-edges (Figure 6*c* [*Alonso et al.*, 1993]), and may terminate abruptly at moundlike deposits at the paleodepositional shelf edge (Figures 4*c* and 6*a* [*Cooper et al.*, 1991*a*]).

The diverse subsurface stratigraphy of seafloor banks (Figures 4 and 6) implies varied histories of deposition and erosion of topset strata (and underlying foreset beds) on the outer shelf, probably due to episodic waxing and waning of marine-based grounded ice sheets. Although the uppermost parts of the topset strata that underlie many banks on the outer shelf have been widely sampled by seafloor cores [e.g., *Anderson*, 1991], Prydz Bay is the only place where the complete section of topset strata has been drilled and underlying

foreset beds have been reached. Here the topset strata consist principally of acoustically opaque, massive diamictites that are believed to be deposited from multiple advances of grounded ice sheets [*Hambrey et al.*, 1991; *Solheim et al.*, 1991]. Ross Sea drilling of topset strata recovered acoustically well-stratified glaciomarine rocks, unlike those of Prydz Bay, from the inner and middle shelf [*Hayes and Frakes*, 1975]. Ice movement on the shelf may be the principal agent of erosion and sediment redistribution for the prograding sequences during glacial times, but cross-shelf currents like those of today [*Anderson*, 1991] may have been important in eroding and depositing sediments during interglacial times. On the basis of Ross Sea and Prydz Bay drilling

[*Hayes and Frakes*, 1975; *Barron et al.*, 1991], the middle to inner shelf areas are likely to be sites of glaciomarine and pelagic sedimentation which may be partly preserved and not eroded away. The large variability in acoustic properties (for example, reflector amplitude and continuity, interval and refraction velocity, and character of acoustic units) of the topset strata suggests that glacial and interglacial strata may exist in the topset strata [*Cochrane and Cooper*, 1991, 1992; *Cooper et al.*, 1991*b*; *Anderson and Bartek*, 1992].

Topset strata beneath unsampled parts of the continental shelf, and its glacially shaped banks, are likely to hold a more complete record than that now available of Neogene and Quaternary glacial and interglacial periods. The preservation of the interglacial record in topset strata will probably vary from nonexistent in highly eroded areas within glacial troughs where few topset strata exist (Figure 4*a*) to possibly good in parts of the middle to outer shelf with relatively thick well-stratified topset strata (for example, Figures 4*c* and 6*b* [*Larter and Barker*, 1991; *Hinz and Block*, 1984; *Eittreim et al.*, 1993]). The innermost and outermost parts of the shelf are commonly subject to persistent ice and current erosion, respectively, and are not likely to be good preservation sites [*Anderson*, 1991; *ten Brink and Cooper*, 1992]. An exception is the good record of pelagic sedimentation that is preserved in deep inner shelf troughs for the last interglacial period (for example, Ocean Drilling Program (ODP) Site 740 [*Barron et al.*, 1991]).

Unfortunately, existing multichannel seismic profiles (for example, Figures 4 and 6) do not have adequate resolution to confirm the presence of interglacial beds that may be only a few meters thick, as has been observed in Prydz Bay drill cores and downhole logs [*Barron et al.*, 1989; *Cooper et al.*, 1991*b*]. Systematic high-resolution seismic studies are needed.

REGIONAL ACOUSTIC HORIZONS

Seismic reflection profiles across the Antarctic continental margin commonly show high-amplitude reflections that can be traced over large areas. The strong reflections denote lithologic contrasts at depositional contacts, unconformities, and special boundaries such as possible diagenetic fronts and gas hydrates. Some reflections are from regional unconformities that outline the upper and lower bounds of the Cenozoic prograding sequences [*Zayatz et al.*, 1990; *Jeffers and Anderson*, 1990; *Alonso et al.*, 1992].

Although the strong reflections are commonly of regional extent, they usually cannot be traced continuously across the entire continental margin from the inner shelf to the ocean basins (Figure 2). Seismic profiles from Wilkes Land and the Ross Sea (Figures 4 and 7) illustrate that at shallow depths (i.e., 1–2 s), reflections within depositional units can confidently be traced over several tens of kilometers. At greater depths, however, reflections along unconformities and within depositional units are not continuous over long distances, especially at reflection times later than the seafloor multiple reflection, particularly from the continental slope and upper rise areas. In these areas, reflections are disrupted or terminate owing to geologic causes, such as basement structures, erosional relief, downlap, toplap, etc., and owing to intrinsic features of seismic data, such as multiple reflections, diffractions, dispersion, attenuation, etc.

In lieu of drilling information, many Antarctic researchers interpret regional reflections to describe the structure, stratigraphy, and evolution of various segments of the Antarctic margin [e.g., *Hinz and Block*, 1984; *Eittreim and Smith*, 1987; *Anderson et al.*, 1990; *Miller et al.*, 1990; *Stagg*, 1985; *Wannesson*, 1990; *Cooper et al.*, 1991*b*]. These investigators acknowledge the difficulties in identifying the nature and continuity of reflections and in determining progressive time gaps across reflections [e.g., *Christie-Blick et al.*, 1990, p. 124]. Nevertheless, stratal ages are commonly inferred by tracing and projecting discontinuous reflections between widely separated structural regimes and depositional environments. Presently, multichannel seismic reflection data are too geographically limited to trace the unconformities around Antarctica. We agree with *Hinz and Kristoffersen* [1987] that some circum-Antarctic and regional events (for example, the opening of ocean pathways, glacial onset and major changes in the volume of the Antarctic ice sheet and in global sea level, and major Antarctic tectonic events) are likely to be recorded in the geologic and seismic records from all depositional environments of the Antarctic margin. However, inferences solely from seismic data are highly speculative without geologic samples from coring and drilling. Our interpretation of the seismic data across the eastern Ross Sea continental shelf and slope areas (Figure 7), although speculative, further illustrates that unconformities can be identified and inferred in these areas from reflections that join, albeit discontinuously. We suspect that individual unconformities, such as U2 and U3, formed during several periods when grounded ice existed on the continental shelf and deep bottom currents eroded the slope and rise.

SCIENTIFIC DRILLING

In our opinion, the thick prograding sequences of the Antarctic margin are likely to hold a detailed proximal record of Cenozoic glacial and interglacial events that are directly linked to fluctuations in the volume of the massive Antarctic ice sheet and in relative sea level. Drilling of the prograding sequences is needed to decipher this record.

Past drilling on the Antarctic continental shelf gives us compelling, albeit scanty, evidence that a glacial history is at least partly preserved and is decipherable if

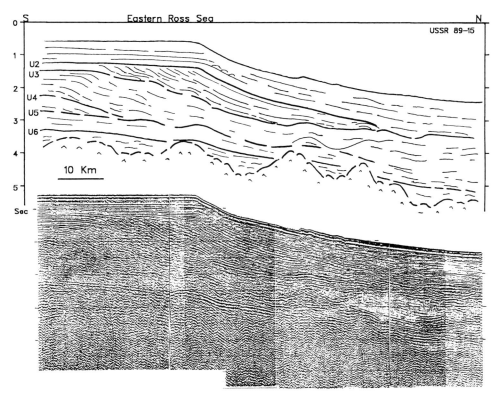

Fig. 7. Seismic reflection profile and interpretive line drawing across the continental margin in the eastern Ross Sea. Unconformities U2 to U6 [*Hinz and Block*, 1984] on the continental shelf may correlate with erosional surfaces beneath the continental rise. Here and elsewhere, reflections are difficult to trace continuously across the entire margin. Modified from *Zayatz et al.* [1990]. See Figure 1 for location.

complete recovery drill cores [e.g., *Barrett*, 1989] can be obtained. In fact, a partial history from early Oligocene time has already been derived at isolated shelf drill sites. The Antarctic ice sheet has been grounded at least three times on the paleo-outer-shelf in Prydz Bay (ODP Site 739 [*Solheim et al.*, 1991]), at least once and probably many times in the central Ross Sea (Deep Sea Drilling Project (DSDP) Site 270 [*Bartek et al.*, 1991]), and many times at the coast at the western Ross Sea (CIROS 1 [*Barrett*, 1989]). Paleo-water-depths have fluctuated greatly at the coast of the western Ross Sea (CIROS 1 [*Barrett*, 1989]) and have increased on the paleo-mid-shelf areas of the Ross Sea (DSDP Site 270 [*Leckie and Webb*, 1983]) and Prydz Bay (ODP sites 739 and 742 [*Hambrey et al.*, 1991]).

Seismic records that have been tied to shelf and slope drill sites have been used to infer that many erosional unconformities and glacial features on the shelf such as troughs, till deltas, and morainal banks are due to grounded ice sheets that developed since early Oligocene time in Prydz Bay [*Cooper et al.*, 1991a, b] and probably in the Weddell Sea [*Kuvaas and Kristoffersen*, 1991] and since at least middle to late Oligocene time in the Ross Sea [*Bartek et al.*, 1991; *Cooper et al.*, 1991a].

Further inferences have been made that the number of glacial and interglacial periods on the shelf has increased since middle to late Miocene time in the eastern Ross Sea [*Cooper et al.*, 1991a; *Anderson and Bartek*, 1992] and has been large since late Miocene time in the Weddell Sea [*Kuvaas and Kristoffersen*, 1991], in the Antarctic Peninsula [*Larter and Barker*, 1989, 1991], and in Prydz Bay [*Cooper et al.*, 1991a, b].

The existing offshore drilling record of the Antarctic continental shelf and slope and rise is sparse. Yet the prior results and the new data compilations show that these areas hold great promise in answering questions about Cenozoic paleoclimates, paleoceanography, and paleobiogeography of the Antarctic and Southern Ocean regions. Drilling across these areas would link the missing proximal continental shelf record with the well-documented distal abyssal basin records of inferred ice volume, temperature, sea level, biostratigraphic, and geochemical changes around Antarctica. Drilling into the prograding sequences and underlying Cenozoic (?) strata from around Antarctica would help establish the extent and history (initiation and synchroneity) of Antarctic ice sheet waxing and waning across the continental shelf around Antarctica and the impact on Cenozoic global sea level fluctuations.

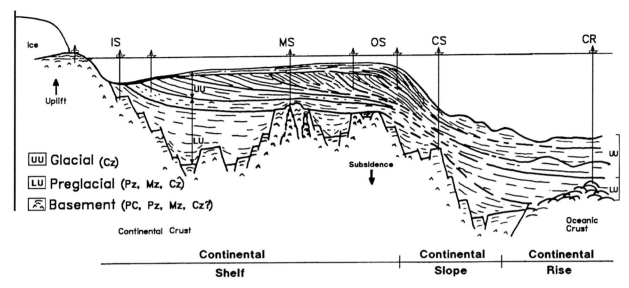

Fig. 8. Generalized section showing possible drill sites along transect across the Antarctic continental margin. The geographic abbreviations (i.e., IS, MS, etc.) and drilling objectives are described in the text.

Drilling would resolve the highly debated late Neogene history of the massive or dispersed Antarctic ice sheet(s) [e.g., *Webb and Harwood*, 1991] and the impact on global climates at this time. Technology exists to drill and core these areas from ships and land-based fast ice drilling platforms. We suggest that future drilling, like past efforts in the Ross Sea [*Hayes and Frakes*, 1975] and in Prydz Bay [*Barron et al.*, 1989], be done along transects across the Antarctic continental margin (for example, Figure 8). The following subsections refer to the generalized Antarctic margin cross section in Figure 8, and although regional variations are likely, the general concepts should be valid.

Coast and Inner Shelf (IS)

Onshore and nearshore sedimentary sections could include late Mesozoic and Cenozoic marine and nonmarine rocks that have been uplifted and/or glacially transported. As the inner shelf is commonly highly eroded, relatively shallow drilling could sample some of the following, in order of increasing age: a thin veneer of Holocene sediments, the oldest parts of the Cenozoic prograding sequences (glacial) that cover the shelf, the preglacial postrift and synrift strata that fill shelf basins, and the prerift and continental basement rocks. Drilling here would provide information on the tectonic and depositional histories of basement and likely Mesozoic through Paleogene nonglacial and glacial rocks.

Midshelf (MS)

The midshelf has prograding sequences in which topset strata, where present, unconformably overlie

gently dipping foreset beds. Drilling could sample the Neogene through Holocene topset strata to acquire a high-resolution record of glacial and interglacial events. Drilling into foreset beds could recover an early record of ice-proximal deposits on the overdeepened paleoshelf and paleoslope (foreset beds), a preglacial history of Mesozoic and Paleogene shelf sedimentation, and possibly a record of continental or volcanic rift basement rocks associated with midshelf structural highs.

Outer Shelf (OS)

As the outer half of the continental shelves are underlain by numerous individual sequences with variable proportions of topset strata and foreset beds, drilling here could give a record of waxing and waning of late Cenozoic (late Oligocene (?) and younger) grounded ice sheets and interglacial periods. At some localities, the glacial to preglacial unconformity could be reached to determine the time of glacial onset and the subsidence and overdeepening history of the continental shelf since glacial onset. Drilling would also help determine possible paleoenvironmental and paleoceanographic events that are responsible for unconformities that extend from the shelf into the abyssal basins.

Continental Slope (CS)

As the upper to middle continental slope is commonly underlain by prograded sequences, drilling here could sample the times of glacial maxima (i.e., times of grounded ice at the paleo-shelf-edge) of late Cenozoic paleoclimatic and paleoceanographic fluctuations. On the lower slope, where canyon systems prevail, relatively shallow drilling could sample Cenozoic unconformities believed to be due to regional and circum-Antarctic glacial/interglacial and paleoceanographic episodes and due to major tectonic events. Here also, preglacial Mesozoic and Cenozoic strata, and possibly basement rocks, deposited during the rifting or convergence history of the margin could be sampled.

Continental Rise (CR)

Drilling beneath the continental rise could use canyons and other erosion sites to sample deep strata and regional unconformities that extend landward beneath the continental slope and onto the shelf. The core samples would date the duration of the suspected hiatuses and would provide paleoenvironmental data for preserved intervals to help establish possible linkages between the deep ocean and continental shelf records. Some sites on the rise could provide ages of oceanic crust and oceanographic and biotic links to Paleogene and older histories of other deep ocean basins including the Arctic.

SUMMARY

Seismic and drilling studies of the Antarctic continental margin provide evidence, much of which is still being assembled and evaluated by the international science community [e.g., *Cooper and Webb*, 1992], that there have been major fluctuations in the lateral extent, and presumably volume, of the grounded Antarctic ice sheet since at least early Oligocene time. The ice sheet and related sea level fluctuations are likely to be recorded in the Cenozoic glacial and interglacial prograding sequences that lie principally between the middle of the continental shelf and the outer continental rise.

Marine-based grounded ice sheets have moved across, flexurally depressed, eroded, and reshaped the bathymetric profile of the continental shelf many times during the Cenozoic. The inferred preglacial seaward dipping profile has been altered to the present generally overdeepened and landward dipping profile that includes deep cross-shelf troughs. Seismic and drill core data suggest to us that many parts of the shelf were overdeepened to water depths greater than 150 m during the initial stages of glaciation when marine-based grounded ice sheets moved across the continental shelf [*Leckie and Webb*, 1983; *ten Brink and Cooper*, 1990, 1993; *Cooper et al.*, 1991*a*, *b*; *Anderson and Bartek*, 1992; *Eittreim et al.*, 1993].

Seismic reflection data show the following on many parts of the margin:

1. A regional reflection horizon can be traced beneath the prograding sequences from at least the middle shelf to continental rise areas. We and others [*Kuvaas and Kristoffersen*, 1991; *Bartek et al.*, 1991; *Cooper et al.*, 1991*a*; *Eittreim et al.*, 1993; *Kuvaas and Leitchen-*

kov, 1993] believe that this horizon is the unconformity that separates glacial and preglacial rocks. We suspect that the unconformity may have been partly cut by initial grounded ice sheets.

2. The position of the paleodepositional shelf edge (PDSE) shifts landward up to 85 km during earliest ice buildup times (i.e., at the aforementioned regional unconformity), and then the PDSE moves unidirectionally seaward and upward to the present shelf edge (Figures 2 and 5*d*).

3. Acoustic geometries indicative of times of low sea level, such as incised fluvial valleys and low-stand fans, like those on low-latitude margins, are rare in the prograding sequences.

4. Bathymetric banks of the middle and outer shelf in less than 500-m water depths are underlain close to the seafloor by varied geometries that are principally flat-lying topset strata. The topset strata are generally thicker where banks are shallower.

We interpret these geometries to indicate the following:

1. Once a part of the shelf was glacially overdeepened, during earliest stages of regional (and circum-Antarctic (?)) glaciation, thereafter that part was kept permanently overdeepened (i.e., never subaerially eroded), principally by later glaciations, particularly the parts with broad cross-shelf troughs. The overdeepening may have occurred at different times for different parts of the shelf, but it did not occur solely during the most recent (i.e., Quaternary) glaciations.

2. The majority of prograding sequences were deposited on or at the seaward edge of an overdeepened shelf either directly beneath or in front of marine-based ice sheets that were grounded on the continental shelf during glacial periods and lowered sea level (Figures 3 and 5). Part of the prograding sequences may be pelagic and glaciomarine strata deposited on the shelf from nearby coastal glaciers during interglacial periods and raised sea level.

3. The outer shelf banks are principally remnants of flat-lying strata that were deposited mostly during Neogene and Quaternary glacial and interglacial periods.

Fluctuations in sea level are believed to partly control the location of ice sheet grounding lines on the overdeepened continental shelf and the acoustic geometries of the Cenozoic prograding sequences [*Jeffers and Anderson*, 1990; *Cooper et al.*, 1991*a*; *Anderson and Bartek*, 1992]. The relative contribution of seismically inferred subglacial (grounded ice) and glaciomarine (nearby ice) deposits appears to vary regionally and temporally around the continental margin. Such variations imply that significant paleoclimatic and paleoceanographic events have occurred and are recorded by the Cenozoic prograding sequences, yet these inferences are largely untested by geologic sampling.

Scientific drilling is needed along transects across the prograding sequences of the Antarctic continental margin (Figure 8) to help establish the detailed Cenozoic paleoclimatic (ice sheet), paleoceanographic (sea level), sedimentologic, and tectonic histories around Antarctica. The prograding sequences hold a proximal and distal record of Cenozoic Antarctic glacial events that could be sampled and resolved by shallow (<500 m) drilling at sites selected using existing seismic data collected and now being compiled by the international community. Such drilling would help correlate deep-ocean, continental shelf, and onshore geologic records and would greatly augment the significant discoveries made over the past 3 decades of shipborne and land-based Antarctic drilling [e.g., *Webb*, 1991; *Kennett and Barron*, 1992].

Acknowledgments. J. Wannesson graciously allowed the use of Institut Français du Pétrole seismic records on the Wilkes Land margin for this study. We appreciate the ongoing discussions with our colleagues in the Antarctic Offshore Acoustic Stratigraphy Project (ANTOSTRAT), which is actively investigating the topics addressed in this paper. We also appreciated the helpful reviews of the paper by M. Fisher, J. Behrendt, R. Larter, J. Anderson, and an anonymous reviewer.

REFERENCES

Alley, R. B., D. D. Blankenship, S. T. Rooney, and C. R. Bentley, Sedimentation beneath ice shelves—The view from Ice Stream B, *Mar. Geol.*, 85, 101–120, 1989.

Alonso, B., J. B. Anderson, J. I. Díaz, and L. R. Bartek, Pliocene-Pleistocene seismic stratigraphy of the Ross Sea: Evidence for multiple ice sheet grounding episodes, in *Contributions to Antarctic Research III, Antarct. Res. Ser.*, vol. 57, edited by D. H. Elliot, pp. 93–103, AGU, Washington, D. C., 1992.

Anderson, J. B., The Antarctic continental shelf: Results from recent marine geologic and geophysical investigations, in *The Geology of Antarctica*, edited by R. Tingey, pp. 285–326, Oxford University Press, New York, 1991.

Anderson, J. B., and L. R. Bartek, Cenozoic glacial history of the Ross Sea revealed by intermediate resolution seismic reflection data combined with drill site information, in *The Antarctic Paleoenvironment: A Perspective on Global Change, Part One, Antarct. Res. Ser.*, vol. 56, edited by J. P. Kennett and D. A. Warnke, pp. 231–263, AGU, Washington, D. C., 1992.

Anderson, J. B., and B. F. Molnia, *Glacial-Marine sedimentation Short Course in Geology*, vol. 9, 127 pp., AGU, Washington, D. C., 1989.

Anderson, J. B., C. Brake, E. Domack, N. Myers, and R. Wright, Development of a polar glacial-marine sedimentation model from Antarctic Quaternary deposits and glaciological information, in *Glacial-Marine Sedimentation*, edited by B. F. Molnia, pp. 233–264, Plenum, New York, 1983.

Anderson, J. B., P. G. Pope, and M. A. Thomas, Evolution and hydrocarbon potential of the northern Antarctic Peninsula continental shelf, in *Antarctica as an Exploration Frontier—Hydrocarbon Potential, Geology, and Hazards*, edited by B. St. John, pp. 1–12, American Association of Petroleum Geologists, Tulsa, Okla., 1990.

Barker, P. F., et al., Leg 113, *Proc. Ocean Drill. Program Initial Rep.*, 113, 785 pp., 1988.

Barrett, P. J., *Antarctic Cenozoic History From the CIROS-1 Drillhole, McMurdo Sound*, 254 pp., Science Information Publishing Center, Wellington, New Zealand, 1989.

Barron, J., et al., Leg 119, *Proc. Ocean Drill. Program Initial Rep.*, 119, 942 pp., 1989.

Barron, J., B. Larsen, and J. G. Baldauf, Evidence for late Eocene to early Oligocene Antarctic glaciation and observations on late Neogene glacial history of Antarctica: Results from Leg 119, *Proc. Ocean Drill. Program Sci. Results, 119*, 869–891, 1991.

Bartek, L. R., P. R. Vail, M. R. Ross, P. A. Emmet, C. Liu, and S. Wu, The effect of Cenozoic ice sheet fluctuations in Antarctica on the stratigraphic signature of the Neogene, *J. Geophys. Res.*, 96, 6753–6778, 1991.

Behrendt, J. C., Multichannel seismic reflection surveys over the Antarctic continental margin relevant to petroleum resource studies, in *Antarctica as an Exploration Frontier—Hydrocarbon Potential, Geology, and Hazards*, edited by B. St. John, pp. 69–76, American Association of Petroleum Geologists, Tulsa, Okla., 1990.

Berger, P., G. Brancolini, C. De Cillia, C. Gantar, A. Marchetti, D. Nieto, and R. Ramella, Acquisition, processing and preliminary results of the Antarctic 1987–88 geophysical survey, in *Proceedings of the 2nd Meeting, Earth Science in Antarctica, Siena 27–28 September, 1988*, pp. 341–379, University of Siena, Siena, Italy, 1989.

Bleil, U., and J. Thiede (Eds.), *Geological History of the Polar Oceans: Arctic Versus Antarctic*, 823, pp., Kluwer Academic, Hingham, Mass., 1990.

Cathles, L. M. (Ed.), *The Viscosity of the Earth's Mantle*, 386 pp., Princeton University Press, Princeton, N. J., 1975.

Christie-Blick, N., G. S. Mountain, and K. G. Miller, Seismic stratigraphic record of sea-level change, in *Sea-Level Change*, edited by R. R. Revelle, pp. 116–140, National Academy Press, Washington, D. C., 1990.

Cochrane, G. R., and A. K. Cooper, Sonobuoy seismic studies at ODP drill sites in Prydz Bay, Antarctica, *Proc. Ocean Drill. Program Sci. Results, 119*, 27–43, 1991.

Cochrane, G. R., and A. K. Cooper, Modeling of Cenozoic stratigraphy in the Ross Sea using sonobuoy seismic refraction data, in *Recent Progress in Antarctic Earth Science*, edited by Y. Yoshida, K. Kaminuma, and K. Shiraishi, pp. 619–626, Terra Scientific, Tokyo, 1992.

Cooper, A. K., and P. N. Webb (Eds.), *International Workshop on Antarctic Offshore Acoustic Stratigraphy (ANTOSTRAT): Overview and Extended Abstracts*, U.S. Geological Survey, Menlo Park, Calif., 1990.

Cooper, A. K., and P. N. Webb, International offshore studies on Antarctic glaciation and sea-level change: The ANTOSTRAT Project, in *Recent Progress in Antarctic Earth Science*, edited by Y. Yoshida, K. Kaminuma, and K. Shiraishi, pp. 655–660, Terra Scientific, Tokyo, 1992.

Cooper, A. K., P. J. Barrett, K. Hinz, V. Traube, G. Leitchenkov, and H. M. J. Stagg, Cenozoic prograding sequences of the Antarctic continental margin: A record of glacio-eustatic and tectonic events, *Mar. Geol.*, 102, 175–213, 1991a.

Cooper, A. K., H. Stagg, and E. Geist, Seismic stratigraphy and structure of Prydz Bay, Antarctica: Implications from ODP Leg 119 drilling, *Proc. Ocean Drill. Program Sci. Results, 119*, 5–25, 1991b.

Cooper, A. K., F. J. Davey, and K. Hinz, Crustal extension and origin of sedimentary basins beneath the Ross Sea and Ross Ice Shelf, Antarctica, in *Evolution of Antarctica*, edited by M. R. A. Thomson, J. A. Crame, and J. W. Thomson, pp. 285–292, Cambridge University Press, New York, 1991c.

Cooper, A. K., et al., A SCAR seismic data library system for cooperative research: Summary report of the international workshop on Antarctic seismic data, Oslo, Norway, April 11–15, 1991, 20 pp., Scott Polar Res. Inst., Cambridge, United Kingdom, 1992.

Davey, F. J., The Antarctic margin and its possible hydrocarbon potential, *Tectonophysics*, 114, 443–470, 1985.

Eittreim, S. L., and G. L. Smith, Seismic sequences and their distribution on the Wilkes Land margin, in *Antarctic Continental Margin: Geology and Geophysics of Offshore Wilkes Land*, edited by S. L. Eittreim and M. A. Hampton, pp. 15–43, Circum-Pacific Council for Energy and Mineral Resources, Houston, Tex., 1987.

Eittreim, S. L., A. K. Cooper, and J. Wannesson, Seismic stratigraphic evidence of ice sheet advances on the Wilkes Land margin of Antarctica, Paleoceanography and Environments of Exploration of the Indian Ocean Basins and Margins, spec. vol., *Sediment. Geol.*, in press, 1993.

Hambrey, M. J., W. U. Ehrmann, and B. Larsen, Cenozoic glacial record of the Prydz Bay continental shelf, East Antarctica, *Proc. Ocean Drill. Program Sci. Results*, *119*, 77–132, 1991.

Hambrey, M. J., P. J. Barrett, W. U. Ehrmann, and B. Larsen, Cenozoic sedimentary processes on the Antarctic continental margin and the record from deep drilling, in *Proceedings of the ICG Symposium No. 5: Glacial and Polar Geomorphology*, edited by M. G. Marcus, H. M. French, and G. Stablein, pp. 73–99, Gebruder Borntraeger, Berlin, 1992.

Haq, B., J. Hardenbol, and P. R. Vail, Chronology of fluctuating sea levels since the Triassic, *Science*, *235*, 1156–1167, 1987.

Haugland, K., Y. Kristoffersen, and A. Velde, Seismic investigations in the Weddell Sea embayment, *Tectonophysics*, *114*, 293–315, 1985.

Hayes, D. E. (Ed.), *Marine Geological and Geophysical Atlas of the Circum-Antarctic to 30°S*, Antarct. Res. Ser., vol. 54, AGU, Washington, D. C., 1991.

Hayes, D. E., and L. A. Frakes, General synthesis: Deep Sea Drilling Project 28, *Initial Rep. Deep Sea Drill. Proj.*, *28*, 919–942, 1975.

Hinz, K., and M. Block, Results of geophysical investigations in the Weddell Sea and in the Ross Sea, Antarctica, in *Proceedings of the Eleventh World Petroleum Congress*, pp. 279–291, John Wiley, New York, 1984.

Hinz, K., and Y. Kristoffersen, Antarctica—Recent advances in understanding of the continental shelf, *Geol. Jahrb. Reihe E*, *37*, 1–54, 1987.

Jeffers, J. D., and J. B. Anderson, Sequence stratigraphy of the Bransfield Basin, Antarctica: Implications for tectonic history and hydrocarbon potential, in *Antarctica as an Exploration Frontier—Hydrocarbon Potential, Geology, and Hazards*, edited by B. St. John, pp. 13–29, American Association and Petroleum Geologists, Tulsa, Okla., 1990.

Kennett, J. P., and J. A. Barron, Introduction, in *The Antarctic Paleoenvironment: A Perspective on Global Change, Part One*, Antarct. Res. Ser., vol. 56, edited by J. P. Kennett and D. A. Warnke, pp. 1–6, AGU, Washington, D. C., 1992.

King, L. H., K. Rokoengen, G. B. J. Fader, and T. Gunleiksrud, Till-tongue stratigraphy, *Geol. Soc. Am. Bull.*, *103*, 637–659, 1991.

Kuvaas, B., and Y. Kristoffersen, The Crary Fan: A trough-mouth fan on the Weddell Sea continental margin, Antarctica, *Mar. Geol.*, *97*, 345–362, 1991.

Kuvaas, B., and G. Leitchenkov, Glaciomarine turbidite and current controlled deposits in Prydz Bay, Antarctica, *Mar. Geol.*, *108*, 367–383, 1992.

Larter, R. D., and P. F. Barker, Seismic stratigraphy of the Antarctic Peninsula Pacific margin: A record of Pliocene-Pleistocene ice volume and paleoclimate, *Geology*, *17*, 731–734, 1989.

Larter, R. D., and P. F. Barker, Neogene interaction of tectonic and glacial processes at the Pacific margin of the Antarctic Peninsula, in *Sedimentation, Tectonics and Eustacy*, edited by D. I. M. MacDonald, pp. 165–186, Blackwell, Oxford, United Kingdom, 1991.

Larter, R. D., and A. P. Cunningham, The depositional pattern and distribution of glacial-interglacial sequences on the Antarctic Peninsula Pacific margin, *Mar. Geol.*, *109*, 203–219, 1993.

Leckie, R. M., and P. N. Webb, Late Oligocene–early Miocene glacial record of the Ross Sea, Antarctica: Evidence from DSDP site 270, *Geology*, *5*, 578–582, 1983.

Miller, H., J. P. Henriet, N. Kaul, and A. Moons, A fine-scale stratigraphy of the eastern margin of the Weddell Sea, in *Geological History of the Polar Oceans: Arctic Versus Antarctic*, edited by U. Bleil and J. Thiede, pp. 131–161, Kluwer Academic, Hingham, Mass., 1990.

National Oceanic and Atmospheric Administration, Environmental data inventory for the Antarctic area, 53 pp., Washington, D. C., 1984.

Posamentier, H. W., and P. R. Vail, Eustatic controls on clastic deposition II—Sequence and systems tract models, in *Sea-Level Changes: An Integrated Approach, Spec. Publ. 42*, edited by C. K. Wilgus, B. S. Hastings, C. G. St. C. Kendall, H. W. Posamentier, C. A. Ross, and J. C. Van Wagoner, pp. 125–154, Society of Economic Paleontologists and Mineralogists, Tulsa, Okla., 1988.

Solheim, A., C. F. Forsberg, and A. Pittenger, Stepwise consolidation of glacigenic sediments related to the glacial history of Prydz Bay, East Antarctica, *Proc. Ocean Drill. Program Sci. Results*, *119*, 169–184, 1991.

Stagg, H. M. J., Structure and origin of Prydz Bay and Mac. Robertson Shelf, East Antarctica, *Tectonophysics*, *114*, 315–340, 1985.

ten Brink, U. S., and A. K. Cooper, Factors affecting the characteristic bathymetry of Antarctic continental margins: Preliminary modeling results from Prydz Bay, in *International Workshop on Antarctic Offshore Acoustic Stratigraphy (ANTOSTRAT): Overview and Extended Abstracts*, edited by A. K. Cooper and P. N. Webb, pp. 274–277, U.S. Geological Survey, Menlo Park, Calif., 1990.

ten Brink, U. S., and A. K. Cooper, Modeling the bathymetry of the Antarctic continental shelf, in *Recent Progress in Antarctic Earth Science*, edited by Y. Yoshida, K. Kaminuma, and K. Shiraishi, pp. 763–771, Terra Scientific, Tokyo, 1992.

Van Wagoner, J. C., H. W. Posamentier, R. M. Mitchum, P. R. Vail, J. F. Sarg, T. S. Loutit, and J. Hardenbol, An overview of the fundamentals of sequence stratigraphy and key defi9nitions, in *Sea-Level Changes: An Integrated Approach, Spec. Publ. 42*, edited by C. K. Wilgus, B. S. Hastings, C. G. St. C. Kendall, H. W. Posamentier, C. A. Ross, and J. C. Van Wagoner, pp. 39–47, Society of Economic Paleontologists and Mineralogists, Tulsa, Okla., 1988.

Wannesson, J., Geology and petroleum potential of the Adelie Margin, East Antarctica, in *Antarctica as an Exploration Frontier—Hydrocarbon Potential, Geology, and Hazards*, edited by B. St. John, pp. 77–87, American Association of Petroleum Geologists, Tulsa, Okla., 1990.

Webb, P. N., A review of the Cenozoic stratigraphy and paleontology of Antarctica, in *Geological Evolution of Antarctica*, edited by M. R. A. Thomson, J. A. Crame, and J. W. Thomson, pp. 599–608, Cambridge University Press, New York, 1991.

Webb, P. N., and D. M. Harwood, Late Cenozoic history of the Ross Embayment, Antarctica, *Quat. Sci. Rev.*, *10*, 215–223, 1991.

Zayatz, I., M. Kavun, and V. Traube, The Soviet geophysical research in the Ross Sea, in *International Workshop on Antarctic Offshore Acoustic Stratigraphy (ANTOSTRAT): Overview and Extended Abstracts*, edited by A. K. Cooper and P. N. Webb, pp. 283–290, U.S. Geological Survey, Menlo Park, Calif., 1990.

(Received March 27, 1992;
accepted December 27, 1992.)

THE ANTARCTIC PALEOENVIRONMENT: A PERSPECTIVE ON GLOBAL CHANGE
ANTARCTIC RESEARCH SERIES, VOLUME 60, PAGES 91–124

CENOZOIC SEDIMENTARY AND CLIMATIC RECORD, ROSS SEA REGION, ANTARCTICA

MICHAEL J. HAMBREY

School of Biological and Earth Sciences, Liverpool John Moores University, Liverpool L3 3AF, United Kingdom

PETER J. BARRETT

Antarctic Research Centre, Victoria University of Wellington, Wellington, New Zealand

The Ross Sea region extends deep into the heart of the Antarctic continent and has received thick accumulations of sediment through Cenozoic times following crustal extension and rifting as the Ross Sea opened. Drilling these sediments since the early 1970s has led to major advances in understanding the history of the Antarctic ice sheet and its role in global climatic and sea level changes. Glaciation in this part of the Antarctic is now known to have begun at least by earliest Oligocene time (36 Ma), with periods of expanded ice cover in late Oligocene, late Miocene, and latest Pliocene times, the latter involving the formation of a polar ice sheet. Major deglacial phases are well established for many intervening periods. Cored lithologies suggest that from early Oligocene to the late Pliocene or even early Pleistocene, the Antarctic ice sheet appears to have been largely temperate, more extensive and more dynamic than the present polar ice sheet. Cores from coastal Victoria Land reveal Pliocene ice advancing periodically from the Transantarctic Mountains and Pleistocene ice grounded across the Ross Sea in response to expansion of the East Antarctic and West Antarctic ice sheets. Outcrops of glacial Sirius Group in the Transantarctic Mountains bear late Pliocene marine diatoms, considered to have been transported from the Antarctic interior and in situ leaf and pollen remains of *Nothofagus*, implying a major recession of the ice sheet at that time. The Quaternary record both onshore and offshore is thin and incomplete, as a result of erosion by successive advances of grounded ice across the shelf. Comparison with drill hole and terrestrial evidence from other parts of the Antarctic continental margin and the deep ocean suggests that the sediments are recording continent-wide ice sheet and climatic fluctuations. Further drilling on the Antarctic margin should seek to extend the climatic history back beyond the Oligocene and to provide more detail and better chronology for critical periods such as the Pliocene and the last glacial cycle.

INTRODUCTION

The Antarctic ice sheet is one of Earth's major physical features, and its history and stability are a matter of continuing interest and contention as recent reviews have indicated [e.g., *Clapperton and Sugden*, 1990; *Webb and Harwood*, 1991; *Wise et al.*, 1991]. Today the ice sheet comprises some 30,000,000 km^3, which, if it all melted, would raise sea level by about 70 m [*Sahagian*, 1987]. The ice sheet has a strong influence on global atmospheric circulation as it represents the major high-altitude topographic heat sink for the atmosphere [*Cattle*, 1991]. The ice sheet also influences oceanic circulation through production of Antarctic Bottom Water, the deepest and coldest in the oceans, coming from the enrichment of brine in seawater as sea ice forms around the continent each winter [*Wadhams*, 1991]. With current concern for possible changes in the Earth's climate, the behavior of the ice sheet and associated phenomena, both short and long term, are particularly important in understanding the Earth-ocean-atmosphere-ice system. An important contribution to that understanding is coming from the documentation of advance and retreat phases of Antarctic ice cover through geological time and associated changes in sea level.

Here we review what is currently known of the history of the Antarctic ice sheet over the last 50 million years or so from the stratal record of the Ross Sea region (Figure 1) concentrating on the offshore portion revealed in seismic surveys and drill holes. The region is well placed for providing a record of the Antarctic ice sheet, as it includes both land and continental shelf that extends closest to the middle of the continent. Thus such questions as the timing and extent of ice grounding across the shelf may be investigated. The review is timely, not only because of recent drilling success, but also because our knowledge of glacial-marine sedimentation in both temperate and polar settings has advanced considerably [*Drewry*, 1986; *Dowdeswell*, 1987; *Powell*

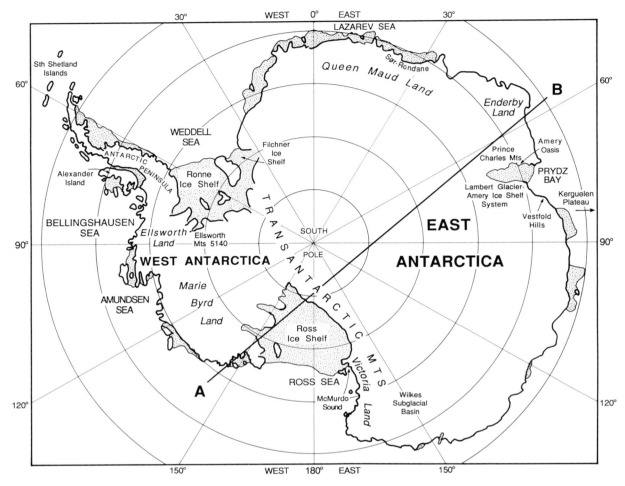

Fig. 1.   Map of Antarctic place names referred to in the text. Line AB is the section for the ice sheet history in Figure 14.

and *Elverhøi*, 1989; *Hambrey et al.*, 1992; *Anderson and Ashley*, 1991]. This paper complements that of *Anderson and Bartek* [1992], who focus on seismic stratigraphy of the Ross continental shelf.

GLACIOLOGICAL, BATHYMETRIC, AND GEOLOGICAL SETTING OF THE ROSS EMBAYMENT

Glaciological Setting

The Antarctic ice sheet behaves not as a single ice mass, but as two: the East Antarctic Ice Sheet and the West Antarctic Ice Sheet, and large parts of both discharge through the Ross Embayment (Figure 2). The much larger East Antarctic Ice Sheet has developed on a continental landmass that even now, despite considerable crustal depression, is mostly above sea level. Flow into the Ross Embayment is largely restrained by the Transantarctic Mountains, rising in several places to

more than 4000 m. However, a number of ice streams exceeding 600 m in thickness drain into the Ross Ice Shelf, including the Reedy, Scott, Beardmore, Nimrod, and Byrd glaciers. To the north, glaciers of similar size, such as the David and Campbell glaciers, drain directly into the Ross Sea and extend beyond the mountain confines as unconstrained, floating ice tongues.

The West Antarctic Ice Sheet, by contrast, is marine based, faster flowing, and inherently much less stable than the land-based East Antarctic Ice Sheet [*Mercer*, 1978; *Bindschadler*, 1991]. At present, ~2/3 of the total ice frontage of the Ross Ice Shelf is derived from West Antarctica (Figure 2). This is largely because the West Antarctic Ice Sheet is unconstrained by topography, discharging mostly as five major ice streams (named A, B, C, D, and E), of the order of 0.8 km thick and 30–60 km across, into the Ross Ice Shelf [*Robin*, 1975]. The ice streams flow at rates of up to 500–800 m/yr, sliding on a wet base of deforming subglacial debris, but Ice

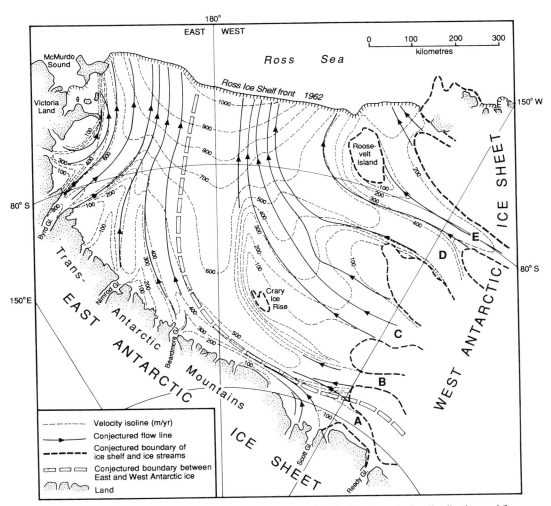

Fig. 2. Present-day glaciological regime of the Ross Ice Shelf, showing velocity distribution and flow lines. Letters A, B, C, D, and E refer to the five major ice streams that feed into the Ross Ice Shelf from the West Antarctic Ice Sheet [after *Robin*, 1975]. Also depicted is the approximate boundary between East Antarctic and West Antarctic ice.

Stream C is now inactive [*Bentley*, 1987; *Shabtaie and Bentley*, 1987]. Today, most of the ice shelf currently gains mass by accumulation on the surface (0.16–0.30 m/yr) but loses it by bottom melting at rates of up to at least 0.50 m/yr, although southern parts may be subject to basal freeze-on [*Robin*, 1975]. Velocities are generally low midshelf and close to grounded zones (100 m/yr or less), but toward the front the ice accelerates as it thins, reaching speeds in excess of 1000 m/yr (Figure 2).

Fundamental questions still remain on factors controlling the behavior of the West Antarctic ice streams and the ice sheet itself, but the potential for rapid collapse and reformation a number of times in the late Quaternary has been recognized [*MacAyeal*, 1992]. Instability and disappearance of West Antarctic ice within the last 400 kyr are supported by geological

evidence. Late Quaternary marine diatoms have been found in mud from beneath Ice Stream B and over 200 km upstream from the present grounding line [*Scherer*, 1991], which implies marine deposition in the West Antarctic interior and hence the absence of an ice sheet there at some time in that period.

Bathymetry

Like most parts of the Antarctic continental shelf, the bathymetry of the Ross Embayment is irregular but, in general, shallows toward the open sea from 500–1000 m beneath the ice shelf to <300 m near the continental shelf edge (Figure 3). Despite the trend the ice shelf is grounded in two shallow areas, the Crary Ice Rise and Roosevelt Island. Several banks lie seaward of the shelf

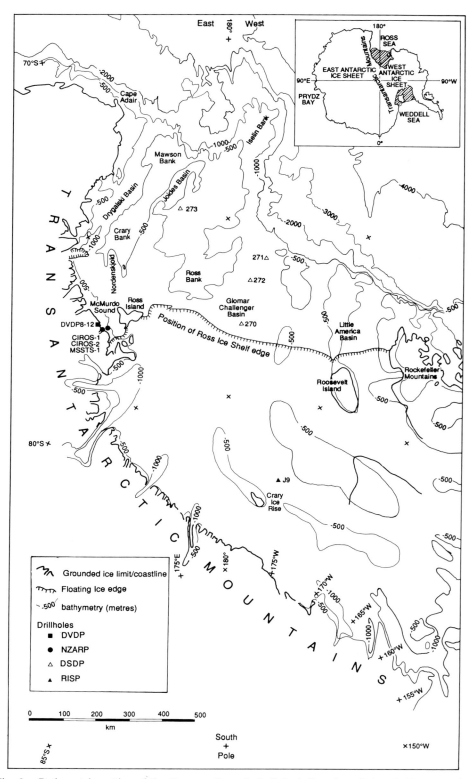

Fig. 3.   Bathymetric setting of the Ross continental shelf, including the sub–ice shelf topography [after *Drewry*, 1983]. Contours are in meters below sea level. Drill holes shown are from the Dry Valley Drilling Project (DVDP), the New Zealand Antarctic Research Programme (NZARP), and the Deep Sea Drilling Project (DSDP). RISP refers to the Ross Ice Shelf Project, which recovered a series of shallow gravity cores from beneath the ice shelf at J9.

separating irregular basins. The deepest is the Drygalski Basin, which attains a depth of more than 1000 m. A number of other troughs more than 1000 m deep extend beyond the edge of the Transantarctic Mountains beneath the ice shelf, turning northward. This trend suggests that grounded ice streams were deflected northward when the troughs were carved. This scenario would require grounded ice in the Ross Sea and hence the presence of both East Antarctic and West Antarctic ice.

Geological Setting

The depressed physiography of the Ross Sea region has resulted from crustal extension over millions of years as part of the West Antarctic Rift System (Figure 4a; *Behrendt et al.* [1991]). Until mid-Mesozoic times the region was the site of slow continental sedimentation [*Barrett*, 1991a], but at ~95 Ma crustal attenuation began, a process associated with the separation of Australia from Antarctica as Gondwanaland began to break up [*Tessensohn and Wörner*, 1991]. The initial phase of extension and sedimentation filled local half-grabens [*Cooper and Davey*, 1985, 1987]. A later more substantial phase of extension and subsidence, accompanied by magmatism and shoulder uplift of the Transantarctic Mountains, was initiated in latest Mesozoic or early Cenozoic time; the timing is quite uncertain but has been estimated to have begun by about 50 Ma [*Tessensohn and Wörner*, 1991]. This led to the formation of three north-south trending sedimentary basins, the Victoria Land basin, the Central Trough, and the Eastern basin [*Davey et al.*, 1982] with sediment thicknesses ranging from 8 to 14 km (Figure 4b). Although the geometry of the basins is now well established from 35,000 km of multichannel seismic profiles [*Behrendt et al.*, 1991], age and lithology are known only from a few drill holes (Table 1). Together these show a complex history of glacial advance and retreat with progressive construction of the shelf in the Ross Embayment at least since Oligocene time [*Bartek et al.*, 1991; *Cooper et al.*, 1991].

The Transantarctic Mountains, like the Ross Sea also a product of crustal extension [*Fitzgerald et al.*, 1986; *Stern and ten Brink*, 1989], mark the western boundary of the rift system, rising on average 5 km through the Cenozoic era from ~60 Ma [*Fitzgerald*, 1989]. The eastern margin of the rift is not well defined, but rifting is evident east of the Ross Embayment through Marie Byrd Land and beyond into the Bellingshausen Sea [*Behrendt et al.*, 1991]. The uplift history, largely determined by fission track analysis, varies somewhat from place to place. In the Scott Glacier area (~86°S), uplift began in late Mesozoic time, with renewed uplift after a period of quiescence from 60 Ma. Similarly, Early Cretaceous uplift is evident in northern Victoria Land, but in the Beardmore Glacier area and southern Victoria

Land only Cenozoic uplift is recorded [*Fitzgerald and Stump*, 1991a]. This uplift is also linked to the separation of Australia and Antarctica, which was relatively slow until early Cenozoic time (55 Ma). Four or five kilometers of erosion are evident along the axis of maximum uplift [*Fitzgerald and Stump*, 1991a]. Lines of evidence such as the high relief onshore and Holocene fault scarps offshore point to recent rapid uplift of the Transantarctic Mountains, in places by several kilometers, in the last 2 or 3 m.y., perhaps forcing a change in the nature of the Antarctic ice sheet, according to *Behrendt and Cooper* [1991].

Basaltic volcanism is a widespread feature of the Ross Embayment, with large edifices such as Mount Melbourne in the Transantarctic Mountains, Mount Erebus in the center of the Victoria Land basin [*Kyle*, 1981], and the volcanoes of Marie Byrd Land [*LeMasurier and Rex*, 1983]. Many more have been identified within the Ross Embayment from magnetic anomalies [*Behrendt et al.*, 1991]. Erebus and Melbourne are still active, and ages from other exposures of volcanic rocks in the Ross Embayment are spread over the last 20 m.y., but older volcanic rocks are likely, as high-velocity strata with volcano geometry have been identified in seismic profiles of the Victoria Land basin [*Cooper and Davey*, 1987]. This volcanism is important in working out glacial history both in allowing direct radiometric dating of ash-bearing strata and of ice-covered or ice-free periods, depending on the character of volcanic deposits.

INTERPRETATION OF SEDIMENTARY SUCCESSIONS

The key to interpretation of sediments in glacial as in other environments is descriptive lithofacies logging [e.g., *Eyles et al.*, 1983]. We have applied a broadly similar approach, along with a consideration of other sedimentary parameters, to the cores from the Ross Sea region. Fuller descriptions of lithofacies are given elsewhere for MSSTS 1 [*Barrett and McKelvey*, 1986], CIROS 1 [*Hambrey et al.*, 1989b], CIROS 2 [*Barrett and Hambrey*, 1992], and ODP drill holes in Prydz Bay [*Hambrey et al.*, 1991]. Interpretations of lithofacies have been based on a mix of conceptual and actualistic models of terrestrial and marine-glacial sedimentation [*Anderson et al.*, 1980, 1991; *Macpherson*, 1987; *Barrett and Hambrey*, 1992; *Hambrey et al.*, 1989b, 1992] and for sea level variation on the classical wave-graded shelf model for nearshore sedimentation [*Elliott*, 1986; *Barrett*, 1989b; *Barrett et al.*, 1990]. A summary of the principal characteristics and interpretation of these lithofacies is given in Table 2.

Some lithofacies are not unique indicators of particular environments of deposition and have had varied interpretations in the past. For example, massive diamictite can be the product of lodgement from grounded ice, continuous rain out of glacial debris close

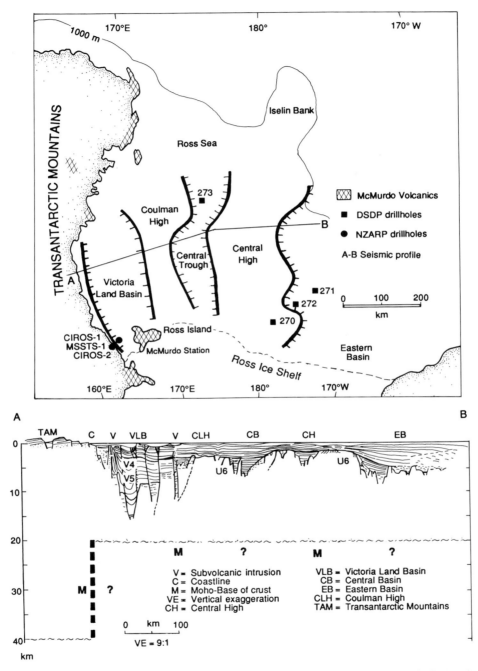

Fig. 4.   Seismically defined structures in the Ross continental shelf. (*a*) Map showing principal basins and highs, together with drill sites [after *Davey et al.*, 1982; *Cooper et al.*, 1987; *Behrendt et al.*, 1991]. (*b*) Geological structure across the Ross continental shelf along the line in Figure 4*a* marked AB. U6 is a widespread unconformity at the base of the late Oligocene [*Hinz and Block*, 1983]. The vertical dashed line represents the Transantarctic Mountain Front, i.e., the change from thick to thin continental crust, as indicated by depth to Moho.

TABLE 1.   Drill Holes of the Ross Continental Shelf and Coastal Victoria Land Which Have Yielded Substantial Data on Cenozoic Sedimentary Successions

Hole	Year	Latitude	Longitude	Elevation, m	Penetration, m	Recovered, %	Oldest Sediment
DSDP 270	1973	77°26.48'S	178°30.19'W	−634	422.5*	62.4	late Oligocene
DSDP 271	1973	76°43.27'S	175°02.86'W	−554	265.0	5.8	Pliocene
DSDP 272	1973	77°07.62'S	176°45.61'W	−629	443.0	36.6	early Miocene
DSDP 273 and 273A	1973	74°32.29'S	174°37.57'E	−495	346.5	24.1	early Miocene
DVDP 10	1974	77°34.72'S	163°30.70'E	+2.8	185.9	83.4	early Pliocene
DVDP 11	1974	77°35.40'S	163°24.67'E	+80	328.0	94.1	late Miocene
DVDP 12	1974	77°38.37'S	162°51.22'E	+75	166*		Pliocene
DVDP 15	1975	77°26.65'S	164°22.82'E	−122	65	52.0	?Pliocene-Pleistocene
MSSTS 1	1979	77°33.43'S	163°23.21'E	−195	229.6	56.1	late Oligocene
CIROS 2	1984	77°41'S	163°32'E	−211	168.1	67.0	early Pliocene
CIROS 1	1986	77°04.91'S	164°29.93'E	−197	702.1	98.0	early Oligocene

*Excludes basement.

to the grounding line, gravity flow recycling of glacial debris, or continuous rain-out from icebergs in a distal glaciomarine setting [*Eyles and Eyles*, 1983; *Forum*, 1984]. However, massive diamictite is now recognized to most likely result from lodgement or grounding line deposition, the rain-out interpretation having now been successfully countered [*Hicock and Dreimanis*, 1989]. Thicknesses of several meters or more increase the likelihood that massive diamictite was deposited subglacially; the interpretation is confirmed by the presence of features such as internal shear surfaces, directed clast fabrics, and overcompaction. *Anderson et al.* [1991] provide modern examples of massive diamictite having been deposited subglacially from cores in front of recently retreating tidewater glaciers and ice shelves.

Still it must be remembered that glacial processes are difficult to study because of the remoteness of the subglacial environment (but see *Powell et al.* [1992]) and the slow accumulation rates. Although we have developed our interpretations from a wide range of evidence and experience and consider them sound, they are not all necessarily definitive.

The Terrestrial Record

Cenozoic terrestrial deposits in the Ross Sea region are limited to the Transantarctic Mountains, with most research to date concentrating in the dry valleys region near McMurdo Sound (Figure 1). This work is reviewed elsewhere [*Clapperton and Sugden*, 1990], but we make some comment on the record of the Sirius Group in view of the differences in scientific opinion on its meaning (see, e.g., *Webb and Harwood* [1991]) and new data from offshore drill cores. The reader is also referred to the Neogene history of Wright Dry Valley, with its alternations of valley glaciation and warmer fjord sedimentation in Neogene times [*Denton et al.*, 1991; *Prentice et al.*, this volume].

The Sirius Group is the name given to the most significant Cenozoic lithological unit on land. The unit was named from Mount Sirius (86°S, Figure 5) by *Mercer* [1972] for a formation of "compact glacial drift" in the Beardmore Glacier area. Its considerable thickness, compactness, and abundant signs of water sorting suggested deposition from temperate ice prior to the present regime. Because of the scattered distribution of outcrops and difficulties in age assignment and correlation, *Mercer* [1981] argued that it was inappropriate to refer to these deposits as a formation. Nevertheless, the name has persisted. The deposits have recently been divided into formations and elevated to group status in the Dominion Range [*McKelvey et al.*, 1991a] but are undifferentiated elsewhere. At the type locality, Mount Sirius (2300 m), the summit area is 100 m of tillite and waterlain sediments capping a Jurassic dolerite [*Mercer*, 1981]. The Sirius Group in most places rests on dolerite or Paleozoic-Mesozoic Beacon Supergroup strata but occasionally on crystalline basement. It is the youngest stratigraphic unit, apart from thin sheets, ribbons, or fans of Quaternary debris.

Sirius outcrops, which extend for over 1500 km along the Transantarctic Mountains (Figure 5), were largely first reported and described by *Mercer* [1972, 1981] and *Mayewski* [1975]. Most outcrops occur above 2000 m and generally form a drape over underlying basement, Beacon Supergroup or Ferrar Dolerite, which commonly is polished, striated, and grooved. Roches moutonnes are developed in places. The diamictite is massive and poorly sorted with boulders up to 1.5 m in a matrix of silt and clay. It is variously described as compact and semilithified. It is a bluff former, but the material can be readily disaggregated in water. Clasts from the diamictite are highly varied, but they are mainly of Ferrar Dolerite, Beacon Supergroup strata, or crystalline basement. Most are subangular and subrounded, and they commonly have striate surfaces.

The thickest recorded succession of the Sirius Group occurs in the Dominion Range in the upper reaches of the Beardmore Glacier [*McKelvey et al.*, 1991a]. The

TABLE 2. Summary of Facies Recovered From Drill Cores in the Ross Embayment and Their Relative Importance and Interpretation [After *Hambrey et al.*, 1992]

FACIES	RELATIVE ABUNDANCE					DESCRIPTION	INTERPRETATION
	DVDP 11	CIROS 2	CIROS 1	DSDP 270	DSDP 272		
Massive diamictite	●●●	●●●	●	●	–	Non-stratified muddy sandstone or sandy mudstone with matrix-supported clasts comprising about 1-20% of rock. Occasional shells and diatoms.	Lodgement till (particularly if having preferred orientation of clast fabric) or waterlain till (random grain orientation).
Weakly stratified diamictite	●●	●●●	●●	●●	●●	As massive diamictite, but with diffuse or wispy stratification. Some bioturbation and slumping. Diatomaceous in part, occasional shells.	Waterlain till to proximal glaciomarine/glaciolacustrine sediment.
Well-stratified diamictite	●	●	●	●	–	As massive diamictite, with prominent but generally discontinuous and commonly contorted stratification. Clasts dispersed with occasional dropstone structures. Significant diatom component and common shells.	Proximal glaciomarine/ glaciolacustrine sediment.
Massive sandstone	●●	●●	●	–	–	Non-stratified, moderately well-sorted to poorly sorted sandstone, commonly with minor mud and gravel component. Some bedding contacts are loaded. Better sorted sands are commonly unconsolidated.	Shoreface with minor ice-rafting in distal glaciomarine setting. Better sorted sands with loaded contacts represent gravity flows on delta slope, and are associated with slumping.
Weakly stratified sandstone	●	●●	●●	●	–	As massive sandstone, but with weak, commonly contorted, irregular, discontinuous, wispy, lenticular stratification. Some brecciation. Some contacts and bioturbation. loaded	Shoreface with minor ice-rafting in distal glaciomarine setting. Better sorted sands are gravity flows and are associated with slumping.
Well-stratified sandstone	●	●	●	–	–	As massive sandstone, but with clear stratification, though commonly contorted.	Shoreface with minor ice-rafting in distal glaciomarine setting. Some post-depositional slumping.
Massive mudstone	–	●	●	●	●	Non-stratified, poorly sorted sandy mudstone with dispersed gravel clasts. Some intraformational brecciation and bioturbation. Dispersed shells and shell fragments.	Offshore with minor ice-rafting in distal glaciomarine setting. Slumping or short-distance debris flowage of unconsolidated material.

Lithofacies	Abundance	Description	Interpretation
Weakly stratified mudstone	• • ••• •• ••	As massive mudstone but with weak discontinuous, sometimes contorted stratification defined by sandier layers. Commonly bioturbated.	Offshore to offshore-transition with minor ice-rafting in distal glaciomarine setting. Subaqueous slumping common.
Well-stratified mudstone	• • •• • –	As massive mudstone, but with discontinuous, well-defined stratification, frequently defined by sandier laminae. Some syn-sedimentary deformation and minor bioturbation.	Offshore-transition with minor ice-rafting in distal glaciomarine setting. Some slumping.
Diatomaceous ooze/diatomite	– – – – –	Weakly or non-stratified siliceous ooze with >60% diatoms. Minor components include terrigenous mud, sand, and gravel.	Offshore with minor ice-rafting in distal glaciomarine setting.
Diatomaceous mudstone	– – • • •••	Massive mud or mudstone with >20% diatoms, and minor sands and gravel.	Offshore with sedimentation predominantly influenced by ice-rafting or underflows in distal glaciomarine setting.
Bioturbated mudstone	– • • • –	As massive mudstone, but stratification highly contorted or almost totally destroyed by bioturbation.	Offshore to offshore-transition with minor ice-rafting. Extensively burrowed.
Mudstone breccia	– • • • –	Non-stratified to weakly stratified, very poorly sorted, sandy mudstone intraformation breccia with up to 70% clasts. Syn-sedimentary deformation and minor bioturbation.	Offshore to offshore-transition slope deposits with minor ice-rafted component, totally disrupted by subaqueous debris flowage.
Breccia	• – • • –	Breccia with sandy matrix	Supraglacial debris dumped from icebergs.
Rhythmite	– • • • –	Graded alternations of poorly sorted muddy sand and sandy mud. Stratification regular on a mm-scale. Some dispersed dropstones.	Subaqueous turbidity underflows derived from subglacial source, with the addition of an ice-rafted component in a proximal glaciomarine setting.
Conglomerate	•• • • • –	Non-stratified to weakly stratified, poorly-sorted, clast-to matrix-supported sandy conglomerate. Both normal and reverse grading evident. Clasts are up to boulder size. Intraclasts of mudstone are frequently incorporated. Loading and other soft-sediment features commonly present.	Proglacial fluvioglacial material or delta-slope debris flows derived directly from subaqueous discharge from glacier. Soft-sediment deformation structures indicate debris flow origin.

••• - greatest abundance; •• - ; • - least abundance; - absent

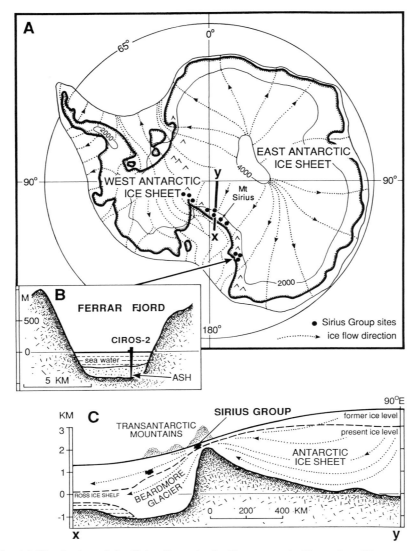

Fig. 5. (a) Distribution of the Sirius Group in the Transantarctic Mountains, (b) a cross section showing the elevation of postulated past and present ice levels, and (c) the flow lines of the present (and most likely past) ice sheets [after *Barrett et al.*, 1992].

sequence, which is horizontally stratified, is divided into two formations with a total thickness of 235 m (Table 3). A feature of the sequence is the occurrence of paleosols, plant stems, and in situ roots at several levels. The material, which includes *Nothofagus*, implies a mean summer temperature of 5°C or more, over 25°C warmer than the present (−20°C). The plant-bearing horizons alternate with massive diamictites, whose fabrics indicate ice flow from the interior and toward the coast by a Beardmore Glacier more extensive and warmer than the present.

Sirius deposits are also reported to contain a varied microfossil assemblage, mostly marine forms, encompassing a number of time periods from Late Cretaceous to Pliocene [*Webb et al.*, 1984; *Harwood*, 1986a]. They include Late Cretaceous and Paleogene foraminifers and diatoms of Oligocene, middle Miocene, late Miocene, early Pliocene, and late Pliocene age [*Harwood*, 1986b; *Webb and Harwood*, 1991]. The diatoms occur both singly and as clasts up to 1 cm across [*Harwood*, 1986a, b]. The assemblage is significant for dating the deposits.

The diatom assemblage also provides an important constraint on the formation of the East Antarctic Ice Sheet, because the microfossils are reported to include planktonic marine diatoms, which from past and present ice flow directions can have come only from the Antarctic interior (Figure 5). The youngest diatoms re-

TABLE 3. Stratigraphy, Facies and Interpretation of the Late Pliocene–?Early Pleistocene Sirius Group in the Beardmore Glacier Area, Central Transantarctic Mountains [After *McKelvey et al.*, 1991a]

Age	Group	Formation	Facies	Interpretation
Late Pliocene to early Pleistocene	Sirius	Mount Mills (50$^+$ m)	Diamictite, poorly sorted breccia, conglomerate, sandstone.	Ablation till, talus and fluvioglacial sediment.
Late Pliocene to early Pleistocene	Sirius	Meyer Desert (185$^+$ m)	Diamictite beds 2 to 7 m thick, with interbeds of laminated siltstone and sandstone with scattered pebbles. *Nothofagus* remains at 3 levels. Paleosols in upper part.	Lodgement till, fluvioglacial and glaciolacustrine sediments representing many phases of glacial advance and retreat.
Devonian-Triassic and Jurassic			*Dominion Erosion Surface* Ferrar Dolerite and Beacon Supergroup	

ported come from six of the seven Sirius localities and include *Actinocyclus actinochilus* and *Nitzschia kerguelensis*, which extend back 3.1 m.y. ago but are still living, and *Thalassiosira vulnifica*, which has a time range from 3.1 to 2.2 m.y. Each of these diatoms is found only at a couple of localities, but *T. lentiginosa* (4.1 m.y. to the present) was extracted from samples of all six localities [*Harwood*, 1986a].

The youthful age of the Sirius diatoms, the perceived antiquity of the Antarctic landscape by geomorphologists, and the varying views on Pliocene ice volume and sea level changes are thorough reconsideration of every aspect of the dating and the glaciological implications [*Sugden*, 1992]. The biostratigraphic ages of Neogene diatom species in the Antarctic region have been confirmed by radiometric dating of ash associated with the late Pliocene Sirius diatoms in drill core from Ferrar Fjord at 2.8 ± 0.3 Ma [*Barrett et al.*, 1992], and the possibility that the diatoms might be wind-blown has also been tested and discounted [*Burckle et al.*, 1988]. Some want better documentation of the biostratigraphic evidence before considering its accommodation with the present body of seemingly contradictory evidence from deep-sea and geomorphological research (e.g., J. P. Kennett, written communication, 1992). We, however, accept the reported occurrences and include a consideration of their implications in the later discussion of climatic history.

The most recent terrestrial deposits in the Transantarctic Mountains differ from Sirius Group strata in forming relatively thin sheets or ribbons of loose debris no more than a few meters thick along the margins of outlet glaciers of the Transantarctic Mountains or sheets on the floors of ice-free valleys. The deposits include sandy tills, lacustrine sediments, and minor amounts of fluvial and eolian sand and gravel and are considered to be late Quaternary in age [*Denton et al.*, 1989a, b]. The dry valleys of the McMurdo Sound region were influenced by at least four major advances of grounded ice from the Ross Sea direction, with a complex interaction between local alpine glaciers and ice moving inland from

the Ross Sea [*Denton et al.*, 1989a; *Moriwaki et al.*, 1991]. Multiple sheets of these deposits are also known from the area of the Beardmore and Hatherton glaciers and may be related to Ross Ice Shelf grounding as far as the continental shelf edge [*Denton et al.*, 1989b].

Other onshore evidence of grounded ice comes from Cenozoic volcanic rocks associated with the West Antarctic Rift System that were erupted subglacially. Rocks of this type, hyaloclastites, provide evidence of an ice sheet thicker than that of today in Marie Byrd Land [*LeMasurier*, 1972]. In northern Victoria Land, subglacial hyaloclastites have been dated as late Miocene (7.3–5.4 Ma) in the neighborhood of the volcanic edifice of Mount Melbourne [*Drewry*, 1981]. Hyaloclastites of Pleistocene age (1.18 Ma) are known from several locations on Ross Island and McMurdo Sound and indicate that the ice became grounded across the continental shelf and was several hundred meters thicker than that of today [*Kyle*, 1981].

Inshore (Fjord) Areas

Sedimentation in fjord settings is represented by CIROS 2, which was drilled in Ferrar Fjord, and by several of the Dry Valley Drilling Project (DVDP) holes in adjacent Taylor Valley, a sediment-filled fjord (Figure 6). The longest core, 328 m, is from DVDP 11 in Taylor Valley. It spans the interval late Miocene to Holocene, but with a major hiatus in the middle Pliocene at 200 meters below seafloor (mbsf). The core has been described and interpreted principally by *McKelvey* [1981], *Powell* [1981], and *Porter and Beget* [1981], with biostratigraphic and paleomagnetic dating refined by *Ishman and Rieck* [1992]. It is dominated by massive diamictite, conglomerate, and pebbly sandstone (Figure 7). According to *Powell* [1981], these sediments were largely deposited subaqueously and are related to fluctuations of the grounding line of an ice shelf, with the diamictites forming by rain-out close to the grounding line (waterlain tills). *Porter and Beget* [1981] noted from microfabric studies that some diamictites, especially those near

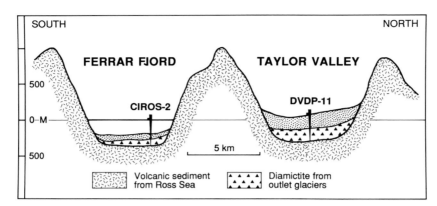

Fig. 6.   Cross section through Ferrar Fjord and Taylor Valley showing the CIROS 2 and Taylor Valley drill holes and the relationship between the cored sequences [after *Barrett and Hambrey*, 1992].

the base of the core, could be basal tills from ice grounding on the floor of the fjord. They further noted from the presence of McMurdo Volcanic Group rocks that the source of the debris for late Pliocene and Pleistocene ice resulted from expansion of the Ross Ice Shelf into the dry valleys.

The CIROS 2 record (Figure 8), drilled 1.2 km east of the terminus of Ferrar Glacier, cored through 165 m of diamictite, sandstone, and mudstone to granitic basement and matches the sequence in Taylor Valley in many respects (Figure 7; *Pyne et al.* [1985]). The lower part, from 165 to 100 mbsf comprises three alternations of thick diamictite and mudstone. Diatom zonation confirmed by radiometric dating of an ash layer at 125 m indicates sedimentation beginning at around 4.5 Ma at the base and extending up to 2.2 Ma at 100 mbsf [*Harwood*, 1986a; *Barrett et al.*, 1992]. The upper 100 m comprises alternations of diamictite and volcanic sand, poorly dated but considered to be Quaternary. The massive clay-rich diamictite units below 100 m we have interpreted as lodgement tills, indicating grounding of temperate ice on the fjord floor. Other diamictites are waterlain tills which are interbedded with proximal and distal glaciomarine and glaciolacustrine sediments. However, the core has no conglomerates like those in Taylor Valley cores, perhaps because the site was in deeper water. The volcanic sand in CIROS 2 is most likely derived from volcanoes to the south and east and blown by the wind onto lake ice before melting through to the floor of the fjord [*Barrett and Hambrey*, 1992], implying a cold climate like the present for the Quaternary of the area.

Nearshore Areas

Drilling beyond the confines of the Victoria Land fjords was intended to uncover the glacial history of the East Antarctic Ice Sheet well back into Cenozoic times by sampling the strata deposited close to its present margin. The first attempt, in 1974, resulted in DVDP 15, 16 km offshore in 122 m of water, but cored to a depth of only 64.6 mbsf with a recovery of less than 50% [*Barrett and Treves*, 1981]. The bulk of the core was well-sorted basaltic sand, like that in the Quaternary part of the CIROS 2 core, and is probably of similar wind-blown origin and age.

Pre-Quaternary strata were cored subsequently by MSSTS 1 to 227 mbsf [*Barrett*, 1986] and then CIROS 1 to 702 mbsf [*Barrett*, 1989a] sited 4 km apart and 12 km offshore (Figure 9), taking the record of glaciation back to the earliest Oligocene. The holes penetrated a sedimentary succession through a glaciomarine delta complex that formed off the coast at the mouth of the ancestral Ferrar Valley. Early Oligocene deepwater foreset beds are overlain by terrestrial to shallow marine topset beds of late Oligocene–early Miocene age. MSSTS 1 penetrated only the topset beds and has a facies association similar to the upper part of CIROS 1, i.e., alternations of diamictite and sandstone or mudstone [*Barrett*, 1986]. However, there are no truly distinctive markers, and correlation is difficult between MSSTS 1 (56% recovery) and CIROS 1 (98% recovery) [*Hambrey et al.*, 1989b, Figure 7]. Strata younger than early Miocene represent a very condensed and poorly dated sequence in both cores, as might be expected from a basin margin.

The early Oligocene sequence from CIROS 1 is 350 m of mainly deepwater mudstone with scattered clasts from floating ice and turbidites, thin sharp-based sandstone, and conglomerate beds [*Hambrey et al.*, 1989b]. The sequence dips seaward at around 10° and is truncated by a 4 m.y. unconformity, on which the late Oligocene–early Miocene sequence of similar thickness rests. The latter offers a relatively detailed record of glacial advance and retreat and sea level change on the East Antarctic margin for the late Oligocene and early Miocene. Core recovery was almost complete (Figure

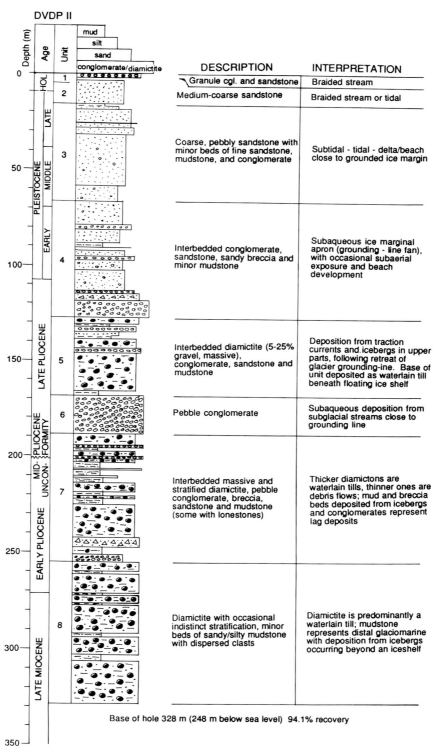

Fig. 7.   Lithological log and interpretation of the sequence cored at DVDP 11 near the mouth of Taylor Valley (lithology after *McKelvey* [1981]; interpretation after *Powell* [1981]).

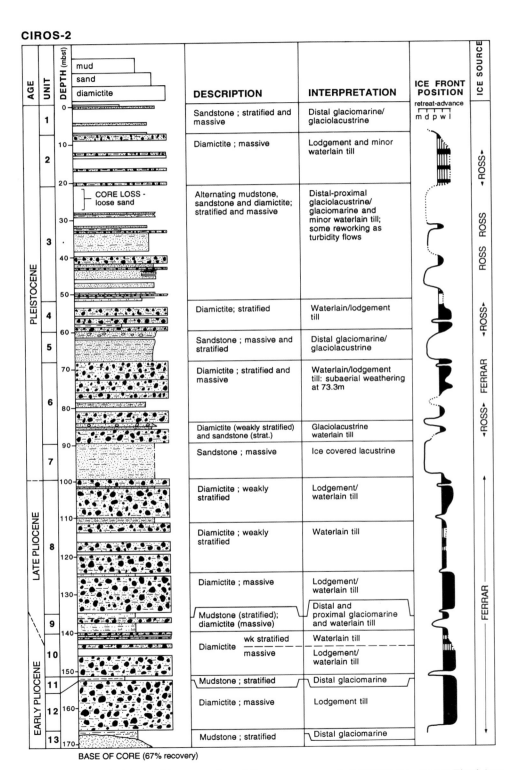

Fig. 8.   Lithological log and interpretation of the sequence cored at CIROS 2 in Ferrar Fjord (see Figure 9 for location).

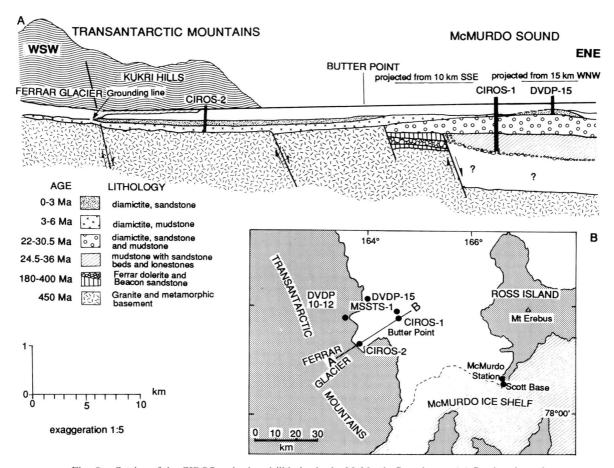

Fig. 9. Setting of the CIROS and other drill holes in the McMurdo Sound area. (*a*) Section through the Transantarctic Mountains down Ferrar Glacier and out into McMurdo Sound, showing the relationships between various bodies of Cenozoic sedimentary strata; controls on age and thickness are based mainly in CIROS 1 and CIROS 2 drill holes [after *Barrett et al.*, 1989]. (*b*) Locations of McMurdo Sound drill holes in relation to the Cenozoic volcano of Ross Island and the Transantarctic Mountains; the line of section in Figure 9*a* is also shown.

10), but the many facies changes represented point to numerous diastems within the sequence [*Barrett et al.*, 1989; *Hambrey et al.*, 1989*b*]. Microfossils, notably diatoms [*Harwood*, 1989] and to a lesser extent benthic foraminifers [*Webb*, 1989], are reasonably common and well preserved and have been combined with the paleomagnetic record [*Rieck*, 1989] to provide a chronology that is perhaps rather optimistically presented with an accuracy of ±0.1 Ma.

The late Oligocene–early Miocene sequence is dominated by seven diamictite units that indicate grounding of ice over the site as well as deposition from a floating glacier tongue. The diamictites are interbedded with sandstone and mudstone units which reflect proximal to distal glaciomarine conditions. The sequence begins at the base with proglacial fluvial conglomerates and continues with a series of marine transgressions and regressions in some but not all cases corresponding to glacial retreats and advances. The frequency of glacial advance and retreat approximates the changes in global ice volume suggested by *Haq et al.* [1987] to account for late Oligocene coastal onlap-offlap cycles and also by *Miller et al.* [1991] for $\delta^{18}O$ events, but chronology is not yet sufficiently refined for confident correlation. These strata have been correlated with widespread reflectors in McMurdo Sound [*Bartek et al.*, 1992] and are probably equivalent to Unit V3 of *Cooper and Davey* [1987] in the western Ross Sea (Figure 4*b*), indicating the regional extent of events described in the core.

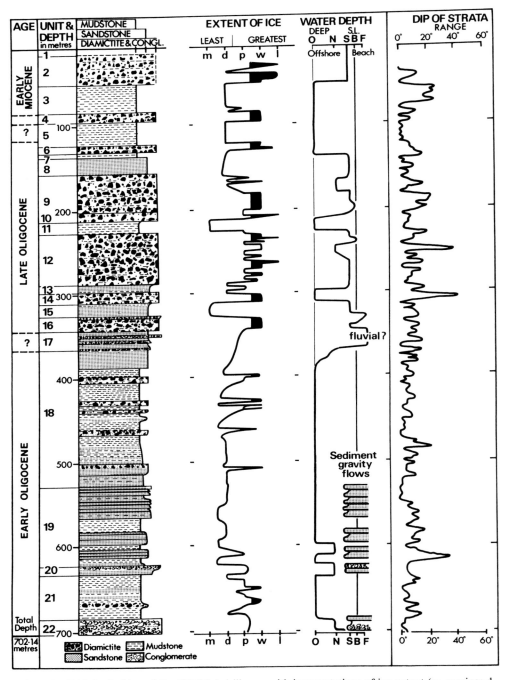

Fig. 10. Lithological log of the CIROS 1 drill core with interpretations of ice extent (m, marine; d, distal; p, proximal; w, waterlain; l, lodgement) and water depth (O, offshore; N, nearshore; S, shoreface; B, beach; F, fluvial) [from *Hambrey et al.*, 1989*b*].

Offshore Areas

The four Deep Sea Drilling Project (DSDP) holes which were drilled on the continental shelf of the central Ross Sea have yielded a record of glacial activity dating back to latest Oligocene time [*Hayes et al.*, 1975].

DSDP sites 270, 272, and 271 lie on a transect running from the southwest near the edge of the Ross Ice Shelf to the northeast, sampling progressively younger strata truncated in Pliocene-Pleistocene times by grounded ice. Site 270 lies on the seismically defined central high,

Site 272 lies at the edge of the eastern basin, and Site 271 is within the basin (Figure 4). Site 273 was drilled in the outer part of the central basin.

The most comprehensive and oldest record comes from Site 270 (Figure 11). The base of the 422-m sequence comprises a sedimentary breccia resting on early Paleozoic marble and calc-silicate gneiss. *Hayes et al.* [1975] interpreted these beds as a talus deposit capped by a paleosol and clearly indicating terrestrial conditions. *Ford* [1991] linked the talus to local uplift of the Transantarctic Mountains around 50 m.y. ago and from geochemical evidence regarded the paleosol as having formed by subaerial weathering. A marine transgression is indicated first by carbonaceous sandstone of probably lagoonal origin, followed by calcareous greensand in which glauconite indicates shallow marine conditions. A K/Ar date of 26 Ma [*McDougall*, 1977] was obtained from the greensand, thereby dating the base of the marine sedimentary sequence as latest Oligocene and soon afterward the first ice at sea level.

The bulk of the sequence consists of mudstone with dispersed gravel clasts, often with a sufficient percentage (1%) for the rock to be classified as a clast-poor diamictite. Chronology from 16 to 25 Ma has been provided by magnetic stratigraphy [*Allis et al.*, 1975] and diatoms and foraminifera [*Leckie and Webb*, 1983]. Both the stratified and nonstratified diamictites in the core were previously interpreted as a product of open water ice rafting rather than subglacial deposition because of their high diatom content [*Hayes et al.*, 1975]. Now we suggest that at least the three intervals of thick massive diamictite are basal tills (Figure 11), though fabric studies need to be undertaken to confirm this. If they are basal tills, then Antarctica several times in the late Oligocene and early Miocene had a more extensive ice cover than today.

Next offshore, Site 272 provides a 443-m section with 37% recovery dating back to the early Miocene but stratigraphically higher than the youngest Tertiary strata in Site 270. *Savage and Ciesielski* [1983] reinvestigated the diatom biostratigraphy and provided revised ages for the sediment, noting a major hiatus between the early and middle Miocene and another before the Pliocene Epoch. The early Miocene part of the core comprises mudstone or diatom-rich mudstone and diatomite, with rare scattered clasts of ice-rafted origin in a distal glaciomarine setting (Figure 12). The middle Miocene part of the core is a diatom-bearing mudstone with dispersed clasts (clast-poor diamictite), suggestive of waterlain till or proximal glaciomarine sediment. Both of the above form part of a prograding sequence. The Pleistocene-Holocene part is flat lying but otherwise similar; this too could be a waterlain till. Compared with Site 270, biogenic productivity was much more pronounced, a reflection of the more distal setting and declining influence of ice. Site 271, on the end of this traverse, yielded only 7% recovery, the oldest sediment

being early Pliocene. As far as could be ascertained, this site has similar diatomaceous sediment to that at Site 272.

Since these Ross Sea sites could have been influenced at different times by either the East Antarctic or West Antarctic ice sheets, establishing the source area of the sediment is of critical importance. *Barrett* [1975b] recorded clast lithologies and found that the majority came from a metasedimentary sequence intruded by granites, such as occurs in Marie Byrd Land today. Typical Transantarctic Mountain lithologies are rare in the eastern Ross Sea cores.

EVIDENCE FOR TEMPERATE OR POLAR ICE SHEETS AND LOCAL OR CONTINENTAL GLACIATION

Temperature is of fundamental importance to the behavior of ice masses [*Drewry*, 1986]. First, ice below the pressure melting point (cold ice) needs higher stresses to deform than ice at the pressure melting point (temperate or warm ice); thus temperate glaciers are more dynamic. Second, significant basal sliding of an ice mass, and thereby subglacial erosion and deposition, will occur only if the basal ice is melting. Third, the release of entrained sediment from the base of an ice mass as melt-out or lodgement till, will occur only if the basal ice is melting. Fourth, sediment distribution and sedimentation rate beneath floating ice shelves, glacier tongues, and icebergs depend on the thermal condition of the ice and the surrounding oceanic waters [*Anderson et al.*, 1991], though the very existence of ice shelves indicates a polar climate and cold ice [*Mercer*, 1972].

Characteristics of temperate and polar glaciomarine sediments have been recently reviewed by *Anderson and Ashley* [1991]. They concluded that there is no clear difference in diamicton facies, representing lodgement or grounding line deposition, deposited from temperate or polar ice. There are, however, major differences in ice marginal and glaciomarine sediments (Table 4) deposited from temperate and polar ice. These differences result largely from lack of meltwater in polar regions, causing low sedimentation rates near the termination of glaciers and in offshore areas. Thus concentrations occur of wind-blown sand off ice-free coasts and of carbonate sand on the outer shelf. However, recycling of interglacial sediment as diamicton by advancing ice may cause polar marine diamictites to be sandier than temperate marine diamictites which are marked by a large component of reworked offshore mud [*Barrett and Hambrey*, 1992].

Although the greater part of today's Antarctic ice sheet is cold, as much as a third may be melting and sliding on its bed and thus temperate at the base [*Zotikov*, 1986]. This trend is due to the increased influence of geothermal heat with thickness. In addition, the ice streams of West Antarctica are now known to be

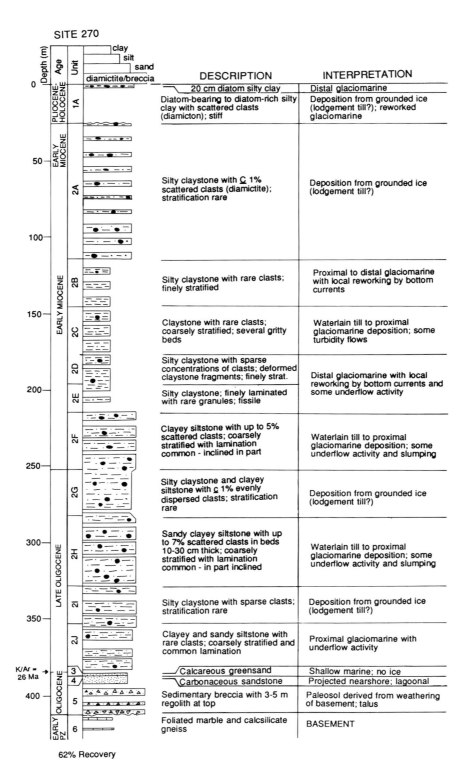

Fig. 11. Lithological log and new interpretation of the sequence cored at DSDP Site 270, central Ross Sea (lithology after *Hayes et al.* [1975]).

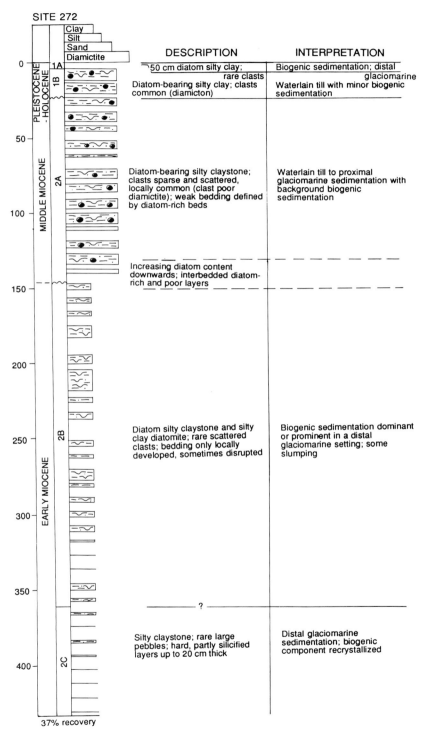

Fig. 12. Lithological log and new interpretation of the sequence cored at DSDP Site 272, central Ross Sea (lithology after *Hayes et al.* [1975]).

TABLE 4.   Facies Differences Between Temperate and Polar Glacial Sediments [After *Anderson and Ashley*, 1991]

Environment	Temperate Facies	Polar/Subpolar* Facies
Subglacial deposition†	diamicton	nil or diamicton under thick ice
Ice marginal terrestrial	conglomerate, sand, silt	nil or minor ablation till or gravel and sand
Fjord	terrigenous mud, minor sand and gravel	sand or biogenic mud
Shelf	terrigenous mud, dropstones rare	biogenic mud, shells and dropstones common

*Temperate summer temperature >10°C; ice at pressure melting. Polar summer temperature <0°C; subpolar summer temperature <10°C; temperature of basal ice <0°C.

†Although Anderson and Ashley report no differences in polar and temperate diamictites, *Barrett* [1989*b*] noted the sandy character of first-cycle polar glacial debris when compared with older debris in the region or with products of the Laurentide ice sheet, presumed to be temperate.

wet based [*Alley et al.*, 1989], and other large ice streams are likely to be also. Therefore subglacial deposition under both present and past polar regimes is quite probable and not distinguishable from temperate subglacial deposition. Discrimination between polar and temperate glaciation must be made on the features described above, along with isotopic measures and paleontological estimates of temperature.

The sedimentary record also allows us to make some judgments on the extent of Antarctic ice cover in the past, as well as its nature. These judgments must be made largely on the distribution of facies, however, rather than their character. Certainly diamictons deposited beneath continental ice sheets will have more varied clast lithologies than those beneath local ice caps, but their texture and fabric will show no critical differences. With our present knowledge the extent of past ice cover at present is best gauged from the record of glacial advance and retreat at various sites both around the Antarctic margin and beneath the ice sheet itself. Basal till at sites beyond the present ice margin indicate more extensive ice at the time of deposition (for example, in the Ross Sea and Prydz Bay in the late Oligocene), and marine sediments beneath the present ice sheet indicate open water at that location in the past (for example, the mud with late Quaternary diatoms beneath the West Antarctic Ice Sheet [*Scherer*, 1991]).

Quantifying variations in ice extent will require much more data from around the continent, and translating ice extent into ice volume will require both more data and criteria for selecting the appropriate ice sheet model. Despite these limitations we believe the lithofacies record does make an important contribution to glacial history, because it is a direct record of glacial advance and retreat and with chronology can be expected to provide the timing for major climatic events on the continent.

THE CLIMATIC AND GLACIAL RECORD

Pre-Oligocene

No sediments of pre-Oligocene age have been cored in the Ross Sea, but there are indirect indications that the earlier Cenozoic climate was sufficiently warm, at least cold to cool temperate, for chemical weathering on land. First, there is the presence of beidellite in the early Oligocene part of the CIROS 1 core [*Claridge and Campbell*, 1989], which they think came from reworking of podzolised soils formed on the pre-Oligocene land surface. Second, many clasts, particularly those of dolerite, have weathering rims up to several millimeters thick, or red staining along fractures, indicative of surface weathering of the clasts in gravel deposits prior to glacial transport and deposition [*Hall*, 1989, 1990].

Deeper drilling on the Antarctic margin is necessary in order to establish when (or whether) glaciers reached the sea prior to the Oligocene [*Barrett and Davey*, 1992]. However, no marine record is likely to provide unequivocal evidence for the onset of glaciation, since a continental ice sheet may have existed for a considerable time on a continent the size of Antarctica prior to reaching the sea. Certainly ice caps and mountain glaciers may well have existed throughout much of the Phanerozoic Eon [*Hambrey et al.*, 1989*b*], a view supported by the recognition of seasonally ice-rafted boulders in Early Cretaceous strata in Australia, which was then attached to Antarctica [*Frakes and Francis*, 1988].

Early Oligocene

The only early Oligocene record from the Ross Sea region comes from the CIROS 1 drill hole in western McMurdo Sound [*Barrett*, 1989*a*]. The shelf there was well below wave base at this time, the dominant sediments being mudstones with ice-rafted pebbles deposited in a distal glaciomarine environment. Sand turbidites occur sporadically. Waterlain tills at several levels indicate that glacier grounding lines approached but did not cover the site. We speculate that early Oligocene conditions varied from limited glacier cover of the rising Transantarctic Mountains, with tidewater glaciers in fjords producing small icebergs, to a situation where ice extended along a broad front beyond the confines of the mountains a short distance across the western Ross continental shelf (Figure 13*a*).

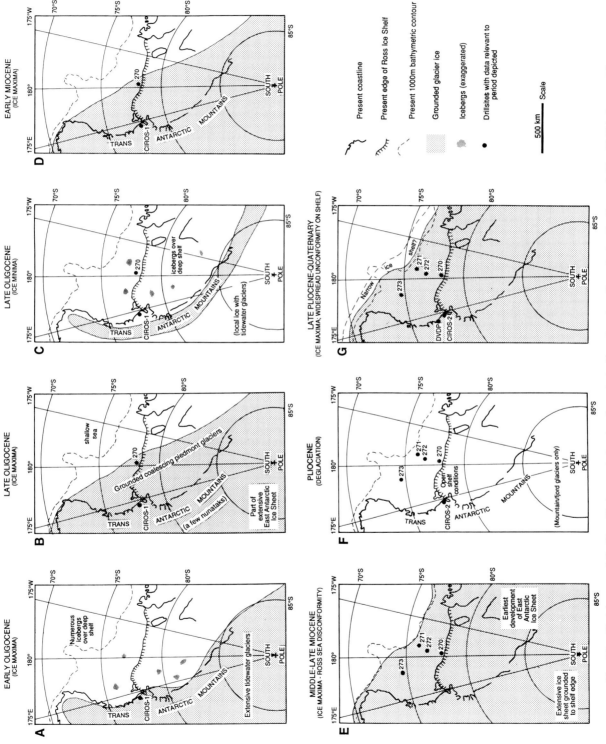

Fig. 13. Sketch maps of the Ross Embayment showing the extent of ice at various times during the Cenozoic Era. Drill holes providing relevant data are also indicated.

Late Oligocene–Early Miocene

Marked marine regression and fluvial deposition with till deposited soon after begin the late Oligocene sequence at the CIROS 1 drill site and follow a period of substantial erosion representing a period of ~4 m.y. The timing of the regression corresponds approximately to that of the largest of the Cenozoic coastal offlaps in the onlap/offlap curve of *Haq et al.* [1987]. A series of marine regressions and transgressions followed through the late Oligocene and earliest Miocene, some corresponding to glacial advance and retreat, suggesting a connection with global sea level change. Correspondence is clearest for the period beginning with grounded ice and shallow water over the site from 29 Ma followed by regression and deposition of deepwater dropstone-free mudstone from 27 to 26.5 Ma followed by shallowing and readvance of the ice from 26.5 to 25.5 Ma. The nearby MSSTS 1 drill core has a similar though more fragmentary record of facies and patterns of sedimentation recording several episodes of glacial advance and retreat with sea level rise and fall in late Oligocene times.

Despite the implications of extensive ice on Antarctica during these periods, both pollen studies of the CIROS 1 core [*Mildenhall*, 1989] and the presence of a *Nothofagus* leaf in the deepwater mudstone [*Hill*, 1989] indicate the persistence of cool temperate vegetation [*Mildenhall*, 1989] on the Victoria Land coast, implying mean annual temperatures of 5°C or more [*Burckle and Pokras*, 1991]. The muddy character of the diamictites and the abundance of mud in the nondiamictite facies also indicates temperate rather than polar glacial sedimentation during this period.

It appears that ice did not reach the middle of the Ross continental shelf until relatively late in the Oligocene, for a thick paleosol, presumably representing early Cenozoic weathering [*Ford*, 1991], survives under shallow marine glauconitic sand at 26 Ma [*McDougall*, 1977] with glaciomarine sediments immediately above [*Barrett*, 1975a]. The subsequent sedimentary record here is one of deepwater deposition largely from floating ice but with diamictite from grounded ice (new interpretation, Figure 11) forming over three intervals in the latest Oligocene and early Miocene. Interglacial periods during at least the early Miocene are recorded by diatomite clasts of this age from a seafloor core near the head of the Ross Embayment at Site J9 [*Harwood et al.*, 1989].

On the basis of the above, we see Antarctic ice cover during this period as varying a number of times from ice sheets of continental dimensions covering most of the Ross Sea continental shelf (Figure 13*b*) to mountain glaciers or local ice caps (Figure 13*c*).

Middle and Late Miocene

A major hiatus separates early Miocene from middle to late Miocene sediments [*Savage and Ciesielski*, 1983]

and suggests a period of extensive erosion, probably by ice, to the edge of the continental shelf. By contrast, diatomite clasts containing middle and late Miocene diatoms in diamictite at Site J9 indicate periods in which the head of the Ross Embayment was ice free [*Harwood et al.*, 1989]. Middle Miocene sediments at Site 273 include diatom-bearing muds with ice-rafted debris, suggesting a distal glaciomarine setting and limited ice at this time. However, approximately equivalent strata at Site 272 include proximal glaciomarine sediments and waterlain tills, which imply an extensive ice cover over much of the Ross Sea.

A late Miocene record has also been obtained from drilling in the dry valleys (DVDP 11). Here diamictites form the oldest sediment cored and may extend to basement. They have been interpreted as waterlain tills [*Powell*, 1981], although fabric studies indicate deposition of some till by lodgement [*Porter and Beget*, 1981]. Other facies include more distal glaciomarine facies. These sediments may represent part of a grounding line fan complex, deposited in the fjord that ultimately filled to create Taylor Valley.

The break between early and late Miocene strata has been termed the Ross Sea disconformity by *Savage and Ciesielski* [1983] and has major significance glaciologically, since it suggests major expansion of the ice sheet. Although its age is poorly constrained, it may be associated with the growth for the first time of the West Antarctic Ice Sheet at about 10 Ma according to *Savage and Ciesielski* [1983] on the basis of West Antarctic-derived clasts in DSDP cores. A combination of East Antarctic and West Antarctic ice reached the edge of the continental shelf at this time (Figure 13*e*) but was still nevertheless subject to major fluctuations.

Pliocene

Evidence of Pliocene climates comes from limited drill hole information and the implications of fossil discoveries described earlier in the Sirius Group in the Transantarctic Mountains. Pliocene polar desert pavements, dated by ash [*Marchant et al.*, 1990], are also considered.

The cores from Taylor Valley and Ferrar Fjord contain a latest Miocene and Pliocene record of the several periods of advance and retreat of glaciers through the Transantarctic Mountains. The diamictite beds in the Taylor Valley cores alternate with gravel and sand considered to be subaqueous outwash; those in Ferrar Fjord alternate with thin mudstone and are presumed to be more distal. The mud, sand, and gravel in the nondiamictite facies and the microfaunas suggest a temperate glacial regime for the Pliocene section of the Taylor Valley cores. Both cores contain a significant proportion of basaltic sand in the upper part, from 202 m in DVDP 11 and from 100 m in CIROS 2. This indicates periodic incursions of Ross Sea ice transporting basaltic debris from volcanoes to the south and east, beginning in the late Pliocene (2.5 Ma in DVDP 11 [*Ishman and Rieck*, 1992] and just under 2.2 Ma in CIROS 2 [*Barrett et al.*, 1992]). We consider the first incursion of Ross Sea ice into the dry valleys to mark the change from temperate to polar ice sheet, as it requires the formation of an ice shelf and then an ice sheet in the Ross Sea.

The discovery of late Pliocene marine diatoms (no older than 3.1 Ma) in high-elevation deposits of the Sirius Group carries the necessary implication that the diatoms were deposited in inland seas in the Antarctic interior in late Pliocene times (2.5–3.1 Ma) [*Webb et al.*, 1984; *Barrett et al.*, 1992]. The extent to which this would require a retreat of the ice is difficult to gauge, but we expect much less ice than the present, perhaps a reduction in ice volume of 50%. A further implication is subsequent erosion of the marine diatoms and deposition in the Transantarctic Mountains by more extensive ice, which some think overrode the Transantarctic Mountains (Figures 6 and 13*g*).

The concept of an expanded "super" ice sheet has been proposed previously by *Mayewski* [1975], who believed that a cold ice sheet with a temperate base covered all of the Transantarctic Mountains during his Queen Maud Glaciation, an event which he argued was never subsequently exceeded. This glaciation was also thought to have carved the dry valleys more than 4.2 m.y. ago, the age of the oldest unglaciated volcanic cones there. Subsequent work reviewed by *Denton et al.* [1991] has provided convincing evidence of glacial overriding events in the form of striated surfaces and erratics on mountains and plateaus at high elevations (3100–4200 m), events which they believe also deposited the diamictites of the Sirius Group. However, a means of directly dating the overriding has yet to be found.

Quaternary

Several distinct phases of expansion of the East Antarctic Ice Sheet from the Ross Sea direction into the Victoria Land dry valleys have been documented from thin spreads of drift and geomorphological evidence. They are known as Ross I, II, III, and IV from youngest to oldest and span the period 1.2 Ma to 12 ka. Late Wisconsin events which, with ^{14}C dating, are the best documented, have been summarized by *Denton et al.* [1989*a*] for the inner Ross Embayment. The dry valleys–McMurdo Sound area has three late Quaternary drift sheets originating alternately from the local Transantarctic Mountain glaciers and the Ross Ice Shelf direction: (1) Marshall Drift (grounding of the Ross Ice Shelf, 180–130 ka), (2) Bonney Drift (advance of Taylor Glacier, 130–90 ka), and (3) Ross Sea Drift (expansion and grounding of the Ross Ice Shelf, 30–10 ka). Final recession of the ice from the area was under way by about 13 ka and was complete by 6600–6020 years ago. Similar drifts occur in the Beardmore and Hatherton

glaciers area [*Denton et al.*, 1989*b*] and imply occasional extension of the Ross Ice Shelf grounding line to the continental shelf edge. Correlation is possible for the Ross Sea drifts but otherwise is difficult.

The grounding of ice to the edge of the Ross Sea continental shelf has recently been recognized in seismic reflection profiles [*Anderson et al.*, 1991; *Bartek et al.*, 1991]. Widespread unconformities and seaward accreting sediments, linked to Pliocene-Pleistocene glacial erosion when sea level dropped, have been identified, although their ages can only be regarded as tentative, as borehole control is limited. The scale of Pliocene-Pleistocene erosion is indicated by the degree of diagenesis in the CIROS 1 core. *Barrett et al.* [1991] estimated that as much as 1 km of strata may have been removed during the Pliocene-Pleistocene advances.

COMPARISON WITH OTHER ANTARCTIC AND SOUTHERN OCEAN SITES

East Antarctica

The only other part of the Antarctic continental shelf that has yielded a long-term record of glaciation is Prydz Bay, selected for Ocean Drilling Program (ODP) Leg 119 as the most likely coastal region of East Antarctica to show the earliest influence of glacier ice. This is because it lies seaward of a long-lived tectonic depression, the Lambert graben, which would have channelled ice from a large part of an early ice sheet to the coast. Today about a fifth of the East Antarctic Ice Sheet drains into Prydz Bay [*Barron et al.*, 1989]. Results from the Prydz Bay drilling included diamictites of early Oligocene age and younger, dominating progradation of the continental shelf and showing that at times ice in this region was more extensive than that of today [*Hambrey et al.*, 1989*a*, 1991].

The Prydz Bay cores also include a major hiatus, apparently lasting from late Oligocene to middle Miocene time. Combined with evidence of ice loading on the underlying sediments [*Solheim et al.*, 1991], it has been argued that this hiatus represents a major advance or series of advances of the East Antarctic Ice Sheet, coinciding with the succession of grounded ice depositional events recorded from CIROS 1 in late Oligocene to early Miocene time [*Hambrey et al.*, 1991] and probably also at DSDP Site 270. The middle-late Miocene hiatus, represented by the Ross Sea disconformity, is interpreted as a period of glacial erosion of the continental shelf and coincides with the gap in the record at Prydz Bay Site 742 (midshelf) and with waterlain till deposition at Site 739 (outer shelf).

In late early Pliocene to early late Pliocene time, it is apparent from Site 742 that there was glacial recession, but not total retreat from coastal areas. However, in the presently ice-free Vestfold Hills, bordering Prydz Bay, high sea levels (75 m) and paleotemperatures have been documented, suggesting major recession of the ice [*Pickard et al.*, 1988]. In the same area, *Quilty* [1991*a*] recorded a mid-Pliocene dolphin in diatomite, dated by diatoms at between 4.2 and 3.5 Ma. In the Larsemann Hills Pliocene faunas and diatoms from 2 to 3 m.y. old have also been reported [*Quilty*, 1991*b*].

These lines of evidence support the Pliocene deglaciation hypothesis, but not total withdrawal of the ice from the Prydz Bay region. In this part of Antarctica, however, major reduction in the size of the East Antarctic Ice Sheet is not incompatible with significant coastal glacier ice according to *Oerlemans'* [1982] model. The area also includes deposits similar to the Sirius Group. They are known informally as the Pagodroma tillites [*Drewry*, 1981; *McKelvey et al.*, 1991*b*] from extensive outcrops in Pagodroma Gorge near Beaver Lake, and they also occur as residual deposits up to 1450 m on the top and flanks of the nearby Fisher Massif. Marine-based grounded ice and other tidewater glacier facies are dominant, and reworked microfossils of Late Cretaceous to late Pliocene age are indicative of a late Pliocene deglacial phase [*McKelvey et al.*, 1991*b*]. In addition, in situ Tertiary mollusks within the Pagodroma tillite at Amery Oasis indicate Pliocene marine sedimentation some 270 km south of the present coastline [*Hart and McKelvey*, 1991]. Geomorphological studies by *Moriwaki et al.* [1991] in the Sør Rondane Mountains of glacigenic sediments and weathering features indicate temperate ice depositing thick till sometime during the Miocene–early Pliocene, followed by deglaciation sometime in the Pliocene–early Pleistocene Epoch.

West Antarctica

The eastern and middle parts of the Ross Sea were influenced by West Antarctic ice from late Miocene time onward [*Savage and Ciesielski*, 1983], as indicated by the DSDP sites. However, a much longer record of glaciation of at least some parts of West Antarctica is evident from intraglacial hyaloclastites, some associated with diamictites, as reviewed by *Burn and Thomson* [1981]. In the Jones Mountains of Ellsworth Land, K/Ar ages between 7 and 24 Ma have been obtained for these deposits, and in Marie Byrd Land ages range from 26 to 29 Ma (late Oligocene) to Quaternary; some sequences may be even older. In Marie Byrd Land the hyaloclastites occur up to 2000 m above the present ice level, but even after allowing for tectonic uplift, this level implies substantially thicker ice at various times over West Antarctica than at present. The volcanic geological record thus provides evidence for a continental ice sheet in late Oligocene time and renewed development of an ice sheet from late Miocene onward [*LeMasurier and Rex*, 1983]. The data are incompatible with the widely held view that the West Antarctic Ice Sheet is a much younger feature than the East Antarctic

Ice Sheet. However, the unstable nature of the West Antarctic Ice Sheet from late Miocene time onward is supported by the volcanic data.

Antarctic Peninsula

The offshore islands of the South Shetlands, beyond the tip of the Antarctic Peninsula, have a much more temperate climate than the Ross Sea but have also yielded a long record of glaciation [*Birkenmajer*, 1988, 1991; *Gazdzicki*, 1989]. The main glacial events are as follows [*Birkenmajer*, 1991]: (1) Krakow Glaciation, glaciomarine sediment of an Antarctic Peninsula (local) provenance, and hyaloclastites; minimum age of 49.4 Ma (middle Eocene) on the basis of a cap of basaltic lava; (2) Polonez Glaciation, lodgement till and glaciomarine sediment of Antarctic continent provenance, capped by andesite-dacite lavas; minimum age of 23.6 Ma (older than early Miocene) on the basis of a cap of andesite-dacite lavas; (3) Legru Glaciation, lodgement till, fluvial sediments, debris flows representing local glaciation; minimum age of 29.5–25.7 Ma (late Oligocene) on the basis of a cap of andesitic lavas; and (4) Melville Glaciation, fossiliferous glaciomarine sediment from an Antarctic continental source; age from fossils of 22–20 Ma (early Miocene).

It was suggested that some of the clasts were derived from East Antarctica, although this is difficult to fit into any glaciological model, nor is it compatible with the deep-sea record. Nevertheless, the South Shetland Islands provide the earliest direct indication of terrestrial ice in Antarctica in having middle Eocene glaciomarine sediments. Evidence of pre-Pleistocene glacial activity in the Antarctic Peninsula is also indicated by hyaloclastites (5–3 Ma) on Alexander Island [*Burn and Thomson*, 1981].

Deep-Sea Record

Various lines of evidence, including the sedimentology, clay mineralogy, paleontology, and isotopic signature of deep-sea sediments, provide a record of climate and ice sheet changes in Antarctica. The reader is referred to other reviews in this volume and also *Ehrmann et al.* [1992].

DISCUSSION

Foci for Initiation and Growth of the Antarctic Ice Sheet

Extensive radio echo soundings over the Antarctic ice sheet have revealed the subglacial topography in broad outline [*Drewry*, 1983]. They show a number of buried mountain ranges which could have provided the necessary foci for the development of the East Antarctic Ice Sheet in addition to those mountains which currently project above the ice surface. However, one cannot be sure that these mountain areas had attained the necessary elevation to initiate glaciation.

Drewry [1975] has examined the possibilities for growth of the East Antarctic Ice Sheet through the Cenozoic Era. Although his conclusion that growth of full-bodied ice sheets postdates early Miocene time is somewhat dated, we believe the style of buildup followed a similar pattern. We suggest that ice buildup takes place in several stages (Figure 14) which have recurred many times since the beginning of the Oligocene Epoch.

Accelerated uplift of the Transantarctic Mountains at around 50 Ma [*Gleadow and Fitzgerald*, 1987], combined with falling global air temperatures, provided the initial conditions for growth of valley and cirque glaciers, fed by cyclonic weather systems entering the Ross Sea, especially on the eastern flanks of the mountains. Ice flowing east calved into the Ross Embayment, while ice flowing down the gentler slopes to the west developed into gradually expanding piedmont glaciers bordering the Wilkes Subglacial Basin, much of which according to the *Drewry* [1983] map of isostatically adjusted bedrock was above sea level (Figure 14a). With continued growth the ice divide moved inland across the crest of the mountains in a manner similar to that which occurred during growth of the Fennoscandian ice sheet (Figure 14b). In a number of places the ice divide was breached, allowing increasingly large volumes of ice to pass eastward through the mountains, although this process may have partly been offset by continuing uplift.

Simultaneously, ice began to build up over the Gamburtsev Subglacial Mountains, although the process there may have been slower because of the drier, more continental setting. Eventually, the ice from these two mountain areas coalesced, filling the intervening basins, and ice began to flow seaward from a point near the center of the continent (Figure 14c). Ice may also have built up over the Prince Charles Mountains and the mountains of Queen Maud Land (if they existed). The breached watersheds in the Transantarctic Mountains, for example, those cut by the Beardmore and Byrd glaciers, provided the main routes for ice to discharge into the Ross Embayment from the inland ice sheet. During peak glacierization, ice extended as a continuous front along the eastern Transantarctic Mountains and across the western part of the Ross continental shelf.

A model of this nature explains how any marine sediments (with diatoms) deposited in the interior basins (such as the Wilkes) during deglacial phases could be transported to high elevations in the Transantarctic Mountains and is further supported by the geomorphological evidence for ice overriding events from pre–early Miocene time onward [*Denton et al.*, 1984].

The West Antarctic Ice Sheet probably developed as a result of expansion of ice from the Transantarctic Mountains into the innermost Ross and Weddell Sea embayments, eventually merging with ice expanding out from Marie Byrd Land and the Ellsworth Mountains. Today, the Ellsworth Mountains include the highest part of Antarctica (5140 m) but are surrounded by deep ice-filled basins. Fission track data suggest that these mountains underwent substantial uplift and denudation in Early Cretaceous time (4 km over 20 million years) [*Fitzgerald and Stump*, 1991b]. Further uplift postdates this early phase but has not yet been resolved. Uplift and denudation occurred during the initial separation of East Antarctica and West Antarctica and continued as the Weddell Sea opened. The Ellsworth Mountains may have been a major growth point for West Antarctic ice long before a full-scale ice sheet developed. Once ice growth had overcome stability problems through accumulating in deep marine basins, it flowed across the Ross and Weddell continental shelves, merging with and deflecting the Transantarctic ice northward, as happens today (Figure 14d).

Role of the Antarctic Ice Sheet in Global Change

It is now widely recognized that global warming will follow increased emission of greenhouse gases by human activity [*Houghton et al.*, 1990]. Carbon dioxide is the most important of these gases, and present trends suggest that an increase in greenhouse gases equivalent to a doubling of CO_2 will occur between 2015 and 2050 A.D., most likely before 2030 [*Department of the Environment, United Kingdom*, 1989]. Forecasts of the temperature rise over the next 40 years, based on global circulation models, range from 0.5° to 2.5°C, the latter figure giving a warming rate more than 10 times higher than anything experienced in the last million years [*Department of the Environment, United Kingdom*, 1989].

The most commonly held perception is that global warming will cause the ice sheets to melt and raise sea level, but this belief is not well founded, and some estimates suggest that warming will increase precipitation and so lower sea level. For example, *Oerlemans* [1982] estimated that the doubling of CO_2 will increase precipitation over Antarctica by 12%, and over the next 250 years this would lower sea level by 30 cm, but this would be countered by a 20-cm rise due to melting of part of the Greenland ice sheet. *Robin* [1986] estimated that changes of sea level due to thickening of the Antarctic ice sheet and melting of the Greenland ice sheet may cancel each other out for a global warming of 3.5°C over the next century. On a longer time scale, *Prentice and Matthews* [1991] have argued a "snow gun" hypothesis, according to which warming increases precipitation and snow accumulation over a cold region.

The short-term data do not take account of the potential instability of the marine-based ice sheets. Rapid deglaciation in West Antarctica might take place

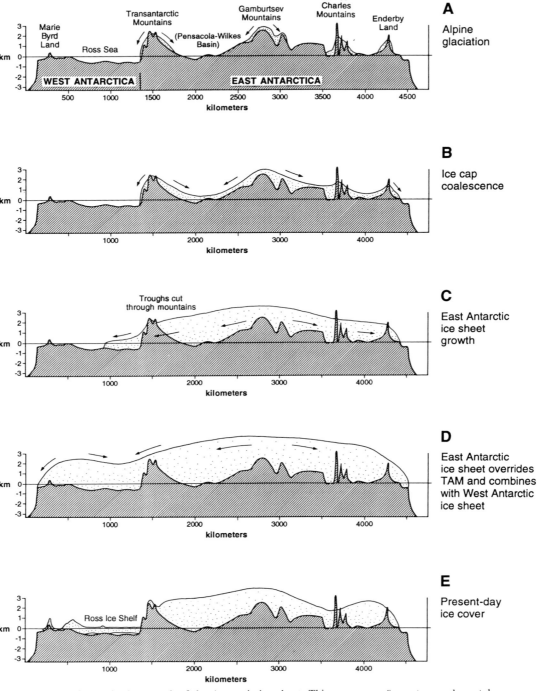

Fig. 14. Stages in the growth of the Antarctic ice sheet. This sequence of events may have taken place many times during the Cenozoic Era prior to the establishment of the present cold ice sheet in late Pliocene or early Pleistocene time. Bedrock and present-day ice sheet profiles are from *Drewry* [1983]. Note that no correction has been made for either tectonic uplift or isostatic loading. Present sea level is indicated for reference only and has little meaning considering the ice volume changes depicted. Ice is lightly stippled; bedrock is striped. Principal ice centers on these profiles include the Transantarctic Mountains, Gamburtsev Subglacial Mountains, and Prince Charles Mountains.

as ocean warming leads to bottom melting of the ice shelves (notably the Ross, Ronne, and Filchner ice shelves). As the grounding lines retreat, rapid sliding could be induced, allowing the ice sheet to collapse. Estimates for the time needed for collapse of the West Antarctic Ice Sheet range from <100 to many hundreds of years, but it is not known whether such a collapse could be started by a global temperature rise of 3.5°C by the end of the next century; even then it is likely that at least 200 years would be necessary to raise sea level by another 5 m [*Robin*, 1986].

Predictions about the future behavior of the Antarctic ice sheet under the influence of global warming, and the associated rise in sea level based on global circulation modeling, need to be counterbalanced with data about real ice sheet behavior in the past. Records of atmospheric composition and temperature come from ice cores for the short term (<200,000 years); these records need to be matched with records of ice extent over this period. For the longer term (>200,000 years), drill holes provide the only data. Those on the continental shelf can offer a direct and specific record of actual ice extent (but not volume), whereas those in the deep sea provide a proxy record of relative ice volume with finer time resolution. Here we are concerned with matching the long-term sedimentary record on the Ross continental shelf with the sea level/temperature curve derived from oxygen isotopes in deep-sea sediments [*Miller et al.*, 1987] and with the global sea level curve derived from sequence stratigraphic analysis by *Haq et al.* [1987] (Figure 15).

Isotopic values on benthic and planktonic foraminifera preserved in the deep oceans reflect variation in size of ice sheets and/or sea water temperature. Opinions differ as to whether the shells with the most reliable guide to ice volume changes are benthic foraminifera [*Miller et al.*, 1991] or tropical surface-dwelling planktonic foraminifera [*Matthews and Poore*, 1980], though some variations, most likely due to ice volume changes, are evident in both. Since northern hemisphere glaciations, as indicated by the first major deep-sea ice rafting in the North Atlantic Ocean [*Shackleton et al.*, 1984] and the North Pacific Ocean [*Rea and Schrader*, 1985], did not begin until 2.4–2.5 Ma, inferred older volume changes almost certainly represent variations in the size of the Antarctic ice sheet alone.

The oxygen isotope record shows that over the last 40 m.y. there has been a stepwise increase in $\delta^{18}O$ values in deep-sea sediments. This was thought to correspond to a decline in temperature and an increase in ice volume [*Shackleton and Kennett*, 1975; *Kennett*, 1977], but not on the scale that we infer from continental shelf sediments. Considerable weight has been placed on the oxygen isotope curve in deriving the glacial record on Earth. For the Quaternary period this curve has proved to be a reliable indicator of changes in global ice volume and sea level [*Williams et al.*, 1988]. However, the curve becomes increasingly difficult to interpret further back in time [*Miller et al.*, 1991]. Some features of the curve do match inferred events on the continental shelf of Antarctica, but others do not. The first major step in Cenozoic cooling near the Eocene/Oligocene boundary coincides with the earliest glacial sediments recovered at CIROS 1, implying major ice buildup at that time, and is further supported by the record from Prydz Bay and the Kerguelen Plateau [*Ehrmann et al.*, 1992] (Figure 15). There is further expansion of ice recorded by upper Oligocene sediments in CIROS 1. The oxygen isotopic record from Maud Rise, Antarctica, contains evidence of some ice growth on East Antarctica during the early to early late Oligocene [*Kennett and Barker*, 1990]. After the middle late Oligocene the isotopic evidence suggests a warming trend, at a time of glacial expansion recorded in CIROS 1. During the early and early middle Miocene, ice expanded and retreated across the Ross continental shelf during a time of relative warmth as inferred from the oxygen isotopic record. The next major indication of ice expansion in the isotope record seems to closely correlate with major ice expansion documented in the development of the Ross Sea disconformity. The inferred development of the West Antarctic Ice Sheet during late Miocene beginning at ~10 Ma and continuing through ~5 Ma is not clearly indicated in the sediment record. A further sharp increase in $\delta^{18}O$ near the early/late Pliocene boundary marks a major development in the northern hemisphere ice sheets (Figure 15). We infer that this also marks the time of development of cold Antarctic ice sheets of the present day.

In some respects, the global sea level curve obtained from seismic sequence stratigraphy by *Haq et al.* [1987] fits the known glacial record on the continental shelf better than the isotope record. From seismic reflection images, Haq et al. have worked out the patterns of transgressions and regressions from the sedimentary strata, and they have been translated into eustatic sea level changes. The resulting curve (Figure 15) illustrates both long-period (10^7–10^8 m.y.) and short-period (10^6 to 5×10^6 m.y.) sea level fluctuations, the former being ascribed to changes in the volume of ocean basins, the latter to global ice volume changes. The global sea level and the isotope curves show several distinct differences. The first indications of a fall associated with known ice sheet development occur in upper Eocene strata, at ~40 Ma, predating the oldest Ross Sea sediments recovered, but coinciding, as far as can be ascertained, with the earliest Prydz Bay sediments cored. A return to the preceding sea level state followed at ~35 Ma, unlike the isotope curve. The next major sea level drop was near the lower/upper Oligocene boundary, not evident in the isotope records, but coinciding with the shallowing and ice expansion documented in CIROS 1. Sea level rise at the end of the Oligocene Epoch was followed by several sharp fluctuations into middle Miocene time, suggesting

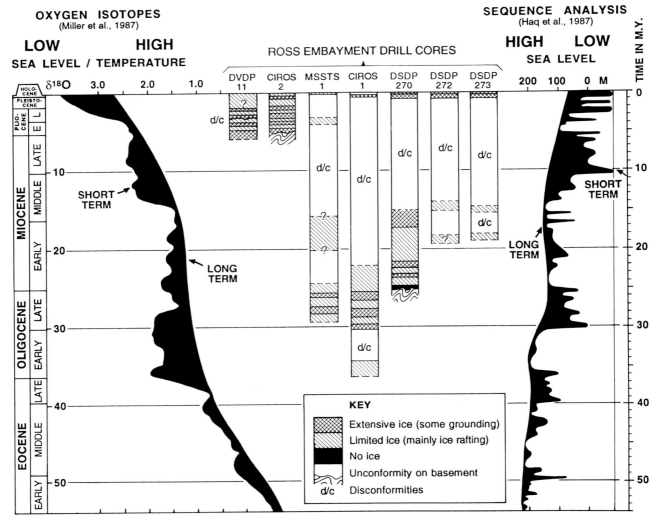

Fig. 15.   Comparison between the Cenozoic sea level/temperature curve based on oxygen isotopic ratios of benthic foraminifera from Atlantic DSDP cores and smoothed with a Gaussian filter [*Miller et al.*, 1987], eustatic sea level changes for the Cenozoic as derived from seismic sequence analysis [*Haq et al.*, 1987], and the stratigraphic record from the Ross continental shelf (adapted from *Barrett* [1991*b*]). Note the broad correspondence between oxygen isotope data and sequence analysis for the last 20 m.y. but significant differences prior to this time. The stratigraphic record from the Ross continental shelf comes from *Ishman and Rieck* [1992] for DVDP 11, *Harwood* [1986*a*] and *Barrett et al.* [1992] for CIROS 2, *Barrett* [1986, 1989] for MSSTS 1 and CIROS 1, *Hayes et al.* [1975] for the lower part of DSDP 270, *Savage and Ciesielski* [1983] for the lower parts of DSDP 272 and 273, and *Kellogg et al.* [1979] for the upper parts of DSDP 270 to 273.

several advances and retreats of the Antarctic ice sheet associated with the time gap in the sedimentary record that is represented by the Ross Sea disconformity. Stepwise falling of sea level thereafter culminated in a sharp fall near the middle/upper Miocene boundary and coinciding with the inferred development of the West Antarctic Ice Sheet. A distinct rise in sea level is evident in the lower Pliocene matching the inferred early Pliocene collapse of the ice sheet of *Webb and Harwood* [1991]. A short-lived phase of early Pliocene low sea

level suggests sharp volume changes of the Antarctic ice sheet. The next major fall occurs near the lower/upper Pliocene boundary at the time of major cryospheric development in the northern hemisphere. This time is also one of inferred ice sheet expansion on land areas bordering the Ross Sea [*Webb and Harwood*, 1991]. Subsequent late Pliocene and Quaternary fluctuations reflect both northern and southern hemisphere ice volume changes, but resolution is inadequate for a detailed comparison. Thus despite problems raised by some authors concerning the magnitude of the sea level changes in the *Haq et al.* [1987] curve [e.g., *Pitman*, 1978; *Summerhayes*, 1986], it does seem to fit our interpretation of the glacial history of the Ross Sea from the Oligocene and early Miocene sedimentary record there better than current interpretation of the isotope record.

CONCLUSIONS AND FUTURE WORK

Twenty years of intermittent drilling activity on the Ross Sea continental shelf and in the dry valleys region, combined with geomorphological and stratigraphic studies in the Transantarctic Mountains, have revolutionized our understanding of the manner in which the East Antarctic and West Antarctic ice sheets have grown and decayed during the last 40 m.y. The principal findings are as follows:

1. Widespread glaciation of East Antarctica was already under way by earliest Oligocene time (36 Ma). Taking into account results from Prydz Bay drilling, ice sheet–scale glaciation is inferred. Accelerated uplift of the Transantarctic Mountains from about 50 Ma may have triggered ice expansion.

2. Many major expansions of East Antarctic ice took place in late Oligocene time and early Miocene time (~30–16 Ma), with grounded ice depositing lodgement tills on the continental shelf following an initial sharp fall in sea level. However, even with glaciers reaching the coast, the mean annual temperature remained sufficiently high to sustain a flora dominated by *Nothofagus* in coastal enclaves.

3. The middle to late Miocene (16–5 Ma) record is poorly represented in the Ross Embayment, and major erosional events by expanded ice are inferred, probably resulting from the development of the West Antarctic Ice Sheet around 10 m.y. ago. However, reworked marine diatoms from the interior of the Ross Embayment and dating within this time indicate several periods of major ice sheet recession.

4. The Pliocene record is best documented, in our view, in cored sequences of fjord sediments in the dry valleys area and by scattered outcrops of Sirius Group tillites and temperate ice-marginal sediments in the Transantarctic Mountains. The drill cores show a number of glacial advance and retreat cycles, with temperate glaciers transporting debris mainly from the Transantarctic Mountains. We accept the hypothesis in which the East Antarctic ice retreated sufficiently to allow seas to penetrate the Antarctic interior ~3 m.y. ago and then expanded as a temperate "super ice sheet" to deposit marine diatom-bearing glacial debris at high elevations in the Transantarctic Mountains. Temperatures were warm enough to support scrubby vegetation (*Nothofagus*) in the Beardmore Glacier area (~85°S) at high elevations but at times had been cold and/or dry enough for the formation of desert pavements at high elevations

in the dry valleys area. The hypothesis we have outlined is disputed by some who point to geomorphic and marine stratigraphic evidence of a persistent polar climate since middle to late Miocene time.

5. The sedimentary and terrestrial plant record suggests to us that the Antarctic ice sheet was temperate and varied greatly in size prior to late Pliocene time. Glaciological models indicate that it would be possible to sustain a significant ice sheet over East Antarctica then, even with temperatures as much as 20°C higher than today, through high elevation and increased snowfall.

6. The Antarctic ice sheet did not take on its present cold thermal character, in our view, until the late Pliocene or early Pleistocene, from which only a thin and patchy record has been produced, either on land or offshore. Ice from East Antarctica and West Antarctica combined to erode the continental shelf further but deposited sediment as a prograding wedge at the shelf edge. Several incursions of grounded Ross Sea ice into the dry valleys of southern Victoria Land took place, apparently when the local glaciers receded.

7. The present regime is one of reduced sedimentation over most of the continental shelf, with a dominance of wind-blown sand and mud deposition near the Victoria Land coast, and biogenic (diatom) sedimentation elsewhere. The Ross Ice Shelf acts as a barrier to sediment transfer, and terrigenous sedimentation is probably most rapid near its grounding line. The role of West Antarctic ice on Ross Sea sedimentation is much more important than that of East Antarctic ice.

8. In a global context, fluctuations of ice in the Ross Embayment appear to match glacial events in other parts of East Antarctica and the Southern Ocean. They only partly correspond to the fluctuations of temperature, sea level, and ice volume inferred from the deep-sea oxygen isotope record, corresponding more closely to sea level variations derived from sequence stratigraphic analysis. Short-term Cenozoic sea level change was probably driven by the Antarctic ice sheets until the late Pliocene, after which the northern hemisphere ice sheets were the main influence, and indeed they may then have increased the extent of Antarctic ice cover through lowering of sea level and grounding of ice shelves to promote ice sheet growth.

Despite these advances in our knowledge in recent years, there remain several major outstanding questions. The question "When did Antarctic glaciation begin?" has still not been answered. Future drilling off the Victoria Land coast, especially near Cape Roberts, will penetrate pre-Oligocene strata to obtain a record of climatic and sea level change possibly back into the Mesozoic Era, as well as the role of the rising Transantarctic Mountains on sedimentation.

The extent and frequency of grounded ice advances across the Ross Sea shelf have still not been satisfactorily determined because many of the sequences cored to date are incomplete through erosion or low recovery. High-resolution seismic profiling, improved biostratigraphic studies and further sedimentological work (especially fabric studies on drill cores) will continue. More specifically, the relative roles of the East Antarctic and West Antarctic ice sheets need to be distinguished, and the important question concerning the stability of the West Antarctic Ice Sheet needs to be addressed. Resolution of these problems will require further information from Ross Sea drilling, especially in the deeper basins where the record is less likely to be interrupted by ice erosional events and in the east where events will clearly relate to West Antarctic ice. Of particular importance would be detailed stratigraphic records of events around the transition from Pliocene to Pleistocene and Pleistocene to Holocene. An equally important task is the reconciliation of conflicting evidence and views on major ice sheet retreat during the late Pliocene. How can Pliocene beech trees be reconciled with Miocene to Pliocene cold desert pavements and old polar soils? Questions on preservation of these features and rates of polar processes have yet to be fully explored.

Future work should in particular be directed toward improving the chronology of late Cenozoic continental shelf sequences around the Antarctic margin, which provide a direct record of ice cover. But we should also be seeking a glacial record for early Cenozoic times for comparison. By combining these data with the proxy data reflecting variations in ice sheet size (especially the isotope record from the Southern Ocean and global seismic sequence data), the validity of the latter may be better assessed and thereby provide a sounder basis for obtaining a detailed global climate record for the Cenozoic Era.

Acknowledgments. This paper is a summary of the results of numerous investigators over many years. We have enjoyed discussions with many of them and have endeavored here to record the range of current opinion, with most weight given to our own views. We thank Jim Kennett, Mike Prentice, and Dietz Warnke for reviewing the manuscript. We also wish to acknowledge the dedication and skill of the drilling teams that have given us access to a major part of the record of Antarctic glacial history. This paper was made possible by a Visiting Fellowship at the Victoria University of Wellington and a travel grant from the Transantarctic Association to M.J.H.

REFERENCES

Alley, R. B., D. D. Blankenship, S. T. Rooney, and C. R. Bentley, Sedimentation beneath ice shelves—The view from Ice Stream B, *Mar. Geol.*, *85*, 101–120, 1989.

Allis, R. G., D. A. Christoffel, and P. J. Barrett, A paleomagnetic stratigraphy for Oligocene and early Miocene marine glacial sediments at Site 270, Ross Sea, Antarctica, *Initial Rep. Deep Sea Drill. Proj.*, *28*, 879–884, 1975.

Anderson, J. B., and G. M. Ashley, Glacial marine sedimentation; paleoclimatic significance; a discussion, *Spec. Pap. Geol. Soc. Am.*, *261*, 223–226, 1991.

Anderson, J. B., and L. R. Bartek, Cenozoic glacial history of the Ross Sea revealed by intermediate resolution seismic reflection data combined with drill site information, in *The Antarctic Paleoenvironment: A Perspective on Global Change, Part One, Antarct. Res. Ser.*, vol. 56, edited by J. P. Kennett and D. A. Warnke, pp. 231–263, AGU, Washington, D. C., 1992.

Anderson, J. B., D. D. Kurtz, E. W. Domack, and K. M. Balshaw, Glacial and glacial marine sediments of the Antarctic continental shelf, *J. Geol.*, *88*, 399–414, 1980.

Anderson, J. B., D. S. Kennedy, M. J. Smith, and E. W. Domack, Sedimentary facies associated with Antarctica's floating ice masses, Glacial Marine Sedimentation: Paleoclimatic Significance, *Spec. Pap. Geol. Soc. Am.*, *261*, 1–26, 1991.

Barrett, P. J., Textural characteristics of Cenozoic preglacial and glacial sediments at Site 270, Ross Sea, Antarctica, *Initial Rep. Deep Sea Drill. Proj.*, *28*, 757–767, 1975*a*.

Barrett, P. J., Characteristics of pebbles from Cenozoic marine glacial sediments in the Ross Sea (DSDP sites 270–274) and the south Indian Ocean, *Initial Rep. Deep Sea Drill. Proj.*, *28*, 769–784, 1975*b*.

Barrett, P. J. (Ed.), Antarctic Cenozoic history from the MSSTS-1 drillhole, McMurdo Sound, *DSIR Bull. N. Z.*, *237*, 174 pp., 1986.

Barrett, P. J. (Ed.), Antarctic Cenozoic history from the CIROS-1 drillhole, McMurdo Sound, *DSIR Bull. N. Z.*, *245*, 254 pp., 1989*a*.

Barrett, P. J., Sediment texture, Antarctic Cenozoic History From the CIROS-1 Drillhole, McMurdo Sound, *DSIR Bull. N. Z.*, *245*, 49–58, 1989*b*.

Barrett, P. J., The Beacon Supergroup of the Transantarctic Mountains and correlatives in other parts of Antarctica, in *Geology of Antarctica*, edited by R. J. Tingey, pp. 120–152, Oxford University Press, New York, 1991*a*.

Barrett, P. J., Antarctica and global climate change: A geological perspective, in *Antarctica and Global Climatic Change*, edited by C. M. Harris and B. Stonehouse, pp. 35–50, Belhaven Press, London, Scott Polar Research Institute, Cambridge, England, and Lewis Publishers, Boca Raton, Fla., 1991*b*.

Barrett, P. J., and F. J. Davey, Cape Roberts Project Workshop report, *R. Soc. N. Z. Misc. Bull.*, *23*, 38 pp., 1992.

Barrett, P. J., and M. J. Hambrey, Plio-Pleistocene sedimentation in Ferrar Fjord, Antarctica, *Sedimentology*, *39*, 109–123, 1992.

Barrett, P. J., and B. C. McKelvey, Stratigraphy, Antarctic Cenozoic History From the MSSTS-1 Drillhole, McMurdo Sound, *DSIR Bull. N. Z.*, *237*, 9–52, 1986.

Barrett, P. J., and S. B. Treves, Sedimentology and petrology of core from DVDP 15, western McMurdo Sound, in *Dry Valley Drilling Project, Antarct. Res. Ser.*, vol. 33, edited by L. D. McGinnis, pp. 281–314, AGU, Washington, D. C., 1981.

Barrett, P. J., M. J. Hambrey, D. M. Harwood, A. R. Pyne, and P.-N. Webb, Synthesis, Antarctic Cenozoic History From the CIROS-1 Drillhole, McMurdo Sound, *DSIR Bull. N. Z.*, *245*, 241–251, 1989.

Barrett, P. J., T. Perrett, and A. R. Pyne, Offshore textural gradient and storm wave base on a prograding sand/mud coast, Pekapeka, New Zealand, in *Abstracts*, International Sedimentological Congress, Nottingham, United Kingdom, 1990.

Barrett, P. J., M. J. Hambrey, and P. H. Robinson, Cenozoic glacial and tectonic history from CIROS-1, McMurdo Sound, in *Geological Evolution of Antarctica*, edited by M. R. A. Thomson, J. A. Crame, and J. W. Thomson, pp. 651–656, Cambridge University Press, New York, 1991.

Barrett, P. J., C. J. Adams, W. C. McIntosh, C. C. Swisher III, and G. S. Wilson, Geochronological evidence supporting Antarctic deglaciation three million years ago, *Nature*, *359*, 816–818, 1992.

Barron, J. A., et al., Leg 119, *Proc. Ocean Drill. Program Initial Rep.*, *119*, 942 pp., 1989.

Bartek, L. R., J. B. Anderson, P. A. Emmet, and S. Wu, Effect of Cenozoic ice sheet fluctuations in Antarctica on the stratigraphic signature of the Neogene, *J. Geophys. Res.*, *96*, 6753–6778, 1991.

Bartek, L. R., S. A. Henrys, and J. B. Anderson, Seismic stratigraphy of Polar Duke-90 data in McMurdo Sound, Antarctica: Implications for glacially influenced early Cenozoic sea level change, *Geol. Soc. Am. Abstr. Programs*, *24*(7), A83, 1992.

Behrendt, J. C., and A. K. Cooper, Evidence of rapid Cenozoic uplift of the shoulder escarpment of the Cenozoic West Antarctic Rift System and a speculation on possible climate forcing, *Geology*, *19*, 315–319, 1991.

Behrendt, J. C., W. E. LeMasurier, A. K. Cooper, F. Tessensohn, A. Trehu, and D. Damaske, The West Antarctic Rift System: A review of geophysical investigations, in *Contributions to Antarctic Research II, Antarctic Res. Ser.*, vol. 53, edited by D. H. Elliot, pp. 67–112, AGU, Washington, D. C., 1991.

Bentley, C. R., Antarctic ice streams: A review, *J. Geophys. Res.*, *92*, 8843–8858, 1987.

Berggren, W. A., D. V. Kent, J. J. Flynn, and J. A. van Couvering, Cenozoic geochronology, *Geol. Soc. Am. Bull.*, *96*, 1407–1418, 1985.

Bindschadler, R., West Antarctic Ice Sheet Initiative, 1, Science and implementation plan, *NASA Conf. Publ.*, *3115*, 54 pp., 1991.

Birkenmajer, K., Tertiary glacial and interglacial deposits, South Shetland Islands, Antarctica: Geochronology versus biostratigraphy (a progress report), *Bull. Pol. Acad. Sci. Earth Sci.*, *36*, 133–144, 1988.

Birkenmajer, K., Tertiary glaciation in the South Shetland Islands, West Antarctica: Evaluation of data, in *Geological Evolution of Antarctica*, edited by M. R. A. Thomson, J. A. Crame, and J. W. Thomson, pp. 629–632, Cambridge University Press, New York, 1991.

Burckle, L. H., and E. M. Pokras, Implications of a Pliocene stand of *Nothofagus* (southern beech) within 500 km of the south pole, *Antarct. Sci.*, *3*, 389–403, 1991.

Burckle, L. H., R. I. Gayley, M. Ram, and J.-R. Petit, Diatoms in Antarctic ice cores: Some implications for the glacial history of Antarctica, *Geology*, *16*, 326–329, 1988.

Burn, R. W., and M. R. A. Thomson, Late Cenozoic tillites associated with intraglacial volcanic rocks, Lesser Antarctica, in *Earth's Pre-Pleistocene Glacial Record*, edited by M. J. Hambrey and W. B. Harland, pp. 199–203, Cambridge University Press, New York, 1981.

Cattle, H., Global climate models and Antarctic climatic change, in *Antarctica and Global Climatic Change*, edited by C. M. Harris and B. Stonehouse, pp. 21–34, Belhaven Press, London, Scott Polar Research Institute, Cambridge, England, and Lewis Publishers, Boca Raton, Fla., 1991.

Clapperton, C. M., and D. E. Sugden, Late Cenozoic glacial history of the Ross Embayment, Antarctica, *Quat. Sci. Rev.*, *9*, 253–272, 1990.

Claridge, G. G. C., and I. B. Campbell, Clay mineralogy, Antarctic Cenozoic History From the CIROS-1 Drillhole, McMurdo Sound, *DSIR Bull. N. Z.*, *245*, 185–193, 1989.

Cooper, A. K., and F. J. Davey, Episodic rifting of Phanerozoic rocks in the Victoria Land basin, western Ross Sea, Antarctica, *Science*, *229*, 1085–1087, 1985.

Cooper, A. K., and F. J. Davey, *The Antarctic Continental Margin: Geology and Geophysics of the Western Ross Sea*,

Earth Sci. Ser., vol. 5B, 253 pp., Circum-Pacific Council for Energy and Mineral Resources, Houston, Tex., 1987.

Cooper, A. K., P. J. Barrett, K. Hinz, V. Traube, G. Leitchenkov, and H. M. J. Stagg, Cenozoic prograding sequences of the Antarctic continental margin: A record of glacioeustatic and tectonic events, *Mar. Geol.*, *102*, 175–213, 1991.

Davey, F. J., D. J. Bennett, and R. E. Houtz, Sedimentary basins of the Ross Sea, Antarctica, *N. Z. J. Geol. Geophys.*, *25*, 245–255, 1982.

Denton, G. H., M. L. Prentice, D. E. Kellogg, and T. B. Kellogg, Late Tertiary history of the Antarctic ice sheet: Evidence from the dry valleys, *Geology*, *12*, 263–267, 1984.

Denton, G. H., J. G. Bockheim, S. C. Wilson, and M. Stuiver, Late Wisconsin and early Holocene glacial history, inner Ross Embayment, Antarctica, *Quat. Res.*, *31*, 151–182, 1989*a*.

Denton, G. H., J. G. Bockheim, S. C. Wilson, J. E. Leide, and B. G. Anderson, Late Quaternary ice surface fluctuations of the Beardmore Glacier, Transantarctic Mountains, *Quat. Res.*, *31*, 183–209, 1989*b*.

Denton, G. H., M. L. Prentice, and L. H. Burckle, Cenozoic history of the Antarctic ice sheet, in *Geology of Antarctica*, edited by R. J. Tingey, pp. 365–433, Oxford University Press, New York, 1991.

Department of the Environment, United Kingdom, Global climate change, 16 pp., Central Off. of Inf., London, 1989.

Dowdeswell, J. A., Processes of glaciomarine sedimentation, *Prog. Phys. Geogr.*, *11*, 52–90, 1987.

Drewry, D. J., Initiation and growth of the East Antarctic Ice Sheet, *J. Geol. Soc. London*, *131*, 255–273, 1975.

Drewry, D. J., The record of late Cenozoic glacial events in East Antarctica (60°–171°E), in *Earth's Pre-Pleistocene Glacial Record*, edited by M. J. Hambrey and W. B. Harland, pp. 212–216, Cambridge University Press, New York, 1981.

Drewry, D. J., *Antarctic Glaciological and Geophysical Folio*, Scott Polar Research Institute, Cambridge, England, 1983.

Drewry, D. J., *Glacial Geologic Processes*, 276 pp., Edward Arnold, London, 1986.

Ehrmann, W. U., M. J. Hambrey, J. G. Baldauf, J. Barron, B. Larsen, A. Mackensen, S. W. Wise, and J. C. Zachos, History of Antarctic glaciation: An Indian Ocean perspective, in *Synthesis of Results From Scientific Drilling in the Indian Ocean, Geophys. Monogr. Ser.*, vol. 70, edited by R. A. Duncan et al., pp. 423–446, AGU, Washington, D. C., 1992.

Elliott, T., Siliciclastic sediments, in *Sedimentary Environments and Facies*, edited by H. G. Reading, pp. 155–188, Blackwell, Oxford, England, 1986.

Eyles, C. H., and N. Eyles, Sedimentation in a large lake: A reinterpretation of the late Pleistocene stratigraphy at Scarborough Bluffs, Ontario, Canada, *Geology*, *11*, 146–152, 1983.

Eyles, N., C. H. Eyles, and A. D. Miall, Lithofacies types and vertical profile model; an alternative approach to the description and environmental interpretation of glacial diamict and diamictic sequences, *Sedimentology*, *30*, 393–410, 1983.

Fitzgerald, P. G., Uplift and formation of the Transantarctic Mountains: Application of apatite fission track analysis to tectonic problems, in *Abstracts, 28th International Geological Congress*, vol. 1, p. 491, Washington, D. C., 1989.

Fitzgerald, P. G., and E. Stump, Uplift history of the Transantarctic Mountains from Victoria Land (~70°S) to Scott Glacier (~86°S): Evidence from fission track analysis, in *Abstracts, Sixth International Symposium on Antarctic Earth Sciences*, p. 144, National Institute of Polar Research, Tokyo, 1991*a*.

Fitzgerald, P. G., and E. Stump, Uplift history of the Vinson Massif, Ellsworth Mountains: constraints from apatite fission track analysis and implications to the tectonic history of West Antarctica, in *Abstracts, Sixth International Symposium on Antarctic Earth Sciences*, p. 145, National Institute of Polar Research, Tokyo, 1991*b*.

Fitzgerald, P. G., T. Sandiford, P. J. Barrett, and A. J. W. Gleadow, Asymmetric extension associated with uplift and subsidence in the Transantarctic Mountains and the Ross Embayment, *Earth Planet. Sci. Lett.*, *81*, 67–78, 1986.

Ford, A. B., Chemical characteristics of greywacke and palaeosol of early Oligocene or older sedimentary breccia, Ross Sea DSDP Site 270, in *Geological Evolution of Antarctica*, edited by M. R. A. Thomson, J. A. Crame, and J. W. Thomson, pp. 293–297, Cambridge University Press, New York, 1991.

Forum, Sedimentation in a large lake: A reinterpretation of the late Pleistocene stratigraphy at Scarborough Bluffs, Ontario, Canada—A discussion, *Geology*, *12*, 185–190, 1984.

Frakes, L. A., and J. E. Francis, A guide to Phanerozoic cold polar climates from high-latitude ice-rafting in the Cretaceous, *Nature*, *339*, 547–549, 1988.

Gazdzicki, A., Planktonic foraminifera from the Oligocene Polonez Cove Formation of King George Island, West Antarctica, *Pol. Polar Res.*, *10*, 47–55, 1989.

Gleadow, A. J. W., and P. G. Fitzgerald, Uplift history and structure of the Transantarctic Mountains and new evidence from fission track dating of basement apatites in the dry valleys area, southern Victoria Land, *Earth Planet. Sci. Lett.*, *82*, 1–14, 1987.

Hall, K. J., Clast shape, Antarctic Cenozoic History From the CIROS-1 Drillhole, McMurdo Sound, *DSIR Bull. N. Z.*, *245*, 63–66, 1989.

Hall, K. J., Palaeoenvironmental reconstruction from redeposited weathered clasts in the CIROS-1 drill core, *Antarct. Sci.*, *1*, 235–238, 1990.

Hambrey, M. J., et al., Forty million years of Antarctic glacial history revealed by Leg 119 of the Ocean Drilling Program, *Polar Rec.*, *25*, 99–106, 1989*a*.

Hambrey, M. J., P. J. Barrett, and P. H. Robinson, Stratigraphy and sedimentology, Antarctic Cenozoic History From the CIROS-1 Drillhole, McMurdo Sound, *DSIR Bull. N. Z.*, *245*, 19–47, 1989*b*.

Hambrey, M. J., W. U. Erhmann, and B. Larsen, The Cenozoic glacial record from the Prydz Bay continental shelf, East Antarctica, *Proc. Ocean Drill. Program Sci. Results*, *119*, 77–132, 1991.

Hambrey, M. J., P. J. Barrett, W. U. Ehrmann, and B. Larsen, Cenozoic sedimentary processes on the Antarctic continental shelf: The record from deep drilling, *Z. Geomorphol.*, *86*, 73–99, 1992.

Haq, B. U., J. Hardenbol, and P. R. Vail, Chronology of fluctuating sea levels since the Triassic, *Science*, *235*, 1156–1167, 1987.

Hart, C. P., and B. C. McKelvey, Tertiary marine molluscs from the Pagodroma Tillite, Prince Charles Mountains, East Antarctica, in *Abstracts, Sixth International Symposium on Antarctic Earth Science*, p. 219, National Institute of Polar Research, Tokyo, 1991.

Harwood, D. M., Diatom biostratigraphy and paleoecology and Cenozoic history of Antarctic ice sheets, Ph.D. dissertation, 592 pp., Ohio State Univ., Columbus, 1986*a*.

Harwood, D. M., Recycled siliceous microfossils from the Sirius Formation, *Antarct. J. U. S.*, *21*, 101–103, 1986*b*.

Harwood, D. M., Siliceous microfossils, Antarctic Cenozoic History From the CIROS-1 Drillhole, McMurdo Sound, *DSIR Bull. N. Z.*, *245*, 67–97, 1989.

Harwood, D. M., R. P. Scherer, and P.-N. Webb, Multiple Miocene marine productivity events in West Antarctica as recorded in upper Miocene sediments beneath the Ross Ice Shelf, *Mar. Micropaleontol.*, *15*, 91–115, 1989.

Hayes, D. E., and L. A. Frakes, General synthesis, Deep Sea Drilling Project, Leg 28, *Initial Rep. Deep Sea Drill. Proj.*, *28*, 919–942, 1975.

Hicock, S. R., and A. Dreimanis, Sunnybrook drift indicates a grounded Wisconsin glacier in Lake Ontario basin, *Geology*, *17*, 169–172, 1989.

Hill, R. S., Fossil leaf, Antarctic Cenozoic History From the CIROS-1 Drillhole, McMurdo Sound, *DSIR Bull. N. Z.*, *245*, 143–144, 1989.

Hinz, K., and M. Block, Results of geophysical investigations in the Weddell Sea and in the Ross Sea, Antarctica, in *Proceedings of the 11th World Petroleum Congress*, pp. 79–91, John Wiley, New York, 1983.

Houghton, J. T., G. J. Jenkins, and J. J. Ephraums, *Climate Change: The IPCC Scientific Assessment, Final Report of Working Party 1*, 364 pp., Cambridge University Press, New York, 1990.

Ishman, S. E., and H. J. Rieck, A late Neogene Antarctic glacio-eustatic record, Victoria Land Basin margin, Antarctica, in *The Antarctic Paleoenvironment: A Perspective on Global Change, Part One, Antarct. Res. Ser.*, vol. 56, edited by J. P. Kennett and D. A. Warnke, pp. 327–347, AGU, Washington, D. C., 1992.

Kellogg, T. B., R. S. Truesdale, and L. E. Osterman, Late Quaternary extent of the West Antarctic Ice Sheet: New evidence from Ross Sea cores, *Geology*, *7*, 249–253, 1979.

Kennett, J. P., Cenozoic evolution of Antarctic glaciation, the circum-Antarctic Ocean, and their impact on global paleoceanography, *J. Geophys. Res.*, *82*, 3843–3860, 1977.

Kennett, J. P., and P. F. Barker, Latest Cretaceous to Cenozoic climate and oceanographic developments in the Weddell Sea, Antarctica: An ocean-drilling perspective, *Proc. Ocean Drill. Program Sci. Results*, *113*, 937–960, 1990.

Kyle, P. R., Glacial history of the McMurdo Sound area as indicated by the distribution and nature of McMurdo Volcanic Group rocks, in *Dry Valley Drilling Project, Antarct. Res. Ser.*, vol. 33, edited by L. D. McGinnis, pp. 403–412, AGU, Washington, D. C., 1981.

Leckie, R. M., and P.-N. Webb, Late Oligocene–early Miocene glacial record of the Ross Sea, Antarctica: Evidence from DSDP Site 270, *Geology*, *11*, 578–582, 1983.

LeMasurier, W. E., Volcanic record of Cenozoic glacial history of Marie Byrd Land, in *Antarctic Geology and Geophysics*, edited by R. J. Adie, pp. 251–260, Universitetsforlaget, Oslo, 1972.

LeMasurier, W. E., and D. C. Rex, Rates of uplift and the scale of ice level instabilities recorded by volcanic rooks in Marie Byrd Land, West Antarctica, in *Antarctic Earth Science*, edited by R. L. Oliver, P. R. James, and J. B. Jago, pp. 663–670, Cambridge University Press, New York, 1983.

MacAyeal, D. R., Irregular oscillations of the West Antarctic Ice Sheet, *Nature*, *359*, 29–32, 1992.

Macpherson, A. J., The Mackay Glacier/Granite Harbour system (Ross Dependency, Antarctica)—A study in near-shore glacial marine sedimentation, Ph.D. thesis, 86 pp., Victoria Univ. of Wellington Libr., Wellington, New Zealand, 1987.

Marchant, D. R., D. R. Lux, C. C. Swisher III, and G. H. Denton, Early Pliocene ash rests on a polar desert pavement, *Antarct. J. U. S.*, *25*, 58–59, 1990.

Matthews, R. K., and R. Z. Poore, Tertiary [18]O record and glacioeustatic sea-level fluctuations, *Geology*, *8*, 501–504, 1980.

Mayewski, P. A., Glacial geology and late Cenozoic history of the Transantarctic Mountains, Antarctica, *Rep. 56*, 168 pp., Inst. of Polar Stud., Ohio State Univ., Columbus, 1975.

McDougall, I., Potassium argon dating of glauconite from a greensand drilled at Site 270 in the Ross Sea, DSDP Leg 28, *Initial Rep. Deep Sea Drill. Proj.*, *36*, 1071–1072, 1977.

McKelvey, B. C., The lithological logs of DVDP cores 10 and 11, eastern Taylor Valley, in *Dry Valley Drilling Project*, *Antarct. Res. Ser.*, vol. 33, edited by L. D. McGinnis, pp. 63–94, AGU, Washington, D. C., 1981.

McKelvey, B. C., P.-N. Webb, D. M. Harwood, and M. C. G. Mabin, The Dominion Range Sirius Group: A record of the late Pliocene–early Pleistocene Beardmore Glacier, in *Geological Evolution of Antarctica*, edited by M. R. A. Thomson, J. A. Crame, and J. W. Thomson, pp. 675–682, Cambridge University Press, New York, 1991*a*.

McKelvey, B. C., M. C. G. Mabin, D. M. Harwood, and P.-N. Webb, The Pagodroma event—A late Pliocene major expansion of the ancestral Lambert Glacier system, in *Abstracts, Sixth International Symposium on Antarctic Earth Science*, p. 403, National Institute of Polar Research, Tokyo, 1991*b*.

Mercer, J. H., Some observations on the glacial geology of the Beardmore Glacier area, in *Antarctic Geology and Geophysics*, edited by R. J. Adie, pp. 427–433, Universitetsforlaget, Oslo, 1972.

Mercer, J. H., West Antarctic Ice Sheet and CO_2 greenhouse effect: A threat of disaster, *Nature*, *271*, 321–325, 1978.

Mercer, J. H., Tertiary terrestrial deposits of the Ross Ice Shelf area, Antarctica, in *Earth's Pre-Pleistocene Glacial Record*, edited by M. J. Hambrey and W. B. Harland, pp. 204–207, Cambridge University Press, New York, 1981.

Mildenhall, D. C., Terrestrial palynology, Antarctic Cenozoic History From the CIROS-1 Drillhole, McMurdo Sound, *DSIR Bull. N. Z.*, *245*, 119–127, 1989.

Miller, K. G., R. G. Fairbanks, and G. S. Mountain, Tertiary oxygen isotope synthesis, sea level history, and continental margin erosion, *Paleoceanography*, *2*, 1–19, 1987.

Miller, K. G., J. D. Wright, and R. G. Fairbanks, Unlocking the ice house: Oligocene-Miocene oxygen isotopes, eustasy, and margin erosion, *J. Geophys. Res.*, *96*, 6829–6848, 1991.

Moriwaki, K., K. Hirakawa, S. Iwata, and M. Hayashi, Glacial history in the Sør-Rondane Mountains, East Antarctica—Thick tills and evidence of warmer climate, in *Abstracts, Sixth International Symposium on Antarctic Earth Sciences*, pp. 418–422, National Institute of Polar Research, Tokyo, 1991.

Oerlemans, J., A model of the Antarctic ice sheet, *Nature*, *297*, 550–553, 1982.

Pickard, J., D. A. Adamson, D. M. Harwood, G. H. Miller, P. G. Quilty, and R. K. Dell, Early Pliocene marine sediments, coastline, and climate of East Antarctica, *Geology*, *16*, 158–161, 1988.

Pitman, W. C., III, Relationship between eustacy and stratigraphic sequences on passive margins, *Geol. Soc. Am. Bull.*, *89*, 1389–1403, 1978.

Porter, S. C., and J. E. Beget, Provenance and depositional environments of late Cenozoic sediments in permafrost cores from lower Taylor Valley, Antarctica, in *Dry Valley Drilling Project*, *Antarct. Res. Ser.*, vol. 33, edited by L. D. McGinnis, pp. 351–364, AGU, Washington, D. C., 1981.

Powell, R. D., Sedimentation conditions in Taylor Valley, Antarctica, inferred from textural analysis of DVDP cores, in *Dry Valley Drilling Project*, *Antarct. Res. Ser.*, vol. 33, edited by L. D. McGinnis, pp. 331–350, AGU, Washington, D. C., 1981.

Powell, R. D., and A. Elverhøi (Eds.), Modern glacimarine environments: Glacial and marine controls on modern lithofacies and biofacies, *Mar. Geol.*, *85*, 101–418, 1989.

Powell, R. D., A. R. Pyne, L. E. Hunter, and N. R. Rynes, Polar versus temperate grounding line sedimentary systems and marine glacier stability during sea level rise by global warming, *Geol. Soc. Am. Abstr. Programs*, *24*(7), A348, 1992.

Prentice, M. L., and R. K. Matthews, Tertiary ice sheet

dynamics: The snow gun hypothesis, *J. Geophys. Res.*, *96*, 6811–6827, 1991.

Prentice, M. L., J. G. Bockheim, S. C. Wilson, L. H. Burckle, D. A. Hodell, and D. E. Kellogg, Late Neogene Antarctic glacial history: Evidence from central Wright Valley, this volume.

Pyne, A. R., P. H. Robinson, and P. J. Barrett, Core log, description and photographs, CIROS 2, Ferrar Fjord, Antarctica, *Antarct. Data Ser. 11*, 80 pp., Victoria Univ. of Wellington, Wellington, New Zealand, 1985.

Quilty, P. G., The geology of Marine Plain, Vestfold Hills, East Antarctica, in *Geological Evolution of Antarctica*, edited by M. R. A. Thomson, J. A. Crame, and J. W. Thomson, pp. 683–686, Cambridge University Press, New York, 1991*a*.

Quilty, P. G., Sources of information on the late Neogene of coastal East Antarctica, in *Abstracts, Sixth International Symposium on Antarctic Earth Science*, p. 488, National Institute of Polar Research, Tokyo, 1991*b*.

Rea, D. K., and H. Schrader, Late Pliocene onset of glaciation: Ice-rafting and diatom stratigraphy of North Pacific DSDP cores, *Palaeogeogr. Palaeoclimatol. Palaeoecol.*, *49*, 313–325, 1985.

Rieck, H. J., Paleomagnetic stratigraphy, Antarctic Cenozoic History From the CIROS-1 Drillhole, McMurdo Sound, *DSIR Bull. N. Z.*, *245*, 153–158, 1989.

Robin, G. de Q., Ice shelves and ice flow, *Nature*, *253*, 168–172, 1975.

Robin, G. de Q., Changing the sea level: Projecting the rise in sea level caused by warming of the atmosphere, in *The Greenhouse Effect, Climatic Change, and Ecosystems (SCOPE 29)*, edited by B. Bolin, B. R. Döös, J. Jäger, and R. A. Warrick, pp. 323–359, John Wiley, New York, 1986.

Sahagian, D. L., Epeirogeny and eustatic sea level changes as inferred from Cretaceous shoreline deposits: Applications to the central and western United States, *J. Geophys. Res.*, *92*, 4895–4904, 1987.

Savage, M. L., and P. F. Ciesielski, A revised history of glacial sediments in the Ross Sea region, in *Antarctic Earth Science*, edited by R. L. Oliver, P. R. James, and J. B. Jago, pp. 555–559, Cambridge University Press, New York, 1983.

Scherer, R. P., Quaternary and Tertiary microfossils from beneath Ice Stream B: Evidence for a dynamic West Antarctic Ice Sheet history, *Palaeogeogr. Palaeoclimatol. Palaeoecol.*, *90*, 395–412, 1991.

Shabtaie, S., and C. R. Bentley, West Antarctic ice streams draining into the Ross Ice Shelf: Configuration and mass balance, *J. Geophys. Res.*, *92*, 1311–1336, 1987.

Shackleton, N. J., and J. P. Kennett, Paleotemperature history of the Cenozoic and the initiation of Antarctic glaciation: Oxygen and carbon isotope analysis in DSDP sites 277, 279, and 281, *Initial Rep. Deep Sea Drill. Proj.*, *29*, 743–755, 1975.

Shackleton, N. J., et al., Oxygen isotope calibration of the onset of ice-rafting in the North Atlantic Ocean, *Nature*, *307*, 620–623, 1984.

Solheim, A., C. F. Forsberg, and A. Pittenger, Stepwise consolidation of glacigenic sediments related to the glacial history of Prydz Bay, East Antarctica, *Proc. Ocean Drill. Program Sci. Results*, *119*, 169–182, 1991.

Stern, T. A., and U. S. ten Brink, Flexural uplift of the Transantarctic Mountains, *J. Geophys. Res.*, *94*, 10,315–10,330, 1989.

Sugden, D. E., Global warming—Ice sheets at risk?, *Nature*, *359*, 775–776, 1992.

Summerhayes, C., Sea level curves based on seismic stratigraphy: Their chronostratigraphic significance, *Paleogeogr., Paleodim., Paleoecol.*, *57*, 27–42, 1986.

Tessensohn, F., and G. Wörner, The Ross Sea rift system, in *Geological Evolution of Antarctica*, edited by M. R. A.

Thomson, J. A. Crame, and J. W. Thomson, pp. 273–277, Cambridge University Press, New York, 1991.

Wadhams, P., Atmosphere-ice-ocean interactions in the Antarctic, in *Antarctica and Global Climatic Change*, edited by C. M. Harris and B. Stonehouse, pp. 65–81, Belhaven Press, London, Scott Polar Research Institute, Cambridge, and Lewis Publishers, Boca Raton, Fla., 1991.

Webb, P.-N., Benthic foraminifera, Antarctic Cenozoic History From the CIROS-1 Drillhole, McMurdo Sound, *DSIR Bull. N. Z.*, *245*, 99–118, 1989.

Webb, P.-N., and D. M. Harwood, The Sirius Formation of the Beardmore Glacier region, *Antarct. J. U. S.*, *22*, 8–12, 1987.

Webb, P.-N., and D. M. Harwood, Late Cenozoic glacial history of the Ross Embayment, Antarctica, *Quat. Sci. Rev.*, *10*, 215–223, 1991.

Webb, P.-N., D. M. Harwood, B. C. McKelvey, J. H. Mercer, and L. D. Stott, Cenozoic marine sedimentation and ice volume variation on the East Antarctic craton, *Geology*, *12*, 287–291, 1984.

Williams, D. F., R. C. Thunell, E. Tappa, R. Domenico, and I. Raffi, Chronology of the Pleistocene oxygen isotope record: 0–1.88 m.y. B.P., *Palaeogeogr. Palaeoclimatol. Palaeoecol.*, *64*, 221–240, 1988.

Wise, S. W., J. R. Breza, D. M. Harwood, and W. Wei, Palaeogene glacial history of Antarctica, in *Controversies in Modern Geology*, edited by J. A. McKenzie, D. W. Müller, and H. Weissert, pp. 133–171, Academic, San Diego, Calif., 1991.

Zotikov, I. A., *The Thermophysics of Glaciers*, 275 pp., D. Reidel, Hingham, Mass., 1986.

(Received March 20, 1992;
accepted March 19, 1993.)

CENOZOIC SOUTHERN MID- AND HIGH-LATITUDE BIOSTRATIGRAPHY AND CHRONOSTRATIGRAPHY BASED ON PLANKTONIC FORAMINIFERA

D. Graham Jenkins

Department of Geology, National Museum of Wales, Cardiff CF1 3NP, Wales, United Kingdom

An essential part of the Cenozoic biostratigraphy of the southern middle and high latitudes has been based on the stratigraphic ranges of planktonic foraminifera. Sequences derived from the Deep Sea Drilling Project and the Ocean Drilling Program (ODP) have been subdivided into biozones, and datum planes based on biostratigraphic events have been used for intersite correlation. The accuracy and reliability of the methodology of using datum planes is discussed. Zonations have been established for the southern mid-latitudes, and these are discussed in relation to the low-latitude zonations. More recently, the ODP has recovered sequences from Maud Rise and the Kerguelen Plateau, and high-latitude zonal schemes have been established; these zones have been correlated with the southern mid-latitude zonation. The sequential zonations of the southern middle and high latitudes are discussed within the Cenozoic chronostratigraphic series boundaries; the identification of these boundaries becomes progressively more difficult with increasing latitude but should improve with the establishment of international boundary stratotypes.

BIOSTRATIGRAPHIC FRAMEWORK

Introduction

Biozones, termed zones in this work, form the biostratigraphic framework of rock sequences, and the accuracy of their identification is dependent on a number of factors, including the recognition of the selected taxa which have been used to define the limits of the zones. Datum planes based on biostratigraphic events are used to correlate the rock sequences. A biostratigraphic framework in a particular area is also dependent on the numbers of fossil species in the sediments which are the result of three main factors: (1) paleotemperature, (2) whether dissolution of tests has taken place, and (3) the ability to extract the maximum numbers of species from the sediments.

There is a rule which both living and fossil species tend to follow: the lower the seawater temperature, the lower the species diversity. Because of the diversity gradient from equator to poles, it follows that normally there are fewer zonal markers available in the middle and high latitudes. This consequence can be easily illustrated by comparing the average duration of Cenozoic zones in the Tropical-Subtropical Province with those in the middle and high latitudes.

According to *Bolli and Saunders* [1985] there are 41 "N" and "P" zones in the Cenozoic Tropical-Subtropical Province with an average duration of 1.6 Ma, and there are 43 named zones with an average duration of 1.5 Ma. In contrast, *Jenkins* [1971, 1975]

named 26 zones in the Cenozoic mid-latitudes of the southwest Pacific with an average duration of 2.5 Ma. *Stott and Kennett* [1990] subdivided the Paleogene sequence of Maud Rise into 15 zones with average duration of 2.5 Ma, and *Berggren* [1992a] subdivided the Neogene of the Kerguelen Plateau into eight zones with an average duration of 3.75 Ma.

And what of the total numbers of species and their durations in the low and middle latitudes? According to H. M. Bolli (personal communication, 1992) there are over 300 species and subspecies in the Cenozoic Tropical-Subtropical Province. This number can be compared with 163 species recorded by *Jenkins* [1971] from New Zealand, to which can be added a further 30 species that have been discovered since then, bringing the total to 193; fortunately a large number of species are common to both areas.

The average species duration in the Cenozoic within the tropical-subtropical areas is 7.1 Ma [*Van Valen*, 1973], and this is also the duration for mid-latitude species of New Zealand [*Jenkins*, 1992c]. Therefore it is the greater number of species in the Tropical and Subtropical Province that makes the finer biostratigraphic subdivision possible.

The recognition of the dissolution of planktonic foraminiferal tests by *Berger* [1967, 1970] led *Jenkins and Orr* [1971, 1972] to suggest that where there has been active dissolution in deep-sea sediments, the more dissolution resistant taxa should be designated as zonal markers. The effects of the dissolution of the tests and reduced diversity resulted in the reduction of the normal nine mid-latitude zones in the Miocene to three zones at Deep Sea Drilling Project (DSDP) Leg 29, Site 278 (latitude 56°S) [*Jenkins*, 1975].

The extraction of fossil tests from a sediment sample is dependent on well established laboratory procedures and this is followed by the patient search for the zonal markers in the washed sediment, because some index species may be present in low numbers.

Accuracy and Reliability of Stratigraphic Markers

Species of planktonic foraminifera which have been chosen to mark zonal boundaries or datum planes have a varying degree of reliability for accurate correlation of fossil sequences, and *Jenkins* [1965] suggested the following order of reliability:

1.   There is evolutionary appearance of a species where the ancestor is also present; the evolution of a species can only occur once, and the event occurs simultaneously in all the oceans.

2.   A cryptogenic species initially appears. If the ancestor is known to exist elsewhere, then it is possible to calculate the time of migration and the degree of diachroneity.

3.   A species becomes extinct. Here two types of extinctions were recognized: (1) the extinction of a species which had existed previously as relatively large populations in the stratigraphic sequence and (2) a species whose numbers decrease slowly and appears to stagger on as low numbers before finally becoming extinct.

4.   Coiling ratio changes are considered to be less reliable for accurate correlation because the sequential changes are probably due to movements of water masses over a fixed point on the seafloor.

Zones and datums used in this work have been tentatively correlated against the time scale of *Berggren et al.* [1985a, b], a correlation which can only be tested as further correlations to the paleomagnetic record become available. A set of zones and selected datums has been tabulated for the Paleocene, the Eocene, the Oligocene, the early, middle, and late Miocene, and the Pliocene-Pleistocene.

Zonal Schemes

Some of the zonal schemes which have been devised for the mid-latitudes of the southwest Pacific area are shown in Figure 1. The first named Cenozoic planktonic foraminiferal zones were a set of 11 zones for an Oligocene-Miocene sequence from southeastern Australia [*Jenkins*, 1960]; *Carter*, [1958, 1964] has produced a set of numbered faunal units from the Aire District of Victoria. The named zones were tested and modified in New Zealand, and additional zones were defined to cover the whole of the Cenozoic [*Jenkins*, 1966a, 1967]. The onshore work in New Zealand by *Finlay* [1939a, b, c, d, 1940, 1947a, b], *Hornibrook* [1958, 1982, 1984], *Jenkins* [1964, 1965, 1966a, b, 1967, 1971], *Kennett* [1966], and *Scott* [1966, 1968a, b, c, 1979, 1980, 1982] has produced an important data set on Cenozoic planktonic foraminifera which has had an essential bearing on the Southern Ocean.

Further work in Australia by *Lindsay* [1967, 1969] and *Ludbrook and Lindsay* [1969] established zones in South Australia, and *Taylor* [1966] (and in the work of *Wopfner and Douglas* [1971]) established zones in Victoria. As can be seen in Figure 1, a number of the zones used in these two areas were the same as those which had been established by *Jenkins* [1960, 1966a, 1967]. In South Australia, *McGowran* [1973] used a set of zones which were different from those of *Lindsay* [1969] and *Ludbrook and Lindsay* [1969] for the middle-late Eocene, and he regarded biostratigraphic events as being more important than zones. Later, *McGowran* [1978a, b] appears to have abandoned the use of local zones in favor of datum planes which he considered to be isochronous between the southern mid-latitudes and the tropical sites. Both *Lindsay* [1986] and *McGowran and Beecroft* [1986] have abandoned the use of zones in favor of selected datum planes; these datum planes have been plotted against local lithostratigraphic units of the middle Eocene–middle Miocene [*Lindsay*, 1986] and against the tropical "P" and "N" zones of the Paleocene-Oligocene [*McGowran and Beecroft*, 1986]. Plotting the biostratigraphic events against the "P" and "N" zones in the mid-latitudes is fraught with difficulties and errors because of (1) lack of local radiometric ages, (2) lack of local paleomagnetic stratigraphy, and (3) the rarity or absence of the 'P" and "N" zonal markers which were originally used to delimit the tropical zones; charts produced in this way have an aura of ill-founded accuracy and are difficult to assess. All that can be derived from the chart by *McGowran and Beecroft* [1986] is that there are 28 selected homotaxial biostratigraphic events in the Paleocene-Oligocene of southern Australia; the supporting data of other biostratigraphic events, paleomagnetic stratigraphy, and radiometric ages are not given.

Kennett [1973] on DSDP Leg 21 produced a set of 12 zones covering the early Miocene-Pleistocene for sites in the Tasman Sea which extended as far south as Site 207 at latitude 37°S, while *Jenkins* [1975] on DSDP Leg 29 used 22 zones to subdivide the late Paleocene-Pleistocene in the southwest Pacific, and 19 zones were recorded in the middle Eocene-Pleistocene of the southeast Atlantic Leg 40 sites 360 and 362 [*Jenkins*, 1978; *Toumarkine*, 1978].

Kaneps [1975] did not recognize any zones in the Cenozoic of the Antarctic sites of DSDP Leg 28, while *Tjalsma* [1977] attempted to recognize a few tropical-subtropical zones at Site 329 (latitude 50°S) drilled on Leg 36 in the southwest Atlantic.

Srinivasan and Kennett [1981] reexamined South Pacific DSDP sites, and of particular importance were sites 206, 207, 281, and 284 in the Tasman Sea, where they produced a set of zones slightly different from those previously published by *Kennett* [1973] and *Jenkins* [1975].

Jenkins and Srinivasan [1985] examined faunas in a north-south traverse from the Equatorial Pacific southward into the Tasman Sea (sites 586–593) together with Site 594 off the east coast of the South Island of New Zealand in Subantarctic waters. Of importance to this work are sites 590–594, south of latitude 31°S.

In Argentina, Malumian and co-workers [*Malumian*, 1970; *Malumian et al.*, 1971] published data on Paleogene faunas, and *Jenkins* [1974] attempted a correlation with the southwest Pacific.

Ocean Drilling Program (ODP) drilling on Leg 113 at Maud Rise and legs 119 and 120 on the Kerguelen Plateau has produced much needed data on the high-latitude Cenozoic faunas, while ODP Leg 114 produced similar results for the southeast Atlantic mid-latitudes.

Stott and Kennett [1990] subdivided the Paleocene–late Oligocene into a set of Antarctic zones which were also used, on the Kerguelen Plateau, by *Huber* [1991] and by *Berggren* [1992a] with minor modifications. *Nocchi et al.* [1991] subdivided the Paleogene on ODP Leg 114 into both "local zones" based on the work of *Jenkins* [1985], *Ludbrook and Lindsay* [1969], and *Orr and Jenkins* [1977] and the "P" zones; *Brunner* [1991] was able to subdivide the late Miocene-Pleistocene using the southern mid-latitude zones [*Jenkins*, 1985].

Zonal schemes are subject to modification as original works are checked and as more data become available; thus *Liska* [1991] has recently found that the low-latitude zone N15 is the equivalent of zone N16 in the late Miocene.

THE CENOZOIC

There are good deep-sea and land mid-latitude Cenozoic planktonic foraminiferal records in the Southern Ocean, and ODP drilling of Maud Rise and the Kerguelen Plateau has provided excellent Paleogene and some Neogene sequences. Future deep-sea drilling will be necessary to obtain high-latitude Cenozoic records in the Pacific sector and Neogene sequences in the southern Indian Ocean. There was a significant reduction in species diversity in the southern high-latitude late Miocene-Pleistocene, and this has resulted in the paucity of datums that can be used for correlation.

Southwest Pacific: Paleocene

The early Paleocene is represented by the *P. eugubina* and *A. pauciloculata* zones and the lower part of

the *G. triloculinoides* Zone in the Waipara Section of New Zealand (latitude 42°S) (Figure 2); the upper *S. triloculinoides*, *P. pseudomenardii*, and *M. velascoensis* zones of the late Paleocene are also found in the Waipara Section. At DSDP Leg 29 Site 277 (latitude 52°S) south of New Zealand, the oldest sediments are from the late Paleocene *S. triloculinoides* Zone.

Indo-Atlantic: Paleocene

McGowran [1964] recorded the *P. pseudomenardii* Zone from the Perth Basin, and *McGowran and Beecroft* [1986] have recorded the ranges of *Planorotalites chapmani* and *Planorotalites haunsbergensis* from southern Australia. *Tjalsma* [1977] apparently identified a Paleocene fauna at DSDP Leg 36 Site 329 (latitude 50°S) when the first appearance of *Globanomalina wilcoxensis* was recorded in Core 32-1 as early Eocene with older Paleocene faunas below in cores 32-4 to 33 core catcher (cc). *Tjalsma* [1977] was able to place faunas in the *P. pseudomenardii* and *M. velascoensis* zones without the diagnostic zone fossils by limiting *Acarinina mckannai* to the *P. pseudomenardii* Zone, following the work of *Bolli* [1957] in Trinidad. This is no longer tenable because *A. mckannai* has a range of late Paleocene to middle Eocene in New Zealand [*Jenkins*, 1971], and in the tropical region it now has a range of *P. pusilla* Zone to *M. velascoensis* Zone [*Bolli and Saunders*, 1985].

Rögl [1976] recorded an early Paleocene *G. edita* Zone fauna from DSDP Leg 35 Site 323 (latitude 63°S) in the Bellingshausen Sea, which is probably equivalent to the *P. eugubina–G. pseudobulloides* zones of *Bolli and Saunders* [1985]; *Rögl* suggested that the lowermost *P. eugubina* Zone could be present.

Stott and Kennett [1990] subdivided the Paleocene at Maud Rise ODP sites 689 and 690 into five zones (Figure 3), and these zones were also used on Kerguelen Plateau ODP Leg 119 sites 738 and 744 by *Huber* [1991] and in part by *Berggren* [1992a] at ODP Leg 120 sites 747–750. *Huber* [1991] redefined the lower boundary Zone AP1a which is now marked at the first appearance of *Globoconusa daubjergensis*. The stratigraphic ranges of taxa at Maud Rise and the Kerguelen Plateau are shown in Figure 4.

Paleocene/Eocene Boundary

Because a boundary stratotype has not been chosen for the Paleocene/Eocene boundary, it is placed by convention at the top of the *M. velascoensis* Zone which is at the midpoint in Chron 24R, at 57.8 Ma [*Berggren et al.*, 1985b; *Berggren and Miller*, 1988].

Jenkins [1964, 1971, 1975] placed the Paleocene/Eocene boundary in New Zealand and in DSDP Leg 29 Site 277 at the first appearance of *Globanomalina wilcoxensis*, but there is a problem with this usage:

Figure 1

Age	New Zealand and Southwest Pacific [*Jenkins*, 1971, 1975]	New Zealand Planktic Foraminiferal Zones [*Jenkins*, 1966a, 1967]	Southwest Pacific DSDP Leg 29 [*Jenkins*, 1975]	Southeast Atlantic DSDP Leg 40 Sites 360, 362 [*Jenkins*, 1978]	Tasman Sea and Southwest Pacific [*Kennett*, 1973]	South Australia [after *Ludbrook* and *Lindsay*, 1969]
Pleistocene	G. truncatulinoides	G. inflata	G. (G.) truncatulinoides	G. truncatulinoides	G. truncatulinoides- G. tosaensis	
Late Pliocene	G. inflata		G. (T.) inflata	G. inflata	G. tosaensis / G. inflata	
Early Pliocene	G. puncticulata		G. (T.) puncticulata	G. puncticulata	G. crassaformis / G. puncticulata / G. margaritae	
Late Miocene	G. conomiozea	G. miozea sphericomiozea	G. (G.) conomiozea	G. conomiozea	G. conomiozea	
Late Miocene	G. miotumida	G. miotumida miotumida	G. (G.) miotumida miotumida	G. miotumida	G. nepenthes / G. continuosa	
Middle Miocene	G. mayeri	G. mayeri mayeri	G. (T.) mayeri mayeri	G. mayeri mayeri	G. mayeri	O. universa
Middle Miocene	O. suturalis	O. suturalis	O. suturalis	O. suturalis	O. suturalis	O. suturalis
Early Miocene	P. glomerosa curva	P. glomerosa curva	P. glomerosa curva	P. glomerosa curva	G. trilobus	P. glomerosa curva
Early Miocene	G. trilobus	G. trilobus trilobus	G. trilobus trilobus	G. trilobus trilobus		G. bisphericus / G. trilobus trilobus
Early Miocene	G. woodi connecta	G. woodi connecta	G. (G.) woodi connecta	G. woodi connecta		G. woodi
Early Miocene	G. woodi woodi	G. woodi woodi	G. (G.) woodi woodi	G. woodi woodi		
Early Miocene	G. dehiscens	G. dehiscens	G. dehiscens	G. dehiscens		G. dehiscens
Late Oligocene	G. euapertura	G. euapertura	G. (G.) euapertura	G. euapertura		G. euapertura / G. stavensis / G. labiacrassata
Early Oligocene	G. angiporoides	G. angiporoides angiporoides	G. (S.) angiporoides angiporoides	G. angiporoides		G. angiporoides angiporoides / C. cubensis
Oligocene	G. brevis	G. brevis	G. (G.) brevis	G. brevis		G. linaperta
Late Eocene	G. linaperta	G. linaperta	G. (S.) linaperta	G. linaperta		G. aculeata
Late Eocene	T. inconspicua	G. inconspicua	G. (T.) aculeata	G. luterbacheri		
Middle Eocene	G. index	G. index index	G. (G.) index	G. index		G. index index
Middle Eocene	A. primitiva	P. primitiva	P. primitiva	P. primitiva		P. primitiva
Middle Eocene						G. australiformis
Early Eocene	M. crater	G. crater crater	G. (M.) crater crater			
Early Eocene	P. wilcoxensis	G. wilcoxensis	G. wilcoxensis			
Late Paleocene	G. triloculinoides	G. triloculinoides	G. (S.) triloculinoides			
Early Paleocene	G. pauciloculata	G. pauciloculata				

Fig. 1. Correlation of sets of Cenozoic

Australia, Victoria [after *Taylor*, 1966; *Sigleton*, 1966; in *Wopfner and Douglas*, 1971]	Australia, Victoria Lakes Entrance Oil Shaft [*Jenkins*, 1960]	Southwest Pacific Site 281 [*Srinivasan and Kennett*, 1981]	Southwest Pacific DSDP Leg 90 [*Jenkins and Srinivasan*, 1985] Site 591	Site 593	Site 594
G. inflata	------------undefined----------	G. truncatulinoides	G. truncatulinoides / G. trun.-G. tosaensis	G. truncatulinoides	G. truncatulinoides
		G. inflata	G. tosaensis	G. inflata	G. inflata
			G. inflata		
		G. puncticulata	G. crassaformis	G. puncticulata	G. puncticulata
			G. puncticulata		
		G. conomiozea	G. sphericomiozea	G. conomiozea	G. sphericomiozea
			G. conomiozea		G. conomiozea
		N. pachyderma	G. nepenthes	G. miotumida	G. miotumida
			N. continuosa	N. continuosa	
G. lenguaensis	G. menardii miotumida	N. challengeri-	G. mayeri	G. mayeri	G. mayeri
G. mayeri	G. mayeri mayeri	G. mayeri	G. fohsi s.l.	G. peripheroacuta	G. peripheroronda
O. universa	O. universa	P. glomerosa-O. suturalis	O. suturalis	O. suturalis	O. conica
G. glomerosa / O. suturalis	C. glom. circularis / C. glomerosa curva	P. glomerosa	P. glomerosa curva	P. glomerosa curva	P. glomerosa curva
G. bisphericus	G. menardii praemenardii	G. miozea	G. miozea	G. miozea	G. miozea
G. trilobus s.s.	G. bisphericus			G. trilobus	
	G. triloba triloba				
G. woodi s.l.	G. woodi	C. dissimilis		G. connecta	
G. kugleri				G. woodi	
G. dehiscens	G. dehiscens dehiscens			G. dehiscens	
G. euapertura / G. (T.) opima s.l.	Pre-G. dehiscens			G. euapertura	
C. cubensis / G. testarugosa				S. angiporoides	
G. (T). gemma / G. angiporoides / G. brevis				G. brevis	
G. ampliapertura/ G. linaperta				G. linaperta	
G. index / H. primitiva / H. australis				G. aculeata	
----------------?------?------?----------					
G. pseudomenardii / G. chapmani					
T. aequa					
T. aff. acuta					
T. champani-ehrenbergi					

zones in the southern hemisphere.

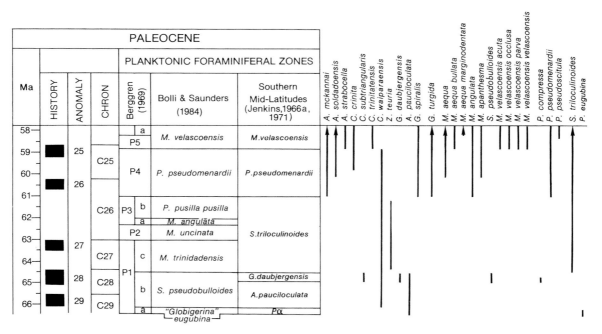

Fig. 2. Paleocene magnetic polarity and time scales after *Berggren et al.* [1985a] plotted against tropical zones [*Berggren*, 1969; *Bolli and Saunders*, 1985] and southern mid-latitude zones [*Jenkins*, 1966a, 1971] and ranges of selected species from the southwestern Pacific.

Planorotalites chapmani, the possible ancestor of *G. wilcoxensis*, has not been found in the southwestern Pacific, and therefore the latter's cryptogenic appearance is later in this region than its evolutionary appearance elsewhere. In southern Australia, *McGowran and Beecroft* [1986] recorded *P. chapmani* becoming extinct well before the appearance of *G. wilcoxensis* (their *P. pseudoiota*).

In the middle Waipara Section, New Zealand, there is a short interval between the extinction of *M. velascoensis* and the appearance of *G. wilcoxensis*, when *P. australiformis* made its appearance. There could therefore be good reason for placing the Paleocene/Eocene boundary at the initial appearance of *P. australiformis* which has a widespread occurrence in the southern hemisphere.

Nocchi et al. [1991] found it difficult to place the Paleocene/Eocene boundary at ODP Leg 114 South Atlantic mid-latitude sites. *Morozovella velascoensis* and other important tropical species were not present, so the boundary has been placed at the first occurrence of *Planorotalites pseudoscitulus* which occurs just before the appearance of *G. wilcoxensis*. According to *Berggren* [1992a], *G. wilcoxensis* occurs simultaneously with the extinction of *M. velascoensis*, which marks the boundary in low latitudes. *Morozovella velascoensis* has not been recorded at any of the high-latitude ODP sites drilled on legs 113, 119, and 120 [*Stott and Kennett*, 1990; *Huber*, 1991; *Berggren*, 1992a]. Unfor-

tunately, our present state of knowledge is such that we do not know enough about the degree of diachroneity of appearances and extinctions of taxa in middle and high latitudes when compared with low latitudes.

Stott and Kennett [1990] used the first appearance of *P. australiformis* to mark the Paleocene/Eocene boundary at Site 690; this taxon also defines the base of Zone AP5. Unfortunately, *P. australiformis* occurs much earlier at Site 689 in Zone AP3. *Huber* [1991] preferred to use the first appearance of *Acarinina wilcoxensis* to mark the boundary at Kerguelen Plateau Site 738, where the first appearance occurs at the AP5/AP6a zonal boundary.

Southwest Pacific: Eocene

Correlation of the tropical zones of *Bolli and Saunders* [1985] with zones established in the southwest Pacific [*Jenkins*, 1985] is not very accurate because very few good datums can be identified in the two areas. A correlation shown in Figure 5 is based on 12 datums; these datums are discussed in descending stratigraphic order: first the New Zealand position followed by the level in the tropical and subtropical areas [*Bolli and Saunders*, 1985] (Figures 5 and 9).

1. The extinction of *Globigerinatheka index* is taken to mark the Eocene/Oligocene boundary which has been placed at 36.6 Ma in New Zealand [*Berggren et al.*, 1985a], but it appears to have become extinct slightly

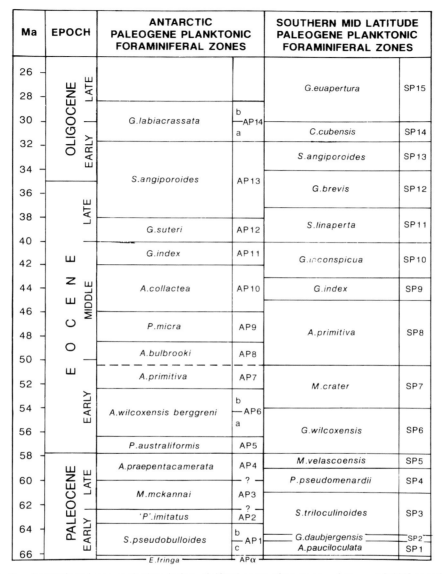

Ma	EPOCH			ANTARCTIC PALEOGENE PLANKTONIC FORAMINIFERAL ZONES		SOUTHERN MID LATITUDE PALEOGENE PLANKTONIC FORAMINIFERAL ZONES	
26 – 28	OLIGOCENE	LATE				G.euapertura	SP 15
30		EARLY		G.labiacrassata	b AP14 a	C.cubensis	SP 14
32						S.angiporoides	SP 13
34 – 36				S.angiporoides	AP 13	G.brevis	SP 12
38	EOCENE	LATE		G.suteri	AP 12	S.linaperta	SP 11
40 – 42		MIDDLE		G.index	AP 11	G.inconspicua	SP 10
44				A.collactea	AP 10	G.index	SP 9
46 – 48				P.micra	AP 9	A.primitiva	SP 8
50				A.bulbrooki	AP 8		
52		EARLY		A.primitiva	AP 7	M.crater	SP 7
54				A.wilcoxensis berggreni	b AP6 a	G.wilcoxensis	SP 6
56							
58				P.australiformis	AP 5		
60	PALEOCENE	LATE		A.praepentacamerata	AP 4	M.velascoensis	SP 5
					?	P.pseudomenardii	SP 4
62				M.mckannai	AP 3	S.triloculinoides	SP 3
64		EARLY		'P'.imitatus	? AP 2		
				S.pseudobulloides	b AP 1 c	G.daubjergensis	SP 2
66						A.pauciloculata	SP 1
				E.fringa	APα		

Fig. 3. Antarctic Paleogene planktonic foraminiferal zones after *Stott and Kennett* [1990] correlated with the southern mid-latitude zones.

earlier in the tropics at the P16/P17 boundary at 37.2 Ma.

2. *Subbotina linaperta* became extinct at the same time as *G. index*, and its range is dotted to the Eocene-Oligocene boundary in the tropics.

3. *Paragloborotalia nana* first appeared in the middle of the *A. aculeata* Zone and toward the top of the *T. rohri* Zone in the tropical areas.

4. *Truncorotaloides collactea* became extinct toward the top of the *A. aculeata* Zone and at the upper boundary at the *T. rohri* Zone.

5. *Globorotaloides suteri* made its first appearance

in the middle of the *A. aculeata* Zone and is at the base of the *O. beckmanni* Zone.

6. *Chiloguembelina cubensis* first appeared at the base of the *A. aculeata* Zone and also at the *M. lehneri/O. backmanni* zonal boundary.

7. *Acarinina primitiva* became extinct in New Zealand at the top of the *G. index* Zone (except for one locality at McCoulloughs Bridge; see *Jenkins* [1971]) and at the upper boundary of the *O. beckmanni* Zone.

8. *Guembelitrioides higginsi* became extinct in the upper half of the *G. index* Zone and at the *M. lehneri/O. beckmanni* boundary.

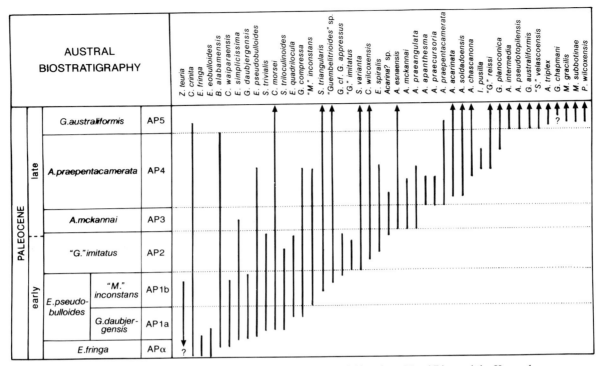

Fig. 4.   Stratigraphic ranges of Paleocene planktonic foraminifera from Maud Rise and the Kerguelen Plateau from a figure supplied by B. Huber.

9. *Acarinina soldadoensis* became extinct in the lower part of the *G. index* Zone and at the *A. penta-camerata/H. nuttalli* zonal boundary.

10. *Morozovella dolabrata* (= *M. lensiformis*) became extinct toward the top of the *M. crater* Zone and in the *M. aragonensis* Zone.

11. *Morozovella crater* (= *M. aragonensis*) made its initial appearance at the base of the *M. crater* Zone and base of the *M. formosa* Zone.

12. *Guembelitrioides higginsi* made its first appearance at the lower part of the *A. primitiva* Zone and at the base of the *A. pentacamerata* Zone.

Indo-Atlantic: Eocene

Toumarkine [1978] described a south to north traverse in the southeast Atlantic at DSDP Leg 40 sites 360, 362, 363, and 364 (Figure 6). It can be seen that the tropical zones were recognized at sites 363 and 364, but at the southernmost Site 360 the zones are grouped together because of the lack of zonal markers. Even so, at latitude 35°S there were still a large number of tropical species present in the Eocene.

Zonal schemes similar to the zonal scheme established in New Zealand have been used in South Australia (Figure 1), and the differences emphasize the faunal differences between the two areas; *Cooper* [1979] used the zones of *Lindsay* [1969] in South Australia. Further

south, *Tjalsma* [1977] recorded a nondescript early Eocene fauna from the southeast Atlantic DSDP Leg 36 Site 329, and *Kaneps* [1975] recorded a late Eocene fauna from DSDP Leg 28 Site 267 in the south Indian Ocean. *Nocchi et al.* [1991] subdivided the ODP Leg 114 sites in the South Atlantic Eocene into six ''local zones'' which had previously been used by *Jenkins* [1985], *Ludbrook and Lindsay* [1969], and *Orr and Jenkins* [1977]. By using the same data on the stratigraphic ranges, the southern mid-latitude zones can be recognized at these sites (Figure 7).

Stott and Kennett [1990] and *Huber* [1991] subdivided the Eocene at the Maud Rise ODP Leg 113 sites 689 and 690 and Kerguelen Plateau ODP Leg 119 sites 738 and 744 into the same set of zones (Figure 3), while *Berggren* [1992a] used similar but slightly modified zones for ODP Leg 120 sites 747, 748, and 749 for the South Kerguelen Eocene. The Antarctic zones used by *Stott and Kennett* [1990] and *Huber* [1991] have been correlated with the southern mid-latitude zones (Figure 3).

Important datums in the high latitudes include the initial appearances and extinctions of (1) *Cassigerinelloita amekiensis* at ODP Site 738 in the early-middle Eocene on the Kerguelen Plateau [*Huber*, 1991] and on Maud Rise Site 689 as *Globigerina* sp. B [*Stott and Kennett*, 1990]; (2) *Jenkinsina triseriata* in the middle Eocene of Site 738 and as *Chiloguembelitria* sp. [*Stott*

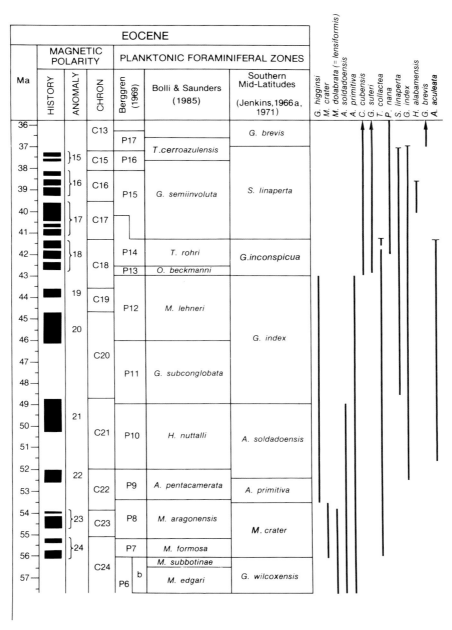

Fig. 5. Eocene magnetic polarity and time scales after *Berggren et al.* [1985a] plotted against tropical zones [*Berggren*, 1969; *Bolli and Saunders*, 1985], southern mid-latitude zones [*Jenkins*, 1966a, 1971], and ranges of selected species from the southwestern Pacific.

and Kennett, 1990] on Maud Rise Site 689; and (3) *Praetenuitella insolita* in the late Eocene at ODP Site 738 and as *T. insolita* on Maud Rise Site 689.

Eocene/Oligocene Boundary

In New Zealand and in the southwestern Pacific the extinction of *G. index* has been taken to mark the Eocene/Oligocene boundary [*Jenkins*, 1971, 1975; *Jen-

kins and Srinivasan*, 1985]; this is also the case in South Australia [*Ludbrook and Lindsay*, 1969; *Lindsay*, 1986]. *Nocchi et al.* [1991] placed the Eocene/Oligocene boundary in ODP Leg 114 sites 699A and 703A in the South Atlantic just after the extinction of the *G. index* in their *P. insolita* Zone (= *G. brevis* Zone).

At the high-latitude sites drilled on Maud Rise and Kerguelen Plateau, *Stott and Kennett* [1990] placed the

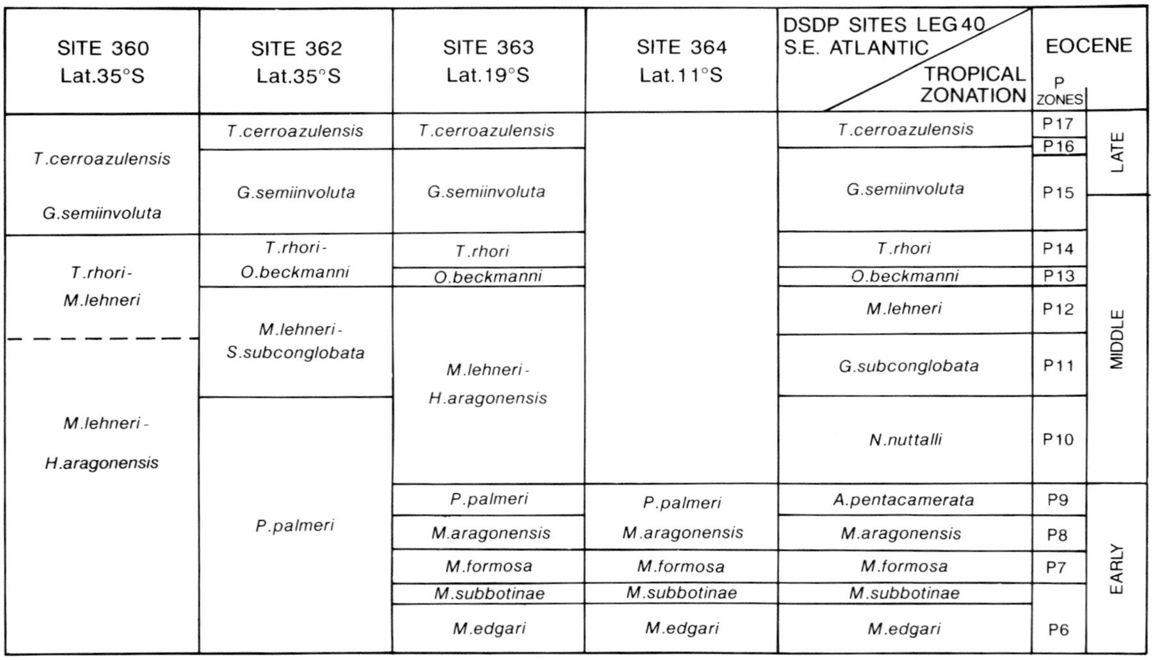

SITE 360 Lat.35°S	SITE 362 Lat.35°S	SITE 363 Lat.19°S	SITE 364 Lat.11°S	DSDP SITES LEG 40 S.E. ATLANTIC TROPICAL ZONATION	P ZONES	EOCENE
T.cerroazulensis	T.cerroazulensis	T.cerroazulensis		T.cerroazulensis	P17	LATE
					P16	LATE
G.semiinvoluta	G.semiinvoluta	G.semiinvoluta		G.semiinvoluta	P15	LATE
T.rhori- M.lehneri	T.rhori- O.beckmanni	T.rhori		T.rhori	P14	MIDDLE
		O.beckmanni		O.beckmanni	P13	MIDDLE
	M.lehneri- S.subconglobata	M.lehneri- H.aragonensis		M.lehneri	P12	MIDDLE
M.lehneri- H.aragonensis				G.subconglobata	P11	MIDDLE
				N.nuttalli	P10	MIDDLE
	P.palmeri	P.palmeri	P.palmeri	A.pentacamerata	P9	EARLY
		M.aragonensis	M.aragonensis	M.aragonensis	P8	EARLY
		M.formosa	M.formosa	M.formosa	P7	EARLY
		M.subbotinae	M.subbotinae	M.subbotinae		EARLY
		M.edgari	M.edgari	M.edgari	P6	EARLY

Fig. 6. Eocene zones in a north-south traverse in the southeastern Atlantic after *Toumarkine* [1978].

boundary in Chron 13N at ODP Leg 113 sites 689 and 690 which is later than the extinction of *G. index*, while both *Huber* [1991] and *Berggren* [1992a] preferred to use the extinction of this taxon to mark the boundary at the Kerguelen Plateau.

Hess et al. [1989], using ^{87}Sr/^{86}Sr stratigraphy, have concluded that the *Globigerinatheka* extinction in the Pacific and Atlantic oceans occurred isochronously at 37 Ma. This is slightly older than the age estimate of 36.6 Ma by *Berggren et al.*, [1985a] for the Eocene/Oligocene boundary and much older than the ~34 Ma at the proposed boundary stratotype at Ancona, Italy, in Chron 13R$_1$ [*Premoli-Silva et al.*, 1988]. The Eocene/Oligocene boundary marked by the extinction of *G. index* at ~37 Ma in the southern hemisphere is stratigraphically too old, and some other fossil markers will have to be found to identify the boundary at ~34 Ma.

Southwestern Pacific: Oligocene

The upper boundary of the Oligocene has been placed at the first appearance of *Globoquadrina dehiscens* following the work of *Jenkins* [1971] and *Kennett and Srinivasan* [1985]. This first appearance seems to be diachronous from south to north. *Jenkins* [1971] and *Bolli and Saunders* [1985], while placing the Oligocene/Miocene boundary in the tropical belt at the base of Zone N4, recorded the first appearance of *G. dehiscens* at the base of N5, which is much later than in the southwestern Pacific. Nevertheless the first appearance

of *G. dehiscens* may be a reasonably good marker within the mid-latitudes.

The stratigraphic ranges of Oligocene planktonic foraminifera have been well documented in New Zealand, and these data have been supplemented with records at DSDP Leg 29 sites 267, 277, 278, and 282 in the southwestern Pacific. The early Oligocene has been subdivided into the upper part of the *G. brevis* Zone, followed by the *S. angiporoides* and *C. cubensis* zones with the *G. euapertura* Zone in the late Oligocene (Figure 8).

Indo-Atlantic: Oligocene

The onshore record from South Australia and Victoria is very good [*Carter*, 1958, 1964; *Jenkins*, 1960; *Ludbrook and Lindsay*, 1969]. The evolutionary appearance of *G. dehiscens* was not recorded in South Australia where the Oligocene/Miocene boundary coincided with the appearance of *Globigerina woodi* [*Ludbrook and Lindsay*, 1969], but *G. dehiscens* with its unnamed ancestor occurred in Victoria [*Jenkins*, 1960]. The South Australian early Oligocene fauna is similar to that in the southwestern Pacific, but there are important differences which have resulted in different zonal schemes (Figure 9).

Tjalsma [1977] recorded an Oligocene fauna from DSDP Leg 36 Site 329 in the southwestern Atlantic and attempted to correlate the fauna with the "P" zones. It is obvious from the recorded species that the sequence

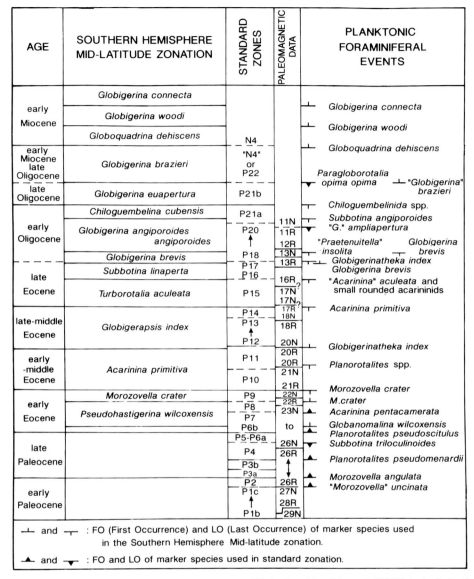

Fig. 7. Planktonic foraminiferal zonal scheme modified after *Nocchi et al.* [1991] to include the southern mid-latitude zones [*Jenkins*, 1985].

is:

28-1, 97-9 - 15-17   *G. euapertura* Zone (?)

← *C. cubensis* extinction

28 cc   *C. cubensis* Zone

← *S. angiporoides* extinction

31 cc - 29-1, 123-125   *S. angiporoides* Zone

The *G. euapertura* Zone is questioned because Tjalsma recorded *G. woodi* in 29 cc and 28 cc, which is stratigraphically too low, but this was possibly due to downhole contamination.

A full record of the Oligocene exists in DSDP Leg 40

sites 360, 362, and 363 from the southeastern Atlantic [*Toumarkine*, 1978]. The fauna at Site 360 (latitude 35°S) is a typical mid-latitude fauna with the addition of the warmer water *Catapsydrax dissimilis*, *Globoquadrina venezuelana*, *G. increbescens*, *G. galavasi*, *G. tripartita*, and *Paragloborotalia opima*. Warmer water taxa are present at sites 362 and 363 (latitude 19°S), including *Subbotina eocenica*, *G. venezuelana*, *Globanomalina barbadoensis*, *Globigerina angulisuturalis*, *G. selli*, *G. gortanii*, and *G. binaiensis*.

P. N. Webb (in the work of *Kaneps* [1975]) recorded a low-diversity fauna from DSDP Leg 28 Site 267 (latitude

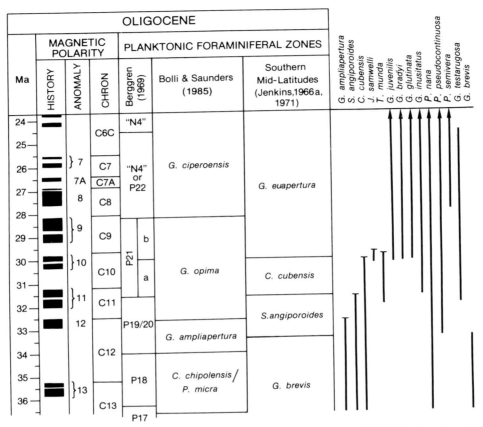

Fig. 8.   Oligocene magnetic polarity and time scales after *Berggren et al.* [1985*a*] plotted against tropical zones [*Berggren*, 1969; *Bolli and Saunders*, 1985], southern mid-latitude zones [*Jenkins*, 1966*a*; 1971], and ranges of selected species from the southwestern Pacific.

59°S) in the southern Indian Ocean and Site 274 (latitude 64°S) in the southwestern Pacific with specimens of *S. angiporoides* and *C. dissimilis*. It is probable that the low diversity of this early Oligocene fauna is due to dissolution, because both recorded species are solution resistant.

Nocchi et al. [1991] recorded the stratigraphic ranges of Oligocene planktonic foraminifera at the South Atlantic ODP Leg 114 sites 699A, 703A, and 704B and subdivided the sequences into six "local zones" previously established by *Jenkins* [1985], *Ludbrook and Lindsay* [1969], and *Orr and Jenkins* [1977]. Using the same data, it has been possible to identify the southern mid-latitude zones of *Jenkins* [1985] (Figure 7).

At the high-latitude sites on Maud Rise ODP Leg 113 sites 689 and 690 and Kerguelen Plateau ODP Leg 119 sites 738 and 744, *Stott and Kennett* [1990] and *Huber* [1991] subdivided the early and part of the late Oligocene into the upper part of the *S. angiporoides* Zone and the *G. labiacrassata* Zone (Figure 3). *Berggren* [1992*a*] subdivided the southern Kerguelen Plateau ODP Leg 120 sites 747, 748, and 749 into the following

zones: *S. angiporoides* at the base, followed by the *C. cubensis*, *G. labiacrassata*, and *G. euapertura* zones.

In the southern mid-latitudes, *Globigerina labiacrassata* became extinct in the early Miocene *G. dehiscens* Zone in New Zealand [*Jenkins*, 1971] and also at DSDP Site 593 in the Tasman Sea [*Jenkins and Srinivasan*, 1985]. This is in contrast to the southern Indian Ocean ODP sites where *G. labiacrassata* became extinct in the late Oligocene: Site 744 in subchron C8N [*Huber*, 1991] and in subchron C9N at sites 689 and 690 [*Stott and Kennett*, 1990] and sites 747 and 748 [*Berggren*, 1992*a*].

The Oligocene/Miocene Boundary

Berggren et al. [1985*b*] quoted *Jenkins* [1966*b*]: "The Oligocene/Miocene boundary is one of the most difficult and controversial boundaries in the Tertiary." The original statement was made in a paper which described for the first time the planktonic foraminiferal fauna in the type Aquitanian, the lowermost stage of the Miocene. The problem of the placement of the boundary has remained unresolved [*Berggren*, 1985*b*], and dif-

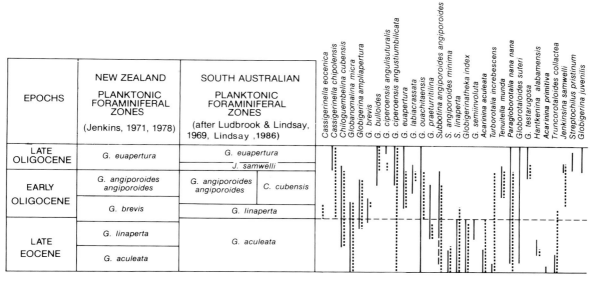

Fig. 9. A comparison of late Eocene–Oligocene zones and selected species from New Zealand (solid lines) and South Australia (dotted lines).

ferent workers have used various biostratigraphic events to mark the boundary.

Bolli and Saunders [1985] used the first appearance of *Globigerinoides primordius* to mark the base of the Miocene in the tropical areas and equated it with the base of Zone N4, while *Berggren et al.* [1985b] have equated the *G. kugleri* Zone with the Zone "N4" and used the base of the zones to mark the boundary. The work of *Bolli and Saunders* [1985] is followed in this publication in order to make use of their published stratigraphic ranges of taxa. To add to the uncertainty of placement of the Oligocene/Miocene boundary at ~23.6 Ma in the southwestern Pacific, the first appearance of *G. dehiscens* is used [*Jenkins and Srinivasan*, 1985], and as was previously noted, this event took place at the base of Zone N5 in the tropical areas at 21.8 Ma; *Berggren et al.* [1983] estimated the first appearance of *G. dehiscens* in the southwestern Atlantic (latitude 30°S) to be at 23.2 Ma. *Nocchi et al.* [1991] also used this species to mark the boundary at the ODP Leg 114 sites in the South Atlantic.

Both *Huber* [1991] and *Berggren* [1992b] have used the extinction of *Globigerina euapertura* to mark the Oligocene/Miocene boundary on the Kerguelen Plateau ODP sites. In southeastern Australia, *G. euapertura* became extinct in the early Miocene *G. dehiscens* Zone [*Jenkins*, 1960], and in New Zealand its extinction occurred later in the early Miocene *G. connecta* Zone [*Jenkins*, 1971].

The problems of the definition of the Oligocene/Miocene boundary will not be resolved until an international boundary stratotype and markers are chosen for the boundary.

Early Miocene

Up until 1981, Berggren placed the early/middle Miocene boundary at the first evolutionary appearance of *Orbulina suturalis*, but *Berggren et al.* [1985b] have positioned the boundary much lower stratigraphically at the base of Zone N.8, at the first appearance of *P. sicana*. In this publication the first appearance of *O. suturalis* marks the top of the early Miocene.

The correlation of the mid-latitude zones with the tropical zonations (Figure 10) is mainly based on the evolutionary appearances of *Globigerinoides trilobus*, *Globorotalia miozea*, and *Praeorbulina glomerosa curva*. In the southwestern Pacific the zonal markers *Globigerinoides primordius* and *Catapsydrax dissimilis* cannot be used because *G. primordius* is too rare and the extinction of *C. dissimilis* appears to be diachronous. *Catapsydrax stainforthi* is not present in the area, and *Globigerinatella insueta* made only one very brief appearance in New Zealand (latitude 39°S) in the *P. glomerosa curva* Zone. In New Zealand, *C. dissimilis* became extinct in the *G. connecta* Zone, while *Kaneps* [1975] thought that further south it ranged nearly to the top of the early Miocene at DSDP Leg 28 sites, but the evidence does not support this. At Site 266 (latitude 56°S) in the southern Indian Ocean, which has the only published reliable stratigraphic range chart, *C. dissimilis* became extinct before the first appearance of *Globorotalia zealandica* and therefore earlier than at sites 279 and 281 [*Kaneps*, 1975]. A low-diversity fauna was recovered from DSDP Leg 29 Site 278 in the southwestern Pacific (latitude 52°S) due mainly to dissolution [*Jenkins*, 1975].

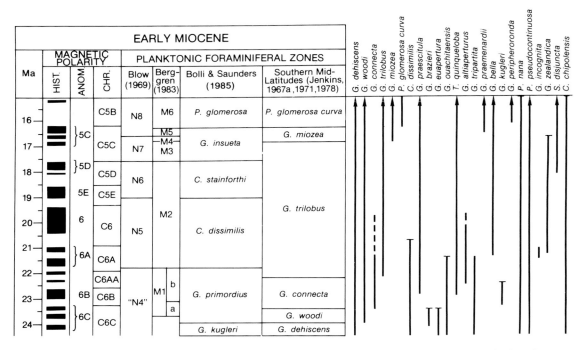

Fig. 10. Early Miocene magnetic polarity and time scales after *Berggren et al.* [1985*b*] plotted against tropical zones [*Blow*, 1969; *Bolli and Saunders*, 1985], temperate-subtropical zones [*Berggren et al.*, 1983], southern mid-latitude zones, and ranges of selected species.

Berggren et al. [1983] worked on material from Site 516 on the Rio Grande Rise of the southwestern Atlantic at latitude 30°S, and this provides an important link between the mid-latitude and the tropical-subtropical areas. From the estimated ages of first appearances and extinctions produced by *Berggren et al.* [1983], a number of events appear to be diachronous when compared with those of the southwestern Pacific, where both *Globorotalia praescitula* and *G. zealandica* appear stratigraphically earlier. The planktonic foraminiferal zones of *Berggren et al.* [1983], namely, M1–M6, appear to be a mixture of mid-latitude and tropical zones which have been termed temperate-subtropical zones by *Berggren et al.* [1985*b*].

Berggren [1992*b*] subdivided the early Miocene in southern Kerguelen Plateau ODP Leg 120 sites 747, 748, and 751 into the upper part of the *G. euapertura* Zone followed by the *G. brazieri*, *P. incognita*, and *G. praescitula* zones. These zones and the later Neogene zones have been correlated with the mid-latitude zones in Figure 11.

Several biostratigraphic early Miocene datums have been correlated to a strontium-isotope-calibrated age model at ODP Site 744 [*Huber*, 1991], and these datums have also been identified at ODP 120 sites [*Berggren*, 1992*b*]. At Site 744 the following were dated by the strontium isotope method: the extinctions of *Globigerina brazieri* (20.5 Ma) and *Catapsydrax dissimilis* (16.7

Ma) and the initial appearances of *Globorotalia praescitula* (18.1 Ma) and *Globorotalia zealandica* (18.1 Ma).

Middle Miocene

General. The top of the middle Miocene has been taken as the extinction of *Neogloboquadrina mayeri* in the mid-latitudes, but instead of having five zones (six if the *G. ruber* Zone is used) as in the tropical area, there are only three zones (Figure 12); in the subtropical area to the north of New Zealand it is possible to use a *G. peripheroacuta* Zone within the lower part of the *G. mayeri* Zone [*Kennett and Srinivasan*, 1983]. The main difference between the southern mid-latitude zonal scheme and that used in the tropical area is the use of the *G. fohsi* keeled taxa to subdivide the sequence.

The *N. nympha* Zone is new in the southern mid-latitudes, and its recognition is an attempt to subdivide the original *N. mayeri* Zone which had a duration of about 3.5 Ma.

N. nympha Zone.

Definition: The base of the zone is defined on the first evolutionary appearance of *N. nympha*, and the top of the zone is based on the extinction of *N. mayeri*.

Age: The age of *N. nympha* is late middle Miocene.

Remarks: Other biostratigraphic events within the zone include the first appearances of *Globorotalia me-*

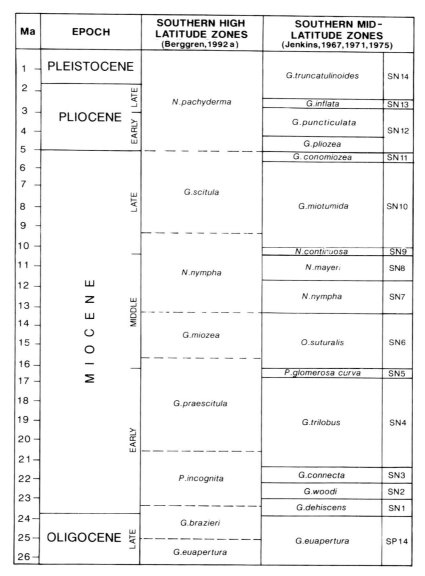

Ma	EPOCH	SOUTHERN HIGH LATITUDE ZONES (Berggren, 1992a)		SOUTHERN MID-LATITUDE ZONES (Jenkins, 1967, 1971, 1975)	
1	PLEISTOCENE			G. truncatulinoides	SN 14
2	PLIOCENE LATE EARLY	N. pachyderma		G. inflata	SN 13
3				G. puncticulata	SN 12
4					
5				G. pliozea	
6				G. conomiozea	SN 11
7	MIOCENE LATE	G. scitula		G. miotumida	SN 10
8					
9					
10				N. continuosa	SN 9
11		N. nympha		N. mayeri	SN 8
12	MIDDLE			N. nympha	SN 7
13					
14		G. miozea		O. suturalis	SN 6
15					
16				P. glomerosa curva	SN 5
17					
18	EARLY	G. praescitula		G. trilobus	SN 4
19					
20					
21					
22		P. incognita		G. connecta	SN 3
23				G. woodi	SN 2
24				G. dehiscens	SN 1
25	OLIGOCENE LATE	G. brazieri		G. euapertura	SP 14
26		G. euapertura			

Fig. 11.   A correlation of the southern high- and mid-latitude zones.

nardii, *G. explicationis*, and *G. scitula* and the extinction of *G. conica*.

Berggren [1992b] has also defined a *N. nympha* Zone from the middle Miocene of the southern Kerguelen Plateau where the zone is based on the total range of the taxon. Berggren found it difficult to place the extinction of *N. nympha* in the Kerguelen area at ODP Leg 120 sites 748 and 751.

In the southwest Pacific at DSDP Leg 29 Site 278 (latitude 56°S) it was possible to recognize the middle Miocene on the presence of the cooler water *G. conica* [*Jenkins*, 1975].

Late Miocene

The late Miocene has been divided into three zones (Figure 12): a very short duration *N. continuosa* Zone followed by the *G. miotumida* Zone which has a duration of about 3.5 Ma and above which is the *G. conomiozea* Zone which lasted about 1 Ma (Figure 13). This is in contrast to the low latitudes where the late Miocene has been subdivided into the three zones N15–N17 [*Berggren et al.*, 1985b], but *Liska* [1991] has recently shown that Zone N15 is the equivalent of Zone N16.

It is possible to further subdivide locally the *G.*

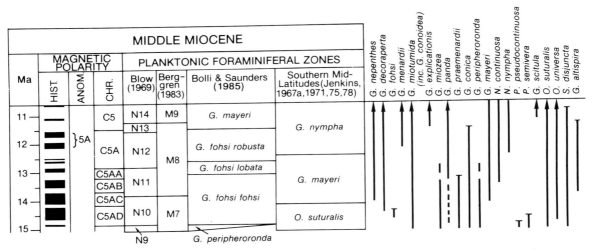

Fig. 12. Middle Miocene magnetic polarity and time scale after *Berggren et al.* [1985b] plotted against tropical zones [*Blow*, 1969; *Bolli and Saunders*, 1985], temperate-subtropical zones [*Berggren et al.*, 1983], southern mid-latitudes zones [*Jenkins*, 1967, 1971, 1975, 1978], and ranges of selected species from the southwestern Pacific.

miotumida Zone using the extinction of *G. dehiscens*, but this event is demonstrably diachronous, becoming extinct in the mid-latitudes in the early part of the late Miocene while ranging to the early Pliocene boundary in the tropical-subtropical areas [*Jenkins*, 1992a].

At ODP Leg 114 Site 704, *Brunner* [1991] recognized the *G. miotumida* and *G. sphericomiozea* zones in the South Atlantic, while *Berggren* [1992b] subdivided the southern Kerguelen Plateau late Miocene into the upper

part of the *N. nympha* Zone, which is succeeded by the *G. scitula* Zone at ODP Leg 120 Site 748.

The Miocene/Pliocene Boundary

The Miocene/Pliocene boundary was placed at the first appearance of both *Globorotalia inflata* and *G. crassiformis* in New Zealand [*Hornibrook*, 1958], but with the documentation of the *G. inflata* lineage it is

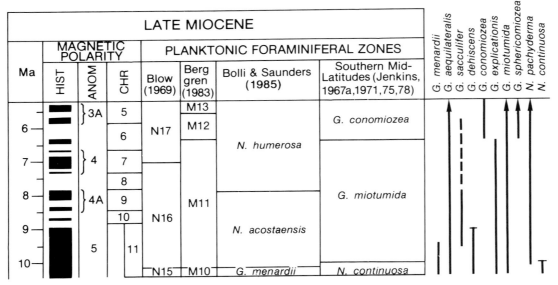

Fig. 13. Late Miocene magnetic polarity and time scale after *Berggren et al.* [1985b] plotted against tropical zones [*Blow*, 1969; *Bolli and Saunders*, 1985], temperate-subtropical zones [*Berggren et al.*, 1983], southern mid-latitude zones [*Jenkins*, 1967, 1971, 1975, 1978], and ranges of selected species from the southwestern Pacific.

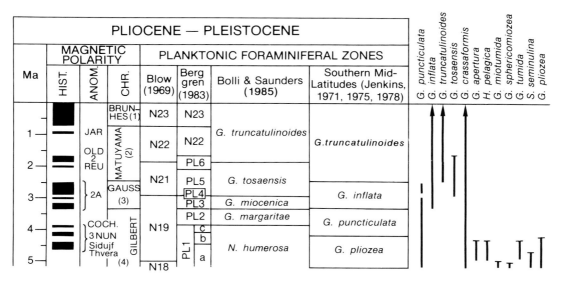

Fig. 14.   Pliocene-Pleistocene magnetic polarity and time scale after *Berggren et al.* [1985*b*] plotted against tropical zones [*Blow*, 1969; *Bolli and Saunders*, 1985], temperate-subtropical zones [*Berggren et al.*, 1983], southern mid-latitude zones [*Jenkins*, 1967, 1971, 1975, 1978], and ranges of selected species from the southwestern Pacific.

now placed at the evolutionary appearance of *G. puncticulata* in the southwestern Pacific [*Hornibrook*, 1982, 1984; *Kennett and Srinivasan*, 1983; *Jenkins and Srinivasan*, 1985].

At ODP Leg 114 Site 704 on Meteor Rise, *Brunner* [1991] did not attempt to place the Miocene/Pliocene boundary, but it can be placed at the first appearance of *G. puncticulata* estimated by Brunner to be at 4.3 Ma. *Berggren et al.* [1985*b*], using paleomagnetic and biostratigraphic data, have estimated the age of the boundary to be much older, at 5.4 Ma, while *Hodell and Kennett*, [1986] also using magnetobiostratigraphic criteria, have estimated the boundary to be at 4.8 Ma.

Pliocene

General.   In the southwestern Pacific the Pliocene can be subdivided into four zones: *Globorotalia pliozea*, *G. puncticulata*, and *G. inflata*, followed by the lower part of the *G. truncatulinoides* Zone (Figure 14). The lower part of the *G. truncatulinoides* Zone is now included in the late Pliocene because the Pliocene/ Pleistocene boundary has been estimated to be at 1.6 Ma [*Berggren et al.*, 1985*b*], but the first evolutionary appearance of this taxon is at 2.5 Ma in the southwestern Pacific [*Jenkins*, 1992*b*].

The *Globorotalia pliozea* Zone is new and is based on the work of *Hornibrook* [1982] on DSDP Leg 29 Site 284 and by *Jenkins and Srinivasan* [1985] on DSDP Leg 90, Tasman Sea Sites 590, 592, 593.

G. pliozea Zone.

Definition: The base of the zone is defined on the

evolutionary appearance of *G. puncticulata*, and the top of the zone is defined at the extinction of *G. pliozea*.

Age: The age of *G. pliozea* is early Pliocene.

Remarks: Events which took place within the zone include the extinction of both *G. sphericomiozea* and *G. miotumida*, which evolved into *G. pliozea* at the base of the zone.

At South Atlantic ODP Leg 114 Site 704, *Brunner* [1991] subdivided the Pliocene into the *G. puncticulata* and *G. inflata* zones, while *Berggren* [1992*b*] used the *N. pachyderma* Zone to identify the Pliocene at ODP Leg 120 sites 748 and 751 on the southern Kerguelen Plateau.

Pliocene/Pleistocene Boundary

The boundary stratotype has been chosen at the Vrica Section in Italy at the base of bed e, an event which has been estimated by *Berggren et al.* [1985*b*] to be at 1.6 Ma. In the southwestern Pacific there appears to be no planktonic foraminiferal marker at this level, and *Brunner* [1991] found it difficult to place the Pliocene/ Pleistocene boundary at Site 704 in the South Atlantic, which was drilled on ODP Leg 114.

Pleistocene

In the New Zealand area the Pleistocene is within the upper part of the *G. truncatulinoides* Zone, and in this interval there are a number of subzones based on the alternating sinistral and dextral populations of *N. pachyderma* [*Jenkins*, 1971]. At Site 704 in ODP Leg

114, the Pleistocene is within the upper part of the *G. inflata* Zone and includes the *G. truncatulinoides* Zone; the latter zone is marked by a nominate taxon which appeared in the area at 0.4 Ma [*Brunner*, 1991].

In the Subantarctic area of the Southern Ocean, *Kennett* [1970] subdivided the middle and late Pleistocene recovered from piston cores into three zones in ascending order: *G. puncticulata*, *G. inflata*, and *G. truncatulinoides* zones. As was previously stated, both *G. inflata* and *G. truncatulinoides* appeared much later in the Subantarctic region than in the mid-latitudes.

At the southern Kerguelen Plateau ODP Leg 120 sites 748 and 751, the Pleistocene is represented in the upper part of the *N. pachyderma* Zone [*Berggren*, 1992*b*].

CONCLUSIONS

Land-based marine sequences in New Zealand and Australia (South Australia and Victoria) were first subdivided into zones based on the ranges of planktonic foraminifera. These mid-latitude zonations were then extended to Deep Sea Drilling Project sites drilled on legs 28, 29, 35, 40, and 90. Sites cored during the Ocean Drilling Program during 1986–1988 on Leg 113 on Maud Rise and on Leg 119 and Leg 120 on the Kerguelen Plateau have yielded high-latitude faunas which have been used to subdivide the Cenozoic; mid-latitude ODP sites cored on Leg 114 in the South Atlantic have been used to supplement southern mid-latitude data.

Further deep-sea drilling is essential to improve our knowledge of high-latitude Cenozoic planktonic foraminiferal faunas in the Pacific sector of the Southern Ocean and Neogene sequences in the southern Indian Ocean.

It is difficult to recognize series boundaries in the high-latitude sites, but this situation will possibly improve when boundary stratotypes with paleomagnetic and isotopic data are chosen for these international boundaries.

Acknowledgments. I would like to thank Carol Whale and Valerie Deisler for typing the manuscript, and Judith Jenkins for checking it; John Taylor, Lin Norton and Gaye Evans drafted the figures and tables; my thanks also go to the referees B. Huber, J. P. Kennett and W. A. Berggren who made valuable suggestions which improved the paper.

REFERENCES

Berger, W. H., Foraminiferal ooze: Solution at depths, *Science*, *156*, 383–385, 1967.

Berger, W. H., Planktonic foraminifera: Selective solution and the lysocline, *Mar. Geol.*, *8*, 111–138, 1970.

Berggren, W. A., Rates of evolution of some Cenozoic planktonic foraminifera, *Micropaleontology*, *15*, 351–365, 1969.

Berggren, W. A., Paleogene planktonic foraminifer magnetobiostratigraphy of the southern Kerguelen Plateau (sites 747–749), *Proc. Ocean Drill. Program Sci. Results*, *120*, 551–568, 1992*a*.

Berggren, W. A., Neogene planktonic foraminifer magnetobio-stratigraphy of the southern Kerguelen Plateau (sites 747, 748 and 751), *Proc. Ocean Drill. Program Sci. Results*, *120*, 631–647, 1992*b*.

Berggren, W. A., and K. Miller, Paleogene tropical planktonic foraminiferal biostratigraphy and magnetobiochronology, *Micropaleontology*, *34*, 362–380, 1988.

Berggren, W. A., M. P. Aubry, and N. Hamilton, Neogene magnetobiostratigraphy of Deep Sea Drilling Project Site 516 (Rio Grande Rise, South Atlantic), *Initial Rep. Deep Sea Drill. Proj.*, *72*, 675–713, 1983.

Berggren, W. A., D. V. Kent, and J. J. Flynn, Paleogene geochronology and chronostratigraphy, The Chronology of the Geological Record, *Geol. Soc. London Mem.*, *10*, 141–186, 1985*a*.

Berggren, W. A., D. V. Kent, and J. A. van Couvering, The Neogene, Part 2, Neogene geochronology and chronostratigraphy, The Chronology of the Geological Record, *Geol. Soc. London Mem.*, *10*, 211–260, 1985*b*.

Blow, W. H., Late Middle Eocene to Recent planktonic foraminiferal biostratigraphy, in R. Bronnimann and H. H. Renz (Eds), *Proc. First Intern. Conf. Plank. Microfossils*, Geneva, *1*, 199–421, E. J. Brill, Leiden, 1969.

Bolli, H. M., Planktonic foraminifera from the Oligocene-Miocene Cipero and Lengua formations of Trinidad, B.W.I., *Bull. U.S. Natl. Mus.*, *215*, 97–123, 1957.

Bolli, H. M., and J. B. Saunders, Oligocene to Holocene low latitude planktic foraminifera, in *Plankton Stratigraphy*, edited by H. M. Bolli, J. B. Saunders, and K. Perch-Nielsen, pp. 155–262, Cambridge University Press, New York, 1985.

Brunner, C. A., Latest Miocene to Quaternary biostratigraphy and paleoceanography, Site 704, Subantarctic South Atlantic Ocean, *Proc. Ocean Drill. Program Sci. Results*, *114*, 201–215, 1991.

Carter, A. N., Tertiary foraminifera from the Aire District, Victoria, *Bull. Geol. Surv. Victoria*, *55*, 1–76, 1958.

Carter, A. N., Tertiary foraminifera from Gippsland, Victoria, and their stratigraphical significance, *Mem. Geol. Surv. Victoria*, *23*, 1–154, 1964.

Cooper, B. J., Eocene to Miocene stratigraphy of the Willunga Embayment, *Rep. Invest. 50*, 101 pp., Dep. of Mines and Energy, Geol. Surv. of South Aust., Adelaide, Australia, 1979.

Finlay, H. J., New Zealand foraminifera: Key species in stratigraphy, No. 1, *Trans. R. Soc. N. Z.*, *68*, 504–533, 1939*a*.

Finlay, H. J., New Zealand foraminifera: The occurrence of *Rzehakina*, *Hantkenina*, *Rotaliatina* and *Zeauvigerina*, *Trans. R. Soc. N. Z.*, *68*, 534–543, 1939*b*.

Finlay, H. J., New Zealand foraminifera: Key species in stratigraphy, No. 2, *Trans. R. Soc. N. Z.*, *69*, 89–128, 1939*c*.

Finlay, H. J., New Zealand foraminifera: Key species in stratigraphy, No. 3, *Trans. R. Soc. N. Z.*, *69*, 309–329, 1939*d*.

Finlay, H. J., New Zealand foraminifera: Key species in stratigraphy, No. 4, *Trans. R. Soc. N. Z.*, *69*, 448–472, 1940.

Finlay, H. J., New Zealand foraminifera: Key species in stratigraphy, No. 5, *N. Z. J. Sci. Technol.*, *B28*, 259–292, 1947*a*.

Finlay, H. J., The foraminiferal evidence for Tertiary trans-Tasman correlation, *Trans. R. Soc. N. Z.*, *76*, 327–352, 1947*b*.

Hess, J., L. D. Stott, M. Bender, J.-G. Schilling, and J. P. Kennett, The Oligocene marine microfossil record: Age assessments using strontium isotopes, *Paleoceanography*, *4*, 655–679, 1989.

Hodell, D. A., and J. P. Kennett, Late Miocene–early Pliocene stratigraphy and paleoceanography of the South Atlantic and southwest Pacific oceans: A synthesis, *Paleoceanography*, *1*, 285–311, 1986.

Hornibrook, N. de B., New Zealand Upper Cretaceous and Tertiary foraminiferal zones and some overseas correlations, *Micropaleontology*, *4*, 25–38, 1958.

Hornibrook, N. de B., Late Miocene to Pleistocene *Globorotalia* (Foraminiferida) from DSDP Leg 29, Site 284, southwest Pacific, *N. Z. J. Geol. Geophys.*, *25*, 83–99, 1982.

Hornibrook, N. de B., *Globorotalia* (planktonic foraminifera) at the Miocene/Pliocene boundary in New Zealand, *Palaeogeogr. Palaeoclimatol. Palaeoecol.*, *46*, 107–117, 1984.

Huber, B., Paleogene and early Neogene planktonic foraminifer biostratigraphy of sites 738 and 744, Kerguelen Plateau (south Indian Ocean), *Proc. Ocean Drill. Program Sci. Results*, *119*, 427–449, 1991.

Jenkins, D. G., Planktonic foraminifera from the Lakes Entrance Oil Shaft, Victoria, Australia, *Micropaleontology*, *6*, 345–371, 1960.

Jenkins, D. G., Location of the Danian–upper Paleocene–Eocene boundaries in North Canterbury, *N. Z. J. Geol. Geophys.*, *7*, 890–891, 1964.

Jenkins, D. G., Planktonic foraminifera and Tertiary intercontinental correlations, *Micropaleontology*, *11*, 265–277, 1965.

Jenkins, D. G., Planktonic foraminiferal zones and new taxa from the Danian to lower Miocene of New Zealand, *N. Z. J. Geol. Geophys.*, *8*, 1088–1126, 1966a.

Jenkins, D. G., Planktonic foraminifera from the type Aquitanian-Burdigalian of France, *Spec. Pub. Cushman Found. Foraminiferal Res.*, *17*, 1–15, 1966b.

Jenkins, D. G., Planktonic foraminiferal zones and new taxa from the lower Miocene to the Pleistocene of New Zealand, *N. Z. J. Geol. Geophys.*, *10*, 1064–1078, 1967.

Jenkins, D. G., New Zealand Cenozoic planktonic foraminifera, *Paleontol. Bull.*, *42*, 278 pp., N. Z. Geol. Surv., Wellington, New Zealand, 1971.

Jenkins, D. G., The present status and future progress in the study of Cenozoic Planktonic foraminifera, *Rev. Esp. Micropaleontol.*, *5*, 133–146, 1973.

Jenkins, D. G., Paleogene planktonic foraminifera of New Zealand and the Austral Region, *J. Foraminiferal Res.*, *4*, 155–170, 1974.

Jenkins, D. G., Cenozoic planktonic foraminiferal biostratigraphy of the southwestern Pacific and Tasman Sea—DSDP Leg 29, *Initial Rep. Deep Sea Drill. Proj.*, *29*, 449–467, 1975.

Jenkins, D. G., Neogene planktonic foraminifers from DSDP Leg 40 sites 360 and 362 in the southeastern Atlantic, *Initial Rep. Deep Sea Drill. Proj.*, *40*, 723–739, 1978.

Jenkins, D. G., Southern mid-latitude Paleocene to Holocene planktic foraminifera, in *Plankton Stratigraphy*, edited by H. M. Bolli, J. B. Saunders, and K. Perch-Nielsen, pp. 263–282, Cambridge University Press, New York, 1985.

Jenkins, D. G., The paleogeography, evolution and extinction of late Miocene–Pleistocene planktonic foraminifera from the southwest Pacific, in *Centenary of Micropaleontology in Japan*, pp. 27–35, Terra Scientific, Tokyo, 1992a.

Jenkins, D. G., The late Cenozoic *Globorotalia truncatulinoides* datum-plane in the Atlantic, Pacific and Indian oceans, in *High Resolution Stratigraphy in Ancient Modern Marine Sequences*, edited by E. Hailwood and R. Kidd, pp. 127–130, Geological Society of London, London, 1992b.

Jenkins, D. G., Predicting extinctions of some extant planktonic foraminifera, *Mar. Micropaleontol.*, *19*, 239–243, 1992c.

Jenkins, D. G., and W. N. Orr, Cenozoic planktonic foraminiferal zonation and the problem of test solutions, *Rev. Esp. Micropaleontol.*, *3*, 301–304, 1971.

Jenkins, D. G., and W. N. Orr, Planktonic foraminiferal biostratigraphy of the eastern equatorial Pacific—DSDP Leg 9, *Initial Rep. Deep Sea Drill. Proj.*, *9*, 1060–1193, 1972.

Jenkins, D. G., and M. S. Srinivasan, Cenozoic planktonic foraminifera from the equator to the Sub-Antarctic of the southwest Pacific, *Initial Rep. Deep Sea Drill. Proj.*, *90*, 795–834, 1985.

Kaneps, G. G., Cenozoic planktonic foraminifera from Antarctic deep-sea sediments, Leg 28, DSDP, *Initial Rep. Deep Sea Drill. Proj.*, *28*, 573–583, 1975.

Kennett, J. P., The *Globorotalia crassaformis* bioseries in North Westland and Marlborough, New Zealand, *Micropaleontology*, *12*, 235–245, 1966.

Kennett, J. P., Pleistocene paleoclimates and foraminiferal biostratigraphy in Subantarctic deep-sea cores, *Deep Sea Res.*, *17*, 125–140, 1970.

Kennett, J. P., Middle and late Cenozoic planktonic foraminiferal biostratigraphy of the southwest Pacific—DSDP Leg 21, *Initial Rep. Deep Sea Drill. Proj.*, *21*, 575–640, 1973.

Kennett, J. P., and M. S. Srinivasan, *Neogene Planktonic Foraminifera, A Phylogenetic Atlas*, 265 pp., Hutchinson Ross, Stroudsburg, Pa., 1983.

Lindsay, J. M., Foraminifera and stratigraphy of the type section of Port Willunga Beds, Aldinga Bay, South Australia, *Trans. R. Soc. South Aust.*, *91*, 93–110, 1967.

Lindsay, J. M., Cainozoic foraminifera and stratigraphy of the Adelaide Plains sub-basin, South Australia, *Bull. Geol. Surv. South Aust.*, *42*, 60 pp., 1969.

Lindsay, J. M., Aspects of South Australian Tertiary foraminiferal biostratigraphy, with emphasis on studies of *Massilina* and *Subbotina*, N. H. Ludbrook Honour Volume, *Spec. Publ. South Aust. Dep. Mines Energy*, *5*, 187–231, 1986.

Liska, R. D., The history, age and significance of the *Globorotalia menardii* Zone in Trinidad and Tobago, West Indies, *Micropaleontology*, *37*, 173–182, 1991.

Ludbrook, N. H., and J. M. Lindsay, Tertiary foraminiferal zones in South Australia, *Proc. Int. Conf. Planktonic Microfossils 1st*, *2*, 366–374, 1969.

Malumian, N., Foraminiferos Danianos de la Formacion Pedro Luro, Provincia de Buenos Aires, Argentina, *Ameghiniana*, *7*, 355–367, 1970.

Malumian, N., and V. Masiuk, Asociaciones foraminiferologicas fossiles de la Republica Argentina, *Actas Quinto Congr. Geol. Argent.*, *3*, 433–453, 1973.

Malumian, N., V. Masiuk, and J. C. Riggi, Micropaleontologia y sedimentologia de la perforacion SC-1 Provincia Santa Cruz, Republica Argentina, *Rev. Assoc. Geol. Argent.*, *26*, 175–208, 1971.

McGowran, B., Foraminiferal evidence for the Paleocene age of the Kings Park Shale (Perth Basin, Western Australia), *J. R. Soc. West. Aust.*, *47*, 81–86, 1964.

McGowran, B., Observation Bore No. 2, Gambier Embayment of the Otway Basin: Tertiary micropaleontology and stratigraphy, *Miner. Resour. Rev. South Aust. Dep. Mines Energy*, *135*, 43–55, 1973.

McGowran, B., Early Tertiary foraminiferal biostratigraphy in southern Australia: a progress report, in "The Crespin Volume: essays in honour of Irene Crespin", *Bull. Bur. Miner. Resour. Geol. Geophys. Aust.*, *192*, 83–95, 1978a.

McGowran, B., Stratigraphic record of early Tertiary oceanic and continental events in the Indian Ocean region, *Mar. Geol.*, *26*, 1–39, 1978b.

McGowran, B., and A. Beecroft, *Guembelitria* in the early Tertiary of southern Australia and its paleoceanographic significance, N. H. Ludbrook, Honour Volume, *Spec. Publ. South Aust. Dep. Mines Energy*, *5*, 247–261, 1986.

Nocchi, M., E. Amici, and I. Premoli-Silva, Planktonic foraminiferal biostratigraphy and paleoenvironmental interpretation of Paleogene faunas from the Subantarctic transect, Leg 114, *Proc. Ocean Drill. Program Sci. Results*, *114*, 233–279, 1991.

Orr, W. N., and D. G. Jenkins, Cainozoic planktonic foraminifera zonation and selective test solution, in *Oceanic Micro-*

palaeontology, vol. 1, edited by A. T. S. Ramsay, pp. 163–203, Academic, San Diego, Calif., 1977.

Premoli-Silva, I., R. Coccioni, and A. Montonari, *The Eocene-Oligocene Boundary in the Marche-Umbria Basin (Italy)*, 268 pp., International Union of Geological Sciences, Ancona, Italy, 1988.

Rögl, F., Late Cretaceous to Pleistocene foraminifera from the southeast Pacific Basin, DSDP Leg 35, *Initial Rep. Deep Sea Drill. Proj.*, *35*, 539–556, 1976.

Scott, G. H., Description of an experimental class within the Globigerinidae (Foraminifera), Parts I and II, *N. Z. J. Geol. Geophys.*, *9*, 513–540, 1966.

Scott, G. H., Comparison of the primary apertures of *Globigerinoides* from the lower Miocene of Trinidad and New Zealand, *N. Z. J. Geol. Geophys.*, *11*, 356–375, 1968a.

Scott, G. H., Comparison of lower Miocene *Globigerinoides* from the Caribbean and New Zealand, *N. Z. J. Geol. Geophys.*, *11*, 376–390, 1968b.

Scott, G. H., Stratigraphic variation in *Globigerinoides trilobus trilobus* (Reuss) from the lower Miocene of Europe, Trinidad and New Zealand, *N. Z. J. Geol. Geophys.*, *11*, 391–404, 1968c.

Scott, G. H., The late Miocene to early Pliocene history of the *Globorotalia miozea* plexus from Blind River, New Zealand, *Mar. Micropaleontol.*, *4*, 341–361, 1979.

Scott, G. H., *Globorotalia inflata* lineage and *G. crassaformis* from Blind River, New Zealand: Recognition, relationship, and use in uppermost Miocene–lower Pliocene stratigraphy, *N. Z. J. Geol. Geophys.*, *23*, 665–677, 1980.

Scott, G. H., Review of Kapitean stratotype and boundary with Opoitian Stage (upper Neogene, New Zealand), *N. Z. J. Geol. Geophys.*, *25*, 475–485, 1982.

Srinivasan, M. S., and J. P. Kennett, Neogene planktonic foraminiferal biostratigraphy: Equatorial to Subantarctic, South Pacific, *Mar. Micropaleontol.*, *6*, 499–534, 1981.

Stott, L. D., and J. P. Kennett, Antarctic Paleogene planktonic foraminifer biostratigraphy: ODP Leg 113, sites 689 and 690, *Proc. Ocean Drill. Program Sci. Results*, *113*, 549–569, 1990.

Taylor, D. J., Upper Cretaceous and Tertiary subsurface biostratigraphic scheme for Gippsland, Bass and Otway Basins, *Rep. 1966/30*, Mines Dep., Victoria, Australia, 1966.

Tjalsma, R. C., Cenozoic foraminifera from the South Atlantic, DSDP Leg 36, *Initial Rep. Deep Sea Drill. Proj.*, *36*, 493–517, 1977.

Toumarkine, M., Planktonic foraminiferal biostratigraphy of the Paleogene of sites 360 to 364 and the Neogene of sites 362A, 363 and 364, Leg 40, *Initial Rep. Deep Sea Drill. Proj.*, *40*, 679–721, 1978.

Van Valen, L., A new evolutionary law, *Evol. Theory*, *1*, 1–30, 1973.

Wopfner, H., and J. G. Douglas, The Otway Basin of southeast Australia, *South Aust. Victoria Geol. Surv. Spec. Bull.*, 1–464, 1971.

(Received April 3, 1992;
accepted August 17, 1992.)

CENOZOIC SOUTHERN OCEAN RECONSTRUCTIONS FROM SEDIMENTOLOGIC, RADIOLARIAN, AND OTHER MICROFOSSIL DATA

DAVE LAZARUS

Geologisches Institut, ETH-Zentrum, 8092 Zürich, Switzerland

JEAN PIERRE CAULET

Laboratoire de Géologie, Musée National d'Histoire Naturelle, 75005 Paris, France

The Antarctic Convergence marks the northern boundary of the Antarctic, or Southern Ocean, a physically distinct region of the world ocean which today contains unique, endemic radiolarian biotas and a distinctive biosiliceous sediment facies on the underlying seafloor. Paleodistributions of these parameters are used to infer Southern Ocean geographic extent and general circulation for Paleocene, Eocene, early Oligocene, late Oligocene to early Miocene, late Miocene, and Pliocene time intervals. Local upwelling along plateaus occurs in the Paleocene to middle Eocene, but no distinct Southern Ocean can be detected. The late Eocene shows increased biosiliceous sedimentation, increasing endemism in biotas, and inferred regional oceanic fronts. A geographically extensive Southern Ocean, circumpolar current, and Polar Front developed in the early Oligocene. Limited data suggest decreased water mass contrasts and more cosmopolitan faunas in the late Oligocene and early Miocene. Southern Ocean environments were again more distinct in the middle Miocene through Pliocene intervals, with strongly endemic radiolarian faunas and widespread deposition of biosiliceous sediments. Carbonate or mixed calc-silica ooze was common on shallow rises in the Paleogene, but by late Miocene/early Pliocene times it had been replaced by siliceous ooze on rises, and clay or biosiliceous clay in deeper basins. The Southern Ocean has increased in geographic extent gradually over the Cenozoic. It has always extended to near 50°S in the Atlantic sector since the early Oligocene but has gradually expanded northward, presumably tracking the mid-ocean ridge system, in the Australian sector.

INTRODUCTION

The Southern Ocean is a distinctive circumpolar water mass and current system which is bounded to the north by the Antarctic Convergence, or Polar Front. This oceanic region (the "Antarctic Ocean" of some authors) plays a central role in the circulation of the world ocean, mixing and exchanging waters between the major ocean basins. Its presence helps maintain the thermal isolation of the Antarctic continent and its ice sheets. Upwelling throughout the Southern Ocean and the resulting growth of plankton sequester carbon and affect the global CO_2 cycle. The biotas of the Southern Ocean are among the most distinctive on Earth, with many endemic species.

The history of this ocean (i.e., its origin and evolution, past geographic extent, and productivity) is only partly understood. Oxygen and carbon stable isotope ratios from lower-latitude benthic foraminifera have served as proxy indicators of Southern Ocean characteristics (*Shackleton and Kennett* [1975] and *Miller et al.* [1987]; see also reviews by *Barrera and Huber* [1991] and *Barron et al.* [1991c], while piston cores and Deep Sea Drilling Project (DSDP) holes have provided more direct data on Southern Ocean history. This work has documented a dramatic change in the characteristics of Cenozoic Antarctic ocean waters, from low-productivity, temperate conditions in the Paleogene to higher-productivity, cold waters and seasonal sea ice conditions by the Pleistocene. These changes are at least partly related to the Cenozoic development of ice sheets on what had previously been a forested, temperate climate Antarctic continent [*Kennett*, 1977; *Kennett and Barker*, 1990; *Barron et al.*, 1991c].

Recent Ocean Drilling Program (ODP) drilling in the Antarctic has substantially enlarged the material available for analysis and has prompted new syntheses of Southern Ocean history. In this paper, we focus on one specific aspect of Southern Ocean development: the origin and subsequent geographic extent of the Southern Ocean as a distinct surface water mass. For this purpose

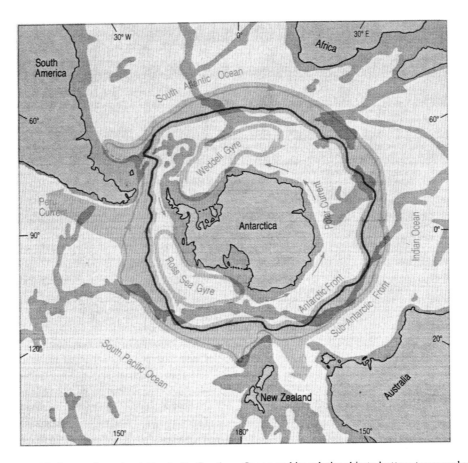

Fig. 1.   Modern surface circulation in the Southern Ocean and its relationship to bottom topography.
Antarctic or Polar Front marking northern boundary of Southern Ocean shown by heavy line. Regions
of ocean bottom shallower than 3 km shown by medium gray bands; other features of southern
hemisphere circulation shown in light gray. Modified from *Brown et al.* [1989].

we make use of two major types of proxy indicator:
radiolarian and other microfossil paleobiogeography
and sediment distributions. Data of this sort have been
used previously by *Kennett* [1977, 1978] to infer the
history of Southern Ocean circulation. New data from
ODP drilling [*Barker et al.*, 1988; *Ciesielski et al.*, 1988;
Barron et al., 1989; *Schlich et al.*, 1989; *Peirce et al.*,
1989], recent syntheses of previous core data [*Lazarus
et al.*, 1987], and continued improvements in our understanding of radiolarian paleobiogeography prompt us to
update this previous work.

Distributional data on proxy indicators is supplemented by predictions obtained from theoretical models
of past ocean circulation. We begin, however, with a
review of those factors which appear to control the
modern circulation and the relation between the modern
circulation and our chosen proxy indicators.

MODERN CIRCULATION

The Southern Ocean (Figure 1) is a circum-Antarctic
ring of water, bounded on the north by the Polar Front.
Its position at any one geographic locality is remarkably
stable from year to year, although its precise position at
any one moment may vary considerably owing to eddies
and other oceanic "weather." However, its latitudinal
position varies considerably with longitude, ranging
from less than 50°S in the South Atlantic to nearly 60°S
south of New Zealand. The surface circulation south of
the Polar Front, driven by strong surface winds, is
predominantly eastward, except for a narrow westward
flowing current near the continental margin. The axis of
the main eastward flow (the Antarctic Circumpolar
Current (ACC)) largely coincides with the maximum
gradient in water density, located at the Polar Front.
The ACC is, by comparison with most other oceanic

currents, unusually deep, extending, albeit with attenuated strength, to a depth of several kilometers, often impinging on the seafloor. This depth is possible in part owing to the unusually weak stratification of upper waters in the Southern Ocean, which in turn reflects the processes of upwelling and deepwater formation in this region. Slow southward flow of waters from the Atlantic, Pacific, and Indian oceans is balanced by two northward flowing water masses formed in the Antarctic. Antarctic Intermediate Water sinks from the surface south of the Polar Front to depths of several hundred meters at the front, as it encounters the much warmer, less dense surface waters to the north. Antarctic Bottom Water forms closer to the continent and sinks to the bottom as it spreads northward throughout much of the world's ocean basins.

The factors controlling the position of the Polar Front, which defines the extent of the Southern Ocean, is of particular interest to our study. Unlike some other oceanic frontal systems, its location does not correspond to a major change in the direction of the dominant surface wind field [Deacon, 1984]. Several factors appear to control the Polar Front position. The immediate determinant of the position of the front is the strong density gradient created by the juxtaposition of colder Southern Ocean waters and warmer, temperate waters to the north. The processes controlling the position of this boundary are less clear but seem to depend at least in part on the bottom topography of the ocean basin, as the surface position of the Polar Front shows a strong tendency to track the position of the underlying oceanic ridge system. The ridge system, which at deeper levels serves to pond in the cold bottom waters of the Southern Ocean, may act to some extent as a stabilizer of the frontal position via its interaction with the deep extending ACC [Gordon, 1967, 1988; Deacon, 1984]. The position of the ACC also follows the general trend of surface wind stress, which largely parallels the distribution of continental masses in the region [Deacon, 1984]. Both the ocean basin topography and the surface wind stress field show somewhat similar patterns, with northward deflections in the Atlantic and Indian oceans and southward deflections in the Drake Passage and south of New Zealand. Thus, in this view, the position of the ACC and Polar Front may have been to a significant degree ultimately controlled by the same geologic process: the spreading of the ocean basins and the resulting distribution of basinal topography and major continental landmasses. We will return to this point later in the paper.

MODERN BIOGEOGRAPHY

We use two major characteristics of the preserved particle flux from Southern Ocean surface waters as tracers of past Southern Ocean geography. The first of these is the taxonomic composition of the radiolarian fauna. The biogeography of microplankton species has long been known to closely approximate surface water mass boundaries, with many species adapted and confined to specific water masses [McGowan, 1971, 1974; Van der Spoel and Pierrot-Bults, 1979; Pierrot-Bults et al., 1986]. The major marine planktonic microfossil groups (diatoms, coccoliths, radiolarians, and planktonic foraminifera) also show this biogeographic specificity, which has been extensively exploited to map past water mass distributions and estimate paleotemperatures [CLIMAP Project Members, 1976]. Radiolarians appear to be the most sensitive of the groups in this regard, with relatively high percentages of endemic or bipolar species in various biogeographic regions [Nigrini, 1967, 1968, 1970; Molina-Cruz, 1977; Moore, 1978; Johnson and Nigrini, 1980; Lombari and Boden, 1985] and particularly so in the Southern Ocean (Figure 2) [Hays, 1965; Lozano and Hays, 1976; Morley, 1979; Boltovskoy and Riedel, 1980]. This endemism makes interlatitudinal stratigraphic correlation difficult with radiolarians but makes radiolarian faunas excellent markers for past surface water masses. Although other microfossil groups can be used for this purpose, they are generally less sensitive indicators of polar water mass boundaries, owing to their more cosmopolitan nature (particularly in high latitudes) and, in the diatoms [Shemesh et al., 1989] and coccolithophores, their relatively incomplete preservation of taxa in the fossil record. Radiolarian assemblages can also be affected by poor preservation, but in the Antarctic region, sufficiently well preserved faunas can be found throughout much of the Cenozoic. The single most limiting factor in using radiolarian faunas to trace past water mass distributions is the limited extent of taxonomic description of many of these past faunas. The fossil record of radiolarians is more diverse than that of all of the other major marine planktonic microfossil groups combined, yet it has received relatively little study to date.

MODERN SEDIMENT DISTRIBUTION

The second tracer of Southern Ocean water mass geography we use is the bulk sediment composition, particularly the abundance of biogenic (mostly diatom) silica. In the modern ocean, widespread upwelling in the Antarctic south of the Polar Front supports high phytoplankton productivity. Throughout the world ocean, relatively high levels of surface productivity produce enhanced relative and absolute rain rates of biogenic silica (mostly diatoms) to the seafloor and a preserved signal in the form of higher biogenic opal content in the sediments [Lisitzin, 1971; Johnson, 1974; Berger, 1976]. The past distribution of biosiliceous sediments can be used to map changes in oceanic productivity and water mass development [e.g., Baldauf and Barron, 1990; Froelich et al., 1991]. Maps of surface sediment type in the Antarctic region show a remarkably good correla-

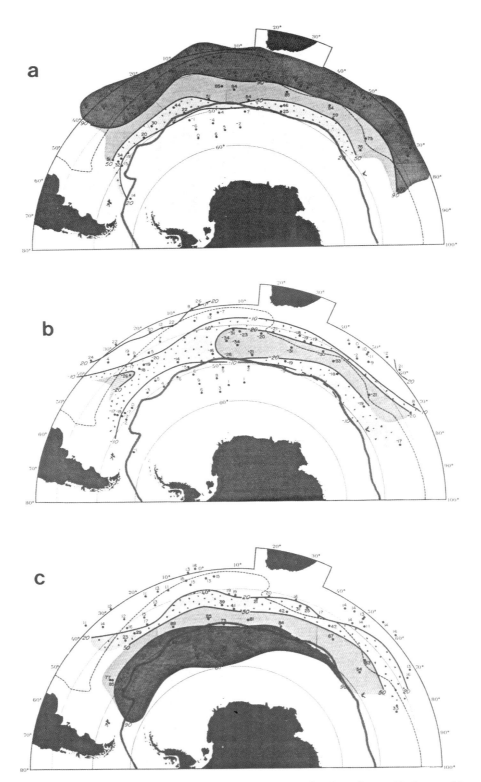

Fig. 2. Distribution of radiolarian assemblages in the modern Southern Ocean. Each assemblage determined from factor analysis of numerous core top samples in the Atlantic and Indian sectors of the Southern Ocean. Contour intervals are relative abundance of each assemblage at 90% (dark gray), 50% (medium gray), and 20% (stippled light gray) of total fauna. Approximate position of the Polar Front indicated by heavy line. (*a*) Subtropical assemblage. (*b*) Subantarctic assemblage. (*c*) Antarctic assemblage. Modified from *Lozano and Hays* [1976].

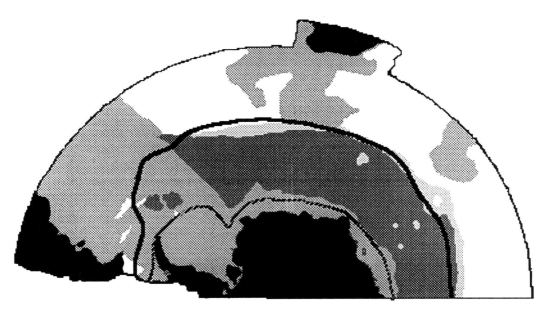

Fig. 3. Distribution of major sediment types in the modern Southern Ocean. Dark gray, siliceous ooze; medium gray, clay and sand; light gray, mixed calc-silica ooze; white, calcareous ooze. Approximate positions of the Polar Front shown by heavy line; minimum sea ice, by dashed line. Redrawn and simplified from *Lozano and Hays* [1976].

tion between biogenic opal-containing sediments and the distribution of productive surface waters of the Southern Ocean (Figure 3) [*Cooke and Hays*, 1982; *Burckle and Cirilli*, 1987]. This map also shows a second characteristic of modern Southern Ocean deep-sea sediments: reduced biogenic silica content in regions near the continent, a pattern created by reduced surface water productivity [*Honjo*, 1990; *Wefer et al.*, 1990] as a result of light limitation by extensive sea ice cover for much of the year [*Cooke and Hays*, 1982; *Tilzer et al.*, 1983; *Burckle and Cirilli*, 1987]. We view these patterns (i.e., upwelling and enhanced relative biosiliceous productivity south of the Polar Front, and reduced productivity in regions of extensive sea ice) as fundamental characteristics of the Southern Ocean today and ones which can be used to map out past Southern Oceans and sea ice development. Other sediment components, such as ice-rafted detritus (IRD) can also be used as indicators of sea ice presence, but much of the IRD in the modern ocean may be transported primarily by icebergs and thus mostly record continental glaciation and the complex peculiarities of iceberg drift patterns within the Southern Ocean [*Cooke and Hays*, 1982; *Wise et al.*, 1991]. The virtual absence of carbonate in modern Southern Ocean sediments might also be seen as a potential tracer for past Southern Ocean water mass distributions, but we believe that carbonate distributions are less useful as a water mass tracer. Carbonate secreting organisms in both plankton and benthos are today underrepresented in the Antarctic, apparently at

least in part owing to physiologic limitations on carbonate secretion in cold environments [*Clarke*, 1990]. The current absence of carbonate in most Antarctic sediments may reflect temperature-controlled exclusion of coccolithophores from modern surface waters, and thus the presence of carbonate in older sediments may be due simply to slightly warmer (>3°C) surface waters [*Burckle and Pokras*, 1991]. Carbonate distribution in deep-sea sediments is also strongly influenced by deep-water dissolution, a complex process whose past history is not well understood.

For purposes of our study, detailed, quantitative values of sedimentary composition, although useful, are not strictly necessary. We have therefore employed a simple classification scheme (similar to that of Figure 3) for past sediments, with assignments being based on previously published sediment descriptions and classifications. Our categories include clay, sand, calcareous ooze, biosiliceous ooze, silica-bearing calcareous ooze, and biosiliceous clay. Sediments were assigned to the last two categories when insufficient biosiliceous material was present to classify them as biosiliceous ooze, but sufficiently well preserved siliceous microfossils were present to be able to date the sediment or determine the biogeographic characteristics of the radiolarian fauna. This definition of biosiliceous is somewhat broader than that employed by most sedimentologists but provides a more sensitive, and we believe more useful, indicator for mapping water mass distributions.

TABLE 1. List of Paleogene Radiolarian Species and Biogeographic Affinities

Paleogene Radiolaria	Biogeog. Affinity	Taxon Type	Mid Eocene	Late Eocene	Early Oligocene	Late Oligocene
Actinomma campilacantha	A	S			X	X
Actinomma kerguelensis	A	S		X	X	X
Actinomma medusa	A	S				X
Amphicraspedum prolixum	T	S	X			
Amphipternis clava	T	S	X			
Amphymenium splendiarmatum	T	S	X	X	X	X
Antarctissa	A	Gr.			X	X
Anthocyrtella mespilus	T	S	X			
Anthocyrtella spatiosa	T	S	X			
Artobotrys auriculaleporis	C	S	X			
Artobotrys biaurita	T	S	X			
Astrophacus inca	T	S	X			
Axoprunum pierinae	T	S	X			
Botryostrobus joides	A	S	X	X		
Botryostrobus kerguelensis	A	S		X	X	X
Botryostrobus rednosus	T	S		X	X	
Calocyclas asperum	T	S		X	X	
Calocyclas semipolita	C	S	X	X	X	
Carposphaera subbotinae	T	S	X			
Cenosphaera (?) oceanica	C	S	X	X	X	X
Ceratocyrtis amplus	A	S			X	X
Ceratocyrtis cuccularis	B	S			X	X
Ceratocyrtis mashae	B	S			X	X
Ceratocyrtis stigi	B	S			X	
Clathrocyclas aurelia	C	S	X			
Clathrocyclas u. nova	C	SS		X		
Cycladophora c. subhumerus	C	SS		X	X	X
Cycladophora conica	B	S		X	X	X
Cymaetron sinolampas	A	S		X	X	
Cyrtocapsella longithorax	A	S			X	X
Cyrtocapsella robusta	A	S		X	X	
Dictyophimus (?) archipilium	A	S	X	X		
Dictyophimus callosus	T	S	X	X	X	X
Dictyophimus pocillum	T	S	X	X	X	X
Dictyophora amphora	T	S	X			
Dictyoprora physothorax	A	S			X	
Dorcadospyris argisca	T	S	X			
Eucyrtidium antiquum	A	S			X	
Eucyrtidium (?) mariae	A	S		X	X	
Eurystomoskevos petrushevskaae	T	S	X	X	X	
Lamprocyclas inexpectata	A	S			X	X
Lamprocyclas prionotocodon	A	S				X
Lithelius aff. L. foremanae	T	Gr.		X	X	X
Lithomelissa cheni	A	S		X	X	
Lithomelissa dupliphysa	A	S		X		
Lithomelissa ehrenbergi	T	S		X	X	
Lithomelissa haeckeli	T	S		X		
Lithomelissa robusta	A	S		X	X	X
Lithomelissa sphaerocephalis	A	S		X	X	X
Lophocyrtis jacchia	T	S	X			
Lophocyrtis dumitricai	A	S	X	X		
Lophocyrtis longiventer	A	S		X	X	X
Lophophaena (?) thaumasia	A	S			X	
Lophophaenoma radians	T	S	X			
Lychnocanoma amphitrite	C	S	X	X	X	
Lychnocanoma babylonis	T	S	X			
Lychnocanoma bellum	T	S	X	X	X	
Lychnocanoma conica	C	S				X
Lychnocanoma tripodium	T	S	X			
Perichlamidium praetextum	T	S		X	X	
Periphaena decora	T	S	X	X	X	
Periphaena heliasteriscus	C	S	X			
Prunopyle frakesi	A	S		X	X	X
Prunopyle hayesi	A	S			X	X
Prunopyle monikae	A	S			X	
Prunopyle polyacantha	C	S		X	X	X
Prunopyle trypopyrena	T	S			X	X
Pseudodictyophimus galeatus	A	S			X	X
Pteropilium contiguum	T	S		X	X	
Rhabdolithis pipa	T	S	X			

TABLE 1. (continued)

Paleogene Radiolaria	Biogeog. Affinity	Taxon Type	Mid Eocene	Late Eocene	Early Oligocene	Late Oligocene
Rhizosphaera antarctica	A	Gr.		X	X	X
Siphocampe imbricata	T	S	X	X	X	X
Siphocampe pachyderma	T	S	X	X		
Siphocampe (?) quadrata	T	S	X			
Spongodiscus americanus	B	S	X			
Spongodiscus cruciferus	T	S	X			
Spongodiscus osculosus	A	Gr.			X	X
Spongodiscus rhabdostylus	T	S	X			
Spongomelissa cucumella	T	S	X			
Stylodictya hastata	T	S	X	X	X	
Stylodictya ocellata	T	S			X	
Stylodictya tainemplekta	A	S	X	X	X	
Stylodictya targaeformis	T	S	X	X	X	X
Stylosphaera coronata	T	S	X	X	X	
Stylosphaera hispida	T	S			X	X
Stylosphaera minor	C	S	X	X	X	
Stylosphaera radiosa	T	S				X
Stylosphaera spinulosa	T	S	X			
Stylotrochus nitidus	T	S	X			
Theocyrtis diabloensis	C	S			X	
Thyrsocyrtis bromia	T	S		X		
Tripilidium clavipes	C	S	X	X	X	X
Velicucullus	B	G	X	X	X	X

Biogeographic affinity codes are as follows: A, Antarctic; T, tropical; C, cosmopolitan; B, bipolar. Taxon type codes are as follows: S, species; SS, subspecies; Gr., group. Remaining columns give approximate stratigraphic range, crosses denoting occurrence in that time interval.

PLATE RECONSTRUCTIONS, PALEOCIRCULATION SIMULATIONS, AND CHRONOLOGY

Although not tracers per se, three other types of information are employed in our analysis. Paleogeographic reconstructions of continental positions for the Paleocene through the middle Miocene are taken from the Terra Mobilis program of *Denham and Scotese* [1988], and mid-ocean ridge system positions are based on those indicated in the program output and reconstructions of *Scotese et al.* [1988]. Terra Mobilis maps are sufficient for the very broad scale analyses of this initial, preliminary set of reconstructions, even though in the future, it would be desirable to use the more accurate recent plate reconstructions that are now available [*Lawver et al.*, 1992]. Core locations were hand plotted using past positions of ocean floor isochrons as a guide. For the late Miocene and early Pliocene, we have used modern locations and commercial mapping software, as the number of data points is too large for manual plotting. Comparison of late Miocene Terra Mobilis maps with modern maps suggests that the geographic errors introduced by this approach are minimal, as very little latitudinal movement occurred between the late Miocene and the Recent in the Antarctic, with the single exception of the Indo-Australian plate,

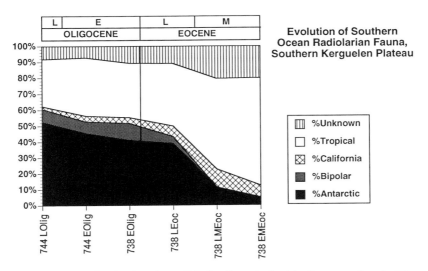

Fig. 4. Development of endemic polar radiolarian fauna in the late Eocene and early Oligocene in ODP Leg 119 sites 738 and 744 on the southern Kerguelen Plateau. Percent refers to percent of species in each time interval, not number of individual specimens. Based on data in Tables 1 and 2.

which has moved northward by more than 5° in latitude. This fact has been taken into account in our reconstructions.

One deficiency in our analysis is our inability to accurately date the opening of key seaways to deep circulation. Despite extensive efforts [*Kennett*, 1977; *Barker and Burrell*, 1977, 1982], the detailed time history of the opening of deep circulation between the Indian and Pacific sectors of the Antarctic and, particularly, the timing of the Drake Passage opening remain poorly constrained by tectonic evidence. However, as we are primarily concerned with surface circulation and water mass boundaries, this imprecision is perhaps not crucial. We implicitly assume shallow water connec-

tions through the Drake passage by the early Oligocene, although we have no direct evidence to support this.

All samples used have been referenced, when possible, to the biostratigraphic schemes developed during the recent suite of ODP drilling in the Antarctic and summarized in the work of *Gersonde et al.* [1990], *Barron et al.* [1991*b*], and *Harwood et al.* [1992]. However, piston core records have been used where no detailed biostratigraphic data are available, only a general age estimate. As some of these estimates were made several years ago, the ages of these samples are less certain. We do not believe that stratigraphic uncertainties, however, materially affect our analyses or conclusions.

TABLE 2. Radiolarian Faunal Composition in Paleogene Antarctic Sections

Site	Age Interval	Faunal Class	Percent Southern	Antarctic	Bipolar	California	Tropical	Unknown	Total
264	middle Eocene	S	5	1	0	4	15	5	25
274	early Oligocene	A	81	13	0	1	2	3	19
278	late Oligocene	S/A	40	10	2	5	13	7	37
281	late Eocene	S/A	45	4	9	5	11		29
511	mid Eocene	S?	0				3		3
511	late Eocene	S	25	3	0	3	6	0	12
511	early Oligocene	A	75	6	0	1	1	1	9
512	middle Eocene	S	12	2	0	3	12	0	17
689	late Oligocene	A	65	20	2	0	12	12	46
690	late Oligocene	A	81	15	2	0	4	8	29
738	middle Eocene	S	12	4	0	4	26	9	43
738	late Eocene	S/A	49	18	2	3	18	5	46
738	early Oligocene	A	58	23	6	19	2	6	56
744	early Oligocene	A	57	25	4	2	20	4	55
744	late Oligocene	A	66	32	5	1	18	5	61

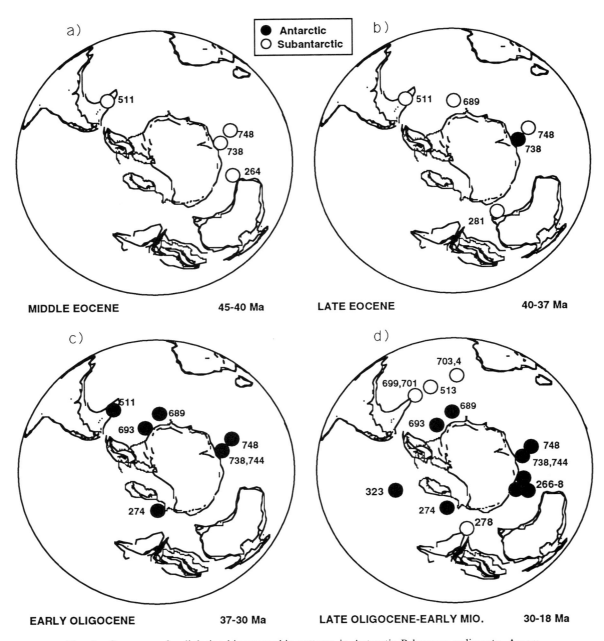

Fig. 5.   Summary of radiolarian biogeographic patterns in Antarctic Paleogene sediments. Assemblages assigned to either Antarctic or Subantarctic on the basis of taxonomic composition of assemblages, as explained in text. Base maps generated by computer program Terra Mobilis [*Denham and Scotese*, 1988].

We have used the oceanic paleocirculation model simulations of *Scotese and Summerhayes* [1986] and *Barron and Peterson* [1991] to guide us in interpreting our faunal and sedimentological data. These simulations are based on many simplifying assumptions, including a simple, basin-shaped ocean bathymetry, but nevertheless give insight into past ocean circulation based on physical oceanographic principles. Used with caution, they are of considerable assistance in interpreting past distributional patterns of biotic and sedimentologic data.

TABLE 3.   List of Neogene Radiolarian Species and Biogeographic Affinities

Neogene Radiolaria	Biogeog. Affinity	Taxon Type	E Miocene	Mid Miocene	Late Miocene	Early Pliocene	Late Pliocene	Pleistocene
Acrosphaera australis	A	S			X			
Acrosphaera labrata	A	S			X	X		
Acrosphaera mercurius	B	S	X	X	X			
Acrosphaera murrayana	C	S	X	X	X	X	X	X
Actinomma golownini	A	S			X	X		
Actinomma haysi	S	S						
actinommids	S	G	X	X	X	X	X	X
Amphymenium challengeri	A	S			X			
Anomalocantha dentata	S	S		X	X			
Antarctissa cylindrica	A	S			X	X	X	
Antarctissa deflandrei	A	S		X	X			
Antarctissa denticulata	A	S			X	X	X	X
Antarctissa sp.	A	S	X					
Antarctissa strelkovi	A	S			X	X	X	X
artostrobids	C	G	X	X	X	X	X	X
cannobotryids	C	G	X	X	X	X	X	X
carpocaniids	S	G	X	X	X	X	X	X
cenosphaerids	S	G	X	X	X	X	X	X
Ceratocyrtis	B	G	X	X	X	X	X	X
cornutellids	C	G	X	X	X	X	X	X
Corythomelissa	B	G	X	X	X	X	X	X
Cycladophora bicornis	S	S		X	X	X	X	X
Cycladophora davisiana	C	S					X	X
Cycladophora golli	A	S	X	X				
Cycladophora humerus	A	S		X	X			
Cycladophora pliocenica	A	S				X	X	
Cycladophora spongothorax	A	S			X			
Cyrtocapsella japonica	S	S		X	X			
Cyrtocapsella longithorax	B	S	X					
Cyrtocapsella tetrapera	S	S	X	X				
Dendrospyris megalocephalis	A	S			X			
Dendrospyris rhodospyroides	A	S	X	X	X			
Dendrospyris stabilis	C	S	X	X	X			
Desmospyris spongiosa	A	S				X		
Dictyophimus	C	G	X	X	X	X	X	X
Diartus/Didymocyrtis	C	G	X	X	X	X	X	X
Druppatractus hastatus	B	S	X	X	X			
Eucyrtidium accuminatum	S	S		X	X	X	X	X
Eucyrtidium calvertense	C	S		X	X	X	X	
Eucyrtidium cienkowski	C	S	X	X				
Eucyrtidium pseudoinflatum	A	S			X			
Eucyrtidum punctatum	A	S	X					
Heliodiscus asteriscus	S	S	X	X	X			
Helotholous? haysi	A	S			X			
Helotholus? vema	A	S				X		
hexastylids	S	G	X	X	X	X	X	X
Lamprocyclas maritalis	S	S					X	X
Lamprocyclas aegles	S	S			X	X		
Lampromitra coronata	A	S			X	X		
Lithatractus timmsi	B	S	X	X	X			
Lithelius minor	C	S	X	X	X	X	X	X
Lithelius nautiloides	A	S			X	X	X	X
Lithomelissa stigi	B	S	X		X			
lithomelissids	S	G	X	X	X	X	X	X
Lychnocanium grande group	B	S			X	X		
prunoids and lithelids	C	G	X	X	X	X	X	X
Prunopyle antarctica	A	S					X	X
Prunopyle hayesi	S	S	X	X	X			
Prunopyle tetrapila	B	S	X	X	X			

TABLE 3.   (continued)

Neogene Radiolaria	Biogeog. Affinity	Taxon Type	E Miocene	Mid Miocene	Late Miocene	Early Pliocene	Late Pliocene	Pleistocene
Prunopyle titan	A	S			X	X		
Pterocanium korotnevi	B	S			X	X		
Pterocanium praetextum	S	S				X	X	X
Pseudocubus warreni	S	S				X		
Rhopalastrum spp.	S	G		X	X			
Saturnalis circularis	C	S	X	X	X	X	X	X
Siphonosphaera vesuvius	A	S			X	X		
spongodiscids	C	G	X	X	X	X	X	X
Spongoplegma antarcticum	A	S			X	X	X	X
Spongotrochus glacialis	B	S			X	X	X	X
Stichocorys spp.	S	G	X	X	X	X		
Stylacontarium bispiculum	S	S	X	X	X			
stylatractids	S	G	X	X	X	X		
Stylatractus neptunus	C	S	X	X	X	X	X	X
Stylatractus universus	C	S			X	X	X	X
stylodiscids	S	G	X	X	X	X	X	X
Triceraspyris antarctica	A	S				X	X	
Triceraspyris coronatus	A	S			X	X		
Velicucullus	B	G	X	X				

An "S" in biogeographic affinity column indicates Subantarctic. Other codes as in Table 1.

MATERIALS

Two main types of marine sediment material are available for reconstruction of past Southern Ocean distributions. DSDP and ODP sections provide the bulk of our material in older time intervals. Piston cores, collected by the ships *Eltanin/Islas Orcadas*, *Marion Dufresne*, *Robert Conrad*, and *Vema*, provide additional data, particularly for the late Miocene to Recent time interval. Core names are indicated in the figures except for the late Miocene and early Pliocene, where the number of data points makes this impractical. All of the samples used here are summarized in the work of *Lazarus et al.* [1987] or in the various initial reports of the Ocean Drilling Program.

RESULTS

Radiolarian Biogeography

Well-preserved Antarctic radiolarian faunas are known from the Late Cretaceous, Paleocene, and early Eocene but are either poorly known taxonomically or consist primarily of cosmopolitan taxa [*Chen*, 1975; *Ling and Lazarus*, 1990; *Ling*, 1991; *Hollis*, 1991]. Middle to late Eocene faunas from the Antarctic, however, are better known and also begin to show increased endemism and provinciality. Between ~48 and 38 Ma, faunal turnover is extensive, with cool water cosmopolitan elements and true Antarctic endemic forms becoming increasingly common (Table 1). Early Oligocene

Neogene Radiolarian Abundances

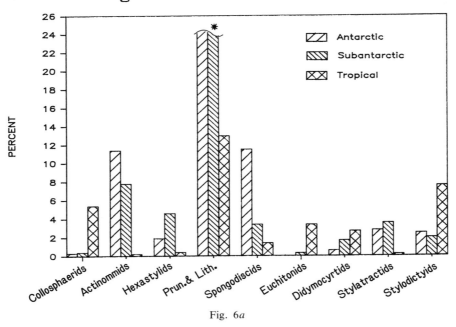

Fig. 6a

Neogene Radiolarian Abundances

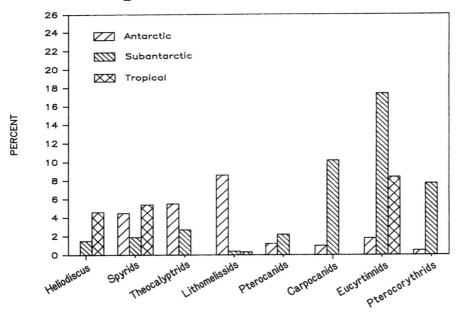

Fig. 6b

Fig. 6. Relative abundance of specimens of radiolarians in (approximate) family level categories in late Miocene and early Pliocene sediments from Antarctic, Subantarctic, and tropical-subtropical environments. Based on published data of *Romine and Lombari* [1985] from the tropical Pacific and unpublished counts of several hundred Antarctic and Subantarctic radiolarian samples by D. B. Lazarus. (*a*) Spumellarian group relative abundances. (*b*) Nassellarian group relative abundances. Antarctic and Subantarctic prunoid-lithelid group abundances (''Prun. & Lith.'') plot off scale at 42% and 31%, respectively. If no bar is shown, abundance is effectively zero.

——— **20 μm**

Fig. 7.   SEM photographs of typical heavily (1, 3) and lightly (2, 4) silicified radiolarian skeletons from south and north of the Polar Front, respectively. Note thicker shells, heavier spines, and smaller pore to bar ratios of Antarctic versus Subantarctic species. (1) *Actinomma popofskii* Petrushevskaya, core MD88-519, 50°01′S, 68°56′E, Kerguelen Plateau, Quaternary. (2) *Actinomma leptodermum* (Jörgensen), core MD83-017, 33°38′S, 47°13′E, Madagascar Plateau, Quaternary. (3) *Eucyrtidium biconicum* (Vinassa), core MD73-028, 49°26′S, 61°45′E, Crozet Basin, early Pliocene. (4) *Eucyrtidium teuscheri* Haeckel, core MD73-017, 33°38′S, 47°13′E, Madagascar Plateau, Quaternary.

radiolarian faunas show even stronger endemic character. While a detailed chronology of this development is not yet available, the general pattern of faunal change from Kerguelen Plateau sites 738 and 744 shows the relatively gradual nature of the transition (Figure 4). Table 2 gives a summary of the major faunal elements at several selected sites in the Paleogene. These data, together with data from other sites, is used to map biogeographic patterns, which also show a relatively gradual change from cosmopolitan or warm water faunas in the middle Eocene (Figure 5a) to cool water, endemic-dominated assemblages in the late Eocene

through early Oligocene. Antarctic endemic forms were most common in the Eocene at Site 738, close to the Antarctic continental margin, where they dominated the assemblage by the late Eocene (Figure 5b). Antarctic forms are present but less common in coeval sites further from the continent (Table 2). Early Oligocene biogeographic patterns were dramatically different (Figure 5c). Antarctic endemic-dominated assemblages were widespread and circumpolar, reaching the northern Kerguelen Plateau (Site 748), Maud Rise in the Weddell Sea (Site 689), the Falkland Plateau (Site 511), and the Pacific sector (Ross Sea Site 274).

Late Oligocene through early Miocene radiolarian faunas, although increasingly common, are less well described than faunas from earlier and later time intervals, as biosiliceous preservation declines during this interval, and many sections are interrupted by extensive hiatuses. However, enough is known of their taxonomic composition to conclude that the general biogeographic pattern established in the early Oligocene persisted throughout this time interval (Figure 5d).

Neogene radiolarian faunas are not yet fully described taxonomically, and the biogeographic affinities of many species are not determined. However, a preliminary listing, even though incomplete (Table 3) gives us a basis for estimating the faunal characteristics of Neogene radiolarian assemblages. At the generic or family group level, there are also clearly visible differences in the taxonomic composition of Antarctic Neogene radiolarian faunas (Figures 6a and 6b). Neogene radiolarians from south and north of the Polar Front also show dramatic differences in the biogenic silica of their skeletons. Those south of the boundary are heavily silicified and give a high refractive contrast when viewed under transmitted light. Species, and even specimens from the same species, north of the boundary tend to have much thinner shells and low refractive contrast (Figure 7). A similar biogeographic variation in refractive index was noted in Holocene lower-latitude faunas by *Goll and Bjørklund* [1971, 1974].

Despite the incompleteness of our taxomomic knowledge, it is possible to readily assign faunas to two categories: Antarctic and Subantarctic. Biogeographic patterns in the Neogene are essentially the same as those of the Oligocene, although the distribution of faunas in younger time intervals is increasingly well defined by increasing numbers of localities. Middle Miocene radiolarian faunas are better preserved, and their biogeography (Figure 8a) is similar to that of the Oligocene. Except for an increasing number of sites, late Miocene–early Pliocene biogeographic patterns (Figure 8b) are the same and, by this point in time, indistinguishable from those of the Recent (Figure 2).

TIME SERIES CHANGES IN RADIOLARIAN FAUNAS FROM SELECTED LOCALES

Previous studies of radiolarian faunas as indicators of water mass characteristics have mostly documented temporal change in faunal composition at a single locality. Although such studies cannot give a geographic picture of paleobiogeographic distribution or the absolute latitudinal extent of temporal shifts in biogeographic pattern, they can determine in much more detail the timing and relative magnitude of temporal change in faunal distributions. Several previous studies of this sort have been done in the Antarctic on pre-Pleistocene material. *Wise et al.* [1991] determined the faunal biogeographic character in early Oligocene sediments from

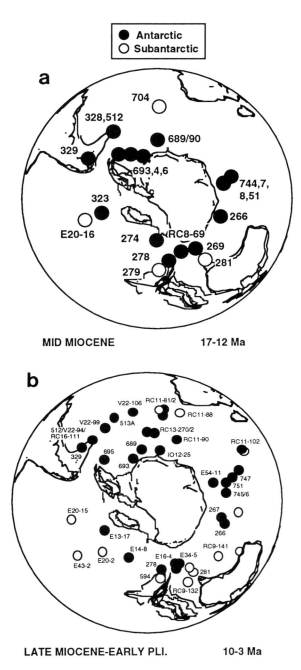

Fig. 8.   Radiolarian biogeographic patterns in Antarctic Neogene sediments. See also caption for Figure 5 and text.

DSDP Site 511 on the Falkland Plateau. Although they divided faunas into three types (cool temperate, Subantarctic, and Antarctic), their results are similar to ours (Table 2; Figures 5b and 5c) for the same time interval. Late Eocene faunas were dominated by "cool temperate" forms, but by the early Oligocene, the overlying waters were occupied by a primarily endemic Antarctic fauna. J. P. Caulet, in a previously unpublished study of Neogene radiolarian assemblages from DSDP Site 594 (Figure 9), showed a gradual increase in colder water radiolarian species in the Subantarctic during the Neogene. Subantarctic species first appeared in the middle Miocene and are the dominant component of the assemblage since the early late Miocene. The shift in assemblage composition may imply a coeval northward shift in the water masses of the Subantarctic and/or cooling of the surface waters. The magnitude of any geographic shift cannot, however, be determined from this site alone. Interestingly, although a shift toward colder climate clay mineral assemblages also occurs at this site [*Robert et al.*, 1986], the clay mineral transition does not occur until several million years after the radiolarian assemblages have become dominated by colder water forms. *Weaver* [1983] documented increasing dominance by Antarctic species at Falkland Plateau DSDP Site 514. Early Pliocene Subantarctic-dominated assemblages are replaced in the late Pliocene by Antarctic assemblages. Again, a northward shift of the front is indicated, although the magnitude of the shift cannot be estimated. A similar study was recently published by *Abelmann et al.* [1990]. We shall argue later that the northward shifts determined in these studies, although significant, represented, at least individually, only relatively minor changes in the overall extent of the Southern Ocean.

CENOZOIC ANTARCTIC DEEP-SEA SEDIMENT DISTRIBUTIONS

Geographic Distribution

Although southern high-latitude Cenozoic biogenic sediments have, except in the later Neogene, been predominantly calcareous, biosiliceous sediments have nearly always been present as well. Most early records are difficult to interpret, as diagenesis to chert has usually occurred, and the original abundance of biogenic silica is thus difficult to estimate. Poor drilling recovery of chert-bearing sections compounds the problem. However, at least one true biosiliceous ooze is known from the Early Cretaceous of the Weddell Sea [*Barker et al.*, 1988], and chert-rich time intervals in the Antarctic include the Campanian-Maastrichtian and much of the Eocene. Paleocene sections from the Antarctic are generally poor in biogenic silica (Figure 10), although local enrichments are present off New Zealand [*Hollis*, 1991], on Kerguelen/Broken Ridge Plateau (Figure 10), and near the Falkland Plateau [*Ciesielski et al.*, 1988]. Carbonate ooze is found in the Weddell Sea and northward toward Africa. Eocene deep-sea sediments have been recovered from various regions around Antarctica (Figure 11). Carbonate ooze is still the dominant lithology, but biosiliceous ooze occurs south of Australia, and silica-bearing carbonate ooze is again present on

the Kerguelen Plateau. Biogenic silica is found in deep-water nannofossil oozes north of the Kerguelen Plateau in *Marion Dufresne* piston cores MD6-31 and MD9-30A, while by the late Eocene, biogenic silica is present in the Weddell Sea as well.

Well-preserved biogenic silica, instead of chert, became more common for the first time in the Oligocene. In the early Oligocene (Figure 12), geographic coverage is somewhat limited, but both biogenic carbonate and biogenic silica deposition appear to be widespread. Biogenic silica became more common in the southern Indian Ocean in the early Oligocene [*Baldauf et al.*, 1992], and true biosiliceous ooze is recorded in the Weddell Sea for the first time since the Late Cretaceous. Geographic coverage is much more extensive for the late Oligocene to early Miocene (Figure 13). Carbonate-bearing sediment is still common throughout the Antarctic, but biosiliceous sediments are the dominant lithology south of 50°S paleolatitude. Biosiliceous clay, first recorded in the early Oligocene in the Weddell Sea, is now common in the deeper basins throughout the region. A similar pattern is seen in the middle Miocene (Figure 14). Late Miocene sediments (Figure 15) also show a similar geographic pattern, but carbonate-bearing sediments south of 50° paleolatitude are increasingly restricted in distribution. Most sediments are clays, biosiliceous clays, or true biosiliceous oozes. Carbonate-bearing sediment is almost absent in the Pliocene Antarctic (Figure 16), being found primarily in one region south of Tasmania. Pleistocene sedimentation patterns in the Antarctic (not shown) are similar to those of the Recent, with biosiliceous ooze (in contrast to the earlier Pliocene and late Miocene) being restricted to a more northward belt near the Polar Front and biosiliceous clay nearer the continent. The northward restriction of biosiliceous productivity is inferred to mark the development of widespread sea ice at this time [*Cooke and Hays*, 1982; *Burkle and Cirilli*, 1987; *Abelmann et al.*, 1990].

Distribution With Depth and Latitude

As Figures 10–16 indicate, there is a characteristic, and quite consistent, pattern over time to the distribution of lithologies with depth and paleolatitude. Biogenic oozes (carbonate and mixed calc-silica oozes in the Paleogene, gradually shifting toward pure biosiliceous oozes in the Neogene) occur on oceanic plateaus and other shallow areas, while biosiliceous clays or other nonbiogenic-dominated lithologies are found throughout the Cenozoic in the deeper basins. Biosiliceous sediments are more common south of ~50°S paleolatitude. This impression is reinforced by the results of *Lazarus et al.* [1987], parts of which are reproduced as Figure 17. In this earlier study the distribution of lithologies in the Pliocene and Miocene were summarized by counting the numbers of cores of each major

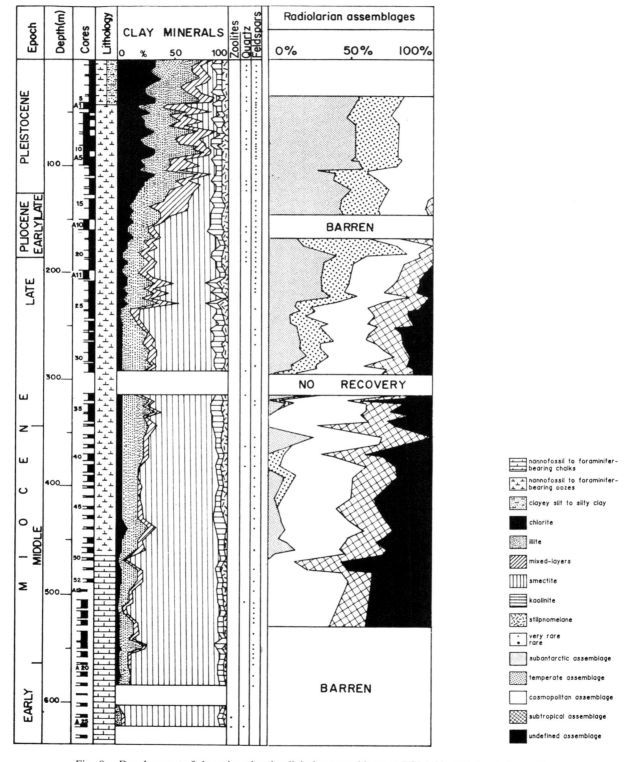

Fig. 9.   Development of clay mineral and radiolarian assemblages at DSDP Site 594, located near the Polar Front southeast of New Zealand. Based on clay mineral results of *Robert et al.* [1986] and unpublished specimen counts by J. P. Caulet. Chronology after *Kennett and von der Borch* [1985].

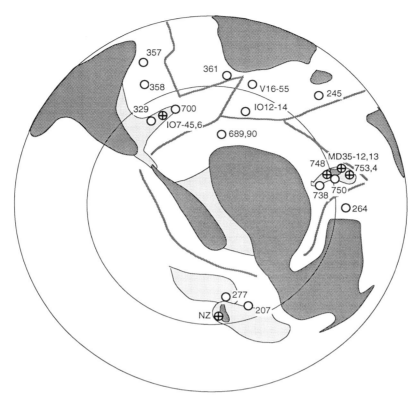

Fig. 10. Paleocene geography and sediment distribution. Key for this and subsequent Figures 11–16: Continent, dark gray; shelf or shallow seas, light gray; ocean ridges and transforms, dark gray lines; white, ice cap. Sediment types: open circles, carbonate ooze; circles with crosses, carbonate ooze with biogenic silica; solid black circles, siliceous ooze; solid black triangles, siliceous clay. Cores without biogenic content not plotted for clarity. Radiolarian faunal assemblage types (from Figures 5 and 8): A, Antarctic; S, Subantarctic; S/A, mixed assemblage. Orthographic southern hemisphere polar projection, latitude circles at 0° and 50°S. Sources of maps and data given in text.

lithologic type by depth or latitudinal interval. This study has the advantage of being based on a larger number of cores (several hundred) than were utilized in the present analysis. However, analyses of this type implicitly assume that the distribution of cores is uniform, which clearly is not so. However, by computing percentages rather than absolute numbers, some of the inherent bias can be avoided. Unlike the present study, no attempt was made in this earlier study to reconstruct either paleodepth or paleolatitude for the sediment material. However, on the basis of their relative distribution of ages for cores with more precise age dates in the Miocene, and on the basis of previous experience, most of the cores compiled by *Lazarus et al.* [1987] are no older than late Miocene in age and thus have been relatively little affected by changes in depth or latitudinal position due to plate motions. It is thus reasonable to compare the patterns described above to the earlier ones. In Figures 17a–17d the distribution of lithologic types versus water depth is shown for Pliocene and Miocene cores from two regions, separated at the

boundary apparent in the earlier paleomaps (Figures 10–16) at ~50°S. During both the Pliocene and the Miocene, the region north of 50°S is dominated by calcareous oozes above ~4-km water depth and by pelagic clays below. Biosiliceous ooze is relatively uncommon at any water depth. South of 50°S the pattern is quite different. Carbonate ooze is relatively rare, and it is mostly seen at depths shallower than 2 km. Biosiliceous ooze is much more common, and it is most common at intermediate water depths, between 2 and 4 km. Biosiliceous ooze is more common in Pliocene sediments at all water depths than it is in the Miocene. Given these results, it is worth plotting the distribution of lithologic types by latitudinal band, restricting, however, the analysis to depths above 4 km, as sediments from depths below this appear, regardless of province, to be primarily nonbiogenic clays. Figures 17e (Miocene) and 17f (Pliocene), again from *Lazarus et al.* [1987], show the results. Sediments north of 40°S are almost entirely calcareous ooze, while those south of 50°S are predominantly biogenic silica and, particularly

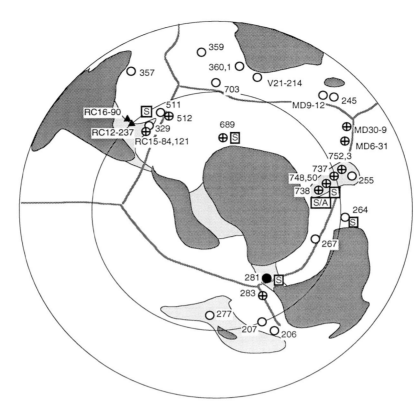

Fig. 11. Eocene geography and sediment distribution.

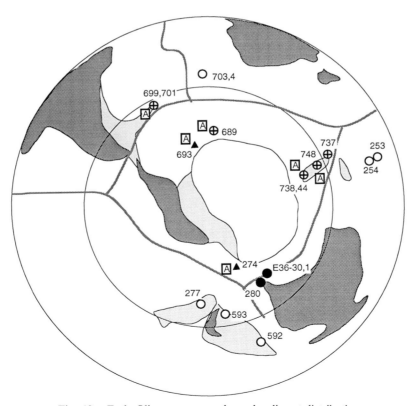

Fig. 12. Early Oligocene geography and sediment distribution.

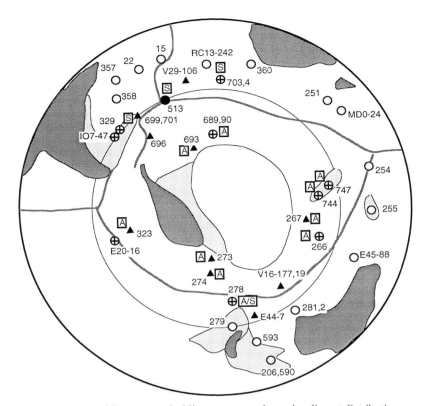

Fig. 13.   Late Oligocene–early Miocene geography and sediment distribution.

in higher latitudes (south of 60°S), clay. Sediments between 40°S and 50°S are transitional: mostly calcareous, but with significant numbers of biogenic silica oozes also being found. Thus in the Pliocene and (late) Miocene, the results of *Lazarus et al.* [1987] confirm the conclusions derived from the paleomaps of Figures 10–16: the presence of a major lithologic boundary at ~50°S.

RECONSTRUCTIONS

Paleocene

The first of our reconstructions, for the Paleocene, is shown in Figure 18. On it are approximate ridge and plate boundaries and inferred paleocirculation patterns. Unlike our other reconstructions, no radiolarian biogeographic data are used since, as mentioned earlier, no extensive radiolarian data are yet available for the Paleocene. Radiolarian assemblages are, however, known from northern Kerguelen Plateau cores MD35-12 and MD35-13 and either are comprised of tropical species or are taxonomically unknown. Biogeographic data from other microfossil groups are used to help interpret circulation patterns. Paleocene calcareous microfossil patterns were dominated by diversification after the K/T extinction event, an evolutionary overprint which makes paleoecologic interpretations more difficult. Nonetheless, all available data [*Boersma and Premoli-Silva*, 1991; *Nocchi et al.*, 1991] suggest that no distinct Southern Ocean biogeographic province existed in the Paleocene or, at most, one defined primarily by the absence of lower-latitude forms [*Sancetta*, 1979]. Planktonic foraminifera and nannofossil assemblages are of moderate diversity and show very little differentiation with latitude.

Biogenic silica in the Paleocene Antarctic appears to be localized to upwelling regions in shallow waters, including the Falkland Plateau [*Fenner*, 1991], New Zealand [*Hollis*, 1991], and Kerguelen (Figure 10). Although the number of data points is limited, there is no evidence for more widespread, open ocean upwelling. The areas of localized upwelling in this and subsequent reconstructions generally match the paleoupwelling predictions of simulations by *Parish and Curtis* [1982] and *Scotese and Summerhayes* [1986]. In these simulations, upwelling is most common along north-south oriented coastlines in paleolatitudes of 40°–50°S. Sedimentation rates of carbonate ooze in the Paleocene are variable but generally of the order of 1 cm/kyr [*Barker et al.*, 1988; *Schlich et al.*, 1989]. Even allowing for compaction, these rates are more typical of oligotrophic environments than upwelling ones today. *Boersma and Premoli-Silva* [1991] also suggest, based on the occurrence

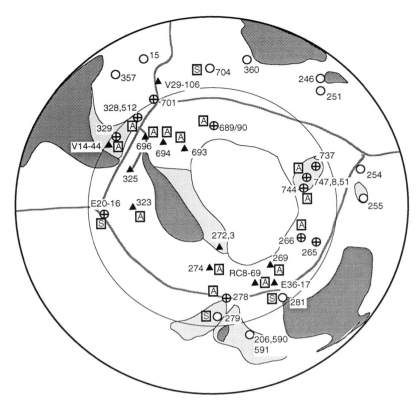

Fig. 14. Middle Miocene geography and sediment distribution.

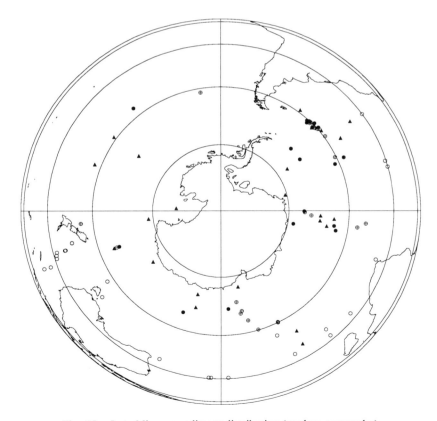

Fig. 15. Late Miocene sediment distribution (modern geography).

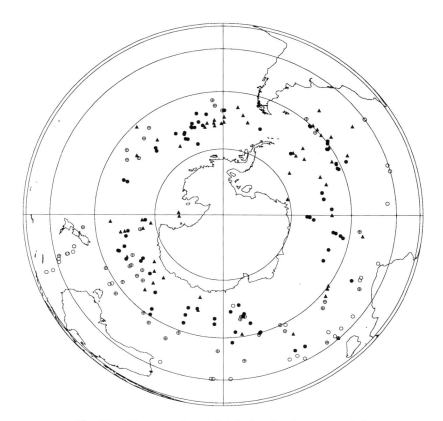

Fig. 16. Pliocene sediment distribution (modern geography).

of spinose planktonic foraminiferal species, that Paleocene Antarctic waters were rather oligotrophic. Isotopic results from the Kerguelen Plateau [*Barrera and Huber*, 1991] and the Weddell Sea [*Stott et al.*, 1990; *Kennett and Stott*, 1990] suggest that the Paleocene Southern Ocean was strongly stratified, and regional upwelling would consequently have been limited. Surface currents would not have penetrated to great depths, as the ACC is able to do today.

Given the limited data available, we have drawn circulation patterns largely following those suggested by *Barron and Peterson*'s [1991] simulation for the Paleocene, with a weak gyre in the Weddell Sea. Barron and Peterson's simulations do not include ocean bathymetry, which probably had a significant influence on ocean circulation in the Paleocene. Most prominent among these bathymetries was the presence of a large, shallow Kerguelen Plateau–Broken Ridge complex [*Peirce et al.*, 1989], which may well have inhibited the formation of the circulation cell depicted in Barron and Peterson's Paleocene map between Austral-Antarctica and India.

Eocene

Eocene data are more extensive and place more constraints on the general circulation regime (Figure

19). Biogenic silica is present, as before, throughout most of the Eocene on oceanic plateaus such as Kerguelen, and it also is found in early Eocene piston cores MD6-31 and MD6-9-30A in the deep Crozet Basin (Figure 11). Assuming that the material is indeed in situ and not reworked or current transported, enhanced productivity and upwelling are suggested at this locality. The radiolarian faunas from these cores, however, show no evidence of provincialism at this time (Figure 5). Indeed, planktonic foraminifera [*Boersma and Premoli-Silva*, 1991; *Nocchi et al.*, 1991], calcareous nannofossil [*Crux*, 1991], and dinoflagellate [*Mao and Mohr*, 1993] assemblages from the early to middle Eocene Antarctic are unusually diverse and contain many tropical elements. These biogeographic data, together with the estimates of near-maximum Cenozoic temperatures at this time in the Antarctic [*Stott et al.*, 1990], suggest that local processes were responsible for biogenic silica accumulation at these sites.

Late Eocene conditions were significantly different and presage those of the following Oligocene. Near Antarctica, radiolarian faunas became increasingly endemic (Table 2; Figure 5c), perhaps within a near-coastal province which has previously been identified elsewhere in the Antarctic by dinoflagellate studies [*Wrenn and Hart*, 1988]. Radiolarian faunas further

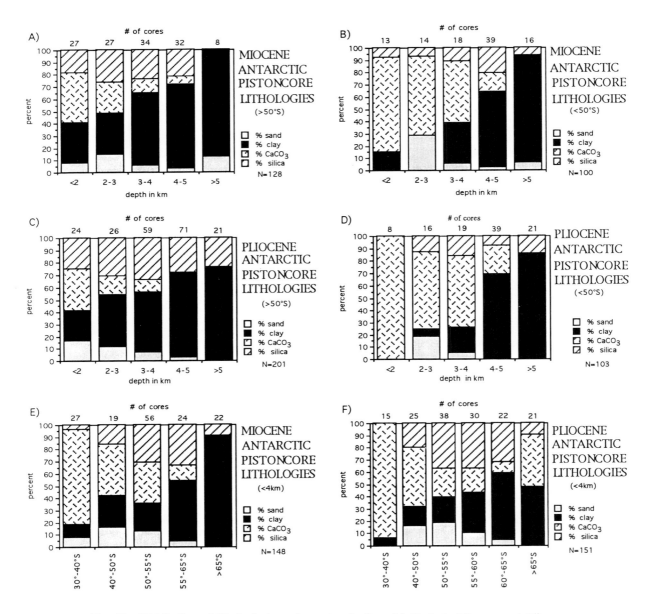

Fig. 17. Distribution of lithologic types by water depth and latitude in Miocene and Pliocene sediments from high southern latitudes. Values are percent of cores within each depth or latitude band falling into each lithologic category. Number of cores used to calculate percent values for each band in each figure given above tops of bars. Modified from *Lazarus et al.* [1987].

from the continent also began to contain significant numbers of endemic species (Table 2). The mixed assemblages of radiolarians in more northerly sites during this time interval may reflect a warmer, shallow water fauna and a deeper, colder water fauna living in a northward moving intermediate water mass developed in surface waters nearer the continent, where the radiolarians were already predominantly of Antarctic type. Biogenic silica deposition became more widespread, and preservation improved significantly. Biogenic silica

deposition and inferred upwelling across the shallow Tasman Rise began, apparently owing to strong currents [*Hampton*, 1975] of the incipient ACC sweeping through the region with the final separation of Australia from Antarctica. Biogeographic gradients in the middle latitudes of the Indo-Pacific region increased through the Eocene, particularly in calcareous nannofossils [*Sancetta*, 1979]. Cooler assemblages of nannofossils became common on the Falkland Plateau, and a decoupling between the biogeography of the Falkland Plateau and

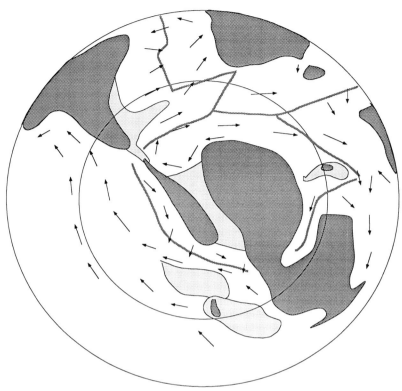

Fig. 18. Paleocene reconstruction of surface circulation in the Antarctic. Arrows give approximate direction of surface currents; other map symbols in this and subsequent Figures 19–24 as described in caption to Figure 10.

eastern Atlantic is apparent [*Crux*, 1991]. Cooler water temperatures and decreasing water column stratification are recorded in middle to late Eocene stable isotopes from both the Weddell Sea [*Stott et al.*, 1990] and Kerguelen [*Barrera and Huber*, 1991].

The circulation simulation of *Barron and Peterson* [1991] for the Eocene shows what they refer to as a "proto-circum-Antarctic" current system, a relatively strong current south of India and north of the Kerguelen Plateau–Broken Ridge complex, and localized upwelling along a frontal system associated with this current may be responsible for the biogenic silica along the northern Kerguelen Plateau and in cores MD6-31 and MD6-9-30A. Less well developed fronts probably existed elsewhere in the late Eocene Antarctic, particularly in the South Atlantic between the Falkland Plateau and the eastern Atlantic. These late Eocene features are shown as isolated lines on our reconstruction of the current systems of the Eocene. *Kennett and von der Borch* [1985, Figures 3a and 3b] infer a poleward flowing current along eastern Australia connecting middle to high latitudes in the Paleogene, reaching beyond 70° paleosouth in the late Eocene. *Barron and Peterson*'s [1991] physical oceanographic simulation for the Eocene shows, by contrast, a moderately well developed west to

east flowing current system throughout much of the Antarctic, with a maximum flow axis well north of the Antarctic continental margin even in the South Pacific near Australia. It should be noted that their simulation for the Eocene did not include any ice on the Antarctic continent. When, as in their Recent simulation, continental ice is included, much stronger currents and oceanic gradients are developed in the Southern Ocean, but again at locations near 50°S, not near the continental margin. There are at present insufficient data to determine which hypothesis for the late Eocene is correct.

Early Oligocene

Our early Oligocene reconstruction (Figure 20) shows a well-developed circum-Antarctic current (ACC) and a geographically extensive Southern Ocean. Local frontal systems of the late Eocene (Figure 19) have coalesced into a true circumpolar current system. The well-known global climatic cooling and biotic turnover of the late Eocene through earliest Oligocene [*Wolfe and Hopkins*, 1967; *Wolfe*, 1971, 1978; *Miller et al.*, 1987; *Prothero*, 1989; *Prothero and Berggren*, 1992] has long been correlated to the development of the circum-Antarctic current, Antarctic glaciation [*Shackleton and Kennett*,

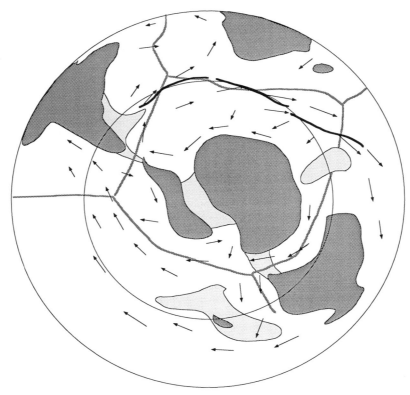

Fig. 19. Eocene reconstruction of surface circulation in the Antarctic. Inferred regional oceanic fronts shown by heavy black lines.

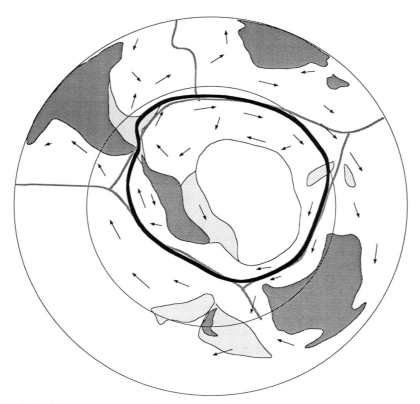

Fig. 20. Early Oligocene reconstruction of surface circulation in the Antarctic. Inferred Polar Front shown by heavy black line.

1975; *Kennett and Shackleton*, 1976; *Kennett*, 1977], and, in recent years, to the formation of major ice sheets [*Matthews and Poore*, 1980; *Wise et al.*, 1991; *Barron et al.*, 1991a; *Zachos et al.*, 1992]. The fragmentation of the Broken Ridge–Kerguelen Plateau in the late Eocene [*Peirce et al.*, 1989] and the separation of Australia from Antarctica allowed deepwater circulation to develop around the Antarctic continent for the first time, despite the seemingly closed Drake Passage [*Barker and Burrell*, 1977, 1982]. It is also possible that circulation through the channel separating East Antarctica and West Antarctica played a significant role. *Barron and Peterson*'s [1991] simulation shows only very weak currents through this channel in the Eocene and Miocene. However, coastal circulation today is partially influenced by the presence of the ice sheet, which creates a strong katabatic coastal wind system [*Deacon*, 1984]. It would be interesting to simulate Southern Ocean paleocirculation between East Antarctica and West Antarctica with an ice sheet on the East Antarctic continent.

Some previous authors have reconstructed the early Oligocene Southern Ocean as being of very limited latitudinal extent. *Kennett and Barker* [1990], for example, conclude that ODP Leg 113 early Oligocene sediments from Maud Rise in the Weddell Sea (paleolatitude ~65°S) "show no indication of the presence of the ACC or Polar Front." Virtually all the available biogeographic data, however, suggest that the early Oligocene Southern Ocean was geographically extensive, with a northern boundary near 50°S in the Atlantic and Indian ocean sectors. The radiolarian biogeographic data discussed above strongly suggest that a well-developed Southern Ocean province existed, with upwelling and a strong oceanographic boundary developed to the north, north of sites 699 and 701 on the Falkland Plateau and Site 747 on the Kerguelen Plateau. Endemic late Eocene and earliest Oligocene dinoflagellates from northern Weddell Sea Site 696 (B. A. R. Mohr, personal communication, 1992) and increased nannofossil floral similarities between northern Antarctic sites 511 and 513 and southern Antarctic Weddell Sea Site 689 in the early Oligocene [*Wise et al.*, 1991] support the concept of a geographically extensive early Oligocene Southern Ocean. ODP Leg 114 results for both calcareous nannofossils [*Crux*, 1991] and planktonic foraminifera [*Nocchi et al.*, 1991] suggest a major biogeographic boundary in the early Oligocene well to the north, between the Falkland Plateau and sites 703 and 704 in the eastern Atlantic. Eastern Atlantic sites appeared in fact to become warmer, even as the Falkland Plateau sites became cooler. A similar decoupling of thermal history was observed across the Eocene/Oligocene boundary in coastal sections of southern Australia [*Kamp et al.*, 1990]. Finally, *Murphy and Kennett* [1985] noted the development of a strong thermal gradient in the early Oligocene at northerly located sites in the southwestern Pacific.

As discussed previously, the modern ACC position

appears to track, and be partly controlled by, deep bottom topography, particularly the mid-ocean ridge system. In our early Oligocene reconstruction, we have drawn the ACC boundary near the paleoridge position in locations where sedimentological or biogeographic data are lacking. In doing this, we make use of the observation [*Stott et al.*, 1990; *Barrera and Huber*, 1991] that water column stratification in the early Oligocene was relatively weak. From this observation we assume that surface circulation patterns were able to penetrate to great depth, in a manner similar to today.

One last feature of our reconstruction of circulation is the development of the western coastal current and Antarctic Divergence. Although today the Antarctic Divergence varies longitudinally in strength, it is a reasonably coherent circumpolar feature. In our early Oligocene reconstruction, we show a western coastal current developed only as part of the Weddell gyre, and (as discussed above) in the East-West Antarctic channel. We assume that the northward extension of Antarctica in the Australian sector, together with a relatively narrow ocean, would prevent a strong circumpolar coastal current and Divergence from developing.

Late Oligocene Through Early Miocene

The late Oligocene through early Miocene reconstruction (Figure 21) is problematical. Biosiliceous sediments are more widespread than at any previous time and are quite circumpolar in distribution. Radiolarian biogeography in general also suggests a geographically extensive southern water mass. However, radiolarian assemblages from Site 278 (near the inferred position of the earlier Oligocene Polar Front) are of mixed affinities, with nearly equal numbers of Antarctic and cosmopolitan forms (Table 2). Also problematical are late Oligocene to early Miocene faunas of the Falkland Plateau region which are reported [*Weaver*, 1983; *Ciesielski et al.*, 1988] as being Subantarctic, rather than Antarctic, in composition. If so, then a weakening and/or a southward shift of the water mass boundary would be indicated. However, it is not clear from the taxa listed by these authors if these assemblages are indeed Subantarctic. Most of the species reported from these sites (*Cyrtocapsella tetrapera*, *Prunopyle hayesi*, etc.) are also common in the southern Weddell Sea [*Abelmann*, 1990] and on the Kerguelen Plateau [*Abelmann*, 1992; *Takemura*, 1992]. A few of these forms are cosmopolitan, a few others bipolar, but many, such as *Cyrtocapsella longithorax* and *Cycladophora golli*, are in our opinion Antarctic endemics. One reason for the assignment of the Falkland Plateau assemblages to the Subantarctic, rather than the Antarctic, province may have been the inability to find *Chen*'s [1975] middle to early Miocene stratigraphic indicator species (e.g., *Spongomelissa dilli*, etc.). *Chen*'s stratigraphic indicator species for the earlier Miocene have, however, recently

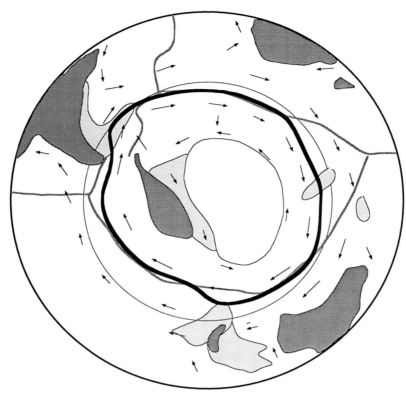

Fig. 21. Late Oligocene to early Miocene reconstruction of surface circulation in the Antarctic.

been found to be extremely rare even within the higher-latitude Antarctic (D. B. Lazarus, unpublished observations). The current biostratigraphic zonation of the late Oligocene to early Miocene time interval is now based on more common taxa [*Abelmann*, 1990, 1992].

Other authors [*Crux*, 1991; *Nocchi et al.*, 1991] have noted a warming in the Antarctic in the latest Oligocene and early Miocene based on various microfossil groups, and warming is also noted in sedimentologic indicators [*Ehrmann*, 1991]. However, the Falkland Plateau planktonic foraminiferal, nannofossil, and diatom assemblages, although "warmer" than in the early Oligocene, are still described as "cold water" or even (diatoms) as belonging to the "circum-Antarctic province" [*Ciesielski et al.*, 1988]. Similarly, calcareous nannofossil floras from the Kerguelen Plateau show no warming in the late Oligocene [*Aubry*, 1992]. A reexamination of the Oligocene Falkland Plateau radiolarian assemblages is clearly needed.

In our reconstruction, we place the boundary of the Southern Ocean along the northern limit of biosiliceous sediments and close to the estimated location of the ridge axes. By the early Miocene, Broken Ridge had moved well north, and deep circumpolar flow was impeded only by the Kerguelen Plateau, which by this time was, on average, probably several hundred meters deeper than in the late Eocene/early Oligocene [*Peirce*

et al., 1989; *Coffin*, 1992], and (possibly) also by a closed Drake Passage.

The gradual movement of the Antarctic continent toward a pole-centered position would have placed a larger band of Indo-Australian Antarctic coastal waters under the influence of polar easterly winds. This movement, and broadening and deepening of the formerly narrow Australian-Antarctic basin, would probably have enhanced the circum-Antarctic development of the western coastal current (or East Wind Drift) and Antarctic Divergence in the Indo-Australian sector by the late Oligocene–early Miocene and possibly led to higher upwelling and productivity near the continent. Unfortunately, the available core data near the continent are too fragmentary in this time interval, owing to both hiatuses and poor recovery, to determine if higher productivity and biogenic sedimentation rates occurred.

Middle Miocene Through Pliocene

Our remaining three reconstructions for the middle Miocene, late Miocene, and Pliocene (Figures 22–24) are all similar both to each other and to those of the Oligocene. A gradual expansion in the inferred extent of the Southern Ocean is visible but may be due to better geographic control, particularly from sedimentologic

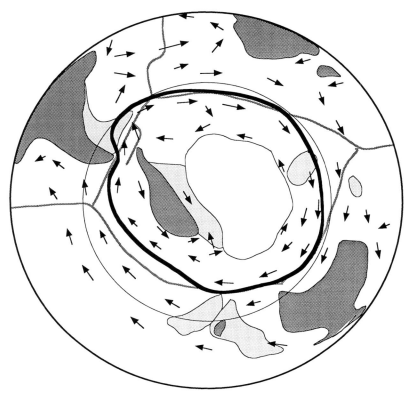

Fig. 22. Middle Miocene reconstruction of surface circulation in the Antarctic.

data in the Pacific sector. Although the Southern Oceans of these time intervals differ little geographically, they are known to have changed substantially in characteristics within the Southern Ocean water mass, such as the extent of biosiliceous versus carbonate sediments, a trend already discussed above. Although they cannot be shown in our maps, the more detailed records available for the later Neogene have allowed other workers to document several short-term fluctuations in Southern Ocean characteristics. The early Pliocene interval, for example, is known to be relatively warm, with higher percentages of nonendemic radiolarian species within the Southern Ocean and high productivity both within the Southern Ocean proper and near the Polar Front [*Elmstrom and Kennett*, 1985; *Abelmann et al.*, 1990; *Grobe et al.*, 1990]. It is worth noting that the biogeographic (Figure 8) and sedimentologic (Figures 14–16) distributions both indicate, with a fair degree of precision, a Southern Ocean northern boundary in most regions at ~50°S and, furthermore, a reasonably close correspondence of this boundary to the inferred position of the ridge system. The geography of our indicators also matches the position of the current axes in *Barron and Peterson*'s [1991] Miocene simulation. Biosiliceous clays in deep basins well north of the estimated position of the Polar Front are clearly evident by the late Miocene, indicating that diatom displace-

ment by strong northward transport of Antarctic Bottom Water [*Booth and Burckle*, 1976; *Burckle*, 1981] was well developed by this time.

DISCUSSION

We conclude from the above analyses that the Southern Ocean first appeared, within the Cenozoic, in the late Eocene or early Oligocene. Since its origin, it has remained a fairly extensive water mass. In the Atlantic and Indian ocean sectors, its northern boundary has been near 50°S for most of its history. In the Australian sector, it has expanded gradually northward with the growth of the ocean basin and its ridge system. The Southern Ocean's geographic extent and long-term geographic stability are probably due to the gradual development of the oceanic basins around the Antarctic continent and relative stability of other major geographic features, such as the Falkland Plateau. Although numerous brief excursions (usually northward) of the Polar Front have been documented by other workers, these excursions appear to be secondary fluctuations superimposed on a longer-term, relatively static or (in the Australian sector) only slowly changing geography. We do not see, except in the Australian sector, a gradual expansion of the Polar Front from a position very near to the continent to its modern position, as hypothesised by several authors

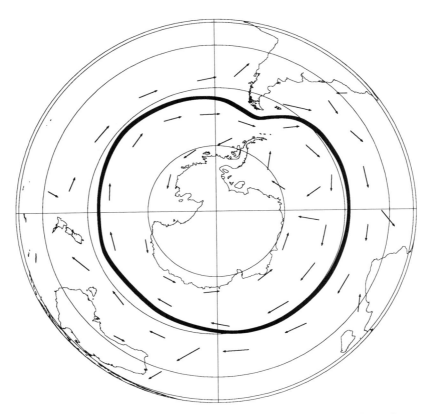

Fig. 23.   Late Miocene reconstruction of surface circulation in the Antarctic.

[*Kemp et al.*, 1975; *Hayes and Frakes*, 1975; *Kennett*, 1977; *Ciesielski and Weaver*, 1983].

Previous workers have traced the Polar Front back in time until the late Miocene to early Pliocene [*Abelmann et al.*, 1990; *Baldauf and Barron*, 1990] or even early Miocene [*Kennett*, 1977]. In this study, we use microfossil biogeography, sediment distributions, and stable isotope data to extend these previous results. We suggest that a well-developed Polar Front existed in the early Oligocene and was present for much of the time between the early Oligocene and the Recent.

Although there are several reasons for these different conclusions by different authors, we discuss two reasons specifically.

Early studies [e.g., *Kemp et al.*, 1975; *Hayes and Frakes*, 1975] were based on a very limited set of cores and a very preliminary stratigraphy, primarily from the one region of the Antarctic (the Australian sector) where the influence of seafloor spreading on frontal position appears to be maximal. The patterns determined by these early data sets do not, in our present analysis, seem to be general features of the Southern Ocean. They particularly do not seem to apply to the Atlantic sector, where we infer a relatively stable geographic position of the Southern Ocean since the early Oligocene. Given the much larger number and greater

geographic distribution of cores now available and the often substantial changes in biostratigraphy, this difference in interpretation is understandable.

A second factor concerns the proxy indicators used to trace water mass development. Some authors have based their interpretations on the timing and growth of the Polar Front using diatom biogeography or the loss of carbonate from sediments, indicators which indeed give much later dates for the origin of a Polar Front. We do not wish to deny the importance of other indicators of water mass conditions and distributions. It is possible, for example, that diatom biogeography in the Antarctic reflects seasonal development of relatively warm surface waters until the late Neogene, while radiolarians record the earlier development of cold average annual conditions in the surface through intermediate water layers. Such differences between proxy indicators will, when better understood, eventually provide more detailed insights into the development of the Cenozoic Southern Ocean. However, as we have stated earlier, we feel that these other indicators are less sensitive markers of general water mass positions than the distribution of radiolarian assemblages and biosiliceous sediments used in this study.

All studies, including this one, run the risk of arguing from negative evidence. Although we do not see any

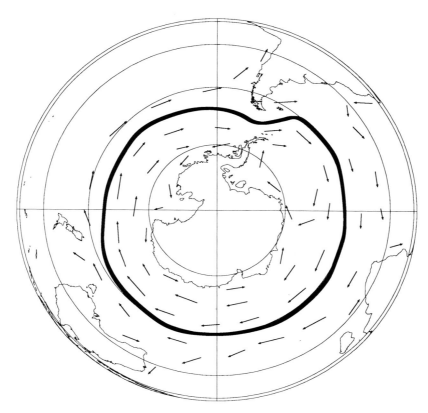

Fig. 24.   Pliocene reconstruction of surface circulation in the Antarctic.

evidence in the present analysis for a distinct Southern Ocean in the early to middle Eocene, for example, this does not prove that it did not exist. It is possible (although we believe it unlikely) that new, more sensitive data sets will at some point be collected that document the presence of at least regional frontal systems and a distinct Antarctic water mass in the early to middle Eocene as well. Thus we can only give a minimum date for the origin of the Southern Ocean as a distinct water mass, together with its geographic extent during those intervals where our data appear to accurately locate its position. During time intervals such as the late Oligocene to early Miocene, or basal Pliocene, where our biogeographic and sedimentologic data are limited or ambiguous, we can only guess at water mass development. Our preference, however, for these intervals, given the relatively stable position of the Polar Front seen during times when our data sets are more reliable, is to assume that some sort of frontal system and ACC was still present, at much the same latitudinal position as always. The presence of higher numbers of Subantarctic forms within the Southern Ocean during such time intervals are interpreted by us as indicators of different (presumably warmer) water conditions within the Southern Ocean, rather than indicators of a Southern Ocean shrunken in extent or entirely absent. This

difference of interpretation may seem to be mostly semantic, but we believe that it has significant consequences.

First, unless paleoceanographic data can be expressed and interpreted in terms that are compatible with physical oceanographic processes, it will be difficult to fully utilize the synergy that can be obtained by comparing physical simulations of past circulation with geologic data that map that past circulation. This is not a criticism of past paleoceanographic studies; until very recently there were no simulations for comparison. However, such simulations will play an increasing role in future paleoceanographic research, and thus we have made a conscious effort in the current study to interpret our results in light of what is known of the underlying physical oceanographic processes. Although present understanding of the mechanisms controlling Polar Front position is limited, to the extent that they are understood (see the introduction section of this paper), a relatively stable position that changes only slowly with geography is expected. Our conclusions may well be wrong, but we have tried to avoid making interpretations which might prove difficult to simulate using known physical oceanographic mechanisms, unless there was a clear pattern in our data which demanded it.

Second, the antiquity, geographic stability, and geographic extent of the Southern Ocean are important pieces of information needed by biologists wishing to understand the origin and evolution of the Antarctic biota. In our view, a distinct biogeographic province, with an oceanographic barrier to dispersal and potential for promoting speciation via interruption of gene flow, has been in existence in the Antarctic for nearly 40 m.y. This is a full order of magnitude longer than if the ACC and Polar Front originated only in the early Pliocene.

Finally, our interpretation of the history of the ACC and Southern Ocean water mass allows us to explain important aspects of the development of this ocean in terms of relatively simple controlling processes, for example, the spreading of the ocean basins and the distribution of continent and ocean. This simplicity of explanation stands in contrast to the more detailed explanations often given, where numerous shifts are individually identified and individually explained. We do not dispute the accuracy of a detailed approach to explanation; we only wish to offer an alternative, which perhaps offers a simpler way to identify general mechanisms in paleoceanography.

Acknowledgments. The authors wish to thank Jim Kennett for his work in organizing the conference and for editing these symposium volumes; Barbara Mohr and Eric Barron for providing preprints of papers; Kurt Meier for assistance in drafting preliminary versions of some of the figures; and Jim Kennett, Ted Moore, and Cathy Nigrini for reviewing the manuscript. This work was supported by the Swiss National Fonds.

REFERENCES

Abelmann, A., Oligocene to middle Miocene radiolarian stratigraphy of southern high-latitudes from ODP Leg 113, sites 689 and 690, Maud Rise, *Proc. Ocean Drill. Program Sci. Results, 113*, 675–708, 1990.

Abelmann, A., Early to mid-Miocene radiolarian stratigraphy of the Kerguelen Plateau (ODP Leg 120), *Proc. Ocean Drill. Program Sci. Results, 120*, 757–784, 1992.

Abelmann, A., R. Gersonde, and V. Spiess, Pliocene-Pleistocene paleoceanography in the Weddell Sea—Siliceous microfossil evidence, in *Geological History of the Polar Oceans: Arctic Versus Antarctic*, edited by U. Bleil and J. Thiede, pp. 729–759, Kluwer, Dordrecht, Netherlands, 1990.

Aubry, M. P., Paleogene calcareous nannofossils from the Kerguelen Plateau, Leg 120, *Proc. Ocean Drill. Program Sci. Results, 120*, 471–492, 1992.

Baldauf, J. G., and J. A. Barron, Evolution of biosiliceous sedimentation patterns—Eocene through Quaternary: Paleoceanographic response to polar cooling, in *Geological History of the Polar Oceans: Arctic Versus Antarctic*, edited by U. Bleil and J. Thiede, pp. 575–607, Kluwer, Dordrecht, Netherlands, 1990.

Baldauf, J. G., J. A. Barron, W. U. Ehrmann, P. Hempel, and D. Murray, Biosiliceous sedimentation patterns for the Indian Ocean during the last 45 million years, in *Synthesis of Results From Scientific Drilling in the Indian Ocean, Geophys. Monogr. Ser.*, vol. 70, edited by R. A. Duncan, D. K. Rea, R. B. Kidd, U. von Rad, and J. K. Weissel, pp. 335–350, AGU, Washington, D. C., 1992.

Barker, P. F., and J. Burrell, The opening of the Drake Passage, *Mar. Geol., 25*, 15–34, 1977.

Barker, P. F., and J. Burrell, The influence upon Southern Ocean circulation, sedimentation, and climate of the opening of the Drake Passage, in *Antarctic Geoscience*, edited by C. Craddock, pp. 377–385, University of Wisconsin Press, Madison, 1982.

Barker, P. F., et al., Leg 113, *Proc. Ocean Drill. Program Initial Rep., 113*, 785 pp., 1988.

Barrera, E., and B. T. Huber, Paleogene and early Neogene oceanography of the southern Indian Ocean: Leg 119 foraminifer stable isotope results, *Proc. Ocean Drill. Program Sci. Results, 119*, 693–717, 1991.

Barron, E. J., and W. H. Peterson, The Cenozoic ocean circulation based on ocean General Circulation Model results, *Palaeogeogr. Palaeoclimatol. Palaeoecol., 83*, 1–28, 1991.

Barron, J., et al., Leg 119, *Proc. Ocean Drill. Program Sci. Results, 119*, 942 pp., 1989.

Barron, J. A., B. Larsen, and J. G. Baldauf, Evidence for late Eocene to early Oligocene Antarctic glaciation and observations on late Neogene glacial history of Antarctica: Results from Leg 119, *Proc. Ocean Drill. Program Sci. Results, 119*, 869–891, 1991*a*.

Barron, J. A., J. G. Baldauf, E. Barrera, J.-P. Caulet, B. T. Huber, B. H. Keating, D. Lazarus, H. Sakai, H. R. Thierstein, and W. Wei, Biochronologic and magnetochronologic synthesis of Leg 119 sediments from the Kerguelen Plateau and Prydz Bay, Antarctica, *Proc. Ocean Drill. Program Sci. Results, 119*, 813–848, 1991*b*.

Barron, J., et al., Leg 119, *Proc. Ocean Drill. Program Sci. Results, 119*, 1003 pp., 1991*c*.

Berger, W. H., Biogenous deep-sea sediments: Production, preservation and interpretation, in *Chemical Oceanography*, vol. 5, edited by J. P. Riley and R. Chester, pp. 266–388, Academic, San Diego, Calif., 1976.

Boersma, A., and I. Premoli-Silva, Distribution of Paleogene planktonic foraminifera—Analogies with the Recent?, *Palaeoecol., 83*, 29–48, 1991.

Boltovsky, D., and W. R. Riedel, Polycystine radiolaria from the southwestern Atlantic Ocean plankton, *Rev. Esp. Micropaleontol., 12*, 99–146, 1980.

Booth, J. D., and L. H. Burckle, Displaced Antarctic diatoms in the southwestern and central Pacific, *Pac. Geol., 11*, 99–108, 1976.

Brown, J., A. Colling, D. Park, J. Phillips, D. Rothery, and J. Wright, *Ocean Circulation*, 238 pp., Pergamon, New York, 1989.

Burckle, L. H., Displaced Antarctic diatoms in the Amirante Passage, *Mar. Geol., 39*, M39–M43, 1981.

Burckle, L. H., and J. Cirilli, Origin of diatom ooze belt in the Southern Ocean: Implications for late Quaternary paleoceanography, *Micropaleontology, 33*, 82–86, 1987.

Burckle, L. H., and E. M. Pokras, Implications of a Pliocene stand of *Nothofagus* (southern beech) within 500 kilometres of the south pole, *Antarct. Sci., 3*, 389–403, 1991.

Chen, P. H., Antarctic radiolaria, *Initial Rep. Deep Sea Drill. Proj., 28*, 437–513, 1975.

Ciesielski, P. F., and F. M. Weaver, Neogene and Quaternary paleoenvironmental history of Deep Sea Drilling Project Leg 71 sediments, southwest Atlantic Ocean, *Initial Rep. Deep Sea Drill. Proj., 71*, 461–477, 1983.

Ciesielski, P. F., et al., Leg 114, *Proc. Ocean Drill. Program Initial Rep., 114*, 815 pp., 1988.

Clarke, A., Temperature and evolution: Southern Ocean cooling and the Antarctic marine fauna, in *Antarctic Ecosystems: Ecological Change and Conservation*, edited by K. R. Kerry and G. Hempel, pp. 9–22, Springer-Verlag, New York, 1990.

CLIMAP Project Members, The surface of the ice-age Earth, *Science, 191*, 1131–1137, 1976.

Coffin, M. F., Subsidence of the Kerguelen Plateau: The Atlantis concept, *Proc. Ocean Drill. Program Sci. Results, 120*, 945–949, 1992.

Cooke, D. W., and J. D. Hays, Estimates of Antarctic seasonal ice cover during glacial intervals, in *Antarctic Geoscience*, edited by C. Craddock, pp. 1017–1026, University of Wisconsin Press, Madison, 1982.

Crux, J. A., Calcareous nannofossils recovered by Leg 114 in the Subantarctic South Atlantic Ocean, *Proc. Ocean Drill. Program Sci. Results*, *114*, 155–178, 1991.

Deacon, G., *The Antarctic Circumpolar Ocean*, 180 pp., Cambridge University Press, New York, 1984.

Denham, C., and C. Scotese, Terra Mobilis (computer program), Earth in Motion Technologies, Austin, Tex., 1988.

Ehrmann, W. U., Implications of sediment composition on the southern Kerguelen Plateau for paleoclimate and depositional environment, *Proc. Ocean Drill. Program Sci. Results*, *119*, 185–210, 1991.

Elmstrom, K. M., and J. P. Kennett, Late Neogene paleoceanographic evolution of Site 590: Southwest Pacific, *Initial Rep. Deep Sea Drill. Proj.*, *90*, 1361–1381, 1985.

Fenner, J. M., Taxonomy, stratigraphy, and paleoceanographic implications of Paleocene diatoms, *Proc. Ocean Drill. Program Sci. Results*, *114*, 123–154, 1991.

Froelich, P. N., P. N. Malone, D. A. Hodell, P. F. Ciesielski, D. A. Warnke, F. Westall, E. A. Hailwood, D. C. Nobes, J. Fenner, J. Mienert, C. J. Mwenifumbo, and D. W. Müller, Biogenic opal and carbonate accumulation rates in the Subantarctic South Atlantic: The late Neogene of Meteor Rise Site 704, *Proc. Ocean Drill. Program Sci. Results*, *114*, 515–550, 1991.

Gersonde, R., A. Abelmann, L. Burckle, N. Hamilton, D. Lazarus, K. McCartney, P. O'Brien, V. Spiess, and J. S. W. Wise, Biostratigraphic synthesis of Neogene siliceous microfossils from the Antarctic Ocean, ODP Leg 113 (Weddell Sea), *Proc. Ocean Drill. Program Sci. Results*, *113*, 915–936, 1990.

Goll, R. M., and K. R. Bjørklund, Radiolaria in surface sediments of the North Atlantic Ocean, *Micropaleontology*, *17*, 434–454, 1971.

Goll, R. M., and K. R. Bjørklund, Radiolaria in surface sediments of the South Atlantic, *Micropaleontology*, *20*, 38–75, 1974.

Gordon, A. L., Structure of Antarctic waters between 20°W and 170°W, *Antarct. Map Folio Ser.*, folio 5, edited by V. Bushnell, Am. Geogr. Soc., New York, 1967.

Gordon, A. L., Spatial and temporal variability within the Southern Ocean, in *Antarctic Ocean and Resources Variability*, edited by D. Sahrhage, pp. 41–56, Springer-Verlag, New York, 1988.

Grobe, H., D. K. Fütterer, and V. Spiess, Oligocene to Quaternary sedimentation processes on the Antarctic continental margin, ODP Leg 113, Site 693, *Proc. Ocean Drill. Program Sci. Results*, *113*, 121–134, 1990.

Hampton, M. A., Detrital and biogenic sediment trends at DSDP sites 280 and 281, and evolution of middle Cenozoic currents, *Initial Rep. Deep Sea Drill. Proj.*, *29*, 1071–1076, 1975.

Harwood, D., D. B. Lazarus, M. Aubry, W. A. Berggren, F. Heider, H. Inokuchi, and T. Maruyama, Neogene stratigraphic synthesis, ODP Leg 120, *Proc. Ocean Drill. Program Sci. Results*, *120*, 1031–1052, 1992.

Hayes, D. E., and L. A. Frakes, General synthesis, Deep Sea Drilling Project Leg 28, *Initial Rep. Deep Sea Drill. Proj.*, *28*, 919–942, 1975.

Hays, J. D., Radiolaria and late Tertiary and Quaternary history of Antarctic seas, in *Biology of the Antarctic Seas II*, *Antarct. Res. Ser.*, vol. 5, edited by G. Llano, AGU, Washington, D. C., 1965.

Hollis, C. J., Latest Cretaceous to late Paleocene radiolaria from Marlborough (New Zealand) and DSDP Site 208, Ph.D. thesis, Univ. of Auckland, Auckland, New Zealand, 1991.

Honjo, S., Particle fluxes and modern sedimentation in the polar oceans, in *Polar Oceanography, Part B: Chemistry, Biology, and Geology*, pp. 687–739, Academic, San Diego, Calif., 1990.

Johnson, D. A., and C. Nigrini, Radiolarian biogeography in surface sediments of the western Indian Ocean, *Mar. Micropaleontol.*, *5*, 111–152, 1980.

Johnson, T. C., The dissolution of siliceous microfossils in surface sediments of the eastern tropical Pacific, *Deep Sea Res.*, *21*, 851–864, 1974.

Kamp, P. J. J., D. B. Waghorn, and C. S. Nelson, Late Eocene–early Oligocene integrated isotope stratigraphy and biostratigraphy for paleoshelf sequences in southern Australia: Paleoceanographic implications, *Palaeogeogr. Palaeoclimatol. Palaeoecol.*, *80*, 311–323, 1990.

Kemp, E. M., L. A. Frakes, and D. E. Hayes, Paleoclimatic significance of diachronous biogenic facies, Leg 28, Deep Sea Drilling Project, *Initial Rep. Deep Sea Drill. Proj.*, *28*, 909–916, 1975.

Kennett, J. P., Cenozoic evolution of Antarctic glaciation, the circum-Antarctic Ocean, and their impact on global paleoceanography, *J. Geophys. Res.*, *82*, 3843–3860, 1977.

Kennett, J. P., The development of planktonic biogeography in the Southern Ocean during the Cenozoic, *Mar. Micropaleontol.*, *3*, 301–345, 1978.

Kennett, J. P., and P. F. Barker, Latest Cretaceous to Cenozoic climate and oceanographic developments in the Weddell Sea, Antarctica: An ocean-drilling perspective, *Proc. Ocean Drill. Program Sci. Results*, *113*, 937–962, 1990.

Kennett, J. P., and N. J. Shackleton, Oxygen isotopic evidence for the development of the psychrosphere 38 myr ago, *Nature*, *260*, 513–515, 1976.

Kennett, J. P., and L. D. Stott, Proteus and proto-oceanus: Ancestral Paleogene oceans as revealed from Antarctic stable isotopic results; ODP Leg 113, *Proc. Ocean Drill. Program Sci. Results*, *113*, 865–880, 1990.

Kennett, J. P., and C. C. von der Borch, Southwest Pacific Cenozoic paleoceanography, *Initial Rep. Deep Sea Drill. Proj.*, *90*, 1493–1517, 1985.

Lawver, L. A., L. M. Gahagan, and M. F. Coffin, The development of paleoseaways around Antarctica, in *The Antarctic Paleoenvironment: A Perspective on Global Change, Part One*, *Antarct. Res. Ser.*, vol. 56, edited by J. P. Kennett and D. A. Warnke, pp. 7–30, AGU, Washington, D. C., 1992.

Lazarus, D. B., A. Pallant, and J. D. Hays, A data-base of Antarctic pre-Pleistocene sediment cores, *Tech. Rep. 87-16*, Woods Hole Oceanogr. Inst., Woods Hole, Mass., 1987.

Ling, H. Y., Cretaceous (Maestrichtian) radiolarians: Leg 114, *Proc. Ocean Drill. Program Sci. Results*, *114*, 311–316, 1991.

Ling, H. Y., and D. B. Lazarus, Cretaceous radiolaria from the Weddell Sea: Leg 113 of the Ocean Drilling Program, *Proc. Ocean Drill. Program Sci. Results*, *113*, 353–364, 1990.

Lisitzin, A. P., Distribution of microfossils in suspension and in bottom sediments, in *The Micropaleontology of Oceans*, edited by B. M. Funnell and W. R. Riedel, pp. 223–230, Cambridge University Press, New York, 1971.

Lombari, G., and G. Boden, Modern radiolarian global distributions, *Spec. Publ. Cushman Found. Foraminiferal Res.*, *16A*, 1985.

Lozano, J. A., and J. D. Hays, Relationship of radiolarian assemblages to sediment types and physical oceanography in the Atlantic and western Indian Ocean sectors of the Antarctic Ocean, Investigation of Late Quaternary Paleoceanography and Paleoclimatology, *Mem. Geol. Soc. Am.*, *145*, 303–336, 1976.

Mao, S., and B. A. R. Mohr, A middle Eocene dinocyst assemblage from Bruce Bank (South Scotia Ridge) off the northern Antarctic Peninsula, *Rev. Palaeobot. Palynol.*, in press, 1993.

Matthews, R. K., and R. Z. Poore, Tertiary $\partial^{18}O$ record and glacio-eustatic sea-level fluctuations, *Geology*, 8, 501–504, 1980.

McGowan, J. A., Oceanic biogeography of the Pacific, in *The Micropaleontology of Oceans*, edited by B. M. Funnell and W. R. Riedel, pp. 3–74, Cambridge University Press, New York, 1971.

McGowan, J. A., The nature of oceanic ecosystems, in *The Biology of the Oceanic Pacific—Proceedings of the 33rd Annual Biology Colloquium*, edited by C. B. Miller, pp. 9–28, Oregon State University, Corvallis, 1974.

Miller, K. G., R. G. Fairbanks, and G. S. Mountain, Tertiary oxygen isotope synthesis, sea level history, and continental margin erosion, *Paleoceanography*, 2, 1–19, 1987.

Molina-Cruz, A., Radiolarian assemblages and their relationship to the oceanography of the subtropical southeastern Pacific, *Mar. Micropaleontol.*, 2, 315–352, 1977.

Moore, T. C. J., The distribution of radiolarian assemblages in the modern and ice-age Pacific, *Mar. Micropaleontol.*, 3, 229–266, 1978.

Morley, J. J., A transfer function for estimating paleoceanographic conditions based on deep-sea surface sediment distribution of radiolarian assemblages in the South Atlantic, *Quat. Res.*, 12, 381–395, 1979.

Murphy, M. G., and J. P. Kennett, Development of latitudinal thermal gradients during the Oligocene: Oxygen-isotopic evidence from the southwest Pacific, *Initial Rep. Deep Sea Drill. Proj.*, 90, 1347–1360, 1985.

Nigrini, C., Radiolaria in pelagic sediments from the Indian and Atlantic oceans, *Bull. 11*, Scripps Inst. of Oceanogr., La Jolla, Calif., 1967.

Nigrini, C., Radiolaria from eastern tropical Pacific sediments, *Micropaleontology*, 14, 51–63, 1968.

Nigrini, C., Radiolarian assemblages in the North Pacific and their application to a study of Quaternary sediments in core V20-130, Geological Investigations of the North Pacific, *Mem. Geol. Soc. Am.*, 126, 139–183, 1970.

Nocchi, M., E. Amici, and I. Premoli Silva, Planktonic foraminiferal biostratigraphy and paleoenvironmental interpretation of Paleogene faunas from the Subantarctic transect, Leg 114, *Proc. Ocean Drill. Program Sci. Results*, 114, 233–280, 1991.

Parish, J. T., and R. C. Curtis, Atmospheric circulation, upwelling, and organic-rich rocks in the Mesozoic and Cenozoic eras, *Palaeogeogr. Palaeoclimatol. Palaeoecol.*, 40, 31–66, 1982.

Peirce, J., et al., Leg 121, *Proc. Ocean Drill. Program Initial Rep.*, 121, 1000 pp., 1989.

Pierrot-Bults, A. C., S. van der Spoel, B. J. Zahuranec, and R. K. Johnson (Eds.), *Pelagic Biogeography, Tech. Pap. in Mar. Sci.*, vol. 49, 295 pp., UNESCO, Paris, 1986.

Prothero, D. R., Stepwise extinctions and climatic decline during the later Eocene and Oligocene, in *Mass Extinctions, Processes and Evidence*, edited by S. K. Donovan, pp. 217–234, Columbia University Press, New York, 1989.

Prothero, D. R., and W. A. Berggren, *Eocene-Oligocene Climatic and Biotic Evolution*, 568 pp., Princeton University Press, Princeton, N. J., 1992.

Robert, C., R. Stein, and M. Acquaviva, Cenozoic evolution and significance of clay associations in the New Zealand region of the South Pacific, Deep Sea Drilling Project, Leg 90, *Initial Rep. Deep Sea Drill. Proj.*, 90, 1225–1238, 1986.

Romine, K., and G. Lombari, Evolution of Pacific circulation in the Miocene: Radiolarian evidence from DSDP Site 289, in *The Miocene Ocean: Paleoceanography and Biogeography*, edited by J. P. Kennett, pp. 273–291, Geological Society of America, Boulder, Colo., 1985.

Sancetta, C., Paleogene Pacific microfossils and paleoceanography, *Mar. Micropaleontol.*, 4, 363–398, 1979.

Schlich, R., et al., Leg 120, *Proc. Ocean Drill. Program Initial Rep.*, 120, 648 pp., 1989.

Scotese, C. R., and C. P. Summerhayes, Computer model of paleoclimate predicts coastal upwelling in the Mesozoic and Cenozoic, *Geobyte*, 1(3), 28–42, 1986.

Scotese, C. R., L. M. Gahagan, and R. L. Larson, Plate tectonic reconstructions of the Cretaceous and Cenozoic ocean basins, *Tectonophysics*, 155, 27–48, 1988.

Shackleton, N. J., and J. P. Kennett, Paleotemperature history of the Cenozoic and the initiation of Antarctic glaciation: Oxygen and carbon isotope analyses in DSDP sites 277, 279 and 281, *Initial Rep. Deep Sea Drill. Proj.*, 29, 743–755, 1975.

Shemesh, A., L. H. Burckle, and P. N. Froelich, Dissolution and preservation of Antarctic diatoms and the effect on sediment thanatocoenoses, *Quat. Res.*, 31, 288–308, 1989.

Stott, L. D., J. P. Kennett, N. J. Shackleton, and R. M. Corfield, The evolution of Antarctic surface waters during the Paleogene: Inferences from the stable isotopic composition of planktonic foraminifers, ODP Leg 113, *Proc. Ocean Drill. Program Sci. Results*, 113, 849–863, 1990.

Takemura, A., Radiolarian Paleogene biostratigraphy in the southern Indian Ocean, ODP Leg 120, *Proc. Ocean Drill. Program Sci. Results*, 120, 735–756, 1992.

Tilzer, M. M., B. von Bodungen, and V. Smetacek, Light-dependence of phytoplankton photosynthesis in the Antarctic Ocean: Implications for regulating productivity, in *Antarctic Nutrient Cycles and Food Webs*, edited by W. R. Siegfried, P. R. Condy, and R. M. Laws, pp. 60–69, Springer-Verlag, New York, 1983.

van der Spoel, S., and A. C. Pierrot-Bults (Eds.), *Zoogeography and Diversity of Plankton*, 410 pp., Halstead, New York, 1979.

Weaver, F. M., Cenozoic radiolarians from the southwest Atlantic, Falkland Plateau region, Deep Sea Drilling Project Leg 71, *Initial Rep. Deep Sea Drill. Proj.*, 71, 667–686, 1983.

Wefer, G., G. Fischer, D. K. Fütterer, R. Gersonde, S. Honjo, and D. Ostermann, Particle sedimentation and productivity in Antarctic waters of the Atlantic sector, in *Geological History of the Polar Oceans: Arctic Versus Antarctic*, edited by U. Bleil and J. Thiede, pp. 363–380, Kluwer, Dordrecht, Netherlands, 1990.

Wise, S. W., J. R. Breza, D. M. Harwood, and W. Wei, Paleogene glacial history of Antarctica, in *Controversies in Modern Geology*, edited by H. Weissert and D. Müller, pp. 133–171, Academic, New York, 1991.

Wolfe, J. A., Tertiary climatic fluctuations and methods of analysis of Tertiary floras, *Palaeogeogr. Palaeoclimatol. Palaeoecol.*, 9, 27–57, 1971.

Wolfe, J. A., A paleobotanical interpretation of Tertiary climates in the northern hemisphere, *Am. Sci.*, 66, 694–703, 1978.

Wolfe, J. A., and D. M. Hopkins, Climatic changes recorded by Tertiary land floras in northwestern North America, in *Tertiary Correlations and Climatic Changes in the Pacific*, edited by K. Hatai, pp. 67–76, Sasaki Publishers, Sendai, Japan, 1967.

Wrenn, J. H., and G. F. Hart, Paleogene dinoflagellate cyst biostratigraphy of Seymour Island, Antarctica, *Mem. Geol. Soc. Am.*, 169, 321–447, 1988.

Zachos, J. C., W. A. Berggren, M. P. Aubry, and A. Mackensen, Isotope and trace element geochemistry of Eocene and Oligocene foraminifers from Site 798, Kerguelen Plateau, *Proc. Ocean Drill. Program Sci. Results*, 120, 839–854, 1992.

(Received April 17, 1992;
accepted October 20, 1992.)

THE EVOLUTION OF THE CENOZOIC SOUTHERN HIGH- AND MID-LATITUDE PLANKTONIC FORAMINIFERAL FAUNAS

D. Graham Jenkins

Department of Geology, National Museum of Wales, Cardiff CF1 3NP, Wales, United Kingdom

The distribution of living species of planktonic foraminifera and the relationship of species diversity to temperature are reviewed with special reference to the Southern Ocean. The average duration of the 28 extant species recorded by Bé (1977) is 10.4 m.y. with the longer surviving species in the cooler waters of 0°–10°C; there is a progressive increase in the durations from 8.1 m.y. in the Tropical Province to 24.5 m.y. in the low-diversity Polar and Subpolar Faunal provinces. These data are compared with an average duration of 7.1 m.y. in both the Cenozoic Tropical-Subtropical Province and the southern mid-latitude province. A sequence of Cenozoic paleoenvironmental changes in the Southern Ocean is discussed, including glaciations in Antarctica. Of great significance to the faunas was the initiation of the circum-Antarctic current in the early Oligocene at ~31 Ma, when the two ocean systems around Antarctica became connected and merged to form the Southern Ocean. Early Paleocene taxa of the Southern Ocean south of 50°S were part of a Temperate Fauna, and these were replaced in the late Paleocene and early Eocene by Subtropical faunas. For the remaining middle and late Eocene, Temperate faunas returned to these high latitudes punctuated by brief incursions of warmer faunas from the north. During the Oligocene the Temperate Fauna expanded northward to ~40°S, and this pattern continued to the early Miocene, when there is some evidence of Subantarctic faunas south of 55°S. The Antarctic and Subantarctic faunas became well established by the late Miocene, and the Southern Ocean faunas acquired the modern appearance in the Pliocene after a possible warming in the middle and high latitudes in the early Pliocene. A sequential review through the Cenozoic is made of data on planktonic foraminifera from Deep Sea Drilling Project and Ocean Drilling Program (ODP) sites from the Southern Ocean and selected mid-latitude areas. The ODP drilling on Maud Rise and on the Kerguelen Plateau has greatly increased our knowledge of Cenozoic high-latitude faunas. A similar pattern of drilling of sites in the Pacific sector of the Southern Ocean is urgently required.

INTRODUCTION

The Southern Ocean Cenozoic planktonic foraminiferal faunas are a mixture of cosmopolitan species, a few species that evolved locally, and some warmer-water species which evolved in the middle and low latitudes and then spread southward [*Jenkins*, 1992a]. The degree of penetration southward of these immigrant species was dependent on a number of factors including the ability of some taxa, such as *Hantkenina* in the late Eocene, to adapt to an increase in water surface temperature in mid-latitudes and the ability of other species, like *Globorotalia truncatulinoides* and *G. inflata*, to adapt to the cooler Subantarctic waters in the late Cenozoic.

The majority of Cenozoic species have been named, are well established in the literature, and are recognized by most authors as valid, but the assignment to phylogenetic classifications and the recognition of iterative evolution have inevitably produced an increasing number of genera and subgenera. Some authors have tried to improve our understanding of Cenozoic phylogeny by producing trinomial (genus, species, and subspecies) and quadrinomial (genus, subgenus, species, and sub-

species) nomenclature [*Jenkins*, 1971a; *Kennett and Srinivasan*, 1983]. One of the problems inherent in an evolutionary approach to classification is the determination of how far back in time a genus can be recognized. Sometimes a consensus can be reached that is acceptable to most workers: for example, *Globigerinatheka* appeared in the middle Eocene, and *Orbulina* appeared in the middle Miocene. Other decisions are more difficult, such as the origin of *Neogloboquadrina* in the middle Miocene. Generic classifications based entirely on wall structures will extend some stratigraphic ranges of Neogene taxa back into the Paleogene [*Hemleben et al.*, 1991].

Only a few species appear to be indigenous to the southern hemisphere, and these include *Antarcticella zeocenica*, *A. coenii*, *A. antarctica*, *A. pauciloculata*, *Globigerina antarctica*, *Hantkenina australis*, *Jenkinsina samwelli*, *Globorotalia amuria*, and *G. conica*.

The living faunas of the Southern Ocean acquired a modern appearance in the Pliocene, and a detailed examination of these taxa is undertaken before examining the sequential faunal changes through the Cenozoic.

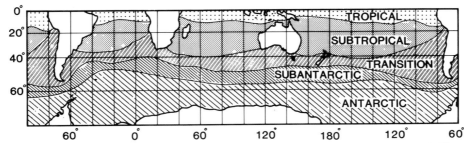

Fig. 1.   Faunal provinces in the southern hemisphere based on living species [*Bé*, 1977].

DISTRIBUTION AND DURATION OF SPECIES

The key to understanding the morphological limits of fossil species and their paleoceanographic distribution is based on our knowledge and understanding of living species mainly from the many papers of Alan Bé, which are reviewed in his 1977 paper.

Modern Distribution of Species

Bé [1977] identified five modern faunal provinces (Figure 1), and the northern limit of the Southern Ocean lies at the boundary between the Subantarctic and the Transition Faunal Province, at the Subtropical Convergence.

When the number of species is plotted for each province from Bé's data, then the diversity gradient becomes obvious (Figure 2). There is a certain amount of asymmetry with four species of the Tropical Province limited to the Indo-Pacific (i.e., *Globoquadrina conglomerata*, *Globorotaloides hexagona*, *Globorotalia cavernula*, and *Globigerinella adamsi*) and only two species in the Arctic as opposed to five in the Antarctic Province. Those species restricted to the Indo-Pacific are not present in the Atlantic because there is no tropical surface oceanic connection between the two oceans.

Species Diversity and Temperature

The cause of the species diversity gradient from the equator to the poles is not known, but it follows a decrease in mean surface water temperature from 30°C at the equator to the near freezing waters in the high latitudes. Species in the tropical waters appear to have adapted to live in various layers of the water column down to below 100 m. In the higher-latitude waters there is no such layering in the water column because the waters are mixed vertically, and consequently there is a rapid decrease in the species diversity between the Transitional and Subantarctic-Antarctic Faunal provinces (Figure 2).

Species Diversity and Paleotemperature

Since there is a decline in species diversity with decrease in temperature, it is reasonable to assume that fossil species diversity would also follow changes in paleotemperature. This hypothesis was tested in the New Zealand Cenozoic and found to be reasonably true [*Jenkins*, 1968], and a more detailed study in the early Miocene of the English Channel confirmed the close link between the two factors [*Jenkins and Shackleton*, 1979]. The plot of the species diversity from Deep Sea Drilling Project (DSDP) Leg 29 sites 277, 279A, and 281, from *Jenkins* [1975], and a paleotemperature curve produced by *Shackleton and Kennett* [1975] is shown in Figure 3 [after *Kennett*, 1978]. The DSDP and Ocean Drilling Program (ODP) sites referred to in the work are shown in Figure 4.

A more detailed study of species diversity has been carried out on the sequence across the Eocene/Oligocene boundary at DSDP Leg 29 Site 277 (Figure 5), and these data are plotted against the oxygen isotope data published by *Kennett and Shackleton* [1976]. There

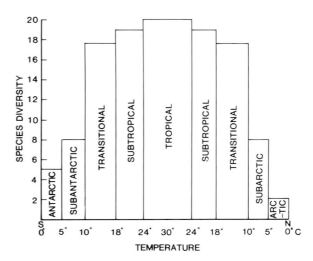

Fig. 2.   Diversity gradient of living species [after *Bé*, 1977].

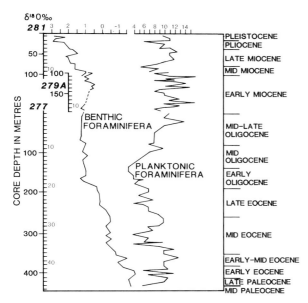

Fig. 3. A paleotemperature curve [*Shackleton and Kennett*, 1975] and species diversity from DSDP Leg 29 sites 277, 279A, and 281 [after *Jenkins*, 1975; *Kennett*, 1978].

seems to be some correlation between the fall in diversity and the isotope record from planktonic foraminifera after the extinction of *G. index*, but this happened well before the 5°C drop in paleotemperature indicated by benthic foraminifera δ^{18}O (Figure 5).

It must be remembered that a number of factors can upset the relationship between paleotemperature and species diversity, and this was brought home to the writer in the analysis of the early Miocene of tropical Pacific DSDP Site 289 where no such relationship existed even with the application of multivariate analyses of the data. The reason for this discrepancy was the selective dissolution of tests which had taken place in the sediments. Where there has been dissolution of tests, a reliable estimate of paleotemperature cannot be made from the species diversity in samples. Thus caution must be used when the species diversity/temperature rule is applied to assess the low-diversity faunas in the Southern Ocean.

Duration of Living Species

Of the 28 living species listed by *Bé* [1977], which includes *Globigerina falconensis* as a species distinct from *G. bulloides*, the average duration is 10.4 m.y., which is in contrast to the mean duration of 7.1 m.y. for the whole middle Eocene–Pleistocene tropical-subtropical species [*Van Valen*, 1973]; this is also true for the New Zealand Cenozoic [*Jenkins*, 1992b]. A possible explanation for the differences could be that the longer-living species found in Arctic-Antarctic to Transition faunal provinces owe their longevity to their adaptation

for living in the cooler waters where there is less interspecific competition, a cooling which has affected the world oceans since the middle Miocene temperature peak at about 15 m.y. The longer-surviving species occupy the cooler water between 0° and 10°C in the Antarctic and Subantarctic provinces (Figure 6).

Survival Rate in the Water Column

In the world oceans the longest-surviving species live in the upper 100 m of the oceans (Table 1), where the average species duration of the 14 species is 16 m.y. with 3 species having survived from the Oligocene, and of the 14 species, 13 have lived longer than 5 m.y.; some of these species have keels. Of the 14 species whose adults live in depths below 100 m, the average duration is 4.9 m.y., with *G. scitula* the longest-surviving species at 15 m.y., and the others have evolved to occupy this depth since the late Miocene; 10 species have only lived at this depth since the early Pliocene at 5.2 Ma.

There is a marked contrast between the average duration of living species in the five faunal provinces (Figure 6) with a progressive increase in duration from 8.1 m.y. in the Tropical Province to 24.5 m.y. in the Antarctic-Arctic provinces. This seems to imply a much more rapid turnover of species in the warmer waters compared with the cooler waters.

PALEOENVIRONMENTAL CHANGES

The sequence of paleoenvironmental changes according to *Kennett* [1978, 1980] that affected Antarctica and the Southern Ocean are shown in Table 2. The major changes that affected the planktonic foraminifera were the development of ice on Antarctica, the consequent cooling of Antarctic waters, and the development of the circum-Antarctic current.

Antarctic Glaciations

There was certainly a major lowering of species diversity at about the Eocene/Oligocene boundary which appears to have occurred about 1 m.y. before the major cooling in the early Oligocene reported from oxygen isotope data on benthic foraminifera [*Shackleton and Kennett*, 1975; *Kennett and Shackleton*, 1976] (see the previous discussion on DSDP Leg 29 Site 277). Thereafter there was a steady increase in species diversity throughout the Oligocene in the mid-latitudes.

The inferred major expansion of ice sheets on East Antarctica between 12 and 14 Ma had very little effect either on the species diversity in New Zealand or on the faunas at DSDP Leg 29 Site 279 (latitude 51°S).

The late Miocene–early Pliocene expansion of the ice sheets had no effect on *N. pachyderma*, and the diversity changes in the upper late Miocene were no greater than those in the early Miocene [see *Kennett and Vella*, 1975]. Nevertheless, *Kennett and Vella* [1975] were able

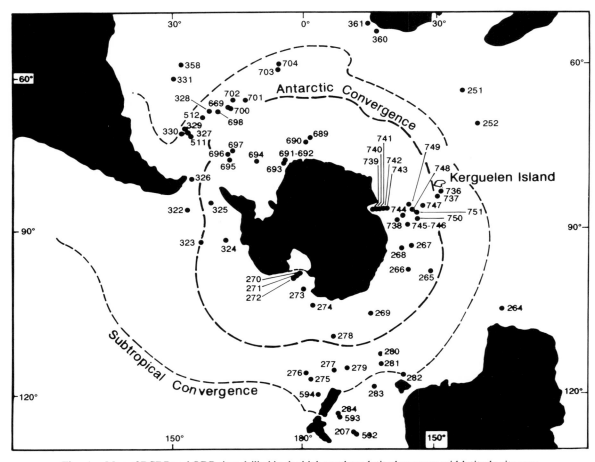

Fig. 4.   Map of DSDP and ODP sites drilled in the high southern latitudes; some mid-latitude sites are also shown.

to state that there was a severe cooling in the late Miocene at DSDP Leg 29 Site 284, and this was confirmed by *Kennett et al.* [1979] by using oxygen isotope data on both planktonic and benthic foraminifera. In New Zealand there was a marked increase in species diversity in the early Pliocene, which suggests a warming at this time [*Jenkins*, 1968].

The major cooling in the late Pliocene–Pleistocene had a more dramatic effect on the faunas with a lowering of diversity and adaptation of the sinistrally coiled *N. pachyderma* to near freezing waters. *Jenkins* [1967] recorded four cooling episodes in New Zealand in the Pleistocene based on the sinistrally coiled populations which had migrated northward along the east coast of the North Island of New Zealand; similar cooling episodes were also recorded by *Devereux et al.* [1970] and *Hornibrook* [1976]. This probably marked north-south fluctuations in the Subtropical Convergence, and a detailed study of the late Miocene–Pleistocene faunas at DSDP Leg 90 Site 594 just south of the Subtropical

Convergence has been made by *Hornibrook and Jenkins* [1993].

Development of the Circum-Antarctic Current

Plate movements in the southern hemisphere during the late Paleogene led to the development of the circum-Antarctic current; South America parted from the Antarctic Peninsula forming the Drake Passage, and a seaway developed between Tasmania and East Antarctica (Figure 7). There are conflicting views as to when these events took place; the evidence appears to have come from three main sources: (1) sedimentation and unconformities, (2) geophysical evidence, and (3) planktonic foraminifera.

Kennett et al. [1975] reviewed the history of the sedimentation and unconformities in the southwestern Pacific and concluded that the circum-Antarctic current developed in the late Oligocene at about 25–30 Ma.

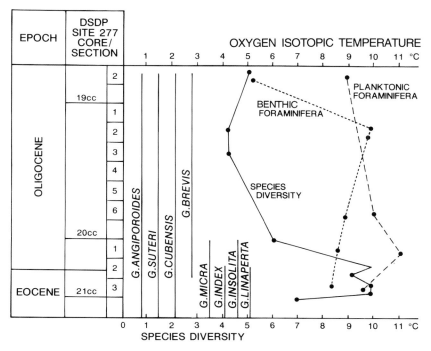

Fig. 5. Species diversity (solid line) and paleotemperature curves at DSDP Site 277; oxygen isotope data on benthic foraminifera (dotted line) and planktonic foraminifera (dashed line) after *Kennett and Shackleton* [1976] and *Jenkins* [1975].

Later, *Kennett* [1977] pointed out that existing geophysical evidence had insufficient resolution to provide an exact date for the opening of a deep seaway south of the Tasman Rise in the period 38–22 Ma. This is supported by the view of *Barker and Burrell* [1977], who concluded from geophysical evidence that the opening of the Drake Passage occurred between 30 and 22 Ma.

Kennett [1980] was of the opinion that there was a shallow water connection between the southern Indian Ocean and the Pacific by the late Eocene, with restricted flow between 38 and 22 Ma, and the development of unrestricted flow of the circum-Antarctic current was in the period 25–22 Ma.

There is no faunal evidence of any connection between the southern Indian Ocean and the southwestern Pacific in the Eocene, but there is some evidence that there were differences between the faunas in South Australia and those in the southwestern Pacific, which will be discussed later. As *Kennett* [1977] pointed out, the important period to examine is the Oligocene, and there is good faunal evidence that the seaway south of Tasmania opened at about 31 Ma and that the plankton which had been partially locked in the Austral Gulf between Australia and Antarctica spread out into the southwestern Pacific. One species, *Jenkinsina samwelli*, was carried around with the current through the Drake Passage to just off the coast of South Africa, and other species such as *C. chipolensis* spread out from the

Austral Gulf into the New Zealand region. P. Gamson (personal communication, 1992) has found *J. samwelli* at ODP Leg 121 Site 754 in the southern Indian Ocean, and *Nocchi et al.* [1991] recorded it at ODP Leg 114 Site 704 in the South Atlantic; at both sites the short stratigraphic range of *J. samwelli* was recorded at the same

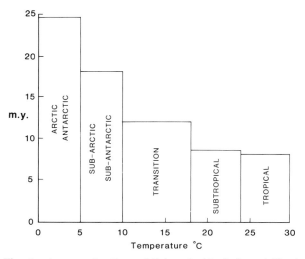

Fig. 6. Average durations of living planktonic foraminiferal species in the five faunal provinces.

TABLE 1. Distribution of Living Species in Water Columns and Faunal Provinces With Durations in Millions of Years
[After *Bé*, 1977]

DEPTH IN WATER COLUMN \ FAUNAL PROVINCES	ARCTIC ANTARCTIC	SUBARCTIC SUBANTARCTIC	TRANSITION	SUBTROPICAL	TROPICAL
0–5 m	T. quinqueloba (22 m.y.)	T. quinqueloba	G. ruber (21.8 m.y.) G. sacculifer (22 m.y.) T. quinqueloba G. conglobatus (5.2 m.y.)	G. ruber G. sacculifer G. conglobatus G. rubescens (3.7 m.y.)	G. ruber G. sacculifer G. conglobatus G. rubescens
50–100 m	G. bulloides (33 m.y.) G. glutinata (28 m.y.) G. bradyi (29 m.y.)	G. bulloides G. glutinata G. bradyi	G. bulloides A. glutinata N. dutertrei (5.2 m.y.) O. universa (15 m.y.) G. aequilateralis (12.2 m.y.) H. pelagica (5.2 m.y.) P. obliquiloculata (5.2 m.y.)	(= G. falconensis (17 m.y.) G. glutinata N. dutertrei O. universa G. aequilateralis H. pelagica P. obliquiloculata	C. nitida (6 m.y.) G. glutinata N. dutertrei O. universa G. aequilateralis H. pelagica P. obliquiloculata
ADULTS BELOW 100 m	N. pachyderma (10.4 m.y.)	N. pachyderma G. scitula (1.5 m.y.) G. inflata (3.2. m.y.) G. truncatulinoides (2.5 m.y.)	N. pachyderma G. scitula G. inflata G. truncatulinoides G. crassaformis (5 m.y.) G. hirsuta (1.8 m.y.) G. menardii (5 m.y.)	G. inflata G. truncatulinoides G. crassaformis G. hirsuta B. digitata (1.8 m.y.)* G. menardii G. tumida (5.2 m.y.)	G. truncatulinoides G. crassaformis G. digitata* G. menardii G. tumida S. dehiscens (4 m.y.) G. conglomerata (4 m.y.) G. hexagona (5.5 m.y.) G. adamsi (0.4 m.y.)*

* spinose taxa below 100 m

TABLE 2. Sequence of Paleoenvironmental Changes in the Southern Hemisphere During the Cenozoic
[After *Kennett*, 1978, 1980]

	Epoch	Changes
		Neogene Environmental Changes
P.D.–1.6 Ma	Quaternary	Further increase in upwelling and biogenic productivity at Antarctic Convergence; increase in ice rafting.
2.5–3 Ma	late Pliocene	Continued global cooling.
3.5 Ma	middle Pliocene	First major glaciation in South America.
4–6 Ma	late Miocene–early Pliocene	Expansion of Antarctic ice sheet, especially on West Antarctica; global cooling; northward expansion of Antarctic surface water; major increase in upwelling and biogenic productivity; increase in ice rafting.
12–14 Ma	middle Miocene	Expansion of major ice sheet on East Antarctica; increased ice rafting; drop in surface water temperatures.
22 Ma	early Miocene	Initial development of Antarctic Convergence, but low degree of upwelling and biogenic productivity; clear differences between Antarctic and Subantarctic microfossil assemblages; increased Subantarctic surface water temperatures to 10°C.
		Paleogene Environmental Changes
22–25 Ma	late Oligocene	Development of unrestricted circum-Antarctic current.
22–38 Ma	Oligocene	Prolonged East Antarctic glaciation including episodic ice sheet; Subantarctic surface water temperatures 7°C; cool temperate vegetation disappearing; increased ice rafting.
34 Ma	early Oligocene	Major, rapid global cooling; development of widespread Antarctic glaciation at sea level; extensive sea ice production and Antarctic bottom water; major ocean between Antarctica and Australia but restricted circum-Antarctic flow.
38–55 Ma	Eocene	Australia moving northward; no important deep circum-Antarctic circulation; shallow water communication between Southern Indian and Pacific by late Eocene; cool temperate climate and vegetation of Antarctica; restricted glaciation and ice rafting; decreasing Subantarctic surface water temperatures (~19°C in early Eocene to ~11°C in late Eocene).
~65 Ma	Paleocene	Tasman Sea open; Antarctica-Australia connected.

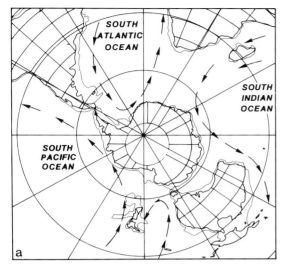

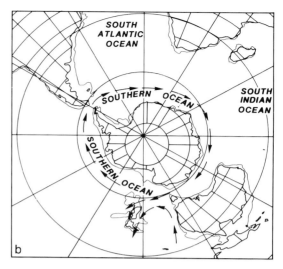

Fig. 7. Two continental reconstructions around Antarctica: (*a*) the late Eocene at 35 Ma and (*b*) the early Oligocene at 31 Ma, showing the oceanic currents and disposition of land [after *Smith et al.*, 1981].

time as the boundary between the early and late Oligocene.

The evidence for the initiation of this circum-Antarctic current was primarily based on the stratigraphic (Figure 8) and paleoceanographic ranges of *J. samwelli*, which was originally misidentified as *G. stavensis* Bandy [*Jenkins*, 1974*b*] and *G.* aff. *stavensis* [*Jenkins*, 1978*a*] before being named *J. samwelli* [*Jenkins*, 1978*b*]. There is still some taxonomic confusion, because *Lindsay* [1986] and *McGowran and Beecroft* [1986] apparently refer to this species as *G. triseriata*. From available evidence, *G. triseriata* was originally

described from the middle Eocene of France and appears to be restricted to this stratigraphic level.

According to records from South Australia [*Lindsay*, 1969, 1986], *J. samwelli* lived in the Austral Gulf for about 6–7 m.y. from the very late Eocene to the lower part of the late Oligocene (Figure 8), although *McGowran and Beecroft* [1986] showed its first appearance to be in the earliest Oligocene. In the lower part of the *G. euapertura* Zone, *J. samwelli* migrated into the southwestern Pacific and has been found at DSDP Leg 29 sites 282, 277, and 276 and Leg 90 Site 593 and also in deposits of the same age in the South Island of New

LOCALITY	SOUTH AUSTRALIA Lat. 35° 15.67' S Long. 138° 26.84' E	DSDP LEG 29 AND LEG 90	WEST OF TASMANIA SITE 282 Lat. 42° 14.76' S Long. 143° 29.18' E	WESTERN EDGE OF CAMPBELL PLATEAU SITE 277 Lat. 52° 13.43' S Long. 166° 11.48' E	EASTERN EDGE OF CAMPBELL PLATEAU SITE 276 Lat. 50° 48.11' S Long. 176° 48.40' E	TASMAN SEA SITE 593 Lat. 40° 30.47' S Long. 167° 40.47' E	NEW ZEALAND SOUTH ISLAND Lat. 40° 46' S Long. 172° 15' E / Lat. 44° 30.6' S Long. 170° 24' E	SOUTH ATLANTIC DSDP LEG 40 SITE 360(a) Lat. 35° 50' S Long. 18° 05' E	ODP LEG 114 SITE 703A (b) Lat. 47°03'S Long. 07°53'
EPOCH	ZONES	ZONES					ZONES	ZONES	
OLIGOCENE L	G. euapertura	G. euapertura					G. euapertura	G. euapertura	
	"G. stavensis"		T	T	T	TT		T	T
OLIGOCENE E	C. cubensis	S. angiporoides angiporoides	C. cubensis / J. samwelli	C. cubensis / J. samwelli	J. samwelli	C. cubensis / J. samwelli	S. angiporoides angiporoides	S. angiporoides angiporoides	C. cubensis / J. samwelli
							C. cubensis / J. samwelli	C. cubensis / J. samwelli	
EOCENE	S. linaperta	G. brevis					G. brevis	G. brevis	

Fig. 8. Late Eocene–Oligocene stratigraphic ranges of *J. samwelli* and *C. cubensis* at localities in the southern hemisphere.

Zealand. It lived only briefly, but long enough to spread through the Drake Passage to DSDP Leg 40 Site 360 off the coast of South Africa, to ODP Site 704 in the South Atlantic [*Nocchi et al.*, 1991] and to the southern Indian Ocean ODP Leg 121 Site 754 (P. Gamson, personal communication, 1992) (Figure 8). Its northern limit in the Tasman Sea has been extended to 40°S at Site 593 [*Jenkins and Srinivasan*, 1986], while in the southeastern Atlantic, *J. samwelli* was found at DSDP Leg 40 Site 360 at 35°S, but it was not present at Site 362 at 19°S.

Confirmation of timing of the short duration of *J. samwelli* has come from the stratigraphic distribution of other taxa including the evolution of *G. munda* into *G. juvenilis*, which occurred during its short stratigraphic range, and the extinction of *C. cubensis*, which has been dated in the Oligocene at 28.8–31.2 ± 1.5 Ma [*Berggren*, 1972] and at 30 Ma [*Berggren et al.*, 1985a].

Hornibrook [1985] in his work on the Heterohelicidae in the New Zealand Oligocene showed nine stratigraphic sections normalized to the extinction of *S. angiporoides*. At two localities, *J. samwelli* was shown to overlap with the uppermost stratigraphic range of *S. angiporoides*: (1) at Trig Z Otiake in the South Island *J. samwelli* was recorded in a greensand just above and with *S. angiporoides* just below a bored surface; (2) at Te Akatea in the North Island *J. samwelli* was again recorded within the uppermost stratigraphic range of *S. angiporoides*. The degree of accuracy and reliability of correlations using datums based on biostratigraphic events in the Cenozoic history of planktonic foraminifera is reiterated by *Jenkins* [this volume]; species extinctions can be unreliable because most of these events are diachronous. Thus *McGowran and Beecroft* [1986] have recorded that the extinction of *S. angiporoides* is diachronous in South Australia, and both *McGowran* [1973] and *Lindsay* [1973] had distinguished two extinction levels as "top good *S. angiporoides*" and a later "top sporadic *S. angiporoides*." The latter type of extinction is present at DSDP Site 593 in the Tasman Sea where the first appearance of *J. samwelli* overlaps the sporadic presence of *S. angiporoides* [*Jenkins*, 1987].

Since publication of this model of the development of the circum-Antarctic current by *Jenkins* [1974b], it has also been tested in the South Atlantic [*Jenkins*, 1978a, b; *Nocchi et al.*, 1991], in the Indian Ocean (P. Gamson, personal communication, 1992), and in the Tasman Sea [*Jenkins and Srinivasan*, 1986]. If the seaway south of Tasmania had been open in the Oligocene prior to 31 Ma, then *J. samwelli* and *C. chipolensis* would have spread out into the southwestern Pacific. At about 31 Ma the seaway was probably only deep enough to let plankton-bearing seawater through, and the deepwater channels developed later, possibly at 22 Ma [*Barker and Burrell*, 1977].

If the cooling between 28 and 31 Ma reported by *Keigwin and Keller* [1984] at Equatorial Pacific DSDP

Leg 9 Site 77 occurred, then there would have been a significant buildup of ice on Antarctica and a consequent fall in sea level. It is predicted that *J. samwelli* spread into the southwestern Pacific just prior to this event and that the seaway may have reclosed for a short while during this cold period.

CENOZOIC FAUNAS

If it is accepted that the circum-Antarctic current began as a relatively shallow water seaway in the early Oligocene at 31 Ma, then up until that time there had been two major faunal provinces in the Paleogene Southern Ocean, namely, the Pacific and the Indo-Atlantic [*Jenkins*, 1985]. In the following discussion the faunas are described separately up to 31 Ma in the early Oligocene, and thereafter, because of the oceanic connection, the two provinces are regarded as one major southern hemisphere faunal province. Although the circum-Antarctic current is regarded as having started at 31 Ma, there is some faunal evidence that there was a seaway connection in the early Paleocene.

Pacific: Paleocene Fauna

The earliest Danian fauna in the *P. eugubina* Zone (the "Pα Zone") has only been described from the New Zealand Waipara Section at 43°S, 20 cm above the K/P boundary [*Strong*, 1984]. In the Waipara Section, there followed the *Antarcticella pauciloculata* and *G. daubjergensis* Zone faunas with a depleted *S. triloculinoides* Zone fauna above. There is an increase in the species diversity in the late Paleocene *P. pseudomenardii* and *M. velascoensis* zones in New Zealand [see *Jenkins*, this volume, Figure 2].

Until recently the only record of *A. pauciloculata* was from the Waipara Section, but now B. T. Huber (personal communication, 1991) has found the taxon in the early Danian of Seymour Island off the Antarctic Peninsula where it was initially identified as *Globoconusa daubjergensis* [*Huber*, 1988, Figures 23, 10, and 11]. This discovery presents us with a major problem: what was the oceanic connection between the South Island of New Zealand and Seymour Island? There are two possibilities: *A. pauciloculata* spread via the waters north of Australia, or there was a southern connection either through Antarctica or via Drake Passage. Because the species of *Antarcticella* are found mainly in shallow water deposits, the latter seaway connection is a distinct possibility.

Late Paleocene faunas found at Leg 29 Site 277 at 52°S were represented by the following species: *Chiloguembelina wilcoxensis*, *Globigerina spiralis*, *Subbotina triloculinoides*, *Acarinina acarinata*, *A. primitiva*, *Morozovella mckannai*, *M. soldadoensis*, *Planorotalites australiformis*, *P. pseudomenardii*, *P.* cf. *reissi*, *Globorotaloides turgida*, *Zeauvigerina parri*, *Z.*

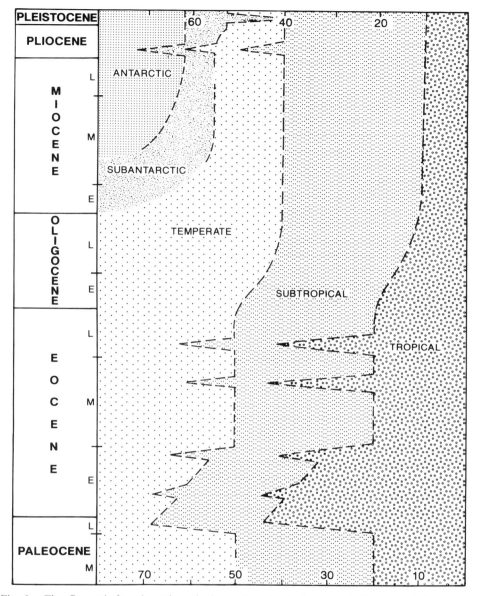

Fig. 9.   Five Cenozoic faunal provinces in the southern hemisphere. Top latitudinal axis from left to right: Antarctic, Subantarctic, Temperate, Subtropical, and Tropical Faunal provinces. Horizontal axis: latitude, °S. Vertical axis: Ma.

teuria, and *Z. zealandica* [*Jenkins*, 1975]. The lack of warm water *Morozovella* species at Site 277 and their presence in the Middle Waipara Section make it possible to place a line separating the Subtropical Fauna from the Temperate Fauna at 50°S (Figure 9).

Indo-Atlantic: Paleocene

An isolated *P. pseudomenardii* Zone fauna was described by *McGowran* [1964] from Western Australia, and two further Paleocene faunas were described from

Western Victoria [*McGowran*, 1965]: *McGowran and Beecroft* [1986] recorded *Planorotalites chapmani* and *P. haunsbergensis* from South Australia.

Tjalsma [1977] recorded a mid-latitude late Paleocene fauna which lacked the warm water keeled *Morozovella* species from DSDP Leg 36 Site 329 and which resembles that of DSDP Leg 29 Site 277. Both mid-latitude faunas lack the diagnostic tropical-subtropical taxa such as *Morozovella angulata* and *M. velascoensis*.

Nocchi et al. [1991] recorded mid-latitude Paleocene

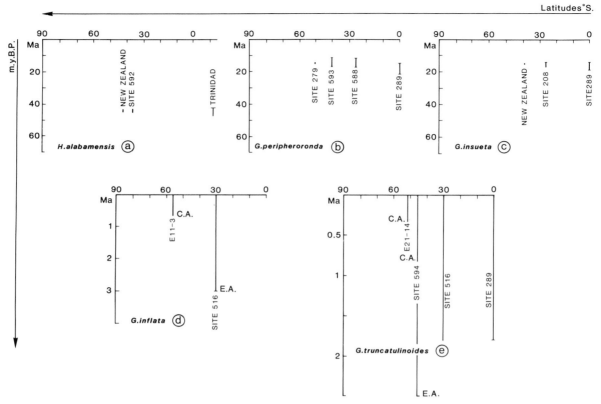

Fig. 10. Latitudinal and stratigraphic distribution of five species in the southern hemisphere and Trinidad. E.A., evolutionary appearance; C.A., cryptogenic appearance [*Jenkins, 1992b*].

faunas from Leg 114 sites 698–702 at 51°–50°S in the South Atlantic. The Pα Zone in the lowermost Danian is missing, and the Paleocene faunas are very much like those of the more southern Maud Rise and Kerguelen Plateau, but they have the additional subtropical taxa which include *Morozovella trinidadensis* and *M. pussila*. Tropical species such as *M. velascoensis* were not recorded.

The following ODP data on the stratigraphic ranges of taxa have recently become available: *Stott and Kennett* [1990] from Maud Rise Leg 113 sites 689 and 690 at 65°S; *Huber* [1991] from Kerguelen Plateau Leg 119 sites 738 and 744 at 61°–62°S, and *Berggren* [1992a] from Kerguelen Plateau Leg 120 sites 747–749 at 55°–59°S. The lowermost Danian *P. eugubina* fauna has not been recognized at these sites, and although there is a difference in species diversity in the early Paleocene between Maud Rise and the southern Kerguelen Plateau, there is a general trend of increasing species diversity toward the late Paleocene in both areas. In spite of this apparent warming trend through the Paleocene, tropical-subtropical taxa such as *Morozovella trinidadensis*, *M. uncinata*, and *M. velascoensis* are not present, but both *Stott and Kennett* [1990] and *Huber* [1991] record a few

warmer-water *Morozovella* species in the late Paleocene, suggesting a possible subtropical incursion of warm water into these high latitudes. A similar late Paleocene increase in diversity has been recorded in New Zealand [*Jenkins*, 1968, 1974a], and this has been interpreted as a significant surface water warming. An abrupt deep-sea warming has also been recorded by *Kennett and Stott* [1991] at Site 690B.

In detail there are differences between the Paleocene faunas recorded at Maud Rise [*Stott and Kennett*, 1990, Site 689] and at the Kerguelen Plateau [*Huber*, 1991, Site 738]. The following number of species were recorded at Maud Rise, with those at Kerguelen in parenthesis: *Acarinina*, 4 (14); *Bifarina*, 0 (1); *Chiloguembelina*, 4 (4); *Eoglobigerina*, 3 (3); *Globoconusa*, 1 (1); *Guembelitrioides*, 0 (1); *Igorina*, 1 (2); *Planorotalites*, 5 (7); *Pseudohastigerina*, 1 (1); "*Morozovella*," 0 (1); *Morozovella*, 7 (2); *Subbotina*, 9 (6); and *Zeauvigerina*, 0 (1). There seems to have been a difference of opinion regarding the designation of some species into the genera *Acarinina*, *Planorotalites*, and *Morozovella*; the much lower species diversity at Maud Rise needs to be checked at sample level (Figure 11).

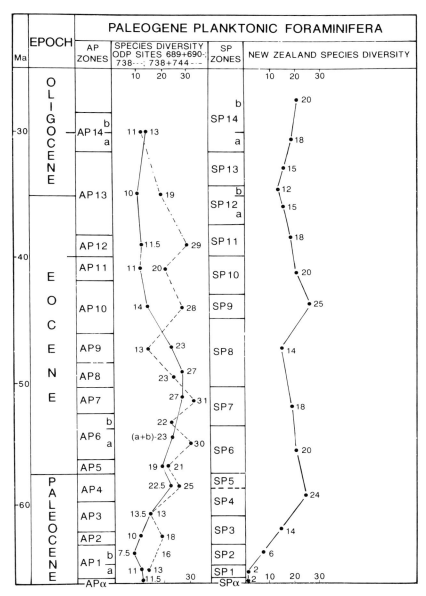

Fig. 11.   Species diversity curves for Maud Rise (ODP sites 689 and 690) and Kerguelen Plateau (ODP sites 738 and 744) plotted against AP zones, and New Zealand diversity plotted against SP Zones (southern mid-latitude Paleogene zones).

Pacific: Eocene

The Eocene mid-latitude fauna has a relatively low diversity compared with the Tropical Province. Of the 90 stratigraphically important species recorded by *Bolli and Saunders* [1985], only 17 occur in New Zealand. A direct consequence of this low diversity is that there are only 6.5 biozones in the New Zealand Eocene (Figure 11) compared with 12 biozones in the Tropical Province [*Bolli and Saunders*, 1985].

Two incursions of warmer-water species occurred in

New Zealand, the first with *Acarinina spinuloinflata* in the middle Eocene lower *G. index* Zone and the second with *H. alabamensis* and *G. cerroazulensis* (Figures 10 and 11) in the lower part of the late Eocene *S. linaperta* Zone of the Port Elizabeth Section, South Island. This latter incursion was fairly widespread, occurring on the west coast of the South Island, the Chatham Islands, and at DSDP Leg 90 Site 592 in the Tasman Sea [*Jenkins and Srinivasan*, 1986]. This warming marked by *Hantkenina* also occurs in South Australia and at DSDP Leg

28 Site 267. It is difficult to say whether this late Eocene warming coincides with one of the late Eocene peaks in temperature recorded by *Shackleton and Kennett* [1975].

A complete Eocene section was drilled at DSDP Leg 29 Site 277 (52°S), and although all the New Zealand planktonic foraminiferal zones were recognized, some of the species stratigraphic ranges are shorter, for example, *Acarinina soldadoensis* became extinct earlier in the *A. primitiva* Zone [*Jenkins*, 1975].

Three species of *Antarcticella* with a total range of early Eocene to middle Miocene have been described in the southern hemisphere. *A. cecionii* [*Canon and Ernst*, 1974] has been recorded from the early middle Eocene of Magellanes Basin (about 53°S), Southern Chile, and *A. zeocenica* [*Hornibrook and Jenkins*, 1965] has been recorded from the middle Eocene *G. index* Zone to the early Oligocene *S. angiporoides* Zone [*Jenkins*, 1971a]. The other species, *A. antarctica* [*Leckie and Webb*, 1985], has a range of late Oligocene–middle Miocene and will be discussed with the Oligocene fauna.

Indo-Atlantic: Eocene

Onshore faunas have been described from South Australia, Argentina, and Chile, and a correlation with the New Zealand zones is shown in the work of *Jenkins* [this volume, Table 1].

Well-documented species ranges have been provided for South Australia by *Lindsay* [1969, 1986] and *Ludbrook and Lindsay* [1969]. The Eocene/Oligocene boundary has been placed at the extinction of *Globigerinatheka index* in New Zealand, while in South Australia it has been placed slightly later at the extinction of *Subbotina linaperta*. There are significant differences in the stratigraphic species ranges between the two areas:

1. *G. index* appears to have become extinct earlier in South Australia.

2. *S. linaperta* ranges above the extinction of *G. index* in South Australia, while in New Zealand both taxa became extinct at the same time.

3. *Acarinina aculeata* (Jenkins) became extinct much earlier in New Zealand.

4. *Cassigerinella winniana* existed in South Australia in the local late Eocene *S. linaperta* Zone but has not been found in New Zealand.

5. In New Zealand there is a distinct overlap between the ranges of *G. index* and *Globigerina ampliapertura*, but no such overlap exists in South Australia.

6. Both *Tenuitella insolita* and *T. gemma* appeared much earlier in the late Eocene of South Australia compared with the records in the southwestern Pacific.

These differences tend to support the concept of there being no oceanic connection in the Eocene via the Tasman Rise between the Indian Ocean and South Australia, and the southwestern Pacific.

Tjalsma [1977] recorded a nondescript fauna from the early Eocene at Leg 36 Site 329 at 50°S. P. N. Webb (in the work of *Kaneps* [1975]) recorded a late Eocene fauna from DSDP Leg 28 Site 267 at 59°S and listed *Chiloguembelina cubensis*, *Catapsydrax martini*, *C. echinatus*, *S. linaperta*, *S. angiporoides*, *G. index*, and a specimen of *Hantkenina* which was tentatively identified as *H. alabamensis compressa*; this last species also occurred briefly in South Australia [*Lindsay*, 1969] in the late Eocene and has been referred to as *H. primitiva* [*McGowran and Beecroft*, 1986].

The early Eocene appears to show a continuation of the warming trend from the late Paleocene in South Atlantic ODP Leg 114 sites 698–702 at 51°–50°S, where *Nocchi et al.* [1991] recorded numerous subtropical *Morozovella* species and also recorded *Cassigerinelloita amekiensis* in the early Eocene *M. crater* Zone. The middle Eocene faunas are very much like those recorded at similar latitudes in the southwestern Pacific [*Jenkins*, 1973; *Jenkins and Srinivasan*, 1986] characterized by numerous *Acarinina primitiva*, *G. index*, and *S. angiporoides*. The main difference is the lack of *Hantkenina* spp. when compared with the southwestern Pacific. The late Eocene is dominated by *G. index* and by *Tenuitella* spp. in the fine fraction. Like most of the Eocene, the low-latitude species tend to be missing, and the taxa can be regarded as belonging to the Temperate faunas (Figure 11).

The early Eocene faunas at Maud Rise and the Kerguelen Plateau are dominated by species of *Acarinina*, *Subbotina*, and *Planorotalites*; *Morozovella* spp. have only been reported at Site 690 (rare *M. aequa*) and at Site 738 (rare *M. gracilis* and *M. subbotinae*) [*Stott and Kennett*, 1990; *Huber*, 1991]. Rare *C. amekiensis*, a possible warmer-water taxon, was recorded at Site 738 and ranged through to the middle Eocene [*Huber*, 1991].

The middle Eocene faunas at Maud Rise and the Kerguelen Plateau are dominated by species of *Acarinina*, *Subbotina*, and toward the top of the interval by *G. index*. *Huber* [1991] recorded a well-documented fauna at Site 738 with a much higher diversity in the late middle Eocene (Figure 11) compared with Maud Rise, with the noticeable addition of *Jenkinsina* species and rare *Morozovella*.

The late Eocene at Maud Rise shows a continued fall in species diversity but with a much higher diversity at the Kerguelen Plateau (Figure 11); again there is an exceptionally well preserved fauna at Site 738 [*Huber*, 1991].

Throughout the Eocene there are low numbers of subtropical taxa in the high-latitude sites, but the peaks in species diversity at Kerguelen Plateau Site 738 in the AP6, AP7, AP10, and AP12 zones (Figure 11) may indicate warm periods. The warming AP12 Zone probably coincides with a similar previously described warming in the southwestern Pacific and South Australia.

Southwestern Pacific: Early Oligocene

With the drop in temperature in the lowermost Oligocene seawater there was a significant drop in species diversity from 15 in the late Eocene to 12 in the early Oligocene in New Zealand (Figure 11). After the extinction of Eocene species the fauna was dominated by nonkeeled genera including *Globigerina*, *Catapsydrax*, *Globorotaloides*, and *Paragloborotalia*. A number of species continued through from the late Eocene including *Globigerina ampliapertura*, *G. brevis*, *G. ouachitaensis*, *G. praeturritillina*, *P. nana*, *S. angiporoides*, *Tenuitella insolita*, and *T. gemma*.

The general rise in surface seawater temperature in the late Oligocene brought immigrant species including *Tenuitella angustiumbilicata*, *C. dissimilis*, and rare *Paragloborotalia opima* southward into the southwestern Pacific.

Indo-Atlantic: Early Oligocene

There are faunal differences between the Indo-Atlantic and the southwestern Pacific, and some of these differences can be seen in the work of *Jenkins* [this volume, Table 9]. An attempt has been made to compare the stratigraphic ranges in South Australia [*Lindsay*, 1969, 1986] with those in New Zealand [*Jenkins*, 1971a], and a number of these have already been referred to. It should be further noted that (1) *Cassigerinella winniana* and *G. angulofficinalis* are recorded in South Australia but have not been found in New Zealand and (2) *S. angiporoides* became extinct slightly earlier than *C. cubensis* in the southwestern Pacific, and *Lindsay* [1986] showed a similar sequence of events, but *McGowran and Beecroft* [1986] showed a much earlier extinction for *S. angiporoides* in South Australia.

Kaneps [1975] recorded only *S. angiporoides* and *C. dissimilis* from the Oligocene DSDP Leg 28 Site 267 (latitude 59°S) and regarded this low diversity as original and not due to dissolution of other foraminiferal tests; the hole was drilled in a water depth of 4564 m. *Tjalsma* [1977] recorded an Oligocene fauna from DSDP Leg 36 Site 329 (latitude 50°S) in the southwestern Atlantic which can be recognized as belonging to the *S. angiporoides* and *G. euapertura* zones; *S. linaperta* was recorded in the early Oligocene. At DSDP Leg 40 Site 360 (latitude 35°S) in the southeastern Atlantic, *Toumarkine* [1978] recorded a transition between mid-latitude and low-latitude Oligocene faunas.

Nocchi et al. [1991] recorded a reduced species diversity in the early Oligocene of ODP Leg 114 sites 698–702 with typical mid-latitude faunas which yielded *S. angiporoides*, *C. cubensis*, and *Tenuitella* spp.; there is a marked increase in diversity in the late Oligocene. *Nocchi et al.* [1991] recorded *J. stavensis* from the lower part of the late Oligocene of Site 704, and from its description and very short stratigraphic range the taxon is most probably *Jenkinsina samwelli*, a species which

has been used to mark the beginning of the circum-Antarctic current [*Jenkins*, 1978a, b; *Jenkins and Srinivasan*, 1986]. P. Gamson (personal communication, 1992) has also found *G. samwelli* at DSDP Leg 121 Site 754 in the southern Indian Ocean at the same stratigraphic level.

At the high-latitude ODP sites, the species diversity of the early Oligocene faunas was found to be low at Maud Rise and on the Kerguelen Plateau (Figure 11; *Stott and Kennett* [1990], *Huber* [1991], and *Berggren* [1992a]). The main taxa recorded include *S. angiporoides*, *S. utilisindex*, *Catapsydrax* spp., and *C. cubensis* in the early Oligocene with *G. labiacrassata*, *G. euapertura*, *T. munda*, and *T. juvenilis* in the late Oligocene.

The low species diversity from the late middle Eocene through the Oligocene at Maud Rise [*Stott and Kennett*, 1990] and from the late Eocene through the Oligocene at the Kerguelen Plateau [*Huber*, 1991] probably reflects the lowering of the surface water temperatures in these areas. There is a slight increase in species diversity into the early Miocene at Site 744 [*Huber*, 1991] which is also recorded in mid-latitudes [*Jenkins*, 1968, 1974a].

According to *Leckie and Webb* [1985], *Antarcticella antarctica* has been recorded only from the Pacific side of the Southern Ocean at DSDP Leg 28 Site 270 (latitude 77°S) [*Leckie and Webb*, 1980, 1983] and Site 273 (latitude 74°S) [*D'Agostino and Webb*, 1980] and in the Ross Sea [*Webb*, 1979] in late Oligocene–middle Miocene sediments; in these high latitudes its occurrence is nearly always monospecific.

In the late Oligocene with the development of the proto circum-Antarctic current a progressive mixing of the Indo-Atlantic with the southwestern Pacific faunas began, and this mixing brought to an end the two-ocean regime. Species which migrated from the Indo-Atlantic via the seaway south of Tasmania include *J. samwelli* and *C. chipolensis*. Species which probably evolved in the southern mid-latitudes include *Globigerina bulloides*, *G. labiacrassata*, *G. brazieri*, *Globigerinoides inusitatus*, *Paragloborotalia pseudocontinuosa*, *Globorotaloides testarugosa*, *Tenuitella juvenilis*, and *T. munda*.

Late Oligocene–Early Miocene

There is a long transition between the late Oligocene and early Miocene faunas of the southern hemisphere, and the Oligocene/Miocene boundary is artificial insofar as there is no major planktonic foraminiferal faunal change [*Jenkins*, 1966; *Berggren et al.*, 1985b]. The late Oligocene–early Miocene faunas first recorded in New Zealand, Victoria, and South Australia are similar to those in DSDP Leg 29 sites 279 and 281 in the southwestern Pacific and to those at DSDP Leg 40 Site 360 in the southeastern Atlantic [*Jenkins*, 1978c]; a link between the mid-latitude faunas and the Tropical-Subtropical provinces exists at DSDP Leg 39 Site 516 of the Rio Grande Rise of the southwestern Atlantic [*Berg-gren et al.*, 1983].

The evolution of taxa took place in the middle and low latitudes, and it is important to discuss some of the aspects of those changes because of their effects on high-latitude faunas when some taxa successfully migrated southward.

There have been two interpretations of the origin of *Globigerinoides trilobus*: (1) it evolved from *Globigerina praebulloides* via *Globigerinoides primordius* [*Blow*, 1969], and (2) *Jenkins* [1965] regarded *G. trilobus* as having evolved from *Globigerina woodi* via *G. connecta*. *Kennett and Srinivasan* [1983] pointed out that *Globigerinoides* is polyphyletic and that there were two lineages with different wall structures (1) *G. praebulloides* → *G. primordius* → *G. altiapertura* which possibly led to *G. obliquus* and (2) *G. woodi* → *G. connecta* → *G. trilobus* which had a number of branches, one of which gave rise eventually to *O. universa* and another to *G. sacculifer* and *G. fistulosus*. The evolution of *G. woodi* → *G. connecta* → *G. trilobus* took place in the mid-latitudes of the southern hemisphere. Although *G. trilobus* became a mid- to low-latitude taxon, *G. woodi* successfully spread southward to the early Miocene higher latitudes at DSDP Leg 29 Site 278 [*Jenkins*, 1975] and at ODP sites 747, 748, and 751 [*Berggren*, 1992*b*].

Jenkins [1973] has long regarded the origin of *Globorotalia praescitula* as cryptogenic and even suggested that it could have been derived from a benthic larval stage in the early Miocene. Alternatively, *Kennett and Srinivasan* [1983] proposed that *G. praescitula* evolved directly from *G. zealandica* in the southwestern Pacific, but this can hardly be true because there are good data which show that *G. praescitula* appeared well before *G. zealandica* in this area [*Jenkins*, 1971*a*, 1975]. *G. praescitula* evolved into *G. miozea* and its descendants in the mid-latitudes, and *G. praescitula* spread into the Tropical Province where it evolved into *G. archaeomenardii*, *G. praemenardii*, and the keeled *Globorotalia*, which *Kennett and Srinivasan* [1983] have placed in the subgenus *Menardella*.

There seems to be some confusion regarding *P. pseudocontinuosa* which evolved from *G. nana* in the southwestern Pacific during the Oligocene. The holotype of *Globorotalia incognita* and specimens illustrated by *Walters* [1965] clearly show that it is closer in morphology to *G. zealandica* than to *P. pseudocontinuosa*. *P. pseudocontinuosa* is the ancestral species from which *G. semivera* and *G. incognita* evolved in the Oligocene and early Miocene, respectively. The main morphological differences between *P. pseudocontinuosa* and *G. incognita* are as follows:

1.   The aperture in *P. pseudocontinuosa* is well rounded and high arched and extends from the umbilicus to about halfway to the periphery, whereas the aperture in *G. incognita* ranges from being low arched

to being a high-arched aperture isolated from the umbilicus in a position about halfway to the periphery.

2.   The chambers are much more inflated globigerine in shape in *P. pseudocontinuosa* with a lobate equatorial periphery compared with the *G. incognita* chambers.

3.   The chamber size increase in the final whorl is much more rapid in *P. pseudocontinuosa* than in *G. incognita* where the final chamber is only slightly larger than the penultimate.

4.   On the spiral side the chamber shape in *P. pseudocontinuosa* is globular, while the ones in *G. incognita* have evolved into a more globorotalid form and have a hemispherical compact shape.

5.   The spiral side is much more flattened in *G. incognita*, and this species eventually evolved into *G. zealandica*, which has a flat spiral side.

According to *Scott et al.* [1990], *G. incognita* is distinguished from *P. pseudocontinuosa* "by only modest architectural changes (weak elongation of chambers in direction of coiling, slight flattening of spiral faces of chambers)." Unfortunately, *Berggren et al.* [1983] regarded the two species as synonymous, and this greatly extended the short stratigraphic range of *G. incognita*. *Berggren et al.* [1983, Plate 5, Figure 1] illustrated one specimen of *P. pseudocontinuosa* and a number of specimens of *G. incognita* from southwestern Atlantic DSDP Site 516 [*Berggren et al.*, 1983, Plate 5, Figures 2–4]; the forms designated as *G. pseudomiozea* are merely thicker-walled varieties of *G. incognita* and *G. zealandica* [*Berggren et al.*, 1983, Plate 5, Figures 10–17].

Kennett and Srinivasan [1983] also regarded *P. pseudocontinuosa* as a junior synonym of *G. incognita* and regarded the ancestor of *G. incognita* as *G. nana*, which became extinct in the lower part of the early Miocene *G. dehiscens* Zone before the appearance of *G. incognita*. *Jenkins* [1971*a*] showed the range of *P. pseudocontinuosa* to be from the Oligocene *G. euapertura* Zone to the middle Miocene *O. suturalis* Zone, while *Kennett and Srinivasan* [1983] showed *G. incognita* (their *G. pseudocontinuosa*) to have an early Miocene range. *Jenkins* [1971*a*] had a more restricted range for *G. incognita* in the lower part of the *G. trilobus* Zone, similar to that of *Walters* [1965] and *Scott et al.* [1990].

Rögl [1976] recorded *G. incognita* in the early Miocene of DSDP Leg 35 Site 325 (latitude 65°S) in the Bellingshausen Sea, with a few specimens of the warm water *G. peripheroronda*; this latter identification needs to be checked because it seems to be well south of its paleogeographic limit (Figure 10).

Tjalsma [1977] recorded *G. zealandica incognita* from the early Miocene of DSDP Leg 36 Site 329 in the southwestern Atlantic. The illustrated specimens are *G. incognita* with the squarer outline than *P. pseudocontinuosa* and with an aperture isolated from the umbilicus.

C. chipolensis became extinct in New Zealand at the top of the late early Miocene *P. glomerosa curva* Zone, but in the Tropical Province it survived nearly to the top of the middle Miocene [*Bolli and Saunders*, 1985].

Jenkins [1975] postulated that *G. conica* was a cool water species which made brief incursions into New Zealand and Southeast Australia, and on DSDP Leg 29 it was recorded at sites 278, 279, and 281, while on Leg 90 it was recorded at southern sites 593 and 594. There is a problem regarding its ancestry, and there are at least two possibilities: (1) it evolved from *G. zealandica* and possible intermediates exist at Site 281 in the *G. miozea* Zone, or (2) it evolved from *G. peripheroronda* and there are similarities including shell structure, and it is conceivable that it evolved by acquiring a flattened spiral side and a vaulted umbilical side with an open aperture. *Scott et al.* [1990] have described *G. amuria*, a heavily encrusted form which closely resembles *G. conica* and having a similar stratigraphic range of late early Miocene to middle Miocene. *Barker et al.* [1988] recorded *G. zealandica* from Leg 113 Site 690 (Core 113-690B-5H) from Maud Rise; specimens supplied by L. Stott from the middle Miocene are *G. amuria*.

G. peripheroronda appeared at about 22 Ma in the Tropical Province and later spread southward with the rise in surface water temperature (Figure 10); it reached latitude 50°S at DSDP Leg 29 Site 289 in the *P. glomerosa curva* Zone at about 15–16 Ma. *Globigerinatella insueta*, another tropical species, showed a similar pattern of migration when spreading southward (Figure 10). Both *G. peripheroronda* and *G. insueta* have distinct southern limits.

Sancetta [1978] used factor analysis of DSDP data from the Pacific and was able to produce successive maps of faunal provinces throughout the Neogene-Holocene. The separation between the Subantarctic, Transition, and Indian Ocean Transitional provinces does not appear to be supported by data from planktonic foraminifera but could very well be substantiated by diatoms, radiolaria, and calcareous nannofossils.

Nocchi et al. [1991] recorded typical southern mid-latitude early Miocene faunas including *G. dehiscens*, *G. woodi*, *G. connecta*, and *C. dissimilis* from South Atlantic sites 703 and 704; the absence of *G. kugleri* again emphasizes the southern mid-latitude position of these sites. *Nocchi et al.* [1991] suggested that dissolution had affected the faunas above the extinction of *C. dissimilis* in the early Miocene.

Berggren [1992*b*] recorded well-preserved late Oligocene–early Miocene faunas from southern Kerguelen Plateau Leg 120 sites 747, 748, 749, and 751 between latitudes 54° and 58°S. Unfortunately, *Berggren* [1992*b*] did not record full faunal ranges from these sites, but from the range charts of key species provided it is obvious that some of the stratigraphic ranges are different from those of the southwestern Pacific. *Berggren* [1992*b*] recorded the extinctions of both *T. gemma* and

T. munda in the early Miocene at Site 748 compared with their Oligocene extinctions in the mid-latitudes. Nevertheless, the first appearances of the key species *G. incognita*, *G. praescitula*, *G. zealandica*, and *G. miozea* are in the normal mid-latitude stratigraphic sequence. Some important early Miocene stratigraphic markers such as *G. kugleri* and *G. trilobus* are missing from these sites.

Middle Miocene

The middle Miocene high species diversity in the southern mid-latitudes was influenced by the temperature peak in the early part of the middle Miocene, which was followed by a steady decline in both diversity and temperature through this interval. The indigenous *Globorotalia miozea–G. miotumida* group dominated the keeled *Globorotalia*, but toward the top of the middle Miocene, *G. menardii* made a brief appearance in sections north of latitude 42°S in New Zealand. *G. altispira*, another relatively warm water species, penetrated as far south as 38°S in the Muddy Creek Section [*Jenkins*, 1971*a*].

Normally, *Neogloboquadrina mayeri* first appears in the middle Miocene of the mid-latitudes, but *Scott et al.* [1990] have recorded very rare species at one locality in New Zealand (Tokomaru Bay, North Island) in the late early Miocene *P. glomerosa curva* Zone.

In the *N. mayeri* Zone, both *N. nympha* and *N. continuosa* evolved from *N. mayeri*. *N. continuosa* evolved into *N. pachyderma* in the lower part of the late Miocene [*Jenkins*, 1967], while in the Tropical-Subtropical provinces *N. continuosa* gave rise to the *N. acostaensis–N. humerosa–N. dutertrei* lineage [*Kennett and Srinivasan*, 1983].

G. conica, whose possible origin in the upper part of the early Miocene has been mentioned, ranges into the middle Miocene *G. mayeri* Zone; its extinction level appears to be diachronous, and its northern limit in New Zealand is 38°S.

At the high-latitude DSDP Leg 28 sites 265 and 266, *Kaneps* [1975] recorded a cool water middle Miocene fauna composed of *G. bulloides*, *G. woodi*, *N. continuosa*, *G. conica*, and *G. bradyi*.

In southeastern Atlantic DSDP Leg 40 Site 360, the fauna is similar to faunas at the same latitude in New Zealand, although there is some indication of slightly warmer water in the upper part of the middle Miocene where *Globigerinoides* cf. *tyrrhenicus* and *G. obliquus* have been found.

Berggren [1992*b*] recorded middle Miocene faunas from southern Kerguelen Plateau ODP Leg 120 sites 744, 748, and 751. There seems to be a significant difference between Site 747 at 54°S and the more southerly Site 751 at 57°S and Site 748 at 58°S. At Site 747, *Berggren* [1992*b*] recorded such forms as *G. panda* and *G. mayeri* which were absent at the two more southerly sites where there is a greatly reduced species diversity. The separation of the Temperate and Subantarctic faunas can be drawn at ~55°C (Figure 9).

N. nympha, after evolving in southern mid-latitudes from N. mayeri, spread quickly southward, and Berggren [1992b] has recorded it at both Site 747 (54°S) and Site 751 (57°S).

Late Miocene

The cooling which began in the middle Miocene continued into the late Miocene, and G. menardii became locally extinct in the lower part of the G. miotumida Zone in New Zealand [Jenkins, 1971a]. Although a few of the warmer-water taxa such as Globigerinella aequilateralis, Globigerina nepenthes, and rare Globigerinoides sacculifer are recorded in the North Island of New Zealand, the species belong to the Temperate Fauna with cooler-water species such as N. pachyderma, G. bulloides, T. quinqueloba, and G. bradyi also being present.

The keeled Globorotalia are represented mainly by the G. miotumida group including the encrusted form G. conoidea, and toward the upper part of the late Miocene the evolution of both the conical-shaped G. conomiozea and the small compact G. sphericomiozea took place; G. sphericomiozea was later to evolve into G. puncticulata in the early Pliocene. Kennett and Srinivasan [1983] recorded G. sphericomiozea from the early Pliocene in the southeastern Pacific, but in New Zealand it also occurs in the late Miocene [Jenkins, 1971a; Scott et al., 1990]; these differences will be resolved once the Miocene/Pliocene international boundary has been defined by the International Union of Geological Sciences.

N. pachyderma which evolved from N. continuosa in the lower part of the late Miocene retained its inherited preference to sinistral coiling throughout this time [Jenkins, 1967]. Like N. nympha in the middle Miocene, N. pachyderma appears to have spread further south than its immediate ancestor; this is a pattern repeated in a number of Neogene lineages [Jenkins, 1992a].

At the high-latitude DSDP Leg 28 Site 265, Kaneps [1975] recorded only G. bradyi from the late Miocene, and he proposed that dissolution was the cause of the low species diversity.

Tjaslma [1977] recorded late Miocene fauna at DSDP Leg 36 Site 329 with N. acostaensis, G. bulloides, G. glutinata, N. continuosa, G. scitula, G. bradyi, G. anfracta, and the keeled G. panda; it is probable that N. nympha has been misidentified as the warm water N. acostaensis at Site 329.

The faunas at DSDP Leg 40 Site 360 in the southeastern Atlantic were consistently warmer than at a similar latitude in New Zealand and yielded G. altispira, G. menardii, Globigerinoides obliquus, G. sacculifer, G. conglobatus, Candeina nitida, and N. humerosa [Jenkins, 1978c].

There is evidence of a late Miocene cooling in the oxygen isotope record, and this evidence of cooling is most pronounced in the planktonic foraminifera at DSDP Leg 29 Site 281 at 47°S [Shackleton and Kennett, 1975]. The decline in temperature of about 2°C occurred between samples 281-9-1 and 281-7-2 and coincides with the evolutionary appearance of N. pachyderma and the local extinction of N. continuosa [Jenkins, 1975].

Brunner [1991] recorded both the G. miotumida and G. sphericomiozea Zone faunas at South Atlantic ODP Leg 114 Site 704 on the Meteor Rise. These faunas closely resemble the faunas of the same age in the southwestern Pacific [Jenkins, 1971a, 1973; Jenkins and Srinivasan, 1986].

The late Miocene faunas at ODP Site 747 at 54°S yielded G. conoidea, which was not recorded at the more southerly sites 748 and 751 at 58°S and 57°S, respectively [Berggren, 1992b]. The presence of N. nympha at all these sites is important because it was originally identified as G. acostaensis by Jenkins [1960] from Southeast Australia and later named N. nympha from the New Zealand late middle Miocene–late Miocene [Jenkins, 1967]; it is a homeomorph of the late Miocene–Recent N. pachyderma.

As in the middle Miocene, the Subantarctic and Temperate faunas can be separated at ~55°S (Figure 9).

Pliocene

There is evidence of a warming in the early Pliocene of New Zealand with Globigerina apertura, Globorotalia tumida, and Hastigerina pelagica in sequences from the North Island; it coincides with a change to dextral in the coiling of N. pachyderma [Jenkins, 1967, 1971a].

The early Pliocene was a time of evolutionary activity both in the Tropical-Subtropical Province and also in the mid-latitudes where G. puncticulata evolved from G. sphericomiozea and G. pliozea evolved from G. miotumida. The keeled Globorotalia tend to become locally extinct in the early Pliocene of the mid-latitudes [Jenkins, 1992a].

The evolution of G. inflata from G. puncticulata is well documented in the mid-latitudes [Walters, 1965; McInnes, 1965; Malmgren and Kennett, 1981; Scott, 1980] and marks the base of the late Pliocene. According to Berggren et al. [1985b] the evolutionary appearance of G. inflata is at 3 Ma, but it made a cryptogenic appearance at 56°S at 0.65 Ma. [Kennett, 1970] (Figure 10). Brunner [1991] examined the Pliocene faunas at Site 704 drilled by ODP Leg 114 in the South Atlantic at 46°S and subdivided the sequence into the lower G. puncticulata Zone followed by the G. inflata Zone. The faunas are typical southern mid-latitude and closely resemble those described in the southwestern Pacific [Jenkins, 1967, 1971a, 1973; Jenkins and Srinivasan, 1986].

Berggren [1992a] recorded low-diversity faunas from the southern Kerguelen Plateau, but at the more northerly Site 747 both *G. sphericomiozea* and *G. puncticulata* were present. In the southwestern Pacific, *G. sphericomiozea* has a very short stratigraphic range in the latest Miocene becoming extinct in the earliest Pliocene [*Scott et al.*, 1990]; at Site 747 it is recorded as ranging from the early to late Pliocene. Similarly, *G. puncticulata* is normally limited to the early Pliocene but ranges from the late Pliocene to Pleistocene at Site 747.

The faunas at sites 748 and 751 are typically Subantarctic in aspect, consisting of *G. bulloides*, *G. scitula*, *G. uvula*, *N. continuosa*, *N. pachyderma*, and *T. quinqueloba*.

According to *Kennett* [1968, 1969] and *Echols and Kennett* [1973] the populations of *N. pachyderma* are 90% or more sinistrally coiled south of the Polar Front, and off New Zealand the separation of the sinistral and dextral populations is a few degrees of latitude south of the Subtropical Convergence. In New Zealand [*Jenkins*, 1967] there are a number of changes in the coiling direction of *N. pachyderma* in the Pliocene with a time sequence of dextral, sinistral, dextrally coiled, and sinistrally coiled populations. Similar coiling changes were recorded by *Devereux et al.* [1970] and by *Hornibrook* [1976].

The coiling in late Miocene–Pleistocene populations of *N. pachyderma* at DSDP Leg 90 Site 594 (45°S) in Subantarctic waters off the east coast of the South Island of New Zealand is mainly sinistral except for some dextral populations in the late Pliocene and Pleistocene. These changes possibly represent northward movement in the Subantarctic waters which carried the sinistrally coiled *N. pachyderma* populations into the mid-latitudes.

The initial discovery by *Berggren et al.* [1967] that *Globorotalia truncatulinoides* had evolved from *G. tosaensis* in the subtropical North Atlantic (26°N) at 1.85 Ma was later followed by a re-evaluation by *Backman and Shackleton* [1983], who re-examined the original piston core CH 61-171; they recalibrated the first appearance of *G. truncatulinoides* to be slightly older than 2.0 Ma. *Weaver* [1986] and *Weaver and Clement* [1986] were able to date its first appearance at DSDP North Atlantic sites between latitudes 37°N and 53°N at 1.80–1.85 Ma using good paleomagnetic records.

In the South Pacific it has been shown that *G. truncatulinoides* evolved much earlier than 1.85 Ma. *Kennett and Geitzenauer* [1969], after examining the *Eltanin* Core 21-5 from the southeastern Pacific at 36°S, concluded that the first appearance of *G. truncatulinoides* had occurred much earlier than the Pliocene/Pleistocene boundary which was then dated at 1.85 Ma. In the southwestern Pacific, *Jenkins* [1971*b*] recorded the first appearance of *G. truncatulinoides* before 2.5 Ma in New Zealand, and *Dowsett* [1988], *Jenkins* [1992*a*], and *Jenkins and Gamson* [1993] recorded its first appearance at DSDP sites 587, 588, 590, and 592 to

be at ~2.5 Ma. *Scott et al.* [1990] also recorded the evolutionary appearance of *G. truncatulinoides* s.l. to be at ~2.5 Ma in New Zealand but stated that populations with 100% keeled tests only appeared at ~1.0 Ma.

The explanation of these time differences can best be resolved by accepting that *G. truncatulinoides* evolved from *G. tosaensis* in the southwestern Pacific at ~2.5 Ma, and at ~1.9 Ma it spread northward into the Central and North Pacific and eastward into the North Atlantic via Drake Passage and the South Atlantic [*Jenkins*, 1992*b*]. *Weaver and Clement* [1986] showed that *G. truncatulinoides* reached the southern part of the North Atlantic by 1.85 Ma and later spread northeastward reaching DSDP Site 611 at 1.35 Ma and Gabon Spur Site 550 by 0.95 Ma [*Pujol and Duprat*, 1985]. Similarly, in the southern hemisphere *G. truncatulinoides* spread southward reaching northern Subantarctic waters in the late Pleistocene [*Kennett*, 1970] (Figure 10).

G. inflata showed a similar pattern of evolution and migration a little earlier in the Pliocene: it evolved from *G. puncticulata* [*Walters*, 1965; *McInnes*, 1965; *Jenkins*, 1975; *Scott*, 1980] in the southwestern Pacific, and this event has been dated at 2.9 Ma by *Malmgren and Kennett* [1981]. *G. inflata* spread northward from the southwestern Pacific and reached the North Atlantic at 2.00–2.26 Ma [*Weaver and Clement*, 1986; *Jenkins et al.*, 1988] and spread southward reaching the Subantarctic region in the Pleistocene [*Kennett*, 1970] (Figure 10).

Pleistocene

The trend toward a pattern similar to the modern distribution of taxa which began in the early Pliocene was concluded in the Pleistocene. The New Zealand fauna is equivalent to *Bé*'s [1977] "Transition Fauna" here described as the Temperate Fauna. Similar faunas have been described by *Jenkins* [1978*c*] from DSDP Leg 40 in the southeastern Atlantic and by *Brunner* [1991] from ODP Leg 114 South Atlantic Site 744.

On the basis of sequences of cold and warmer faunas, *Kennett* [1970] was able to recognize eight intervals of warming in Subantarctic cores of middle late Pleistocene age. The appearance of *G. inflata* at 0.7 Ma was interpreted as its adaptation to the colder Subantarctic water, and *G. truncatulinoides* appeared at ODP Leg 114 Site 744 in the South Atlantic, at 46°S at 0.40 Ma [*Brunner*, 1991; *Kennett*, 1970] after having evolved in the southwestern Pacific mid-latitudes at 2.5 Ma.

Keany and Kennett [1972] were able to recognize 10 warmer intervals in Subantarctic cores in the Matuyama Reversed Epoch and six in the Brunhes Normal Epoch. The cold intervals were recognized by sinistrally coiled *N. pachyderma* and warmer intervals by dextrally coiled populations. *Keany and Kennett* [1972] also described as new, *Globigerina antarctica* which ranged throughout the Matuyama Reversed Epoch; its paleo-geographic and stratigraphic ranges need to be tested.

Berggren [1992*b*] recorded low-diversity Subantarctic faunas from Leg 120 ODP sites 747, 748, and 751 at 54°–58°S. Again the faunas at Site 747 at 54°S are slightly different because they include *G. bulloides* and *G. puncticulata*. Further south, *Kennett* [1968] examined bottom samples from the Ross Sea south of 70°S and reported mainly *N. pachyderma* with a few *G. inflata* and *G. megastoma* in this Antarctic Fauna.

CONCLUSIONS

Up until early Oligocene at ~31 Ma there were two oceans around Antarctica: the Indo-Atlantic and the South Pacific. With the development of the circum-Antarctic current at ~31 Ma these oceans were joined and the Southern Ocean probably became fully developed by early Miocene at ~22 Ma.

The Cenozoic planktonic foraminiferal faunas in the oceanic waters around Antarctica show definite responses to changes in both paleotemperature and paleoceanography. In the high latitudes south of 50°S a warming in the late Paleocene saw Temperate faunas being replaced by Subtropical faunas in the late Paleocene and early Eocene. These were subsequently replaced by Temperate faunas south of 50°S which later spread northward to ~40°S in the Oligocene. The Miocene showed a gradual cooling, and by late Miocene Subantarctic faunas are found south of 55°S. After an early Pliocene warming, the high-latitude faunas took on a modern aspect with Antarctic and Subantarctic faunas south of the Subtropical Convergence (Figure 9).

Future ODP drilling should aim at a better Cenozoic planktonic foraminiferal record from the high latitudes of the Pacific sector of the Southern Ocean.

Acknowledgments. The writer wishes to thank J. M. Jenkins and J. P. Kennett for their encouragement; Valerie Deisler, who typed the manuscript; John Taylor and Lin Norton, who drafted the figures and tables; and the two referees, N. de B. Hornibrook and J. P. Kennett, who made useful suggestions for improving the manuscript. N. de B. Hornibrook also provided specimens and samples from New Zealand.

REFERENCES

Backman, J., and N. J. Shackleton, Quantitative biochronology of Pliocene and early Pleistocene calcareous nannoplankton from the Atlantic, Indian and Pacific oceans, *Mar. Micropaleontol.*, 8, 141–170, 1983.

Barker, P. F., and J. Burrell, The opening of the Drake Passage, *Mar. Geol.*, 25, 15–34, 1977.

Barker, P. F., et al., Leg 113, *Proc. Ocean Drill. Program Initial Rep.*, 113, 785 pp., 1988.

Bé, A. W. H., An ecological, zoogeographic and taxonomic review of Recent planktonic foraminifera, in *Oceanic Micropaleontology*, vol. 1, edited by A. T. S. Ramsay, pp. 1–88, Academic, San Diego, Calif., 1977.

Berggren, W. A., A Cenozoic time-scale, some implications for regional geology and paleobiology, *Lethaia*, 5, 195–215, 1972.

Berggren, W. A., Paleogene planktonic foraminifer magneto-biostratigraphy of the southern Kerguelen Plateau (sites 747–749), *Proc. Ocean Drill. Program Sci. Results*, 120, 551–568, 1992*a*.

Berggren, W. A., Neogene planktonic foraminifer magnetobiostratigraphy of the southern Kerguelen Plateau (sites 747, 748 and 751), *Proc. Ocean Drill. Program Sci. Results*, 120, 631–647, 1992*b*.

Berggren, W. A., J. D. Phillips, A. Bertels, and D. Wall, Late Pliocene-Pleistocene stratigraphy in deep sea cores from the south-central North Atlantic, *Nature*, 216, 253–255, 1967.

Berggren, W. A., M. P. Aubry, and N. Hamilton, Neogene magnetobiostratigraphy of Deep Sea Drilling Project Site 516 (Rio Grande Rise, South Atlantic), *Initial Rep. Deep Sea Drill. Proj.*, 72, 675–713, 1983.

Berggren, W. A., D. V. Kent, and J. J. Flynn, Jurassic to Paleogene, Part 2, Paleogene geochronology and chronostratigraphy, in *The Chronology of the Geological Record*, *Mem. 10*, edited by N. Snelling, pp. 141–186, Geological Society of London, London, 1985*a*.

Berggren, W. A., D. V. Kent, and J. A. van Couvering, The Neogene, Part 2, Neogene geochronology and chronostratigraphy, in *The Chronology of the Geological Record*, *Mem. 10*, edited by N. Snelling, pp. 211–260, Geological Society of London, London, 1985*b*.

Blow, W. H., Late middle Eocene to Recent planktonic foraminiferal biostratigraphy, *Proc. Int. Conf. Planktonic Microfossils 1st*, 1, 199–421, 1969.

Bolli, H. M., and J. B. Saunders, Oligocene to Holocene low latitude planktic foraminifera, in *Plankton Stratigraphy*, edited by H. M. Bolli, J. B. Saunders, and K. Perch-Nielsen, pp. 155–262, Cambridge University Press, New York, 1985.

Brunner, C. A., Latest Miocene to Quaternary biostratigraphy and paleoceanography, Site 704, Subantarctic South Atlantic Ocean, *Proc. Ocean Drill. Program Sci. Results*, 114, 201–215, 1991.

Canon, A., and M. Ernst, Part II, Magallanes Basin foraminifera, A System of Stages for Correlation of Magallanes Basin Sediments, *Mem. Geol. Soc. Am.*, 139, 61–117, 1974.

D'Agostino, A. E., and P. N. Webb, Interpretation of mid-Miocene to Recent lithostratigraphy and biostratigraphy at DSDP Site 273, Ross Sea, *Antarct. J. U. S.*, 15(5), 118–120, 1980.

Devereux, I., C. H. Hendy, and P. Vella, Pliocene and early Pleistocene sea temperature fluctuation, Mangaopari Stream, New Zealand, *Earth Planet. Sci. Lett.*, 8, 163–168, 1970.

Dowsett, H. J., Diachrony of late Neogene microfossils in the southwest Pacific Ocean: Application of the graphic correlation method, *Paleoceanography*, 3, 209–222, 1988.

Echols, R. J., and J. P. Kennett, Distribution of foraminifera in surface sediments, in *Marine Sediments of the Southern Ocean*, *Antarct. Map Folio Ser.*, folio 17, edited by V. Bushnell, Am. Geogr. Soc., New York, 1973.

Hemleben, C., D. Muhlen, R. K. Ollson, and W. A. Berggren, Surface texture and the first occurrence of spines in planktonic foraminifera from the early Tertiary, *Geol. Jahrb.*, A128, 117–146, 1991.

Hornibrook, N. de B., *Globorotalia truncatulinoides* and the Pliocene-Pleistocene boundary in northern Hawkes Bay, New Zealand, in *Progress in Micropaleontology*, pp. 83–102, American Museum of Natural History, New York, 1976.

Hornibrook, N. de B., Heterohelicidae (Foraminiferida) in the New Zealand Oligocene, *N. Z. Geol. Surv. Rec.*, 9, 67–69, 1985.

Hornibrook, N. de B., and D. G. Jenkins, *Candeina zeocenica* Hornibrook and Jenkins, a new species of foraminifera from the New Zealand Eocene and Oligocene, *N. Z. J. Geol. Geophys.*, 8, 839–842, 1965.

Hornibrook, N. de B., and D. G. Jenkins, DSDP Site 594, Chatham Rise, New Zealand: Late Neogene planktonic foraminiferal biostratigraphy revised, *J. Foraminiferal Res.*, in press, 1993.

Huber, B. T., Upper Campanian–Paleocene foraminifera from the James Ross Island region, Antarctic Peninsula, Geology and Paleontology of Seymour Island, Antarctica, *Mem. Geol. Soc. Am.*, 169, 163–245, 1988.

Huber, B. T., Paleogene and early Neogene planktonic foraminifer biostratigraphy of sites 738 and 744, Kerguelen Plateau (South Indian Ocean), *Proc. Ocean Drill. Program Sci. Results*, 119, 427–449, 1991.

Jenkins, D. G., Planktonic foraminifera from the Lakes Entrance oil shaft, Victoria, Australia, *Micropaleontology*, 6, 345–371, 1960.

Jenkins, D. G., Planktonic foraminifera and Tertiary intercontinental correlations, *Micropaleontology*, 11, 265–277, 1965.

Jenkins, D. G., Planktonic foraminiferal zones and new taxa from the Danian to lower Miocene of New Zealand, *N. Z. J. Geol. Geophys.*, 8, 1088–1126, 1966.

Jenkins, D. G., Recent distribution, origin and coiling ratio changes in *Globorotalia pachyderma* (Ehrenberg), *Micropaleontol.*, 13, 195–203, 1967.

Jenkins, D. G., Variations in the numbers of species and subspecies of planktonic Foraminiferida as an indicator of New Zealand paleotemperatures, *Palaeogeogr. Palaeoclimatol. Palaeoecol.*, 5, 309–313, 1968.

Jenkins, D. G., New Zealand Cenozoic planktonic foraminifera, *N. Z. Geol. Surv. Paleontol. Bull.*, 42, 278 pp., 1971a.

Jenkins, D. G., The reliability and accuracy of some Cenozoic planktonic foraminiferal "datum-planes" used in biostratigraphic correlation, *J. Foraminiferal Res.*, 1, 82–86, 1971b.

Jenkins, D. G., The present status and future progress in the study of Cenozoic planktonic foraminifera, *Rev. Esp. Micropaleontol.*, 5, 133–146, 1973.

Jenkins, D. G., Paleogene planktonic foraminifera of New Zealand and the Austral region, *J. Foraminiferal Res.*, 4, 155–170, 1974a.

Jenkins, D. G., Initiation of the proto circum-Antarctic current, *Nature*, 252, 371–373, 1974b.

Jenkins, D. G., Cenozoic planktonic foraminiferal biostratigraphy of the southwestern Pacific and Tasman Sea—DSDP Leg 29, *Initial Rep. Deep Sea Drill. Proj.*, 29, 449–467, 1975.

Jenkins, D. G., *Guembelitria* aff. *stavensis* Bandy, a paleoceanographic marker of the initiation of the circum-Antarctic current and the opening of Drake Passage, *Initial Rep. Deep Sea Drill. Proj.*, 40, 687–693, 1978a.

Jenkins, D. G., *Guembelitria samwelli* Jenkins, a new species from the Oligocene of the southern hemisphere, *J. Foraminiferal Res.*, 8, 132–137, 1978b.

Jenkins, D. G., Neogene planktonic foraminifers from DSDP Leg 40 sites 360 and 362 in the southeastern Atlantic, *Initial Rep. Deep Sea Drill. Proj.*, 40, 723–739, 1978c.

Jenkins, D. G., Southern mid-latitude Paleocene to Holocene planktonic foraminifera, in *Plankton Stratigraphy*, edited by H. M. Bolli, J. B. Saunders, and K. Perch-Nielsen, pp. 263–282, Cambridge University Press, New York, 1985.

Jenkins, D. G., Forum on Tertiary unconformities in Otago: Oligo-Miocene unconformities in North Otago and Tasman Sea, *J. R. Soc. N. Z.*, 17, 177–180, 1987.

Jenkins, D. G., The paleogeography, evolution and extinction of late Miocene–Pleistocene planktonic foraminifera from the southwest Pacific, in *Centenary of Micropaleontology in Japan*, pp. 27–35, Terra Scientific, Tokyo, 1992a.

Jenkins, D. G., Predicting extinctions of some extant planktonic foraminifera, *Mar. Micropaleontol.*, 19, 239–243, 1992b.

Jenkins, D. G., Cenozoic southern mid- and high-latitude biostratigraphy and chronostratigraphy based on planktonic foraminifera, this volume.

Jenkins, D. G., and P. Gamson, The late Cenozoic *Globorotalia truncatulinoides* datum-plane in the Atlantic, Pacific and Indian oceans, in *High Resolution Stratigraphy in Marine Sequences*, edited by E. Hailwood and R. Kidd, pp. 127–130, Geological Society of London, London, 1993.

Jenkins, D. G., and N. Shackleton, Parallel changes in species diversity and paleotemperature in the lower Miocene, *Nature*, 278, 50–51, 1979.

Jenkins, D. G., and M. S. Srinivasan, Cenozoic planktonic foraminifera from the equator to the sub-Antarctic of the southwest Pacific, *Initial Rep. Deep Sea Drill. Proj.*, 90, 795–834, 1986.

Jenkins, D. G., D. Curry, B. M. Funnell, and J. E. Whittaker, Planktonic foraminifera from the Pliocene Coralline Crag of Suffolk, eastern England, *J. Micropaleontol.*, 7, 1–10, 1988.

Kaneps, G. G., Cenozoic planktonic foraminifera from Antarctic deep-sea sediments, Leg 28, DSDP, *Initial Rep. Deep Sea Drill. Proj.*, 28, 573–583, 1975.

Keany, J., and J. P. Kennett, Pliocene–early Pleistocene paleoclimatic history recorded in Antarctic-Subantarctic deep sea cores, *Deep Sea Res.*, 19, 529–548, 1972.

Keigwin, L., and G. Keller, Middle Oligocene cooling from equatorial Pacific DSDP Site 77B, *Geology*, 12, 16–19, 1984.

Kennett, J. P., Latitudinal variation in *Globigerina pachyderma* (Ehrenberg) in surface sediments of the southwest Pacific Ocean, *Micropaleontology*, 14, 305–318, 1968.

Kennett, J. P., Distribution of planktonic foraminifera in surface sediments southeast of New Zealand, *Proc. Int. Conf. Planktonic Microfossils 1st*, 2, 307–322, 1969.

Kennett, J. P., Pleistocene paleoclimates and foraminiferal biostratigraphy in Subantarctic deep-sea cores, *Deep Sea Res.*, 17, 125–140, 1970.

Kennett, J. P., Middle and late Cenozoic planktonic foraminiferal biostratigraphy of the southwest Pacific—DSDP Leg 21, *Initial Rep. Deep Sea Drill. Proj.*, 21, 575–640, 1973.

Kennett, J. P., Cenozoic evolution of Antarctic glaciation, the circum-Antarctic ocean, and their impact on global paleoceanography, *J. Geophys. Res.*, 28, 3843–3860, 1977.

Kennett, J. P., The development of planktonic biogeography in the Southern Ocean during the Cenozoic, *Mar. Micropaleontol.*, 3, 301–345, 1978.

Kennett, J. P., Paleoceanographic and biogeographic evolution of the Southern Ocean during the Cenozoic, and Cenozoic microfossil datums, *Palaeogeogr. Palaeoclimatol. Palaeoecol.*, 31, 123–152, 1980.

Kennett, J. P., and K. R. Geitzenaur, Pliocene-Pleistocene boundary in a South Pacific deep-sea core, *Nature*, 224, 899–901, 1969.

Kennett, J. P., and N. J. Shackleton, Oxygen isotope evidence for the development of the psychrosphere 38 m.y. ago, *Nature*, 260, 513–515, 1976.

Kennett, J. P., and M. S. Srinivasan, *Neogene Planktonic Foraminifera, a Phylogenetic Atlas*, 265 pp., Hutchinson Ross, Stroudsburg, Pa., 1983.

Kennett, J. P., and L. D. Stott, Abrupt deep-sea warming, paleoceanographic changes and benthic extinctions at the end of the Paleocene, *Nature*, 353, 225–229, 1991.

Kennett, J. P., and P. Vella, Late Cenozoic planktonic foraminifera and paleoceanography at DSDP Site 284 in the cool subtropical south Pacific, *Initial Rep. Deep Sea Drill. Proj.*, 29, 769–799, 1975.

Kennett, J. P., and N. D. Watkins, Late Miocene–early Pliocene paleomagnetic stratigraphy, paleoclimatology and biostratigraphy in New Zealand, *Geol. Soc. Am. Bull.*, 85, 1385–1398, 1974.

Kennett, J. P., R. E. Houtz, P. B. Andrews, A. R. Edwards,

V. A. Gostin, M. Hajos, M. Hampton, D. G. Jenkins, S. V. Margolis, A. T. Ovenshine, and K. Perch-Nielsen, Cenozoic paleoceanography in the southwest Pacific Ocean, Antarctic glaciation, and the development of the circum-Antarctic current, *Initial Rep. of the Deep Sea Drill. Proj.*, *29*, 1155–1169, 1975.

Kennett, J. P., N. J. Shackleton, S. V. Margolis, D. E. Goodney, D. E. Dudley, and P. M. Kroopnick, Late Cenozoic oxygen and carbon isotopic history and volcanic ash stratigraphy: DSDP Site 284, South Pacific, *Am. J. Sci.*, *279*, 52–69, 1979.

Leckie, R. M., and P. N. Webb, Foraminifera of DSDP Site 270 as indicators of the evolving Ross Sea in the late Oligocene–early Miocene, *Antarct. J. U. S.*, *15*(5), 117–118, 1980.

Leckie, R. M., and P. N. Webb, Late Oligocene–early Miocene glacial record of the Ross Sea, Antarctica: Evidence from DSDP Site 270, *Geology*, *11*, 578–582, 1983.

Leckie, R. M., and P. N. Webb, *Candeina antarctica* n.sp. and the phylogenetic history and distribution of Candeina spp. in the Paleogene–early Miocene of the Southern Ocean, *J. Foraminiferal Res.*, *15*, 65–78, 1985.

Lindsay, J. M., Cainozoic foraminifera and stratigraphy of the Adelaide Plains sub-basin, South Australia, *Bull. Geol. Surv. South Aust.*, *42*, 1–60, 1969.

Lindsay, J. M., Oligocene in South Australia, in *ANZAAS 45th Congress, Abstracts*, section 3, pp. 105–106, Australian and New Zealand Association for the Advancement of Science, 1973.

Lindsay, J. M., Aspects of South Australian Tertiary foraminiferal biostratigraphy, with emphasis on studies of *Massilina* and *Subbotina*, N. H. Ludbrook Honour Volume, *Spec. Publ. South Aust. Dep. Mines Energy*, 187–231, 1986.

Ludbrook, N. H., and J. M. Lindsay, Tertiary foraminiferal zones in South Australia, *Proc. Int. Conf. Planktonic Microfossils 1st*, *2*, 369–375, 1969.

Malmgren, B. A., and J. P. Kennett, Phyletic gradualism in a late Cenozoic planktonic foraminiferal lineage: DSDP Site 284, southwest Pacific, *Paleobiology*, *7*, 230–240, 1981.

McGowran, B., Foraminiferal evidence for the Paleocene age of the Kings Park Shale (Perth Basin, Western Australia), *J. R. Soc. West. Aust.*, *47*, 81–86, 1964.

McGowran, B., Two Paleocene foraminiferal faunas from the Wangerip Group, Pebble Point coastal section, western Victoria, *Proc. R. Soc. Victoria*, *79*, 9–74, 1965.

McGowran, B., Observation bore no. 2, Gambier Embayment of the Otway Basin: Tertiary micropaleontology and stratigraphy, *Miner. Resour. Rev. South Aust. Dep. Mines Energy*, *135*, 43–55, 1973.

McGowran, B., and A. Beecroft, *Guembelitria* in the early Tertiary of southern Australia and its paleoceanographic significance, N. H. Ludbrook Honour Volume, *Spec. Publ. South Aust. Dep. Mines Energy*, *5*, 247–261, 1986.

McInnes, B. A., *Globorotalia miozea* Finlay as an ancestor of *Globorotalia inflata* (d'Orbigny), *N. Z. J. Geol. Geophys.*, *8*, 104–108, 1965.

Nocchi, M., E. Amici, and I. Premoli Silva, Planktonic fora-

miniferal biostratigraphy and paleoenvironmental interpretation of Paleogene faunas from the Subantarctic Transect, Leg 114, *Proc. Ocean Drill. Program Sci. Results*, *114*, 233–279, 1991.

Pujol, C., and J. Duprat, Quaternary and Pliocene planktonic foraminifers of the northeastern Atlantic (Goban Spur), DSDP Leg 80, *Initial Rep. Deep Sea Drill. Proj.*, *80*, 683–723, 1985.

Rögl, R., Late Cretaceous to Pleistocene foraminifera from the Southeast Pacific Basin, DSDP Leg 35, *Initial Rep. Deep Sea Drill. Proj.*, *35*, 539–556, 1976.

Sancetta, C., Neogene Pacific microfossils and paleoceanography, *Mar. Micropaleontol.*, *3*, 347–376, 1978.

Scott, G. H., *Globorotalia inflata* lineage and *G. crassaformis* from Blind River, New Zealand: Recognition, relationship, and use in uppermost Miocene–lower Pliocene stratigraphy, *N. Z. J. Geol. Geophys.*, *23*, 665–677, 1980.

Scott, G. H., S. Bishop, and B. J. Burt, Guide to some Neogene Globorotalids (Foraminiferida) from New Zealand, *N. Z. Geol. Surv. Paleontol. Bull.*, *61*, 1–135, 1990.

Shackleton, N. J., and J. P. Kennett, Paleotemperature history of the Cenozoic and the initiation of the Antarctic glaciation: Oxygen and carbon isotope analysis in DSDP sites 277, 279 and 281, *Initial Rep. Deep Sea Drill. Proj.*, *29*, 743–756, 1975.

Smith, A. G., A. M. Hurley, and J. C. Briden, *Phanerozoic Paleocontinent World Maps*, 102 pp., Cambridge University Press, New York, 1981.

Stott, L. D., and J. P. Kennett, Antarctic Paleogene planktonic foraminifer biostratigraphy: ODP Leg 113, sites 689 and 690, *Proc. Ocean Drill. Program Sci. Results*, *113*, 549–569, 1990.

Strong, C. P., Cretaceous-Tertiary boundary, Mid-Waipara River section, North Canterbury, New Zealand, *N. Z. J. Geol. Geophys.*, *27*, 231–234, 1984.

Tjalsma, R. C., Cenozoic foraminifera from the South Atlantic, DSDP Leg 36, *Initial Rep. Deep Sea Drill. Proj.*, *36*, 493–517, 1977.

Toumarkine, M., Planktonic foraminiferal biostratigraphy of the Paleogene of sites 360 to 364 and the Neogene of sites 362A, 363 and 364, Leg 40, *Initial Rep. Deep Sea Drill. Proj.*, *40*, 679–721, 1978.

Van Valen, L., A new evolutionary law, *Evol. Theory*, *1*, 1–30, 1973.

Walters, R., The *Globorotalia zealandica* and *G. miozea* lineages, *N. Z. J. Geol. Geophys.*, *8*, 109–127, 1965.

Weaver, P. P. E., Late Miocene to Recent planktonic foraminifera from the North Atlantic: DSDP Leg 94, *Initial Rep. Deep Sea Drill. Proj.*, *94*, 815–829, 1986.

Weaver, P. P. E., and B. M. Clement, Synchroneity of Pliocene planktonic foraminiferal datums in the North Atlantic, *Mar. Micropaleontol.*, *10*, 295–307, 1986.

Webb, P. N., Initial reports on geological materials collected at RISP Site J9, 1978–79, *Ross Ice Shelf Proj. Tech. Rep.*, *79-1*, 127 pp., 1979.

(Received April 3, 1992;
accepted August 17, 1992.)

THE ANTARCTIC PALEOENVIRONMENT: A PERSPECTIVE ON GLOBAL CHANGE
ANTARCTIC RESEARCH SERIES, VOLUME 60, PAGES 195–206

UNUSUAL SILICOFLAGELLATE SKELETAL MORPHOLOGIES FROM THE UPPER MIOCENE–LOWER PLIOCENE: POSSIBLE ECOPHENOTYPIC VARIATIONS FROM THE HIGH-LATITUDE SOUTHERN OCEANS

Kevin McCartney

Micropaleontology Undergraduate Research Laboratory, University of Maine at Presque Isle, Presque Isle, Maine 04769

Sherwood W. Wise, Jr.

Department of Geology, Florida State University, Tallahassee, Florida 32306

Six unusual morphotypes of the silicoflagellate subspecies *Distephanus speculum speculum* consti-tute the "*pseudofibula* plexus," the distribution of which is concentrated in uppermost Miocene–lowermost Pliocene sediments centered around the Antarctic continent. The occurrence of this plexus in Antarctic and Subantarctic waters seems to correlate with the late Miocene–earliest Pliocene glaciations of the continent. Therefore some type of ecologic control is suspected to account for its distribution in time and space. Distribution maps show that members of the plexus dominate silicoflagellate assemblages closest to the continent (class I province) but that their numbers diminish away from the continent (through class II and class III provinces), as does the thickness of the interval they occupy; thus the stratigraphic boundaries of the plexus may be diachronous. Only at Ocean Drilling Program Site 704 on Meteor Rise does the plexus occur in a continuous pelagic carbonate sequence with well-developed stable isotope and magnetobiostratigraphy. Its first abundance peak at that site, just above the base of the Messinian Stage, corresponds with a major interglacial event recorded in the planktonic and benthic foraminiferal isotopic records, an episode believed to have produced significant melting of the continental ice sheet and the injection of low-salinity meltwaters across the surface of the Southern Ocean. The major abundance peak of the plexus at this site occurs farther upsection, where it follows closely another anomalous negative excursion in the planktonic oxygen isotopic record. At the conclusion of the latest Miocene–earliest Pliocene glaciations, however, the *Pseudofibula* plexus disappeared abruptly from the Southern Ocean, suggesting that its existence was closely tied to glacial/interglacial events on the continent. Injections of meltwaters into a nutrient-rich upwelling environment may have triggered its blooms.

INTRODUCTION

Silicoflagellates are autotrophic protists with simple geometric skeletons consisting of hollow rods. They are the least abundant and studied of the major protist groups. The silicoflagellates are especially noteworthy for an extraordinary skeletal variability found in fossil and modern populations (see *McCartney and Wise* [1990] for a review), although specific skeletal morphol-ogies can have extensive fossil histories. The combina-tion of population variability and geologic stability of silicoflagellate skeletal design has led to a variety of species concepts. Some workers, particularly biolo-gists, "lump" a variety of morphologies into a single species, while paleontologists tend to finely subdivide the morphologies into separate taxa because of potential biostratigraphic utility.

An excellent example of silicoflagellate variability is the "*pseudofibula* plexus," which is used as a biostratigraphic zone in upper Miocene/lower Pliocene high latitudes [*Mc-Cartney and Wise*, 1990]. The plexus consists of a combi-nation of several unusual variants of *Distephanus specu-lum speculum* that occur together and can be extremely abundant and dominant. The various skeletal morpholo-gies that make up this group have been known for some time and were first used as a biostratigraphic zone by *Ciesielski* [1975], but little was said about their variability or interrelationships. Detailed examination did not occur until they were found in extreme abundance during Ocean Drilling Program (ODP) Leg 113. *McCartney and Wise* [1990] coined the name "*pseudofibula* plexus" to cover the entire variety of closely related morphologies that straddle the Miocene/Pliocene boundary in the Southern Ocean. A plexid interval of similar age also occurs in high northern latitudes but is poorly known.

195

TABLE 1.   Occurrence of the *Pseudofibula* Plexus Morphologies in Southern Ocean DSDP/ODP Holes

Hole	Latitude, Longitude	Plexid Interval, m	Plexids Dominant, m	Proportion of Class I Samples	Water Depth, m	Age of Plexid-Dominant Interval, Ma
		Leg 28 [Ciesielski, 1975]				
Hole 266	56°24.13′S, 110°06.70′E	1.75 ± 1.25			4173	
Hole 267B	59°14.55′S, 104°29.94′E	insufficient data	insufficient data		4564	
Hole 269	61°40.57′S, 140°04.21′E	72.0 ± 22.0	47.5 ± 0.5	0/10	4285	
Hole 274	68°59.81′S, 173°25.64′E	40.5 ± 8.5	6.0 ± 0.5	0/11	3326	
		Leg 36 [Busen and Wise, 1977]				
Hole 328	49°48.67′S, 36°39.53′W	4.5 ± 05			5013	
Hole 329	50°39.31′S., 45°05.73′W	plexids rare	plexids rare		1519	
		Leg 71 [Shaw and Ciesielski, 1983]				
Hole 513A	47°34.99′S, 24°38.40′W	plexids rare	plexids rare		4383	
		Leg 90 [Locker and Martini, 1986]				
Hole 591	31°35.06′S, 164°26.92′E	plexids rare	plexids rare		2131	
Hole 594	45°31.41′S, 174°56.88′E	plexids rare	plexids rare		1204	
		Leg 113 [McCartney and Wise, 1990]				
Hole 689B	64°31.01′S, 03°05.99′E	10.5 ± 1.5	10.5 ± 1.5	3/3	2080	4.7 to 8.0
Hole 690B	65°09.63′S, 01°12.30′E	3.0 ± 1.5	3.0 ± 1.5	2/2	2914	includes hiatus
Hole 693A*	70°49.89′S, 14°34.46′W	55.1 ± 5.1	55.1 ± 5.1	4/6	2359	4.4 to 7.8
Hole 695A	62°23.48′S, 43°27.10′W	34.65 ± 5.65	34.65 ± 5.65	0/5	1305	4.6 to 4.7+
Hole 696B	61°50.96′S, 42°56.00′W	59.1 ± 21.0	59.1 ± 21.0	0/4	605	4.4 to 7.9?
Hole 697B	61°48.63′S, 40°17.73′W	incomplete section	incomplete section		3483	
		Leg 114 (Figure 3)				
Hole 699A	51°32.54′S, 30°40.62′W	3.0 ± 1.5			3705	
Hole 704A	46°52.76′S, 07°25.25′E	12.0 ± 1.5			2532	5.8 to 6.2
		Leg 119 (Figure 4)				
Hole 745B	59.59°S, 85.85°E	37.5 ± 1.5	21.0 ± 1.5	0/11	4093	4.7 to 5.8+
Hole 746A	59.57°S, 86.87°E	23.25 ± 0.75	23.25 ± 0.75	0/10	4070	5.8− to 6/8+
		Leg 120 [McCartney and Harwood, 1992]				
Hole 747A	54°48.68′S, 76°47.64′E	incomplete interval	incomplete interval		1697	5.6
Hole 751A	57°43.55′S, 79°48.89′E	5.5 ± 0.75	5.5 ± 0.75		1634	5.7 to 8.3

* For additional Hole 693A data, see Figure 2.
    Plexid interval is the thickness of the interval where plexids exceed 2% of the silicoflagellate assemblage. Plexids dominant is the interval in which plexids are >50% of the silicoflagellate assemblage. Age of plexid-dominant interval is the estimated age of the interval in which plexids are dominant, based on paleomagnetic information (if available). Class I samples are the proportion of samples in which plexids are very abundant (>100 silicoflagellates per slide) and dominant (>50% of assmblage). Intervals given in meters cover the range from first to last occurrence of plexids in the specified abundance range plus half the distance to the sample above and below this range.

Members of the *pseudofibula* plexus are found in all deep-sea sites drilled in the Southern Ocean that include a Miocene-Pliocene interval, although the abundance of the skeletons and the duration of the interval varies considerably (Table 1). After being formally recognized as a group in the ODP Leg 113 material, the plexus was noted to form a pronounced interval in Leg 120 cores [*McCartney and Harwood*, 1992]; it also occurs in Leg 114 and Leg 119 sequences, although absolute abundance counts were not provided in the *Proceedings of the Ocean Drilling Program Scientific Results* for those

legs. The purpose of this study is to review the occurrence of this interesting group in Antarctic sediments recovered by Deep Sea Drilling Project (DSDP) and ODP drilling and to discuss their paleoenvironmental implications in view of new data we provide.

In this paper we use the term "plexid" when referring to individuals or morphologies that constitute the *pseudofibula* plexus. Previously, members of this group have often been referred to as "pseudofibulid," but we prefer to reserve that term for the morphology represented by *Distephanus speculum speculum* f. *pseudofibula*.

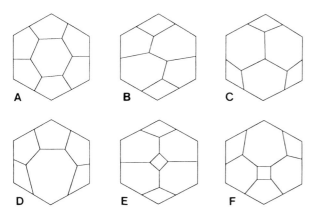

Fig. 1. Apical structures of silicoflagellate skeletal morphologies with six basal sides [from *McCartney and Wise*, 1990, Figure 5]. (*a*) *Distephanus speculum speculum*. (*b*) *D. speculum speculum* f. *pseudofibula*. (*c*) *D. speculum speculum* f. *varians*. (*d*) *D. speculum speculum notabilis*. (*e*) *D. speculum speculum* f. *pseudocrux* (centered ring). (*f*) *D. speculum speculum* f. *pseudocrux* (uncentered ring). Morphologies in Figures 1*b*–1*f* are members of the *pseudofibula* plexus.

GENERAL DESCRIPTION OF THE *PSEUDOFIBULA* PLEXUS ASSEMBLAGE

The *pseudofibula* plexus is an assemblage of silicoflagellates that generally have a six-sided basal ring and are found in upper Miocene/lower Pliocene sediments at high latitudes. The morphologies are unusual in that they have apical structures that do not consist of a six-sided ring. These morphologies are generally rare in the geologic record but, where they are present, can be very abundant. They appear to be polymorphic within single populations. Besides the Neogene *Distephanus speculum speculum* discussed here, plexid morphologies also occur for *Distephanus boliviensis* in the Pliocene [*McCartney and Harwood*, 1992], for *Distephanus speculum* in the Eocene [*Locker and Martini*, 1986], and for *Dictyocha grandis* in the Eocene [*Shaw and Ciesielski*, 1983].

The *pseudofibula* plexus includes three major and several minor morphologies (Figure 1); there are also five- and seven-sided variants. The three most abundant plexid morphologies are the formae *pseudofibula*, *notabilis*, and *varians*. These morphologies consistently occur together and have similar geologic ranges. They also have a similar abundance, although *varians* is generally somewhat more abundant and *pseudofibula* is often slightly less abundant than the other two. Other morphologies included in the plexus are the formae *pseudocrux*, the five-sided *pseudopentagonus*, and the unnamed seven-sided morphologies; these are all relatively rare, even within the plexid interval.

The plexid morphologies differ from all other commonly occurring *Distephanus*, which have skeletons with similar-sided apical and basal rings. The plexid

morphologies would generally be considered aberrant, but the more characteristic teratologies, such as fused or forked spines or struts or distorted shapes, appear to be no more abundant in the plexid interval than elsewhere. As was stated previously, the plexid morphologies can also be very abundant and predominate over the more typical skeletal designs, as in the upper Miocene/lower Pliocene Southern Ocean sediments. There is, however, little question that the plexid morphologies are closely related to more typical *Distephanus*. In the Miocene/Pliocene interval the plexids generally have the small size and fragile appearance of co-occurring *Distephanus speculum speculum*; indeed, some specimens found in the plexid interval have a centered six-sided apical ring verging on the *notabilis* form.

The plexid morphologies are also unusual in light of mathematical modeling by *McCartney and Loper* [1989], which suggests that the minimization of apical surface area is the most important known factor influencing silicoflagellate skeletal design; the six-sided *Distephanus* with an apical ring is very efficient at minimizing this surface area. The double-ring morphology, however, uses a relatively large amount of skeletal material, leading *McCartney and Loper* [1989] to suggest that simpler skeletal morphologies such as *Dictyocha* might predominate where silica is a more limiting nutrient. The morphologies of the *pseudofibula* plexus are apparently also more silica efficient than typical *Distephanus speculum speculum*. The unusual plexid morphologies suggest that they result from unusual environmental conditions [*McCartney and Wise*, 1990].

METHODOLOGY

This paper discusses the geographic distribution of the "*pseudofibula* plexus" using in part the published literature of different silicoflagellate researchers. Such comparisons are fraught with difficulty, since the technique of slide preparation and examination can vary from one worker to another and, even for a single researcher, can vary from one study to another. The recorded absolute abundances of silicoflagellates, for example, can vary according to the amount of sample material on the microscope slide. There are also differences in taxonomic nomenclature and species concept among various workers. Only previous literature that has absolute abundance counts, therefore, is used for this study. For most of the high southern latitude deep-sea legs in which the plexid morphologies are known to occur, sediment samples have been prepared independently and examined by the first author, who also did all of the counts on the Leg 113 and Leg 120 materials; absolute abundance counts by the various silicoflagellate workers appear to be comparable, although the counts should not be taken too literally.

The slides examined expressly for this study, and included in Figures 2–4, were prepared using a method-

Age	Pseudofibula plexus	Core, section, interval (cm)	Depth (mbsf)	Number of slides examined	Total number of specimens	% pseudofibula plexus	Bachmannocena diadon	Dictyocha aspera (s. ampl.)	D. fibula (s. ampl.)	D. stapedia	Distephanus boliviensis	D. boliviensis (7 sides)	D. crux (s. ampl.)	D. speculum pentagonus	D. speculum speculum (6 sides)	D. speculum speculum (7 sides)	D. speculum speculum f. notabilis	D. speculum speculum f. pseudocrux	D. speculum speculum f. pseudofibula	D. speculum speculum f. pseudopentagonus	D. speculum speculum f. varians
Early Pliocene		12R-2, 28-30	100.68	1.0	32	0	.	.	.	.	16	.	12	.	4	.	.	.	.	.	.
		12R-7, 43-45	108.33	0.1	300	0	.	.	.	.	286	5	1	8	.	.	.	.	.	.	.
		13R-1, 43-45	109.03	0.2	300	70	.	.	.	.	77	.	.	4	9	.	70	.	54	.	86
		13R-2, 43-45	110.53	0.3	300	95	.	.	.	.	9	.	.	.	7	.	76	1	63	4	140
		13R-3, 43-45	112.03	0.2	300	75	.	.	.	.	66	.	.	.	9	.	71	1	41	5	107
		13R-4, 43-45	113.53	1.0	11	91	.	.	.	.	.	.	.	.	1	.	2	.	4	.	4
		13R-5, 43-45	115.03	1.0	35	66	.	.	.	.	11	.	.	.	1	.	3	.	8	.	12
		13R-6, 14-16	116.24	1.0	122	89	.	.	.	.	.	.	.	.	13	1	30	.	22	.	56
		14R-1 44-46	118.64	1.0	128	77	.	2	2	1	22	.	.	.	2	1	26	.	25	1	46
		14R-2, 98-100	120.68	1.0	146	86	.	4	.	1	5	.	.	.	7	1	46	.	16	.	64
		14R-3, 98-100	122.18	1.0	166	77	.	2	1	7	19	.	.	2	8	.	45	.	20	.	62
		14R-4, 97-99	123.67	1.0	132	91	.	.	.	1	2	.	.	.	9	.	39	.	28	1	52
		14R-5, 97-99	125.17	1.0	211	85	.	.	.	.	24	.	.	.	8	.	43	.	39	.	97
		15R-1, 101-103	128.81	1.0	134	96	.	.	.	.	.	.	.	.	15	.	41	.	25	1	52
		15R-2, 101-103	130.31	1.0	210	75	.	.	.	.	.	.	.	.	3	.	30	.	47	1	79
		15R-3, 101-103	131.81	1.0	115	98	.	.	.	.	.	.	.	.	2	.	46	.	25	.	42
		15R-4, 61-63	132.91	1.0	46	80	.	2	.	.	7	.	.	.	.	.	15	.	6	.	16
		16R-1, 97-99	138.47	0.9	300	100	.	.	.	.	.	.	.	.	.	.	114	.	44	4	138
		17R-1, 97-99	148.17	1.0	144	100	.	.	.	.	.	.	.	.	.	.	44	.	27	3	70
		17R-2, 97-99	149.67	1.0	66	100	.	.	.	.	.	.	.	.	.	.	27	.	5	.	34
		17R-3, 97-99	151.17	1.0	31	100	.	.	.	.	.	.	.	.	.	.	9	.	6	1	15
		18R-1, 102-104	157.92	1.0	2	100	.	.	.	.	.	.	.	.	.	.	1	.	.	.	1
		18R-2, 102-104	159.42	1.0	26	85	1	.	.	.	.	.	.	1	.	.	6	.	5	.	11
		19R-2, 30-32	168.40	0.1	16	0	.	.	.	.	6	.	1	8	1	.	.	.	.	.	.

Fig. 2.   Distribution of the *pseudofibula* plexus and other silicoflagellates in the assemblage from ODP Leg 113 Hole 693A in the Weddell Sea.

ology similar to that of *McCartney and Wise* [1990] and *McCartney and Harwood* [1992]. The entire slide is counted unless 300 silicoflagellate specimens are found first. To provide the reader with a better understanding of the absolute silicoflagellate abundance, the percentage of a slide that was counted is included with the counts. This percentage is an estimate determined visually. We recommended that future silicoflagellate work use this or similar methods for expressing absolute abundance.

For purposes of this study, plexids are considered to be "very abundant" when the total silicoflagellate count exceeds 100 specimens on a slide. The plexids are considered "dominant" when their number exceeds 50% of the total silicoflagellate count. The bottom and top of the plexid interval is defined as the lowest and highest samples that have plexid abundances greater than 2% of the silicoflagellate count; the interval thicknesses given in Table 1 are calculated from the mid-

points between the boundary and the next counted sample. The estimated age of the plexid-dominant interval is determined using age versus depth plots with paleomagnetic control when available.

Taxa considered in this paper are listed in Appendix A; taxonomic problems and distinctions are discussed in Appendix B.

GEOGRAPHIC DISTRIBUTION OF THE PLEXID MORPHOLOGIES

DSDP and ODP sites in the Southern Ocean where plexids are known to occur are shown in Figure 5, which is based on data summarized in Table 1 and the new information provided in Figures 2–4. The Southern Ocean drill holes that yielded plexids can be divided into three categories, or classes, as follows: (class I) sites where the plexids are both very abundant (commonly 100 silicoflagellates per microscope slide) and dominant

Age	Pseudofibula plexus	Core, section, interval (cm)	Depth (mbsf)	Number of slides examined	Total number of specimens	% pseudofibula plexus	Bachmannocena diadon	Dictyocha aspera (s. ampl.)	D. fibula (s. ampl.)	D. pentagona	D. stapedia (asperid)	D. stapedia (fibulid)	Distephanus boliviensis	D. boliviensis (7 sides)	D. boliviensis (multiwindowed)	D. crux (s. ampl.)	D. speculum pentagonus	D. speculum speculum (6 sides)	D. speculum speculum f. notabilis	D. speculum speculum f. pseudofibula	D. speculum f. varians
Hole 699A																					
early Pliocene		6-1, 33-35	46.93	1.0	8	0	.	.	.	.	.	1	6	1	.	.	.	.	.	.	.
		6-3, 57-59	50.17	0.8	300	0	.	1	.	.	2	5	150	3	2	125	11	.	.	.	1
		6-5, 58-60	53.18	0.1	300	0	.	.	.	1	.	.	269	5	2	.	23	.	.	.	.
		7-1, 58-60	56.68	1.0	210	1	29	1	9	.	2	3	152	1	1	.	8	2	1	1	.
	▨	7-2, 58-60	58.18	1.0	35	14	.	1	.	.	8	10	6	.	.	1	.	4	3	.	2
late Miocene	▨	7-3, 56-58	59.66	1.0	28	32	.	1	.	.	.	1	6	1	1	.	1	8	4	2	3
		7-4, 56-58	61.16	1.0	2	0	.	.	.	.	.	.	.	5	.	.	.	2	.	.	.
		7-5, 56-58	62.66	1.0	7	0	2	1	.	.	.	.	.	.	.	.	.	4	.	.	.
		8-1, 56-58	66.16	1.0	42	0	.	.	.	.	.	.	.	.	.	5	1	36	.	.	.
Hole 704B																					
		25-4, 43-45	228.10	1.0	18	0	.	5	.	.	.	.	13	.	.	.	.	2	.	.	.
		25-5, 43-45	229.60	1.0	17	0	.	3	.	.	.	.	12	.	.	.	.	2	.	.	.
		25-6, 43-45	231.10	1.0	102	1	.	15	14	.	3	32	25	.	.	.	1	11	.	.	1
	▨	26-1, 43-45	233.10	1.0	25	12	.	.	.	.	.	.	12	.	.	.	.	10	.	.	3
	▨	26-2, 48-50	234.70	1.0	93	38	.	.	.	.	.	.	36	1	.	.	.	21	1	14	20
	▨	26-3, 48-50	236.20	1.0	42	10	.	.	.	.	.	.	11	.	.	.	.	27	2	1	1
late Miocene	▨	26-4, 48-50	237.70	1.0	21	10	2	2	.	.	.	8	.	.	.	1	.	6	.	.	2
	▨	27-1, 48-50	242.70	1.0	9	22	.	.	.	.	.	.	1	.	.	.	1	5	1	.	1
	▨	27-2, 61-63	244.30	1.0	11	18	.	.	.	.	.	.	1	3	.	2	.	3	1	.	1
	▨	27-3, 61-63	245.80	1.0	23	9	.	.	.	.	.	.	2	.	.	.	.	19	.	1	1
	▨	27-4, 61-63	247.30	1.0	65	11	4	17	.	.	.	.	.	8	.	1	.	28	.	2	5
		27-5, 48-50	248.70	1.0	12	0	8	1	.	.	.	.	.	.	.	.	1	2	.	.	.
		27-6, 61-63	250.30	1.0	70	1	54	1	.	.	.	.	.	.	2	.	3	9	.	.	1
		27-7, 61-63	251.80	1.0	7	0	7	1	.	.	.	.	.	.	.	.	.	.	.	.	.

Fig. 3.   Distribution of the *pseudofibula* plexus and other silicoflagellates in the assemblage from ODP Leg 114 Hole 699A and Hole 704B in the Subantarctic Atlantic Ocean.

(>50% of silicoflagellates), (class II) sites where plexids are commonly dominant but not very abundant, and (class III) sites where silicoflagellates are present but rare.

As is shown in Figure 5, the geographic distribution of the *pseudofibula* plexus seems to show a high correlation between abundance and dominance of the morphologies and proximity to the Antarctic continent. In general, three geographic provinces can be recognized on the basis of distribution of sites by class.

Sites 689, 690, 693, and 751 fall into the class I province. At these locations the plexid morphologies are often extremely abundant; indeed, 300 specimens are sometimes found in one or two transects across a slide. The plexids can be so predominant that nonplexid morphologies are uncommon. The four sites form a narrow band adjacent to the Antarctic continent in the area of the Weddell Sea and the Kerguelen Plateau. The

Weddell Sea sites are located along the Antarctic margin or on the offshore Maud Rise, where topographic upwelling produces seasonal polynya within the present-day ephemeral sea ice [*DeFelice and Wise*, 1981]. The high abundance and dominance of plexids in Hole 751A on the Kerguelen Plateau might also be attributed to topographic upwelling, yet it appears to be somewhat anomalous in that it is surrounded by other holes in which the plexus is not as abundant or dominant. Hole 751A, however, was drilled in much shallower water than the holes that recovered the interval nearby or south of it. Numerous workers [e.g., *Schrader*, 1972] have commented that silicoflagellates may be especially susceptible to dissolution because of their fragile construction of hollow skeletal elements; this characteristic may explain the higher abundance and dominance of the plexus at this relatively shallow water site.

Alternatively, silicoflagellate occurrences within the

Age	Pseudofibula plexus	Core, section, interval (cm)	Depth (mbsf)	Number of slides examined	Total number of specimens	% pseudofibula plexus	Bachmannocena diodon	Dictyocha aspera (s. ampl.)	D. fibula (s. ampl.)	Distephanus boliviensis	D. crux (s. ampl.)	D. pentagona	D. speculum speculum (6 sides)	D. speculum speculum (7 sides)	D. speculum speculum (8 sides)	D. speculum speculum (multiwindowed)	D. speculum speculum f. notabilis	D. speculum speculum f. pseudocrux	D. speculum speculum f. pseudofibula	D. speculum speculum f. pseudopentagonus	D. speculum speculum f. varians
Hole 745B																					
early Pliocene (NSOD 13)		20H-6, 60-62	175.60	1.0	6	0	.	.	.	.	.	4	2	.	.	.	.	.	.	.	.
		21H-2, 60-62	179.10	1.0	42	5	.	1	.	.	5	4	27	1	.	2	.	1	.	.	1
		21H-3, 60-62	180.60	1.0	10	0	.	.	.	.	.	2	8	.	.	.	.	.	.	.	.
		21H-4, 60-62	182.10	1.0	76	7	.	.	.	23	.	10	37	1	.	.	1	.	1	.	3
		21H-6, 60-62	185.10	1.0	70	3	.	.	.	.	.	4	61	2	.	.	.	.	.	.	2
		22H-2, 60-62	188.60	1.0	40	20	.	.	.	3	.	6	23	.	.	.	1	.	3	.	4
		22H-4, 60-62	191.60	1.0	26	39	.	4	.	1	.	.	11	.	.	.	1	.	3	.	6
late Miocene (NSOD 12)		22H-6, 60-62	194.60	1.0	32	59	.	2	5	.	.	.	6	.	.	.	1	.	5	.	13
		23H-2, 60-62	198.10	1.0	28	79	.	.	.	.	.	.	5	.	1	.	8	.	6	.	8
		23H-4, 60-62	201.10	1.0	6	33	.	.	.	.	.	1	3	.	.	.	.	.	1	.	1
		23H-6, 60-62	204.10	1.0	21	48	.	.	.	.	.	.	11	.	.	.	.	.	4	.	6
		24H-2, 60-62	207.60	1.0	15	73	.	.	.	.	.	.	4	.	.	.	5	.	.	1	5
		24H-4, 60-62	210.60	1.0	3	100	.	.	.	.	.	.	.	.	.	.	1	.	1	.	1
		24H-6, 60-62	213.60	1.0	10	90	.	.	.	.	.	.	1	.	.	.	6	.	.	.	3
Hole 746A																					
		4H-1, 60-62	165.40	1.0	63	56	.	1	.	.	.	2	3	1	1	.	15	.	10	.	30
		4H-3, 60-62	168.40	1.0	17	34	.	1	3	.	.	.	9	.	.	.	1	.	1	.	2
		4H-5, 60-62	171.40	1.0	2	100	.	.	.	.	.	.	.	.	.	.	.	.	1	.	1
		5H-1, 60-62	174.90	1.0	9	22	.	4	2	.	.	.	.	1	.	.	1	.	.	.	1
late Miocene (NSOD 12)		5H-3, 60-62	177.90	1.0	29	45	10	2	.	.	.	.	3	1	.	.	4	.	2	.	7
		5H-5, 60-62	180.90	1.0	0	0	.	.	.	.	.	.	.	.	.	.	.	.	.	.	.
		5H-7, 60-62	183.90	1.0	3	33	1	.	.	.	.	.	1	.	.	.	.	.	1	.	.
		6H-1, 65-67	184.45	1.0	8	38	4	.	.	.	.	.	1	.	.	.	2	.	1	.	.
		6H-3, 65-67	187.45	1.0	8	25	5	.	.	.	.	.	1	.	.	.	1	.	1	.	.
		6H-4, 65-67	188.95	1.0	7	43	1	.	.	.	.	.	2	.	.	.	2	.	1	.	.
		6H-5, 65-67	190.45	1.0	4	0	.	.	.	.	.	.	4	.	.	.	.	.	.	.	.
		6H-6, 68-70	191.98	1.0	2	0	1	.	.	.	.	.	.	.	.	.	.	.	.	.	.

Fig. 4. Distribution of the *pseudofibula* plexus and other silicoflagellates in the assemblage from ODP Leg 119 Hole 745B and Hole 746A on the Kerguelen Plateau in the southern Indian Ocean. NSOD, Neogene Southern Ocean diatom zone of *Baldauf and Barron* [1991].

plexid interval of Hole 693A, the thickest interval for any class I site, are given in Figure 2 (see *McCartney and Wise* [1990, Table 7] for additional counted samples from this interval). The plexids are very abundant throughout most of the interval, although abundances decline considerably where there are large quantities of terrigenous clastic grains. The plexid morphologies are extremely abundant near the top of the interval. Silicoflagellates also remain extremely abundant a few meters above the plexus (see sample 113-693A-12-7, 43–45 cm) but then decline to more normal levels. In the topmost sample of the interval (693A-13-1, 43–45 cm) the relative abundance of the plexids declines in favor of *Distephanus boliviensis*, which is larger in size; this also

occurs in sample 114-699A-7-1, 58–60 cm (Figures 3 and 4). This pattern of increasing plexid abundance upsection with larger, abundant *D. boliviensis* near the top appears to be a characteristic trend for the *pseudofibula* plexus, as similar patterns are also found in holes 690B and 695A and elsewhere.

Class II sites generally occur farther from the continent than the four sites mentioned above and are characterized by lower absolute abundances and diminished plexid dominance. Most of the Southern Ocean sites drilled through the upper Miocene/lower Pliocene during DSDP (through 1980) encountered this province (Figure 5), which explains why the plexus did not attract greater attention before ODP Leg 113 encountered three

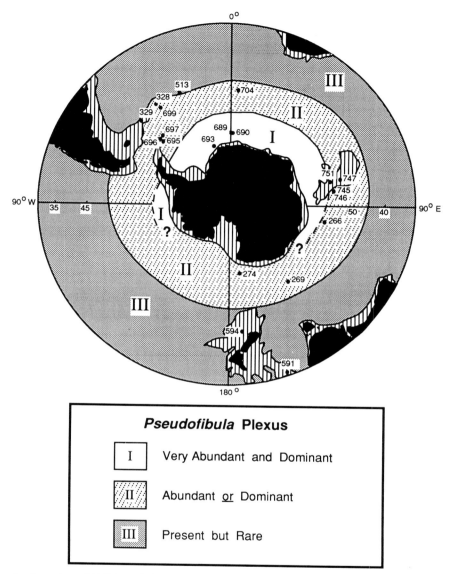

Fig. 5. Location of DSDP and ODP drill holes where the *pseudofibula* plexus has been found. Vertical hachured pattern denotes submerged features less than 3000 m deep. Holes where plexids are generally both abundant (>100 specimens per slide) and dominant (>50% of total silicoflagellates) are designated class I. Holes were plexids are commonly found but are not abundant and dominant are designated class II. In class III holes, plexids are present but rare.

class I sites in 1987. Data for holes 699A, 704B, 745A, and 746A, for which abundance counts and distribution charts have not been published previously, are given in Figures 3 and 4. Plexids are generally uncommon (often fewer than 10 specimens per slide) and are outnumbered by other *Distephanus* or *Dictyocha* taxa. While the silicoflagellates are often very abundant immediately above the plexid interval, they are not as abundant within the interval; this is in strong contrast to the class I sites.

Plexids are rare at locations north of 45°S latitude,

and the interval is usually unrecognizable. This we call the class III province, which is a region where members of the *Pseudofibula* plexus are not useful for biostratigraphy.

Aside from the general trend discussed above of decreasing abundance and dominance of the *pseudofibula* plexus away from the continent, there are also concomitant changes in the overall silicoflagellate assemblage within the interval. Near the continent the plexids are associated almost exclusively with typical six-sided *Distephanus*, although *Bachmannocena* can be sporadically abundant. Further away from the continent, the plexids are commonly associated with various *Dictyocha* skeletons, *Distephanus crux*, and other taxa. *Dictyocha* is considered a warm water indicator and has been used in ratio with other taxa to estimate Pliocene Southern Ocean paleotemperatures [*Ciesielski and Weaver*, 1974].

In general, the total abundance of silicoflagellates and the relative abundance of the plexids decrease rapidly away from the Antarctic continent. In addition, the thickness of the stratigraphic interval and the time duration represented by the *pseudofibula* plexus appear to decrease away from the continent, although this trend is difficult to judge because of variations in sediment accumulation rates, core recovery, and the geologic completeness of the sections. This variation suggests that the upper and/or lower boundaries of the plexus, which are sometimes used to delineate a biostratigraphic zone, are not time synchronous.

PALEOENVIRONMENTAL INTERPRETATIONS

The occurrence of the *pseudofibula* plexus in ODP Leg 113 sediments has been described in detail by *McCartney and Wise* [1990]. They noted that in the Weddell Sea, the plexus is closely associated with the presence of dropstones and ice-rafted debris. It appears to have developed toward the end of the late Miocene–earliest Pliocene Antarctic glaciations and to have disappeared shortly thereafter. These glaciations appear to have overlapped in time the Messinian salinity crisis. Thus the morphologic variability within the plexus may be an ecophenotypic response to one or a combination of environmental factors, such as reduced salinity, dissolved silica, or rapid changes in upwelling or climate.

As was mentioned above, precise correlation of the plexus among sites is difficult, particularly in class I and II provinces, owing to the prevalence of disconformities and the lack of a carbonate stratigraphy at most sites. Age dates assigned to cores from these provinces in the *Proceedings of the Ocean Drilling Program Scientific Results* volumes are based primarily on diatom stratigraphy. Pliocene occurrences are noted only at sites relatively near the continent, such as sites 689/690 and 745/746. The absence of Miocene occurrences of the plexus at sites 689/690 is due to disconformities [*McCartney and Wise*, 1990], but a hiatus would not explain the absence of Pliocene occurrences farther away from the continent, such as at Site 704. This trait indicates that the upper boundary of the plexus is diachronous.

Only at Site 704, on Meteor Rise near the northern extremity of the class II province (Figure 5), has the *pseudofibula* plexus been captured in a continuous pelagic-carbonate section. Sedimentation rates across the interval of the plexus vary widely between 10.0 and 62.5 m/m.y. [*Müller et al.*, 1991], sufficient nonetheless

to yield excellent high-latitude calcareous and siliceous biostratigraphies plus stable isotope and magnetostratigraphies. Although this occurrence of the plexus represents only a distal feather edge of the plexus, the site provides an opportunity to calibrate its stratigraphic boundaries in the Subantarctic and to correlate abundance peaks of the plexids with regional and global events.

Stratigraphic and paleogeographic events during the late Miocene–earliest Pliocene at Site 704 have been described and discussed in detail in the ODP Leg 114 scientific reports by *Froelich et al.* [1991], *Hodell et al.* [1991], *Mead et al.* [1991], and *Müller et al.* [1991]. The major events and their regional and global correlations are outlined in Figures 6 and 7, which are modified from *Müller et al.* [1991], their Figures 6 and 4, respectively, to include the occurrence of the *pseudofibula* plexus at Site 704. According to the Leg 114 authors cited above, between 9.8 and 6.4 Ma, carbonate contents were high with little variability and sustained productivity by calcareous plankton. Decreased carbonate (40%) along with the first significant biogenic opal between 8.45 and 8.2 Ma signaled the first increased cooling at this site that accompanied the late Miocene West Antarctic glaciations (Figure 6). The interval from 6.3 to 4.5 Ma is characterized by low carbonate values with high variability (61.17%), suggesting strong fluctuations in the production and/or dissolution of carbonate. The onset of this interval is marked by a decrease in carbonate values that divides the carbonate record into two distinct intervals as indicated in Figure 7. This delineation coincides with a well-defined 1.0°/oo decrease in $\delta^{13}C$ mean values for both planktonic and benthic foraminifers in Chron C3AR, the well-known "carbon shift" recorded at this site as well as in carbon isotope stratigraphies worldwide. The benthic $\delta^{18}O$ record indicates a strong glacial interval from 5.8 to 5.4 Ma (Figure 7). This interval is followed by an interval of intense carbonate dissolution and low surface productivity from 5.35 to 4.77 Ma during the earliest Gilbert Chron, which ended abruptly when carbonate values increased to 75.3 ± 5.2% at this site (Figure 6). The low carbonate contents and highly variable surface water conditions between 6.3 and 4.8 Ma correlate with the Messinian salinity crisis in the Mediterranean.

If we apply the age model developed by the Leg 114 authors for the section at Site 704, the *pseudofibula* plexus first appears at about 6.3 Ma (Figure 7), near the beginning of the Messinian and during the global "carbon shift." The first abundance peak, although modest (seven specimens), occurs at 247.31 m. The plexids are accompanied by a substantial number of *Dictyocha*, the warm water silicoflagellate indicator (Figure 3). Thereafter, the number of plexids remains low until a major peak is developed at 234.68 m (Figures 3 and 7). Although plexids vie with *Distephanus boliviensis* for dominance of the assemblage, no *Dictyocha* are present

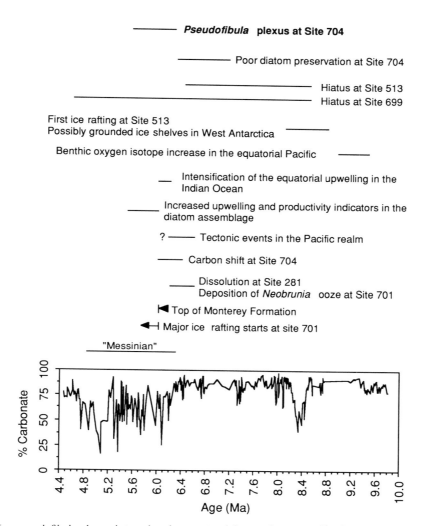

Fig. 6. *pseudofibula* plexus interval and percent calcium carbonate at Site 704 plotted against the temporal distribution of late Miocene–earliest Pliocene depositional, isotopic, and erosional events in the Southern Ocean and elsewhere [after *Müller et al.*, 1991, Figure 6].

in this sample. Dictyochids do dominate the assemblage at 231.13 m, however, which is considered to lie just above the plexid interval, although it contains one specimen of the *pseudofibula* plexus assemblage.

We can interpret these occurrences in Hole 704B in light of the extensive paleoenvironmental information provided by the Leg 114 authors cited above. The first minor peak at 247.31 m, near the base of the Messinian, corresponds to the first in the series of high-amplitude changes in several sedimentary and geochemical parameters plotted in Figures 6 and 7 that includes a major increase in ice rafting at this site. These strong fluctuations indicate instabilities in a paleoceanographic/paleoclimatic system [*Hodell et al.*, 1991]. In addition to the decrease in percent carbonate, the beginning of the period of extreme instability is denoted by anomalously high-amplitude planktonic $\delta^{18}O$ variations (<1‰). Al-

though our sample intervals are slightly different, two closely spaced samples with low planktonic $\delta^{18}O$ values closely bracket the initial plexid peak at 247.31 m (horizontal arrows in Figure 7). As these plexids are matched by a *Dictyocha* peak (Figure 3), a warm event is clearly indicted. Indeed, *Hodell et al.* [1991] speculated that these low oxygen isotope values (0.33°/oo and 0.48°/oo) and others farther upsection may denote brief deglacial events that injected large volumes of meltwater into the Southern Ocean. If so, then the *pseudofibula* plexus may well be a response to reduced salinities as suggested by *McCartney and Wise* [1990]. *Hodell et al.* [1991] and *Müller et al.* [1991] further correlate the anomalous low planktonic $\delta^{18}O$ minimum at 6.15 Ma with the deposition of laminated organic-rich *Neobrunia* ooze in ODP Leg 114 Hole 701 and with an *Ethmodiscus* ooze at DSDP Site 520 in the South Atlantic. They

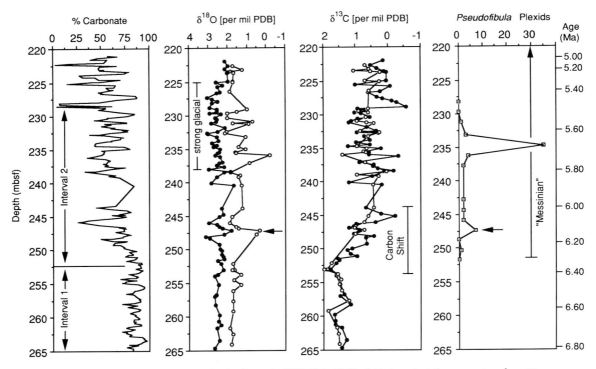

Fig. 7. Occurrence of *pseudofibula* plexus in ODP Hole 704B plotted against the percent carbonate and stable isotope records for the uppermost Miocene–lowermost Pliocene [after *Müller et al.*, 1991, Figure 4]. Stable isotope measurements are for planktonic (open circles) and benthic (solid circles) foraminifers. Note the variability in the percent carbonate (interval 2) and stable isotope values after 6.3–6.4 Ma. Horizontal arrows mark the initial *pseudofibula* plexid peak and a $\delta^{18}O$ anomaly centered at 247.5 m.

speculate that suboxic bottom waters may have formed in response to a meltwater lid that temporarily halted the production of Antarctic Bottom Water during rapid deglacial events in West Antarctica. This interpretation appears to be reasonable because the isotopic minimum in the planktonic record at 6.15 Ma is also mirrored in the benthic record, which suggests that a possible ice volume reduction may have accompanied the major warming event.

The major *pseudofibula* plexus peak up column at 234.68 m falls within the intense glacial episode recorded by benthic $\delta^{18}O$ (Figure 7). The average values during this episode are nearly equal to those of the Holocene. This is also within an interval of upwelling according to diatom analyses (see J. Fenner cited in the work of *Müller et al.* [1991]). An anomalously low planktonic $\delta^{18}O$ value is recorded at 235.91 m, just 1.23 m below the major *pseudofibula* plexus peak. Again, our sample intervals do not coincide with those selected for stable isotope analyses. In addition, no *Dictyocha* occur with the plexids at this level, and the planktonic anomaly is not mirrored in the benthic record.

It is not clear, therefore, whether or not the major plexid peak in Figure 4 represents a warm meltwater

event(s), as appears to be the case with the initial minor peak at 247.31 m. The difficulty in interpreting the major peak could be due to the fact that the stable isotope and silicoflagellate analyses were run on different samples. Closer, coordinated sampling would probably be necessary to record brief, but significant, fluctuations in silicoflagellate abundances produced by meltwater events. The close proximity of the major plexid peak to the most extreme planktonic stable isotope anomaly within the interval displayed in Figure 7, however, suggests a possibility that these unusual silicoflagellate morphologies are likely caused by environmental factors. We can only speculate that major plexid blooms may well have been triggered by low salinities caused by meltwater events, which in turn resulted from an unstable Antarctic ice sheet in a state of flux. Nutrient upwelling, as indicated by diatoms through the strong glacial interval, might also provide a second necessary condition for a major plexid bloom. If the plexids represent an opportunistic taxon group, as their numbers indicate, then meltwater injection into a high-nutrient environment might have produced the bizarre array of morphotypes that constitute the *pseudofibula* plexus. Closely spaced sampling for silicoflagellates coordinated with stable isotope analyses would be necessary to further test this hypothesis.

SUMMARY AND CONCLUSIONS

With a few exceptions, the distribution of the upper Miocene–lowermost Pliocene silicoflagellate *pseudofibulid* plexus in time and space is concentrated in waters proximal to the Antarctic continent, which suggests some form of ecologic control. Because the morphologies exhibited occur elsewhere in the geologic record dating back to the Eocene, the forms are not accorded species rank, but they are classified as variants of species or subspecies. Their occurrence in Antarctic waters seems to correlate with the late Miocene–earliest Pliocene glaciations, which involved the formation of an apparently unstable West Antarctic Ice Sheet that suffered episodes of deglaciation as well as advance. High-amplitude fluctuations in many sedimentary and geochemical indices during this interval denote a time of transition for the Antarctic paleoceanographic/paleoclimatic system. It was in this rapidly changing environment that the *pseudofibula* plexus flourished.

Site 704 is the only site yet available in which the *pseudofibula* plexus occurs in a continuous, well-dated pelagic carbonate sequence. Its first abundance peak at that site, just above the base of the Messinian Stage, corresponds with a major interglacial event recorded in the planktonic and benthic foraminiferal isotopic record. This interglacial episode is believed to have produced significant melting of the ice sheet and the injection of low-salinity meltwaters across the surface of the Southern Ocean. The major abundance peak for the plexus, which occurred about 450 kyr after the first, follows closely another anomalous negative excursion in the planktonic isotopic record during a time of otherwise intense glaciation and upwelling. Both the upwelling of nutrients and the injection of low-salinity meltwater may have fostered opportunistic blooms of the *pseudofibula* plexus. At the conclusion of the latest Miocene–earliest Pliocene glaciations, however, conditions apparently stabilized and the *pseudofibula* plexus disappeared abruptly from the Southern Ocean.

APPENDIX A: TAXA IN THIS PAPER

Taxa considered in this paper are given in alphabetical order of generic epithets. Discussions of taxonomy are given in Appendix B and in the work of *McCartney and Wise* [1990], who provide bibliographic references for the taxa below:

Bachmannocena diodon (Ehrenberg) Bukry, 1987; *Dictyocha fibula* var. *aspera* Lemmermann, 1901; *Dictyocha fibula* Ehrenberg, 1939, fide Loeblich et al., 1968; *Dictyocha pentagona* (Schultz) Bukry and Foster, 1973; *Dictyocha stapedia* Haeckel, 1887; *Distephanus boliviensis* (Frenguelli) Bukry, 1975; *Distephanus crux* (Ehrenberg) Locker, 1974; *Distephanus speculum pentagonus* (Lem-

mermann) Bukry, 1976; *Dictyocha speculum speculum* Ehrenberg, 1840; *Distephanus speculum speculum* f. *pseudofibula* Schulz, 1928; *Distephanus speculum speculum* f. *varians* Gran and Braarud, 1935; *Distephanus speculum speculum* f. *notabilis* (Locker and Martini) McCartney and Wise, 1990; *Distephanus speculum speculum* f. *pseudocrux* Schulz, 1928; *Distephanus speculum speculum* f. *pseudopentagonus* McCartney and Wise, 1990.

APPENDIX B: TAXONOMIC DISTINCTIONS

One problem in examining the plexid assemblage is the distinction between *Distephanus speculum speculum* and *D. boliviensis*. There is often some difficulty in distinguishing between these taxa, since both are quite variable and have similar six-sided shapes. These taxa present a special problem, however, in the plexid interval found in the Southern Ocean. *Distephanus boliviensis* generally has a more circular basal ring with more equant spines, but the wide variability in both taxa causes overlaps. The easiest distinguishing characteristic is size, with *D. boliviensis* generally being larger and having more robust elements. In the *pseudofibula* plexus interval, however, both taxa are smaller than usual, and the size ranges can overlap more than usual; the *D. speculum* also tend to have circular basal rings and equant spines. This variation makes classification somewhat subjective; in this study the largest, most robust skeletons are counted as *D. boliviensis*, but some members of that species are probably included with the *D. speculum* count. Interestingly, the size ranges of the two taxa become considerably more distinct near the top of the plexid interval where the relative abundance of *D. boliviensis* increases dramatically.

The problems in distinguishing between *Distephanus speculum* and *D. boliviensis* in the Antarctic Miocene/Pliocene apply also to the plexid morphologies. Plexids occur for *D. boliviensis* (see *McCartney and Harwood* [1992], Table 5, sample 120-751A-3H-1, 10–11 cm), although they are seldom abundant. Plexid morphologies do appear to occur for *D. boliviensis* in the *pseudofibula* plexus interval, but they are much less abundant than the smaller skeletons associated with *D. speculum*. Because of the general difficulties in distinguishing between the two taxa in the first place, and for convenience, all of the plexid morphologies are counted as varieties of *Distephanus speculum speculum*.

Acknowledgments. We thank the organizers of the JOI/USSAC conference on Southern Ocean Paleoceanography and Climate Change for the opportunity to present these results. Silicoflagellate slides or abundance information were generously provided by John Barron, Paul Ciesielski, and Jim Ling. Daniel W. Müller kindly provided the age assignments for the percent carbonate data plotted in Figure 6, and Wuchang Wei read an early version of the manuscript and made helpful suggestions. The manuscript was further improved by the conscientious and constructive reviews by Paul Ciesielski, Jim Ling, David A. Hodell, and Sigurd Locher and careful editing by Diana M. Kennett. Li Liu prepared most of the samples, which were provided by the Ocean Drilling Program. Liu and James J. Pospichal skillfully assisted in preparing the figures. Support was provided by NSF grant 7-5-33500 to K. M., which made possible the microscope work, and by NSF grant DPP 91-18480 to S. W. W. This is publication 003 of the Undergraduate Micropaleontology Research Laboratory at the University of Maine, Presque Isle.

REFERENCES

Baldauf, J. G., and J. A. Barron, Diatom biostratigraphy: Kerguelen Plateau and Prydz Bay regions of the Southern Ocean, *Proc. Ocean Drill. Program Sci. Results*, *119*, 547–598, 1991.

Busen, K. E., and S. W. Wise, Silicoflagellate stratigraphy, Deep Sea Drilling Project Leg 36, *Initial Rep. Deep Sea Drill. Proj.*, *36*, 697–743, 1977.

Ciesielski, P. F., Biostratigraphy and paleoecology of Neogene and Oligocene silicoflagellates from cores recovered during Antarctic Leg 28, Deep Sea Drilling Project, *Initial Rep. Deep Sea Drill. Proj.*, *28*, 625–691, 1975.

Ciesielski, P. F., Relative abundances and ranges of select diatoms and silicoflagellates from sites 699 and 704, Subantarctic South Atlantic, *Proc. Ocean Drill. Program Sci. Results*, *114*, 753–778, 1991.

Ciesielski, P. F., and F. M. Weaver, Early Pliocene temperature changes in the Antarctic seas, *Geology*, *2*, 511–516, 1974.

DeFelice, D. R., and S. W. Wise, Surface lithofacies, biofacies, and diatom diversity patterns as models for delineation of climatic change in the southeast Atlantic Ocean, *Mar. Micropaleontol.*, *6*, 29–70, 1981.

Froelich, P. N., P. N. Malone, D. A. Hodell, P. F. Ciesielski, D. A. Warnke, F. Westall, E. A. Hailwood, D. C. Nobes, J. Fenner, J. Mienert, C. J. Mwenifumbo, and D. W. Müller, Biogenic opal and carbonate accumulation rates in the Subantarctic South Atlantic: The late Neogene of Meteor Rise Site 704, *Proc. Ocean Drill. Program Sci. Results*, *114*, 515–550, 1991.

Hodell, D. A., D. W. Müller, P. F. Ciesielski, and G. A. Mead, Synthesis of oxygen and carbon isotopic results from Site 704: Implications for major climatic-geochemical transitions during the late Neogene, *Proc. Ocean Drill. Program Sci. Results*, *114*, 475–480, 1991.

Locker, S., and E. Martini, Silicoflagellates and some sponge spicules from the southwest Pacific, Deep Sea Drilling Project, Leg 90, *Initial Rep. Deep Sea Drill. Proj.*, *90*, 887–924, 1986.

McCartney, K., and D. M. Harwood, Silicoflagellates from Ocean Drilling Program Leg 120 on the Kerguelen Plateau, Southwest Indian Ocean, Ocean Drilling Program, Scientific Reports, Leg 120, *Proc. Ocean Drill. Program Sci. Results*, *120*, 811–831, 1992.

McCartney, K., and D. E. Loper, Optimized skeletal morphologies of silicoflagellate genera *Dictyocha* and *Distephanus*, *Paleobiology*, *15*, 283–298, 1989.

McCartney, K., and S. W. Wise, Jr., Silicoflagellates and ebridians from Ocean Drilling Program Leg 113: Biostratigraphy and notes on morphologic variability, *Proc. Ocean Drill. Program Sci. Results*, *113*, 729–760, 1990.

Mead, G. A., D. A. Hodell, D. W. Müller, and P. F. Ciesielski, Fine-fraction carbonate oxygen and carbon isotope results from Site 704: Implications for movement of the Polar Front during the late Pliocene, *Proc. Ocean Drill. Program Sci. Results*, *114*, 437–458, 1991.

Müller, D. W., D. A. Hodell, and P. F. Ciesielski, Late Miocene to earliest Pliocene (9.8–4.5 Ma) paleoceanography of the Subantarctic southeast Atlantic: Stable isotopic, sedimentologic, and microfossil evidence, *Proc. Ocean Drill. Program Sci. Results*, *114*, 459–474, 1991.

Schrader, H. J., Kieselsaure-Skelette in Sedimenten des iberomarokkanischen Kontinentalrandes und angrenzender Tiefsee-Ebenen, *"Meteor" Forschungsergeb. Reihe A/B*, *8*, 10–36, 1972.

Shaw, C. A., and P. F. Cielsielski, Silicoflagellate biostratigraphy of middle Eocene to Holocene Subantarctic sediments recovered by Deep Sea Drilling Project Leg 71, *Initial Rep. Deep Sea Drill. Proj.*, *71*, 687–737, 1983.

(Received August 6, 1992;
accepted February 16, 1993.)

THE ANTARCTIC PALEOENVIRONMENT: A PERSPECTIVE ON GLOBAL CHANGE

ANTARCTIC RESEARCH SERIES, VOLUME 60, PAGES 207–250

LATE NEOGENE ANTARCTIC GLACIAL HISTORY: EVIDENCE FROM CENTRAL WRIGHT VALLEY

M. L. Prentice,[1] J. G. Bockheim,[2] S. C. Wilson,[2] L. H. Burckle,[3] D. A. Hodell,[4] C. Schlüchter,[5] and D. E. Kellogg[1]

As a test of the divergent hypotheses for Late Neogene Antarctic climate and East Antarctic Ice Sheet behavior, we examined the surficial geology in central Wright Valley, a major ice-free valley cut into the seaward flank of the Transantarctic Mountains in the McMurdo Sound region. The four major climate episodes that are in evidence involve climates at least slightly warmer than at present and infrequent large-scale glaciation. The oldest deposit on the floor of Wright Valley is the Jason glaciomarine diamicton. On the basis of marine diatoms and the $^{87}Sr/^{86}Sr$ ratio of a shell fragment from the Jason glaciomarine deposit, we suggest that a fjord occupied Wright Valley at 9 ± 1.5 Ma. From the negative $\delta^{18}O$ of the shell fragment, we infer that the Jason Fjord was both warmer and less saline than modern fjords in this region. Hence the Jason Fjord episode represents a warmer-than-present interval with reduced local ice extent. We propose a shallow fjord which implies that the local mountains were less than 400 m below their present elevation at 9 ± 1.5 Ma. Hence East Antarctic Ice Sheet expansions sufficient to invade and overdeepen Wright Valley prior to the Jason Fjord episode achieved at least present ice sheet dimensions. At 5.5 ± 0.4 Ma, Wright Valley was largely ice free and occupied by a fjord in which were deposited the Prospect Mesa gravels. The presence of marine diatoms from the Antarctic Convergence and a thick-shelled pecten with negative $\delta^{18}O$ coupled with the absence of coccolithophores implies that fjord waters were 0°–3°C. Because fjord influx derives from the Ross Sea, the sea level climate of the Ross Sea was at least slightly warmer than it is today. The age range is based on the $^{87}Sr/^{86}Sr$ ratios of the pectens and marine diatoms. We estimate that the maximum uplift of this area since the Prospect Fjord episode is less than 400 m. During the Neogene before 3.9 Ma, a largely wet-based glacier draining the East Antarctic Ice Sheet filled Wright Valley. The distribution and character of the resulting Peleus till suggest that the concurrent climate was warmer than at present. The Peleus glacial episode probably reflects a larger-than-present East Antarctic Ice Sheet. Since deposition of the Peleus till and the Prospect Fjord episode (3.9 Ma), the central valley has experienced both alpine glacier fluctuations and colluviation. The late Neogene evidence is not consistent with extreme climate warming or a continuous polar desert climate such as prevails today. Rather, the evidence is consistent with occasional mild warming intervals and dominance of polar desert conditions.

INTRODUCTION

The Antarctic ice sheet has been a dominant factor in global climate since its inception at at least 35 Ma [*Barrett et al.*, 1987; *Prentice and Matthews*, 1988; *Zachos et al.*, 1992]. Whereas the behavior of this ice sheet during late Quaternary global ice ages and the controls of this behavior are reasonably well known [*Denton et al.*, 1991], little is known about the Tertiary history of the ice sheet. At least three different hypotheses for the late Neogene history of the Antarctic ice sheet and climate have been proposed. One is that the

East Antarctic Ice Sheet has been relatively stable at about its present size under a polar desert climate since ~14 Ma. This hypothesis is based in part on interpretations of (1) deep-sea oxygen isotopic data [e.g., *Savin et al.*, 1975; *Shackleton and Kennett*, 1975], (2) Southern Ocean planktonic microfossil distributions [e.g., *Kennett*, 1978], (3) semiconsolidated tills from the Transantarctic Mountains referred to as the Sirius Group [e.g., *Mercer*, 1978], and (4) polar desert gravel pavements buried by well-preserved airfall ash in the "dry valleys" region of McMurdo Sound [e.g., *Wilch et al.*, 1989; *Marchant et al.*, 1993]. Another hypothesis is that the late Neogene East Antarctic Ice Sheet was dynamic, generally smaller than at present, and was driven by large climatic oscillations. At times during the Pliocene to early Quaternary, climate was very warm in the Antarctic, and the ice sheet was very small. Support for this hypothesis includes interpretations of (1) Antarctic marine diatoms and southern beech (*Nothofagus*) twigs and leaves in some outcrops of the Sirius Group [*Webb*

[1]Department of Geological Sciences and Institute for Quaternary Studies, University of Maine, Orono, 04469.

[2]Department of Soil Science, University of Wisconsin, Madison, 53706.

[3]Lamont-Doherty Earth Observatory of Columbia University, Palisades, New York 10964.

[4]Department of Geology, University of Florida, Gainesville, 32611.

[5]The University of Bern, Bern, Switzerland.

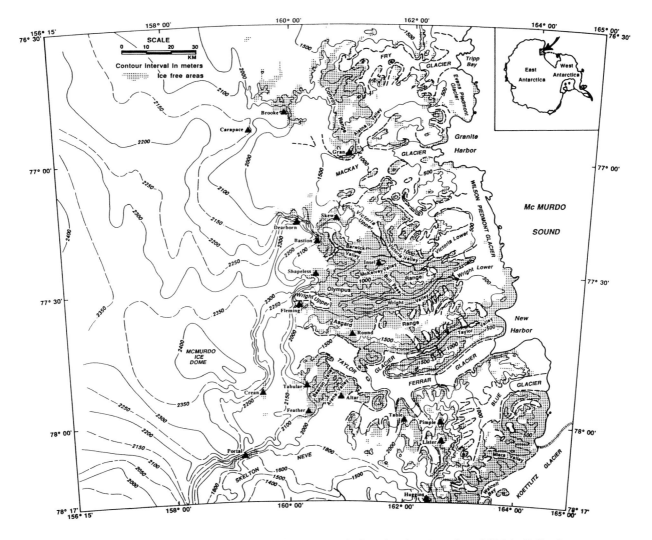

Fig. 1. Generalized topographic map of the McMurdo Sound region. Location of Wright Valley is shown. The Wright Upper Glacier flows from the northeast portion of the McMurdo Ice Dome. The Wright Lower Glacier flows westward from the Wilson Piedmont Glacier.

et al., 1984; Harwood, 1991; Webb and Harwood, 1991], (2) a vertebrate fauna and δ¹⁸O of bivalves from the Vestfold Hills [Quilty, 1991], (3) deep-sea silicoflagellate populations [Ciesielski and Weaver, 1974], and (4) high early Pliocene stands of the sea [e.g., Haq et al., 1988]. The third hypothesis is that the East Antarctic Ice Sheet was dynamic and occasionally slightly larger than at present and driven by a dynamic cool climate [Prentice and Denton, 1988]. This hypothesis draws on interpretations of (1) glacial erosional forms and sediments at the highest Transantarctic Mountain elevations [e.g., Denton et al., 1984; Prentice et al., 1986; Sugden et al., 1991], (2) thick steep-sided wedges of Sirius Group drift [Denton et al., 1991; McKelvey et al., 1991], (3) deep-sea δ¹⁸O data [Prentice and Matthews, 1991], and (4)

climate and ice sheet modeling studies [Prentice et al., 1993]. Each of the three hypotheses has numerous strengths and weaknesses. Late Neogene Antarctic climate and ice sheet history are still poorly understood. In this paper, we test these three hypotheses using the record from the central portion of Wright Valley.

Wright Valley is the central "dry valley" facing the McMurdo Sound sector of the Ross Sea (Figures 1 and 2). This valley is slightly more than 80 km long, 2 km deep, and 9 km wide. It is overdeepened with a minimum elevation of 3 meters above sea level (masl) in the central Lake Vanda basin, a valley-mouth threshold at about 270 masl [Calkin, 1974] and a two-step headwall with 1900 m of relief. Because Pleistocene advances of the trunk glaciers, Wright Upper [Bockheim, 1983] and

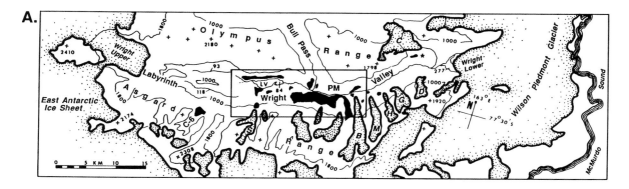

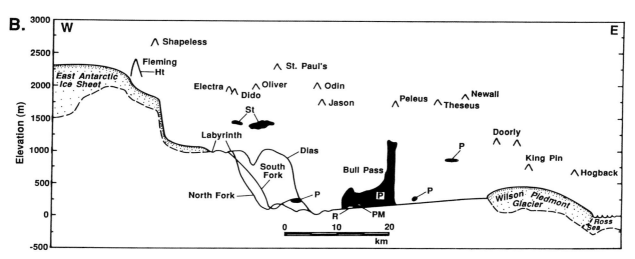

Fig. 2. (a) Sketch map of Wright Valley. Shaded black areas depict outcrops of Peleus till. Alpine glaciers are as follows: S, Sykes; B, Bartley; M, Meserve; H, Hart; G, Goodspeed; D, Denton; and C, Clark. PM is Prospect Mesa. LV is Lake Vanda. C-6 is Njord Valley in the Asgard Range. Just south of Clark Glacier is a patch of Peleus till with an excavation, 8444, marked by an asterisk. Location of Figure 3 is shown. (b) Longitudinal topographic profile of Wright Valley. Shaded black areas marked P represent Peleus till locations. Names along the top are mountain peaks in the Olympus and Asgard ranges on the north and south, respectively. The largest patch of Peleus till exists below Mount Peleus.

Wright Lower [*Nichols*, 1971; *Bockheim*, 1979], as well as the alpine glaciers [*Calkin and Bull*, 1972] did not cover central Wright Valley, Neogene glacial sediments are still exposed there [*Prentice*, 1982]. A major hanging valley, Bull Pass, intersects the steep northern wall of Wright Valley connecting it to the McKelvey-Victoria valley system to the north (Figure 1). A large debris fan on the floor of Wright Valley heads into Bull Pass (Figure 2). Water and sediment gravity flows from Bull Pass have dissected the fan, isolating, in one area, an elongate mesa with 18 m of relief, referred to as Prospect Mesa. The glacial sediments exposed in Prospect Mesa and to the west around Lake Vanda have provided most of the evidence concerning the glacial history of this area (Figure 2).

An important stratigraphic unit exposed in Prospect Mesa is stratified sandy gravel that contains pecten

shells [*Nichols*, 1961, 1965]. We refer to this unit as the Prospect Mesa gravels. The Prospect Mesa gravels are encased in both stratified and massive pebbly mud. *Nichols* [1961, 1971] inferred that all Prospect Mesa sediments were glacially transported from McMurdo Sound [*Calkin et al.*, 1970]. *Webb* [1972, 1974] identified a rich but low-diversity foraminiferal fauna in the Prospect Mesa gravels (his Pecten gravels). Because of the excellent preservation and wide size range of the Prospect gravels microfossils, *Webb* [1972], *Vucetich and Topping* [1972], *Brooks* [1972], and *McSaveney and McSaveney* [1972], among others, reinterpreted Prospect Mesa strata as having been deposited in a fjord on which floated an expanded Wright Upper Glacier. *Vucetich and Topping* [1972] referred to the entire sequence of Prospect Mesa sediments, including the pebbly mud, as the Prospect Formation. The pebbly mud of the Prospect Formation was traced across much of the southern wall of the central valley [*Vucetich and Topping*, 1972; *Webb*, 1972] and underneath alpine glacier drift [*McSaveney and McSaveney*, 1972] (Figure 2).

The age of the Prospect Mesa gravels has long been debated. *Webb* [1972] and *Brady* [1979] proposed that the Prospect Mesa gravels were middle and early Pliocene in age, respectively. Subsequently, *Burckle et al.* [1986] recovered marine diatoms indicating a late Pliocene age from the Prospect gravels.

The Prospect Mesa gravels and the enclosed fossils provide an indication of coastal Antarctic climate. *Webb* [1972] suggested that the waters of the Prospect Fjord were warm, possibly as high as 10°C, and that the fjord existed under "interglacial" climate. *Prentice et al.* [1987] inferred that the waters were no warmer than 5°C on the basis of enclosed marine diatoms. The warmth of the Prospect Fjord and, by extension, the adjacent sector of the Ross Sea has important implications for Antarctic ice extent and global climate. *Ciesielski and Weaver* [1974] proposed that the interglacial conditions reflected by the Prospect Mesa gravels were correlative with a Pliocene interval of exceptionally warm Southern Ocean surface waters that they inferred from silicoflagellate data. *Ciesielski et al.* [1982] hypothesized that the marine incursion that resulted in the Prospect Fjord reflected partial deglaciation of West Antarctica in response to early Pliocene warming. *Pickard et al.* [1988] suggested, on the basis of diatomaceous sands presently just above sea level in the Vestfold Hills, that the volume of the East Antarctic Ice Sheet was much reduced in relation to today during the existence of the Prospect Fjord.

The pebbly mud of the Prospect Formation has been given many interpretations. *Vucetich and Topping* [1972] considered all of it as glaciomarine. *Calkin et al.* [1970] suggested an interglacial mudflow origin. *Prentice* [1982, 1985] and *Prentice et al.* [1985] interpreted the diamicton that overlies the Prospect gravels as basal till. *Prentice* [1982, 1985] and *Denton et al.* [1984] suggested that this till, referred to as Peleus till, was deposited beneath thick ice that submerged the local Transantarctic Mountains. More recently, *Prentice et al.* [1987] presented counter arguments that Peleus till could have been deposited by a temperate valley glacier draining the East Antarctic Ice Sheet.

Glaciomarine sediments, older than the Prospect Mesa gravels, were recovered in a Dry Valley Drilling Project (DVDP) core (Hole 4A) from the bottom of Lake Vanda (Figure 2). Using marine diatoms, *Brady* [1977, 1979, 1982] and *Brady and McKelvey* [1983] inferred that the basal pebbly mud in DVDP Hole 4A was deposited in a fjord in middle to late Miocene time. *Prentice* [1982] referred to this unit as the Jason glaciomarine diamicton and suggested that it was deposited between 15 and 9 Ma also on the basis of marine diatoms. *Burckle et al.* [1986] further revised the age of

Jason diamicton to latest Miocene or early Pliocene time. This deposit is the oldest yet discovered from the floor of a major valley in the Transantarctic Mountains. Hence the Jason diamicton places important constraints on the age of the Antarctic ice sheet [*Denton et al.*, 1984].

The bedrock geology of Wright Valley is pertinent to the glacial history. Precambrian basement, composed of interbedded marbles, hornfelses, and schists, crops out extensively in Wright Valley east of Bartley Glacier [*McKelvey and Webb*, 1962] (Figure 2). Lower Paleozoic acidic plutonic rocks, including granites, gneisses, and diorites, intrude the Precambrian metasediments and crop out throughout the valley as do heterogeneous felsic-to-mafic dike rocks. The gently dipping Devonian-to-Jurassic Beacon Supergroup, a largely nonmarine sequence of sandstones, siltstones, and conglomerates, unconformably overlies the basement complex and is exposed almost exclusively in the mountain ranges bordering the valley west of Bartley Glacier. Sills and dikes of the Jurassic Ferrar Dolerite intrude all rocks mentioned above and crop out on the valley walls much more to the west of Bartley Glacier than to the east.

Numerous cones of subaerially erupted basalt, assigned to the Cenozoic McMurdo Volcanics, are scattered over the southern valley wall east of Bartley Glacier. Scoria from two small basalt cones just below the Goodspeed Glacier [*Nichols*, 1962] (Figure 2) have yielded K/Ar ages of 4.2 Ma [*Fleck et al.*, 1972] and 3.5 Ma [*Armstrong*, 1978]. These cones were inferred to be in situ [*Fleck et al.*, 1972]. A cone in the accumulation zone of Meserve Glacier has a K/Ar age of 2.5 Ma [*Fleck et al.*, 1972] (Figure 2). Cones, inferred to exist under the Bartley and Meserve glaciers [*Nichols*, 1962, 1965; *Denton and Armstrong*, 1968], have yielded basalt with K/Ar ages of 3.4 Ma [*Fleck et al.*, 1972] and 3.75 Ma [*Armstrong*, 1978].

METHODS

We mapped central Wright Valley using aerial photography, ground surveys, observations from over 300 excavations averaging 1–12 m in depth, and information from DVDP Hole 4A (Figure 3). We collected samples of bulk sediment finer than -4ϕ, averaging 1.7 kg, for grain size analysis. To avoid contamination, separate samples were collected for microfossil analysis. Samples of gravel between -4ϕ and -6ϕ were collected for lithologic and shape analysis using sieves. Soil development in each deposit was examined at numerous sites in pits excavated to a depth of 1 m. Morphologic properties of these soils, including depth of oxidation, of ghosts (pseudomorphs), of visible salts, and of cohesion, were measured in the field. In addition, a morphogenetic salt stage [*Bockheim*, 1979] and a weathering stage [*Campbell and Claridge*, 1975, 1987] were assigned.

Sediment samples were sieved in the laboratory at -1ϕ, and sand/mud splits of 30–100 g were extracted, desalted, and dispersed [*Jackson et al.*, 1949]. Gravel and sand were sieved on a Ro-tap shaker [*Folk*, 1974]. Mud was analyzed by the pipette technique [*Galehouse*, 1971]. Frequency curves were constructed according to *Brotherhood and Griffiths* [1947]. Size frequency statistics were calculated by the method of moments [*McBride*, 1971]. Replicate analyses indicate a precision in any size interval better than 3%.

We measured the long, intermediate, and short axes as well as the roundness of washed and numbered gravel according to *Krumbein* [1941]. Maximum projection sphericity was calculated after *Sneed and Folk* [1958], and oblate-prolate index was after *Dobkins and Folk* [1970]. We rated each stone according to the development of eight different characteristics: weathering (state of preservation), broken faces (bruises), glacial marks (polish, striation, and molding), pitting (hollows), carbonate crusts, weathering rinds (quartzite), desert varnish (stain or crust), and ventifaction (flutes and facets). The ratings are zero, one, or two for negligible, significant, and exceptional development, respectively. The sample rating is the sum of individual stone ratings divided by the total number of stones in the sample and multiplied by 100. Stones were assigned to lithologic associations which are related to the bedrock units mapped in Wright Valley. The lithologic composition of each sample is reported in volume percentages. Volume percentages are less subject to the effects of differential stone resistance than are number percentages and so better reflect the amount of rock present.

Shell samples were cleaned and polished, washed in dilute (1 *M*) acetic acid, and ground to a fine powder. Approximately 20 mg of the carbonate powder was dissolved in 0.1 *N* HCl, which is sufficiently dilute to minimize leaching of strontium from noncarbonate material. The sample solution was filtered and the volume reduced by evaporation before passing through standard Dowex 50 × 12 ion exchange columns. The sample was eluted in 2 *N* HCl and dried. The strontium fraction was then loaded as a nitrate onto a single oxidized Ta filament. Isotope ratios were measured on the University of Florida VG 354 triple collector mass spectrometer in the dynamic mode with mass fractionation normalized to $^{87}Sr/^{86}Sr = 0.1194$. For the entire procedure, the blank was less than 1-ng total strontium. Because we collect 100 ratios, our internal precision is usually equal to or better than 1×10^{-5}. Intrarun precision is estimated at 2×10^{-5} on the basis of repeated analyses of SRM-987, which have a mean value of 0.710235. Sample reproducibility based on seven analyses of the same pecten sample is 1.6×10^{-5} (Table 7). All samples were normalized to an average SRM-987 value of 0.710235 by correcting each run for the difference between its SRM-987 values and 0.710235.

The Sr/Ca ratio of pecten shells was determined at the University of Florida and replicated at the University of Maine by atomic absorption. Shell powders remaining after Sr analyses were reacted on line in 100% orthophosphoric acid at 90°C and analyzed in a V G Prism Series II mass spectrometer in the University of Maine Stable Isotope Laboratory according to standard procedures [*Prentice et al.*, 1993]. Precision on carbonate standards is 0.1‰ for $\delta^{18}O$ and $\delta^{13}C$. Calibration to PDB is through both NBS-19 and NBS-20. For the period of analysis, measurements of NBS-20 yield -4.08 ± 0.1 ($\delta^{18}O$) and -1.01 ± 0.09 ($\delta^{13}C$); those of NBS-19, -2.25 ± 0.12 ($\delta^{18}O$) and 1.87 ± 0.1 ($\delta^{13}C$). Diatom slides were prepared at Maine (D. Kellogg) and Lamont-Doherty (L. Burckle) after *Schrader* [1974].

RESULTS

Numerous stratigraphic units were resolved within central Wright Valley and named informally (Table 1 and Figures 3 and 4). Unconsolidated till crops out in patches throughout the central valley and is regarded as the same unit, the Peleus till. Peleus till overlies the Prospect Mesa gravels and a number of physically separated water-laid diamictons. Peleus till is overlain by colluvium, alpine glacier drift, and drift deposited by the Wright Lower Glacier. The principal deposits are presented below in stratigraphic order from oldest to youngest.

Jason Glaciomarine Diamicton

Jason glaciomarine diamicton (JGD) (Table 1) crops out extensively along the north shore of Lake Vanda and was penetrated at DVDP Hole 4A (Figure 3). Stratigraphic sections are exposed in the channels of the debris fans that head into the Olympus Range beneath Mount Jason (Figure 5). The surface of JGD has been reworked in the Holocene by lake ice which formed well-developed beaches up to 50 m above the present lake level. Basal till, interpreted as Peleus till, outcrops higher on the north valley wall and presumably overlies JGD, although the contact between these units was not observed. A 40-cm-thick sliver of JGD crops out just north of the outer moraine of Sykes Glacier at excavation 794 nearly 250 masl (Figure 3). At 794, JGD overlies massive pebbly muddy sand.

Jason glaciomarine diamicton varies from pebbly muddy sand to sandy mud but is commonly slightly pebbly sandy mud (Figure 6). Three different grain size distributions are apparent (Figure 7 and Table 2). At the western end of Lake Vanda, a fine-skewed diamicton with a primary mode at 1.5ϕ (medium sand) crops out at excavation 8352 above a more size-symmetric facies with a primary mode at 4.5ϕ (coarse silt). At the eastern end of the lake, JGD does not exhibit a strong mode and is more poorly sorted.

Gravel in the JGD consists predominantly of granite

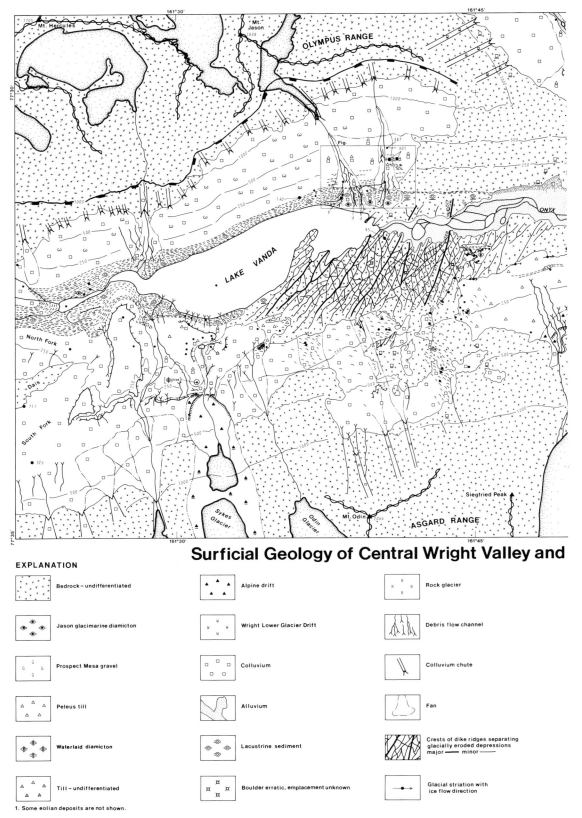

Surficial Geology of Central Wright Valley and

EXPLANATION

Bedrock – undifferentiated		Alpine drift		Rock glacier	
Jason glacimarine diamicton		Wright Lower Glacier Drift		Debris flow channel	
Prospect Mesa gravel		Colluvium		Colluvium chute	
Peleus till		Alluvium		Fan	
Waterlaid diamicton		Lacustrine sediment		Crests of dike ridges separating glacially eroded depressions major —— minor ——	
Till – undifferentiated		Boulder erratic, emplacement unknown		Glacial striation with ice flow direction	

1. Some eolian deposits are not shown.

Fig. 3.   Surficial geologic map of central Wright Valley and southern Bull Pass. The location of inset maps for Prospect Mesa dots represent excavations identified in Figure 4.

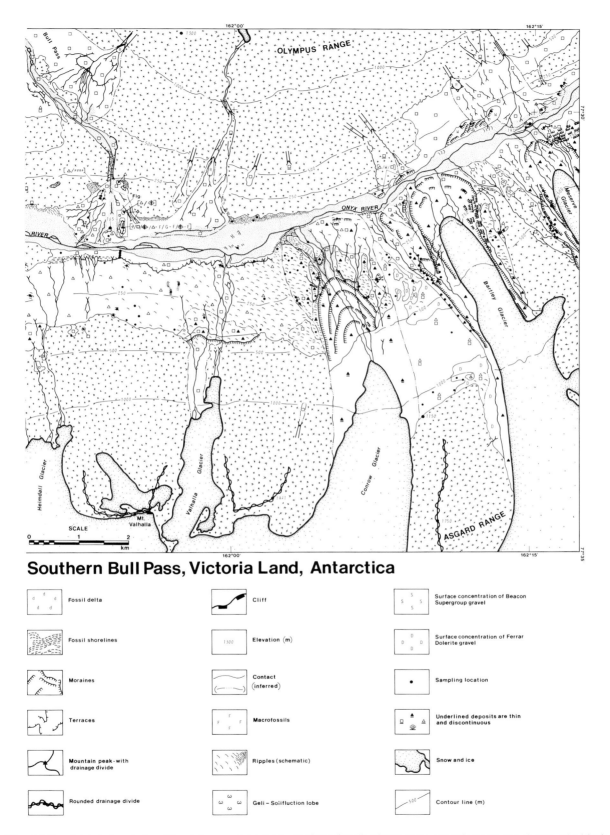

Southern Bull Pass, Victoria Land, Antarctica

Fossil delta	Cliff	Surface concentration of Beacon Supergroup gravel
Fossil shorelines	Elevation (m)	Surface concentration of Ferrar Dolerite gravel
Moraines	Contact (inferred)	Sampling location
Terraces	Macrofossils	Underlined deposits are thin and discontinuous
Mountain peak - with drainage divide	Ripples (schematic)	Snow and ice
Rounded drainage divide	Geli - Solifluction lobe	Contour line (m)

(Figure 14) and debris fans below Mount Jason, where Jason glaciomarine diamicton crops out (Figure 5), are shown. The black

TABLE 1. Lithologic And Biologic Characteristics Of Primary Glacial And Periglacial Deposits In Central Wright Valley.

Deposit/Lithofacies (Max. Observed Thickness)	Color* Texture†	Structure§	Gravel** Lithology and Shape††	Fossil Content§§
Jason glaciomarine diamicton (JGD); (8 m)	pale yellow (5y 7/3); slightly pebbly mud; unimodal: more commonly coarse silt than medium sand	mostly massive minor stratification	mostly granite, dolerite metamorphics; trace limestone; granite: compact-bladed, sub-rounded; slightly weathered; moderately chipped; few glacial features; desert surface features rare	fragmented marine diatoms common; few non-marine diatoms; sponge spicules common; few calcareous worm tubes, small shell fragments; rare radiolaria and silicoflagellate fragments
Onyx ponds waterlaid diamicton (OPWD); (2.5 m)	light olive gray (5y 6/2); slightly pebbly mud; unimodal: fine to coarse silt; gravel distributed unevenly	mostly massive; some crude to moderate stratification; lenticular bedding; some sand beds deformed by pebbles; worm burrows	mostly granite, dolerite; some dike rock; granite: bladed, sub-rounded; slightly weathered, slightly chipped; glacial features common; no desert surface features	marine and non-marine diatoms rare; few sponge spicules
Heimdall glaciomarine diamicton (HGD); (0.75 m)	white (5y 8/2); slightly pebbly mud; unimodal: coarse silt	indistinctly stratified	none recovered	whole and fragmented marine diatoms common; silicoflagellate fragments rare
Prospect Mesa lower waterlaid diamicton:(PMLWD); stratified lithofacies; (SL) (10 m)	pale yellow (5y 7/3); slightly pebbly mud; unimodal: commonly coarse silt	flaser-lenticular bedding; symmetric-asymmetric ripples; graded beds scour-fill cross-bedding; ball-pillow, flame structures; convolutions, faulting; mud laminae conform to boulders, worm burrows	mostly granite; some dolerite, dike rock; granite: compact-bladed; sub-rounded; slightly weathered; slightly chipped; few glacial features; no desert surface features	marine diatoms rare ; few non-marine diatoms; sponge spicules rare; few small shell fragments
Prospect Mesa lower waterlaid diamicton:(PMLWD); massive lithofacies; (ML) (1.5 m)	pale yellow (5y 7/3); slightly pebbly mud; unimodal: commonly coarse silt	generally massive; local indistinct stratification;	same as PMLWD-SL	same as PMLWD-SL
Prospect Mesa lower waterlaid diamicton:(PMLWD); lag lithofacies (LL) ; (1 m)	pale yellow (5y 7/3); sandy granular gravel; unimodal: granule clast-supported	massive to indistinctly stratified	same as PMLWD-SL	pollen spores
Prospect Mesa gravel (PMG); gravel lithofacies (2.5 m)	brownish grey (2.5y 6/4); muddy sandy pebble gravel; bimodal: med sand, pebble: matrix and clast-supported;	massive to stratified; beds: horizontal - inclined; thick to very thickly bedded; normal grading rare; unconformable contacts; 8 beds with shells stacked + parallel to bedding inclined beds dip up to 30°; minor imbrication	mostly granite, dolerite; some dike rocks; granite: compact-bladed, sub-rounded; slightly weathered; slightly chipped; glacial features rare; no desert surface features	few marine, non-marine diatoms; sponge spicules, radiolarian frags rare; rich, low-diversity foram fauna dominated by *Ammoelphidiella antarctica* Webb (1974) from -1φ to 4φ; abundant *Chlamys (tuftensis* (Turner, 1967); valve hts: 47-71 mm, lengths: 44-69 mm;
Prospect Mesa gravel (PMG);	pale yellow (5y 7/3); pebbly mud;	massive to crudely stratified; medium to thick bedding;	same as PMG-GL	marine diatoms rare; few non-marine diatoms;

Unit	Texture/Color	Structure	Lithology / Clasts	Fossils
mud lithofacies (ML); (1 m)	bimodal: medium sand to coarse silt; matrix-supported	2 shell beds, shells in low concentration		sponge spicules rare; few small shell fragments
Peleus till (PT); (6 m)	pale yellow (5y 7/3); pebbly muddy sand; bimodal:fine sand, pebble; matrix-supported; bullet-shaped boulders exhibit strong fabric; moderately compact	not stratified except at base; sandy beds in basal 25 cm; subhorizontal fissility; augen structures; vertical and subhorizontal joint systems.	mostly granite, dolerite;some dike rocks metamorphics,limestone NE of Lake Vanda; minor Vida granite east of Conrow Gl; trace sandstone; granite: compact-bladed,sub-rounded - rounded; slightly weathered;moderately chipped; glacial features common to abundant; desert surface features rare	few marine diatoms; few non-marine diatoms; few shell fragments
Prospect Mesa upper waterlaid diamicton (PMUWD); (2 m)	pale yellow (5y 7/3); slightly pebbly mud; unimodal:coarse silt	massive to moderately ; stratified; flaser bedding; pockets pebbly sand	mostly granite;some dolerite,dike rocks; granite: compact-bladed,rounded; slightly weathered;moderately chipped; glacial features rare; no desert surface features	non-marine diatoms rare
Prospect Mesa colluvium (PMC); (3.5 m)	light olive gray (5y 6/2); muddy sandy pebble gravel; unimodal: pebble clast-supported	massive	mostly dolerite granite;some dike rock; granite: compact- bladed, sub-rounded ; slightly weathered;moderately chipped; no glacial features; desert surface features common	unfossiliferous
Hart ash; (30 cm)	white (2.5y 8/0) slightly pebbly mud	upper 25 cm w/ some gravel basal 5 cm is pure ash	some granite desert surface features common	unfossiliferous
Odin colluvium (OC); (2 m)	pale yellow (5y 7/3); pebbly muddy sand; unimodal: medium sand clast-supported	massive to crudely stratified; medium bedded	mostly dolerite;some granite, sandstone, quartzite, and dike rock; granite: bladed-platy,sub-angular - sub-rounded; no glacial features; desert features common moderately weathered; moderately chipped;	unfossiliferous
Alpine III drifts (A III); (1.2 m)	brownish yellow (10yr 6/6); pebbly muddy sand; bimodal: med sand, pebble clast-supported	massive	mostly granite,basalt highly variable; granite: bladed - platy ;angular ; slightly weathered; slightly chipped; no glacial features;desert features:few-common	unfossiliferous
Wright Lower drift (WLD); (1 m)	brownish yellow (10yr 6/6); sandy pebble gravel; clast-supported	massive	mostly granite; some dike and metamorphic rocks; granite:bladed,sub-angular - sub-rounded	unfossiliferous
Ripples (RIP); (15 cm)	pale yellow (5y 7/3); pebbly sandy mud; bimodal:coarse sand, pebble matrix-supported	massive to crudely stratified; beds: 10-30 cm thick	mostly dike rocks; some granite,dolerite; trace sandstone quartzite; granite: bladed to platy,sub-angular - sub-rounded; moderately weathered; moderately chipped; no glacial features; desert surface features common	unfossiliferous

Note : Stratigraphic order is generally oldest to youngest. Contacts between all units are sharp.
* Munsell soil colors recorded as hue value/chroma.
† See Figures 7, 11, 15, 18, 21, 23, 28; Table 2
§ Sedimentary structure terminology follows Reineck and Singh (1975).

** See Figure 8
†† See Figures 8, 9
§§ See Table 3

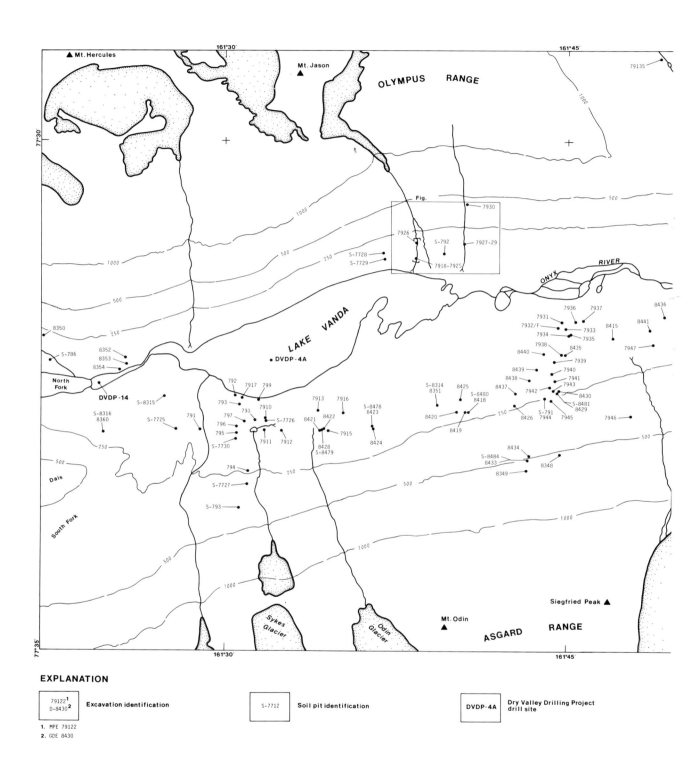

Fig. 4.   Identification of hand-dug excavations in central Wright Valley.

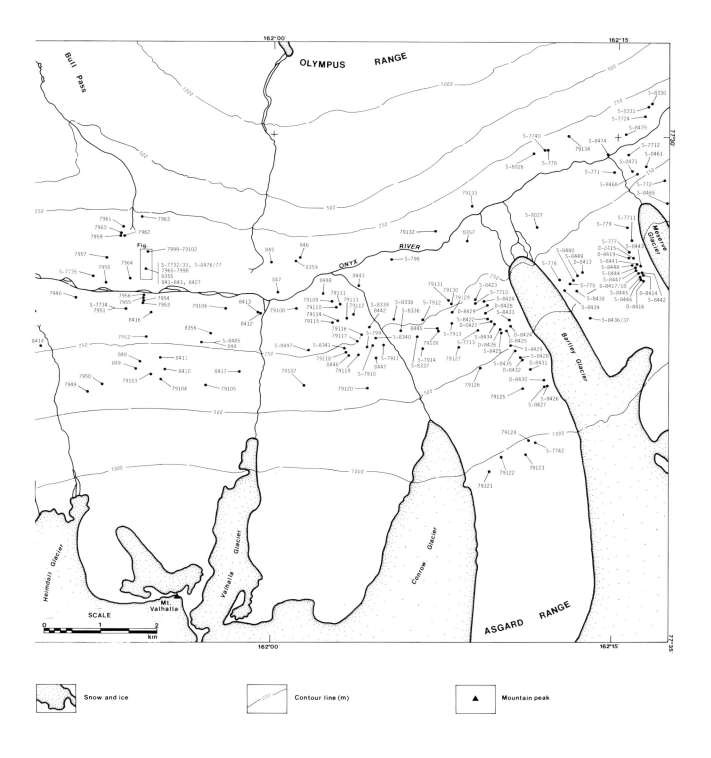

Fig. 4. (continued)

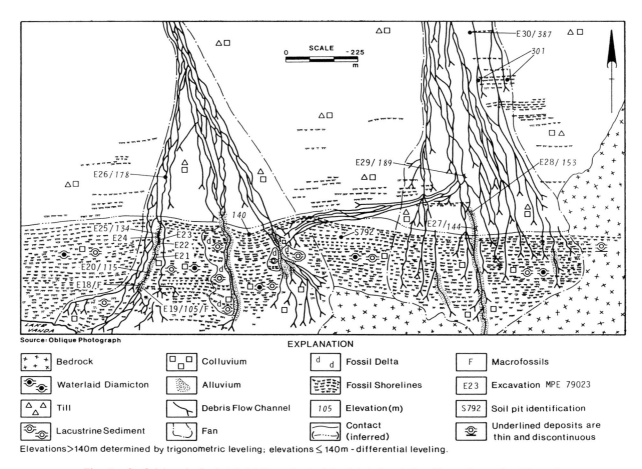

EXPLANATION

Bedrock	Colluvium	Fossil Delta
Waterlaid Diamicton	Alluvium	Fossil Shorelines
Till	Debris Flow Channel	Elevation (m)
Lacustrine Sediment	Fan	Contact (inferred)

Macrofossils	F
Excavation MPE 79023	E23
Soil pit identification	S792
Underlined deposits are thin and discontinuous	

Elevations >140m determined by trigonometric leveling; elevations ≤ 140m - differential leveling.

Fig. 5.   Surficial geologic sketch (oblique view) of the debris fans below Mount Jason. See Figure 3 for location within valley. Water-laid diamicton, Jason glaciomarine diamicton. Till, Peleus till. The Holocene-age shorelines of Lake Vanda are prominent up to 140 m. Linear horizontal features etched into drift above 140 m and up to 387 m may reflect older shorelines. Elevations above 140 m were determined by trigonometric leveling. Elevations below 140 m were determined by differential leveling. An oblique photograph by G. Denton served as the base for this sketch.

and dolerite lithologies (Figure 8). Small percentages of sandstone and a carbonaceous mudstone occur in JGD from the western end of Lake Vanda but not from the eastern end. Two of the samples from the eastern end of the lake contain small percentages of metamorphic lithologies. All lithologies except the metamorphic rocks crop out locally. The metamorphic rocks are only known to crop out in the eastern end of Wright Valley. Their presence suggests an ice advance from the east.

The surface texture of the gravel from the JGD provides insights into transport and depositional processes as well as weathering history. A significant portion of the gravel from all three eastern outcrops exhibits glacial markings indicating that these clasts are not far traveled from the point of release from wet-based glacier ice (Figure 8). A few clasts exhibit desert varnish and ventifaction, features indicative of polar desert conditions. However, the very low abundance of clasts

with these features suggests that local reworking is a more plausible explanation than the attainment of a polar desert climate at the time of Jason deposition. In local reworking, clasts either slumped off the vertical walls of the debris flow channel during peak summer thaw or were eroded by mudflow. These clasts were likely exposed to polar desert conditions before being reworked back into the JGD on the channel floor to be exposed later as the channel migrated.

The plutonic clasts in the JGD exhibit compact-bladed to bladed form (Figure 9). Because lithology strongly influences form, we restrict our examination to the plutonic rocks. JGD contains only a few plutonic clasts that are very platy or very bladed. These forms are indicated by a maximum projection sphericity (mps) below 0.5 with an oblate-prolate index (opi) below -10 for very platy and between -15 and 5 for very bladed. On the other hand, there are numerous compact clasts

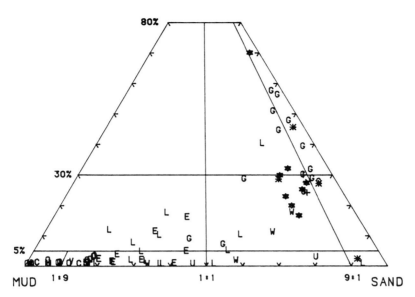

Fig. 6.   Ternary diagram for gravel, sand, and mud percentages of central Wright Valley water-laid diamictons and ripples. Letters are for water-laid diamictons and are as follows: E, Jason glaciomarine diamicton from excavations 7918–7930; W, Jason glaciomarine diamicton from excavations 8352–8354; G, Prospect Mesa gravel from excavations 7966–7995; H, Heimdall glaciomarine diamicton from excavation 7941; L, Prospect Mesa lower water-laid diamicton from excavations 7972–79102; O, Onyx ponds water-laid diamicton from excavations 7931–7937; and U, Prospect Mesa upper water-laid diamicton from excavations 7978, 7997, and 8427. Symbols are for ripples and are as follows: plus sign, ripples from excavations 8356, 841, and 842; asterisk, ripples from excavations 8453–8454, and star, ripples from excavations 8426–8442.

(mps > 0.8), a mature form toward which other forms converge.

Whole and fragmented marine diatoms are abundant in JGD (Tables 3–6). The age ranges of the marine diatoms overlap in the late Miocene [*Baldauf and Barron*, 1991]. On the basis of the overall abundance and

preservation of JGD marine microfossils, we suggest that they are nearly in situ and conclude that the depositional environment of JGD was a fjord. The high concentration and variable preservation of marine diatoms are typical of Antarctic continental shelf glaciomarine sediments [*Elverhøi et al.*, 1983; *Anderson et al.*,

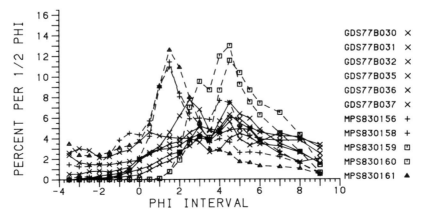

Fig. 7.   Grain size frequency distributions for Jason glaciomarine diamicton. The gravel/sand boundary is at -1ϕ (2 mm) and the sand/mud boundary is at 4ϕ (63 μm). GDS77B030–GDS77B037 are from west-facing stream-channel wall at excavation 7923 shown in Figure 5 just below an elevation of 134 masl. MPS830156 and MPS830158 are from excavation 8352 north of the western end of Lake Vanda. MPS830159 and MPS830160 are from excavation 8353. MPS830161 is from excavation 8354.

TABLE 2.   Moment Statistics and Size Modes for Central Wright Valley Deposits

Deposit*	Mean Size, ϕ	Standard Deviation, ϕ	Skewness	Kurtosis	Delta†	Mode§ Primary, ϕ	Secondary, ϕ	N
AIII	0.8	3.0	0.8	4.6	0.17	1.5	−4.0	10
OC	1.5	3.2	0.8	3.8	−0.05	1.5	· · ·	7
PMC	−0.7	2.8	1.6	6.6	−0.05	−3.0	· · ·	1
PMUWD	3.7	3.2	0.7	4.4	0.18	4.5 (1.5)	· · ·	2
PT	2.4	3.5	0.6	3.0	−0.18	2.0	−4.0 (4.5)	46
PMG GL	0.0	2.9	1.4	6.5	0.12	1.0	−4.0	4
PMG-ML	3.9	3.7	0.4	2.7	−0.19	1.5	5.0	2
PML WD-SL, ML	4.4	3.3	0.1	3.4	0.12	4.5 (2.5)	· · ·	4
PML WD-LL	0.8	3.2	1.1	4.2	−0.17	−2.0	3.0	1
OPWD	7.5	2.8	0.1	2.3	−0.27	4.5-7.0	· · ·	2
HGD	6.6	2.7	0.5	2.7	· · ·	−0.27	4.5	1
JGD	4.8	3.4	0.4	2.9	−0.12	4.5 (1.5)	· · ·	11

*See Table 1 for deposit codes.
†Delta = $(2K − 3Sk^2 − 6)/(K + 3)$, where Sk is skewness and K is Kurtosis [*Leroy*, 1981].
§Modes in parentheses are less common.

1992; *Kellogg and Truesdale*, 1979] and are atypical of terrestrial subglacial deposits laid down in a marine setting [*Elverhøi*, 1984]. Benthic diatoms dominate the JGD flora and suggest that this fjord was shallow. The absence of Pliocene to Pleistocene diatoms suggests that JGD diatoms were not reworked into Wright Valley by Pleistocene advances of ice from the Ross Sea [*Denton et al.*, 1991]. We rely on diatom abundance and preservation to preclude earlier reworking from the Ross Sea,

the Transantarctic Mountains, or even interior East Antarctica. The JGD diatom flora is unlike that recovered from the Sirius Group [*Webb et al.*, 1984] or basal Wright Upper Glacier debris [*Harwood*, 1986] which these authors consider to come from interior East Antarctica. We prefer the simpler interpretation that JGD marine diatoms are from the Wright Valley Fjord rather than being from seaways interior to East Antarctica.

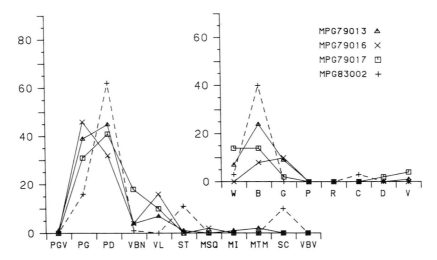

Fig. 8.   Lithology and surface texture of gravel (*b* axis: 16–64 mm) from the Jason glaciomarine diamicton. Left-hand plot shows volume percentage of lithologic associations. PGV, Vida Granite; PG, granites other than Vida Granite; PD, Ferrar Dolerite; VBN, dark dike rock; VL, light dike rock; ST, Beacon sandstone; MSQ, quartzite; MI, metaigneous rock; MTM, schists and marbles; SC, limy mudstone; and VBV, dark vesicular basalt. Right-hand plot depicts development of surface features. W, weathering; B, broken faces; G, glacier marks; P, pitting; R, rind; C, carbonate crust; D, desert varnish; and V, ventifaction. MPG79013 (excavation 7925, *n* = 144 clasts), MPG79016 (excavation 7920, *n* = 39 clasts), and MPG79017 (excavation 7919, *n* = 50 clasts) are from channel exposures below Mount Jason (Figure 5). MPG83002 (excavation 8352, *n* = 30 clasts) is from north of the western edge of Lake Vanda.

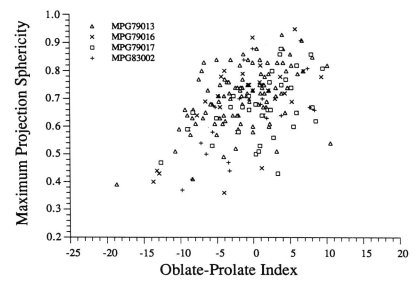

Fig. 9. Maximum projection sphericity (mps) and oblate-prolate index (opi) of granite (PG, PGV) and dolerite (PD) gravel (*b* axis: 16–64 mm) from Jason glaciomarine diamicton. Values for these two variables define one of ten possible clast forms in Folk's form triangle [*Dobkins and Folk*, 1970]. Compact forms are indicated by mps > ~0.85. Clasts with mps < ~0.85 are either platy, bladed, or elongate, depending upon whether opi < −3, −3 < opi < 3, or opi > 3, respectively. If mps < ~0.4, clasts are either very platy, very bladed, or very elongate. Samples are described in Figure 8.

The fine grain size of JGD is consistent with deposition in standing water. JGD is finer-grained than typical tills deposited by wet-based ice [*Mills*, 1977; *Lawson*, 1979] as well as colluvium [*Landim and Frakes*, 1968]. On the other hand, JGD particle size distribution and lack of sedimentary structure typify glacial deposits accumulating in ice-distal glaciomarine environments where ice-rafted detritus mixes with fine-grained sediment settling out of turbid meltwater plumes [*Vorren et al.*, 1983; *Powell*, 1990; *Elverhøi*, 1984; *Barrett and Hambrey*, 1992]. The presence of glacially processed gravel suggests that glacier ice extended into the fjord. The ice must come from the inland end of the valley for the valley mouth to remain open to the sea.

Two unidentifiable shell fragments recovered from JGD were analyzed for $^{87}Sr/^{86}Sr$ and Sr/Ca (Table 7 and

TABLE 3. Diatom Assemblages in Wright Valley Deposits and the Sirius Group

Jason Glaciomarine Diamicton	Heimdall Glaciomarine Diamicton	Prospect Mesa Gravels	Peleus Till	Wright Upper Glacier*	Sirius Group†
Actinocyclus ingens	*A. ingens*		*A. ingens*	*Actinocyclus*	*A. actinochilus*
Denticulopsis dimorpha			*D. dimorpha*		
Denticulopsis hustedtii			*D. hustedtii*		
Denticulopsis lauta					
Eucampia antarctica	*E. antarctica*	*E. antarctica*	*E. antarctica*		
		Nitzschia kerguelensis	*N. kerguelensis*		*N. kerguelensis*
			Nitzschia praeinterfrigidaria		
Thalassionema nitzschioides		*T. nitzschioides*	*T. nitzschioides*	*Thalassionema*	*T. nitzschioides*
		Thalassiosira insigna	*T. insigna*		*T. insigna*
Thalassiosira cf torokina					
		Thalassiosira lentiginosa	*T. lentiginosa*	*Thalassiosira*	*T. lentiginosa*
			Thalassiosira vulnifica	*Thalassiosira*	*Thalassiosira vulnifica*

Harwood and Webb [1986].
†*Harwood* [1991].

TABLE 4. Occurrence of Microfossils in Jason Glaciomarine Diamicton at DVDP-4a and E794.

Location	DVDP-4a							E794	E794
Sample ID	16-2 10-12	17-2 3-5	17-2 12-14	17-2 25-26	19-1 5-6	19-3 45-46	21-1 10-11	MPF79 Y198	MPF79 Y197
Depth (m)	5.1	6.2	6.3	6.5	8.6	9.7	9.8	0.05-0.15	0.20-0.25
Actinocyclus cf ehrenbergi								R	
Actinocyclus ingens	F	R	R	R	R	R	R	C	
Cestodiscus								R	
Cocconeis				F	R	R			
Coscinodiscus sp.	R		R	R	R				
Coscinodiscus furcatus								R	
Coscinodiscus marginatus								R	
Coscinodiscus cf obscurus								R	
Cosmiodiscus cf intersectus								R	
Denticulopsis dimorpha								R	
Denticulopsis hustedtii	R	R			R			C	
Denticulopsis lauta	F	R	R	R	R				
Diploneis			R		F				
Eucampia antarctica								R	
Hyalodiscus sp.								R	
Melosira sol	F	F	F	F	F				
Melosira sulcata	R	R							
Nitzschia sp.	R								
Rhizosolenia sp.								R	
Rhizosolenia cf barboi	R	R						R	
Rhizosolenia habetata								R	
Rhizosolenis styliformis	R							R	
Rouxia naviculoides								R	
Stephanopyxis cf turris		R							
Thalassionema sp.	R	R	R	R	R	R	R	R	
Thalassionema nitzschioides	F	R	R	R					
Thalassiosira antarctica	R								
Thalassiosira oestrupii								R	
Thalassiosira cf torokina								R	
Thalassiothrix sp.								R	
Trachyneis aspera	R						R		
Trinacria sp.	R	R	R	R	R			R	
Non-marine diatoms								R	
Diatom fragments	C	C	C	C	C	C	C	F	R
Sponge spicules								F	
Radiolarian fragments	P		P	P	P				

A = abundant, C = common, F = few, R = rare, P = present.

Figure 10*b*). The mean $^{87}Sr/^{86}Sr$ ratio of JGD shells, 0.708875, suggests an age of about 9 ± 1.5 Ma for this deposit according to the compilation of *Hodell et al.* [1991]. This is consistent with the marine diatom stratigraphy, particularly the abundance of *Denticulopsis hustedtii*, which suggests a middle to late Miocene age. We assume that the JGD shells were in equilibrium with the oceanic $^{87}Sr/^{86}Sr$ reservoir. This assumption is supported by the $^{87}Sr/^{86}Sr$ of the modern Antarctic scallop, *Adamussium colbecki*, from the present shoreline adjacent to Wright Valley, which is statistically indistinguishable from modern seawater $^{87}Sr/^{86}Sr$ (Table 7 and Figure 10*b*).

A number of potential flaws in the Sr-based age estimate deserve mention. One is that JGD shell $^{87}Sr/^{86}Sr$ may be diagenetically altered. We suggest that this is not the case because the Sr/Ca ratio of the shell, a good measure of the extent of recrystallization, equals that of modern pristine pectens (Table 7). Another possible flaw is that the $^{87}Sr/^{86}Sr$ of the Jason Fjord may not have been in equilibrium with the oceanic reservoir but, rather, was contaminated by excessive continental Sr. Given typical continental Sr with $^{87}Sr/^{86}Sr$ of 0.716, the effect would increase Jason Fjord water $^{87}Sr/^{86}Sr$ relative to coeval seawater values and artificially decrease the age of the JGD shells, making our age a

TABLE 5. Occurrence of Microfossils in Jason Glaciomarine Diamicton at E7918, 7920-24 from the Mt. Jason Region.

Excavation	E79 18	E79 18	E79 18	E79 20†	E79 21¥	E79 22§	E79 22	E79 23	E79 23	E79 23	E79 23	E79 24
Sample ID	MPF 79 Y142	MPF 79 Y143	MPS 79 Y144	MPF 79 Y141	MPF 79 Y283	MPS 79 G283	MPS 79 G282	GDS 77 B38	GDS 77 B34	GDS 77 B31	GDS 77 B30	MPS 79 G281
Depth (m)	0.4-0.5	0.7-0.8	1.0-1.1	1.0-1.1	1.6-1.7	0.9-1.0	1.5-1.6	0.9-1.0	1.9-2.0	2.9-3.0	3.1-3.2	1.5-1.6
Actinocyclus cf ehrenbergi		R										
Actinocyclus ingens					R	R	R					R
Cocconeis sp.								R				
Coscinodiscus sp.											F	
Denticulopsis dimorpha				R								
Denticulopsis hustedtii	R	R		R	R		R					
Denticulopsis lauta				R	R		R	R				
Eucampia antarctica				R	R							
Grammatophora sp.		R										
H. amphioxys		R	R									
Melosira sol										R		
Melosira sulcata							R					
Rhabdonema sp.								R				
Rhizosolenia cf barboi						R			R			
Stephanopyxis cf turris							R					
Thalassionema sp.								R				
Thalassionema nitzschioides										R		
Thalassiosira cf torokina					R			R				
Trachyneis aspera								R				
Non-marine diatoms				R								
Diatom fragments	A	C	F	F	C	A	F	F	R	R	F	A
Sponge spicules				F	F	R	R	R	R	R	F	

A = abundant, C = common, F = few, R = rare.
† Samples with fragments only = MPF79Y138 (2.2-2.3 m) and MPF79Y139 (1.7-1.8 m).
¥ Samples with fragments only = MPS79G285 (0.7-0.8 m) and MPF79Y284 (1.0-1.1 m).
§ Samples with fragments only = MPS79Y282 (1.2-1.3 m).

minimum. A third potential problem is that the late Neogene history of seawater $^{87}Sr/^{86}Sr$ is not yet fully defined and may shift with additional data.

The oxygen isotopic composition of one JGD shell fragment ranges between 1.8 and 0.76‰, which is about 3‰ more negative than the $\delta^{18}O$ of modern scallops from nearby McMurdo Sound (Figure 10*a*). Given the high Sr/Ca ratio and the small range in $^{87}Sr/^{86}Sr$ of one of the Jason shells, we suggest that the $\delta^{18}O$ data are unaltered and reflect the Jason Fjord environment. To interpret these data, we assume equilibrium $\delta^{18}O$ precipitation [*Epstein et al.*, 1953]. *Barrera et al.* [1990] demonstrated equilibrium $\delta^{18}O$ precipitation for *A. colbecki* in McMurdo Sound waters characterized by temperatures of $-1.9°C$ to $-1.0°C$ and $\delta^{18}O$ of $-0.2‰$ to $-0.7‰$ (SMOW). If we are correct, JGD shell $\delta^{18}O$ indicate that Jason Fjord water was at least slightly warmer and more isotopically negative (less saline) than water in McMurdo Sound today. We discuss this issue in a later section.

Onyx Ponds Water-Laid Diamicton

Onyx ponds water-laid diamicton (OPWD) consists of a small isolated patch of drift less than 1 km south of the Onyx River midway between Prospect Mesa and Lake Vanda (Figure 3). The underlying and surrounding bedrock exhibits a corrugated morphology in which the dikes stand above host granite. Peleus till overlies this diamicton. Stratigraphic sections are exposed in the sides of a water-eroded channel through this drift outlier. The maximum observed thickness is 2.5 m at excavation 7931.

OPWD is extremely fine grained (Figure 6) and massive except for a few unevenly distributed pockets of gravel. Primary grain size modes are 4.5ϕ, like the JGD, and 7ϕ (fine silt) (Figure 11). These textural characteristics suggest that OPWD was deposited in standing water. This is supported by a complex network of worm burrows observed in excavation 7932. Although all clasts examined consist of local lithologies, many ex-

TABLE 6. Occurrence of Microfossils in Jason Glaciomarine Diamicton at E7919 in the Mt. Jason Region

Sample ID	GDS 77 B29	GDS 77 B28	GDS 77 B27	GDS 77 B26	GDS 77 B25	GDS 77 B24	GDS 77 B23	GDS 77 B21	GDS 77 B20	MPS 79 G134	MPS 79 G135	MPS 79 G136	MPS 79 Y136	MPS 79 Y137
Depth (m)	0.4-0.5	0.7-0.8	0.9-1.0	1.1-1.2	1.2-1.3	1.4-1.5	1.8-1.9	1.9-2.0	2.0-2.1	2.0-2.1	2.1-2.2	2.3-2.4	2.3-2.4	2.3-2.4
Actinocyclus cf *ehrenbergi*														R
Actinocyclus ingens	R	F	R	R		R	R	R	R	R	R	R		R
Cocconeis sp.	R		R	F	R		R	R	R			R		
Coscinodiscus sp.	R		R	R				R				R		
Coscinodiscus marginatus												R		R
Cosmiodiscus cf *intersectus*												R		
Denticulopsis dimorpha	F		R	R	R		C	F		R	C	R		R
Denticulopsis hustedtii				R	R	R	R				C	C	F	C
Denticulopsis lauta				R	R	R	R			R				
Denticulopsis speculum	F	F	R	F	R	R	F	R	R	R		F		F
Diploneis sp.											R	R		
Eucampia antarctica	R	R	R	R	R		R	R				R		
Hyalodiscus sp.														
Melosira sol	R	R	R	R	R	R	R	R				F		R
Melosira sulcata							C	R				R		
Nitzschia sp.	C	C	F	F	F	R	R	R	R					
Rhabdonema sp.	R		R	R			C							
Rhizosolenia cf *barboi*	R								R			R		R
Rouxia naviculoides														
Stephanopyxis cf *turris*	F									R		R		
Thalassionema	F	R	R	R	R	R	R	R	R			R		
Thalassionema nitzschoides		R	R	R			R					R		
Thalassiosira cf *torokina*														
Trachyneis aspera	C	C	F	F	R	R	F	F	R					
Trinacria sp.														R
Non-marine diatoms	C	C	C	C	C	A	A	A	A	A	R	C		F
Diatom fragments	F	F	C	F	F	F	F	A	A	A	R	A		C
Sponge spicules					C	F	F	F	C					R
Radiolarian fragments											R	R		
Silicoflagellates														R

A = abundant, C = common, F = few, R = rare.

TABLE 7. ^{Sr}/Ca, ^{87}Sr/^{86}Sr, and δ^{18}O Data for Carbonates From Wright Valley and Vicinity

Deposit*	Excavation	Sample Identification	Sr/Ca	^{87}Sr/^{86}Sr†	δ^{18}O	δ^{13}C
BS§	NH§	MPF86001a	2.13	0.709122	4.56	2.41
BS		MPF86001b	2.15	0.709108	4.54	2.48
BS		MPF86001c	2.14¶		4.70	2.59
PMG	7972	GDS77B87/88a		0.709058		
PMG	7972	GDS77B87/88b	1.72	0.709094		
PMG	7972	GDS77B87/88c	1.66¶	0.709171	0.09	1.23
PMG	7973	MPS79G296a	1.82	0.708981		
PMG	7973	MPS79G296b	1.74¶	0.708993	0.79	1.74
PMG	7973	GDS77B90a	2.04	0.708953	2.23	1.70
PMG	7973	GDS77B90b	1.99¶	0.709010		
PMG	7977	MPS79G262a	2.10	0.708957	2.13	2.78
PMG	7977	MPS79G262b	2.10¶	0.708983		
PMG	7985	MPS79Y290a	1.79	0.709238	−3.50	1.40
PMG	7985	MPS79Y290b	1.72¶	0.709137		
PMG	7985	MPS79Y290c		0.709351		
PMG	7985	MPS79Y290d		0.709164		
PMG	7985	MPS79Y290e		0.709263		
PMG	7985	MPS79Y290f		0.709252		
PMG	7989	MPS79G273a	1.84	0.709052	0.52	2.36
PMG	7989	MPS79G273b	2.24¶	0.709022		
PMG	7989	MPS79G273c		0.709021		
PMG	7989	MPS79G273d		0.709028		
PMG	7989	MPS79G273e		0.709024		
PMG	7989	MPS79G273f		0.708995		
PMG	7989	MPS79G273g		0.709036		
JGD	7919	GDS77B14a	2.14	0.708900	1.80	3.09
JGD	7919	GDS77B14b	2.18	0.708846	0.76	2.15
JGD	7919	GDS77B14c	2.22¶	0.708818		
JGD	7919	GDS77B14d		0.708898		
JGD	7919	GDS77B14e		0.708872		
JGD	7919	MPS79Y136		0.708880		

Note: We take the ^{87}Sr/^{86}Sr of modern seawater to be 0.709123.
*See Table 1.
†Data were corrected such that mean of standards (NBS-987) is 0.710235.
§Modern beach in New Harbor region (Figure 1).
¶Data collected at University of Maine; otherwise, University of Florida.

hibit glacial marks suggesting some transport at the sole of a wet-based glacier (Figure 12). The clasts exhibit bladed to compact-bladed form as they do in JGD (Figure 13). Diatoms and sponge spicules are rare in this diamicton (Table 8), preventing differentiation of marine from lacustrine environments.

Heimdall Glaciomarine Diamicton

Heimdall glaciomarine diamicton (HGD) was seen at excavation 7941 at an elevation of 240 masl, directly south of the OPWD (Figure 3). Here HGD is assumed to be overlain by Peleus till. Sedimentologically, HGD closely resembles OPWD (Figure 11). The principal difference is that marine diatoms are abundant in one HGD sample (Tables 3 and 8). Given this abundance, HGD marine diatoms are inferred to be in situ for the same reasons as the JGD diatoms. Hence HGD was also probably deposited in a fjord. The absence of the marine diatoms diagnostic of JGD, including *Denticulopsis hustedtii*, *D. lauta*, and *D. dimorpha*, suggests that

HGD was not deposited in the Jason Fjord. The presence of diatoms *Actinocyclus ingens* and *Eucampia antarctica* suggests that HGD has a maximum age of late Miocene [*Baldauf and Barron*, 1991]. The diatom assemblage suggests a nearshore low-salinity environment. Benthic diatoms dominate the HGD flora and suggest that this fjord was shallow.

Prospect Mesa Lower Water-Laid Diamicton

The Prospect Mesa lower water-laid diamicton (PMLWD) is the lowest unit exposed at Prospect Mesa (Figures 3 and 14). PMLWD consists of three complexly interbedded lithofacies (Table 1). A fine-grained stratified lithofacies dominates the unit. In this lithofacies, beds dominated by a fine sand mode alternate with beds exhibiting a coarse silt mode (Figure 11). Mud laminae commonly bend below and drape striated pebbles and boulders. The massive pebbly mud lithofacies is more poorly sorted. The lag lithofacies is bimodal with strong primary mode in the granule size range (Figure 11). The

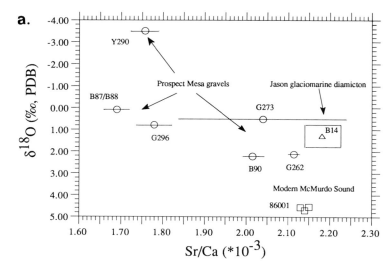

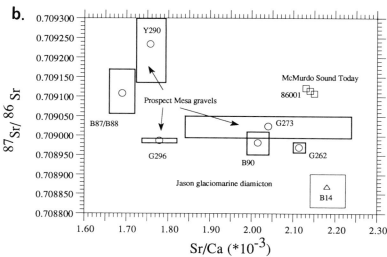

Fig. 10. Sr/Ca, δ^{18}O, and ^{87}Sr/^{86}Sr of marine carbonates from central Wright Valley from Table 7. (*a*) Sr/Ca versus δ^{18}O. Circles are mean values for *Chlamys tuftsensis* from the Prospect Mesa gravels. Triangle is mean for carbonates from Jason glaciomarine diamicton. Small boxes are for modern *Adamussium colbecki* from New Harbor shown in Figure 1. Horizontal lines show the range of variability of individual Sr/Ca measurements. (*b*) Sr/Ca versus ^{87}Sr/^{86}Sr. Symbols as in Figure 10*a* except that boxes define variability in ^{87}Sr/^{86}Sr and Sr/Ca.

gravel in PMLWD is dominated by plutonic lithologies and exhibits glacial markings (Figure 12). The clasts exhibit a distinctive elongate form (Figure 13). PMLWD contains a few small shell fragments and nonmarine diatoms as well as rare marine diatoms (Table 8). Extensive worm burrows were observed in excavation E7992.

Stratification, worm burrows, and fine grain size imply that PMLWD was deposited in standing water. Sedimentary structures around some of the larger pebbles indicate that these pebbles were dropped, presumably from floating glacier ice. Abundant penecontempo-

raneous deformation structures attest to rapid sedimentation and an unstable arrangement of sediments. The fossils do not indicate whether the water body was marine or nonmarine.

Prospect Mesa Gravels

The Prospect Mesa gravels (PMG) are a wedge-shaped unit that overlies PMLWD (Figure 14). PMG also crops out just to the west of the mesa near excavation 7964 (Figure 3). A gravel lithofacies exhibiting some stratification dominates the gravels (Table 1).

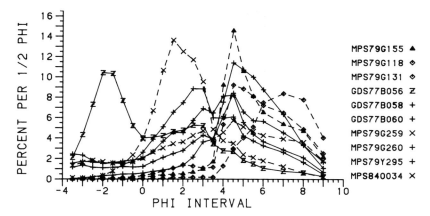

Fig. 11. Grain size frequency distributions for water-laid diamictons, other than Jason glaciomarine diamicton, in central Wright Valley. MPS79G155 (excavation 7941) represents Heimdall glaciomarine diamicton. MPS79G118 (excavation 7931) and MPS79G131 (excavation 7933) are from Onyx ponds water-laid diamicton. GDS77B056 (excavation 7972) represents the lag lithofacies of the Prospect Mesa lower water-laid diamicton. GDS77B058 and GDS77B060 are from the stratified lithofacies of the Prospect Mesa lower water-laid diamicton at excavation 7972. MPS79G259 (excavation 7992), MPS79G260 (excavation 7992), and MPS79Y295 (excavation 7999) are from the massive lithofacies of the Prospect Mesa lower water-laid diamicton. MPS840034 (excavation 8427) represents Prospect Mesa upper water-laid diamicton.

This lithofacies is bimodal with small pebble and medium sand modes (Figure 15). A massive mud lithofacies with strong silt modes is complexly interbedded with the gravels in the eastern face of Prospect Mesa. The clasts are all of local lithologies (Figure 16); some exhibit very bladed and very elongate forms (Figure 13). A 1-m-long block of PMLWD occurs within the basal layers of PMG. Contacts between many beds within this unit show evidence of scour and channeling.

PMG contains an abundance of pecten shells and foraminifers. The shells belong solely to an extinct species of thick-walled pecten, *Chlamys tuftsensis* [*Turner*, 1967]. The foraminiferal fauna is dominated by the benthic species *Ammoelphidiella antarctica* (equal to *Trochoelphidiella onyxi* of *Webb* [1974]) but also contains *Rosalina globulosa*, *Patellina antarctica*, and *Cibicides* sp. (P.-N. Webb, personal communication, 1985). Gravel lithofacies beds at excavations 7972, 7973,

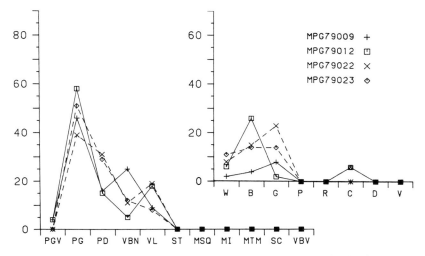

Fig. 12. Lithology and surface texture of gravel (16–64 mm) from water-laid diamictons. Axes labels as in Figure 8. MPG79009 (*n* = 50 clasts) is from Prospect Mesa lower water-laid diamicton at excavation 7972. MPG79012 (*n* = 50 clasts) is from Prospect Mesa upper water-laid diamicton at excavation 7978. MPG79022 (*n* = 13 clasts) and MPG79023 (*n* = 35 clasts) are from Onyx ponds water-laid diamicton at excavations 7931 and 7932, respectively.

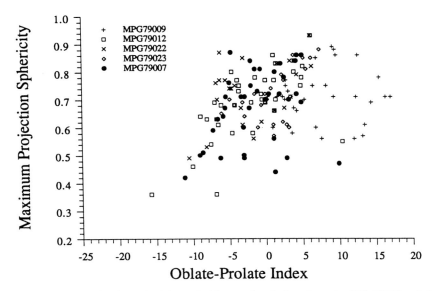

Fig. 13. Maximum projection sphericity and oblate-prolate index of granite (PG, PGV) and dolerite (PD) gravel (16–64 mm) from central valley water-laid diamictons and the Prospect Mesa gravels (MPG79007 at excavation 7972, n = 50 clasts). Water-laid diamictons are as follows: Onyx ponds (MPG79022, MPG79023), Prospect Mesa lower (MPG79009), and Prospect Mesa upper (MPG79012). Samples are described in Figure 12.

and 7985 are less than 10 cm thick and contain well-preserved articulated valves in high concentration. The shells are commonly stacked parallel to bedding. Less well-preserved shells occur in moderate-to-low concentration in five other beds of the gravel lithofacies and two beds of the mud lithofacies.

Marine and nonmarine diatoms as well as sponge spicules and a radiolarian fragment (Tables 3 and 8) occur in the gravel lithofacies. The critical marine diatoms for stratigraphic purposes are *Nitzschia kerguelensis* and *Thalassiosira insigna* (Table 3). The co-occurrence of these two species suggests an age of about 3 to 2.5 Ma based on the Subantarctic ocean diatom stratigraphy [*Gersonde and Burckle*, 1990; *Baldauf and Barron*, 1991; *Barrett et al.*, 1992]. This flora is quite different from and younger than that in either JGD or HGD.

We agree with previous workers [e.g., *Webb*, 1972] that the macrofossils and foraminifers are in situ and that PMG was deposited in a fjord. The high concentration of these fossils in the gravels as well as the variety of ontogenetic stages represented and their excellent preservation constitute the strongest evidence. The high density of stacked pecten valves is typical of mature bivalve communities on the floor of McMurdo Sound [*Bullivant*, 1959, 1961; *Dayton et al.*, 1970]. The high density probably reflects a lack of disturbance which allowed the community to build up vertically as live pectens at the surface died and formed a protective substrate for succeeding pectens. According to the studies of *Menard and Boucot* [1951] and *Brenchley and*

Newall [1970], the preservation, stacking, and orientation of the pecten valves imply insignificant movement.

An alternative interpretation for PMG is that the fossils were frozen onto the base of ice advancing through McMurdo Sound, transported 50 km to Prospect Mesa, and deposited there in proglacial outwash [*Nichols*, 1965, 1971]. Evidence against this is the absence in the gravels of metamorphic lithologies which are restricted to eastern Wright Valley. If PMG was deposited by ice from McMurdo Sound, it should contain some of these lithologies. Also, complex glaciological mechanisms must be invoked to protect the delicate shells.

The sedimentology of the gravels is consistent with deposition in a fjord. Coarse grain size, cut-and-fill structures, and clast-supported frameworks are all typical of subaqueous flows of high velocity [*Rust and Romanelli*, 1975; *Middleton and Hampton*, 1976; *Stanley et al.*, 1978]. Two graded beds on the eastern side of Prospect Mesa are good evidence for turbidity currents. We prefer this sedimentation model to that of *Webb* [1972, 1974], who invoked vertical sedimentation from floating ice. But what is the source of these flows? Do they reflect mass wastage on the Prospect Mesa debris fan [*Vucetich and Topping*, 1972]? Alternatively, could they have emanated from a nearby ice margin and be a grounding line morainal bank deposit?

The fossil content of the gravels is more compatible with a debris fan environment than a grounding line environment. Sporadic calm intervals would clearly be required of each alternative to allow pectens and fora-

TABLE 8. Occurrence of Siliceous Microfossils in Central Wright Valley Deposits Other Than Jason Diamicton.

Deposit	OPWD†	HGD¥	PMLWD§	PMG		Peleus till ¶				
Excavation	E79 31	E79 44	E79 102	E79 72	E79 72	E79 72	E79 72	E79 72	E79 72	E79 126
Sample ID	MPF 79Y 116	DMF 87	MPS 79 G295	HCF 84 4	HCF 84 5	GDS 77 B84	MPS 84 51	GDS 77 B76	MPS 84 50	MPS 79 G204
Depth (m)	2.25-2.40	0.1-0.2	1.2-1.3	2.0-2.3	3.0-3.3	3.5-3.6	4.5-4.7	5.25-5.35	5.5-5.7	0.9-1.05
Actinocyclus actinochilus									R	
Actinocyclus ingens		R		R			R		R	
Actinocyclus sp.				R	R					
Azpeitia nodulifera								R		
Cocconeis sp.		R								
Coscinodiscus endoi						R				
Coscinodiscus marginatus					R			R		
C. oculus-iridus				R		R				
C. symbolophorus			R					R		
Coscinodiscus tabularis						R				
Coscinodiscus sp.				R						
Denticulopsis sp.							R			
Denticulopsis dimorpha					R	R	R		R	
Denticulopsis hustedtii						R	R			
Eucampia antarctica		R			R	R	R	R	R	
Hemidiscus cuneiformis				R				R		
Hemidiscus karstenii								R		
Melosira sol	R		R							
Nitzschia kerguelensis				F	F	F	C	F	R	
Nitzschia lanceolata						R	R			
Nitzschia praeinterfrigidaria									R	
Nitzschia sp				R	R					
Rhizosolenia sp.		R		R	R	R				
Rhizosolenia cf barboi				F		R	R		R	
Stephanopyxis cf turris						R				
Stephanopyxis sp.				R	R					
Thalassionema sp.	R		R	R	F					
Thalassionema nitzschioides	R		R	F	F	F	R	F	R	
Thalassiosira antarctica								R		
Thalassiosira insigna				F	F	R			R	
Thalassiosira lentiginosa				F	F	R	R	R	R	
Thalassiosira oestrupii							R			
Thalassiosira vulnifica							R			
Non-marine diatoms										R
Diatom fragments	R	C	R	C	C			P		F
Sponge spicules	P									F
Radiolarian fragments				R	R			P		
Spores								P		

Samples examined A = abundant, C = common, F = few, R = rare, P = present
Samples examined and found barren.

† Onyx ponds waterlaid diamicton. Barren samples= GDS77B045, GDS77B046, MPF79Y114 all from excavation E7931. MPF79Y169 and MPF79Y170 from E7932. MPF79Y128 and MPF79Y131 from E7933. MPF79Y133 (E7937).

¥ Heimdall glaciomarine diamicton. Barren samples= MPF79Y155 and MPF79Y156 from E7941.

§ Prospect Mesa lower waterlaid diamicton. Barren samples= MPS79G289 (excavation E7969). GDS77B55, -B56, GDS77B57, -B58, -B59, -B60, -B61 from E7972. MPS79G246 (E7979). GDS77B93 (E79100). GDS77B91 (E79101).

¶ Peleus till. Barren samples= MPS79210 (E798). MPF79Y280 (E7926). MPF79Y150 (E7928). MPS79G127 (E7933). MPS79G221 (E79124).

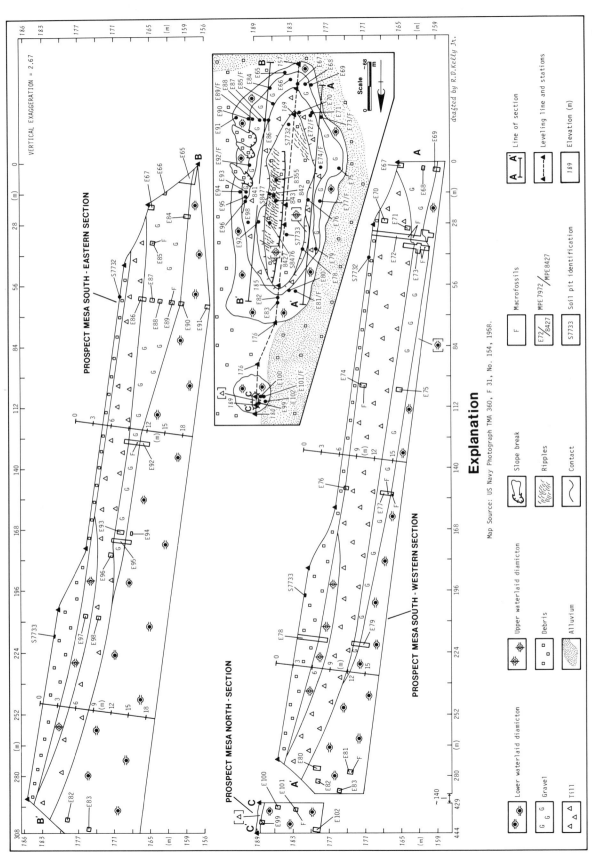

Fig. 14. Surficial geologic map (center, right) and stratigraphic sections of Prospect Mesa. Figure 3 shows the location of the map within the central valley. Prospect Mesa consists of the primary mesa south and a much smaller mesa north. Stratigraphic sections A-A', B-B', and C-C' are shown on the map. Deposit contacts are inferred where they were not exposed. Till, Peleus till. Debris, Prospect Mesa colluvium. All other deposit names refer to Prospect Mesa.

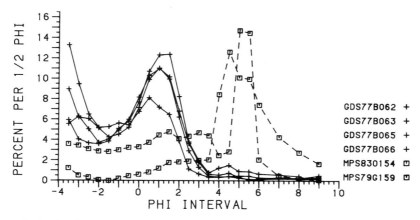

Fig. 15. Grain size frequency distributions for Prospect Mesa gravels. GDS77B062, GDS77B063, GDS77B065, and GDS77B066 are from the gravel lithofacies at excavation 7972. MPS79G159 and MPS83154 represent the mud lithofacies and are from excavations 7992 and 7995, respectively.

minifers to colonize the gravels. We consider such intervals as uncommon in the grounding line environment. Moreover, we consider the ice-proximal environment as restrictive because of the high potential for turbulence, turbidity, and reduced salinity. The sedimentology of PMG is consistent with both glacial [*Elverhøi et al.*, 1983; *Powell*, 1984] and colluvial [*Stanley and Unrug*, 1972; *Walker*, 1975; *Kelling and Holroyd*, 1978; *Stanley et al.*, 1978] alternatives.

The absence of gravel deposits similar to PMG elsewhere in Wright Valley (Figure 3) favors a submarine colluvial origin. If the glacier alternative were correct, deposits similar to PMG should litter the valley as the conditions necessary for their formation would have existed throughout. This would be consistent with the commonly reported "cluster" occurrence of grounding line gravels [e.g., *Banerjee and Mcdonald*, 1975; *Rust and Romanelli*, 1975]. On the other hand, the uniqueness of the PMG in Wright Valley and the uniqueness of their location adjacent to a major tributary valley, Bull Pass, suggest a cause-and-effect relationship that favors the colluvial alternative.

We measured the $^{87}Sr/^{86}Sr$ of six different pecten samples from PMG (Table 7). We suspect that four of the six samples (Y290, B87/88, G296, and G273) are altered by diagenesis on the basis of low and highly variable Sr/Ca ratio (Figure 10b). We accept the $^{87}Sr/^{86}Sr$ ratios of the two samples, B90 and G262, as unaltered. To infer age from these ratios, we assume that we have avoided the three potential problems stated previously for JGD. However, there is also a fourth potential complication related to local Pliocene volcanism. Such volcanism may have biased Prospect Fjord waters with lowered $^{87}Sr/^{86}Sr$. We assume that any volcanics erupted during this time of intermittent volcanism formed subaerially and so did not bias Prospect Fjord water $^{87}Sr/^{86}Sr$. On the basis of the mean

value, 0.708975, we suggest an age of 5.5 ± 0.4 Ma using the seawater $^{87}Sr/^{86}Sr$ record of *Clemens et al.* [1993] from Ocean Drilling Program Site 758. This is significantly older than the diatom-based age estimate.

The $\delta^{18}O$ of six different *C. tuftsensis* shells from six different excavations varies between 2.2‰ and −3.5‰ (Figure 10a). As with JGD carbonates, to interpret *C. tuftsensis* $\delta^{18}O$, we assume equilibrium $\delta^{18}O$ precipitation. The two shells with the most positive $\delta^{18}O$ also have a relatively high and invariant Sr/Ca ratio near that for modern pristine scallops. This fact suggests that they are reliable. The four shells that we considered altered diagenetically on the basis of low or highly variable Sr/Ca ratios have a relatively low $\delta^{18}O$. One way to derive these lower $\delta^{18}O$ values is by solution/reprecipitation in the presence of meteoric waters, isotopically very negative at first but becoming increasingly positive through evaporation [e.g., *Allan and Matthews*, 1982]. Overall, we regard the $\delta^{18}O$ of B90 and G262 as primary and suggestive of a fjord paleoenvironment warmer than McMurdo Sound today.

Peleus Till

Peleus till crops out discontinuously throughout central and eastern Wright Valley up to an elevation of 1150 masl west of Bartley Glacier and up to 1000 masl on the north valley wall by Clark Glacier (Figures 2 and 3). All patches of Peleus till lack moraines and exhibit feather edges. Peleus till satisfies the principal criteria for subglacial till [*Prentice*, 1985] established by *Dreimanis* [1988] and *Dreimanis and Schlüchter* [1985] (Table 1). *Prentice et al.* [1987] previously inferred that this till could represent a single advance of largely wet-based glacier ice down the valley from west to east. This inference is probably an oversimplification. The physically separated diamicton units assigned to Peleus till

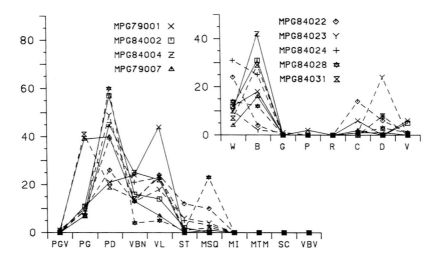

Fig. 16. Lithology and surface texture of gravel (16–64 mm) from the Prospect Mesa gravels and central Wright Valley colluvium. Axes labels as in Figure 8. Prospect Mesa gravels are represented by MPG79007 (n = 50 clasts) from excavation 7972. Prospect Mesa colluvium is represented by MPG79001 (excavation 7972, n = 50 clasts), MPG84002 (excavation 841, n = 85 clasts) and MPG84004 (excavation 841, n = 72 clasts). Odin colluvium is represented by MPG84022 (excavation 8418, n = 116 clasts), MPG84023 (excavation 8418, n = 120 clasts), MPG84024 (excavation 8419, n = 108 clasts), MPG84028 (excavation 8425, n = 165 clasts), and MPG84031 (excavation 8426, n = 90 clasts) from the northeastern corner of Odin colluvium.

might represent multiple glaciations from the west and, possibly, from the east or local alpine areas. Given the variety of source areas for glaciers that could traverse the valley, the variety of mass-wastage processes that could generate diamictons, and the great length of time available for sedimentation, the simplification represented by the single glaciation hypothesis cannot be overemphasized. We now present our evidence for Peleus till on an outcrop-by-outcrop basis.

Peleus till in the Lake Canopus region between Lake Vanda and Sykes Glacier to the south is very poorly sorted (Figure 17) and has a multimodal size frequency distribution (Table 2 and Figure 18a). This till exhibits a primary mode in the medium-to-fine sand sizes and a secondary mode in the coarse pebble sizes (Figure 18a). Gravel percentages are high and uniform; mud percentages are low. Peleus gravel from this region shows a significant degree of glacial marking (Figure 19a) and no signs of weathering in a polar desert climate. Clast form is compact bladed to bladed (Figure 20a). The lithologic distribution of Peleus gravel is dominated by granite but contains a moderate percentage of dolerite and a small percentage of dike rocks. Importantly, the till contains some Vida Granite and no metamorphic or sedimentary rocks. Given that Vida Granite crops out on or near the valley floor only to the west of this region in the North Fork and South Fork (Figure 2), the presence of Vida Granite in Peleus till south of Lake Vanda suggests that Peleus ice flowed from west to east. If the Peleus ice had advanced from

the east, metamorphic rocks, which crop out east of Bartley Glacier, should be a constituent of the till.

Peleus till from below Mount Jason contains more gravel than that around Lake Canopus and has a slightly finer sand fraction mode. Gravel from this Peleus till exhibits significant glacial markings, carbonate crust, and some ventifaction. This gravel also contains a significant percentage of metamorphic lithologies, as well as half the granite and twice the dolerite as samples from Lake Canopus. Further, Vida Granite occurs only in trace quantity in one of three samples examined. The presence of the metamorphic rocks might indicate that this till was deposited by glacier ice from the eastern end of Wright Valley. Alternatively, these rocks may have come from xenoliths in granite to the west of this location. Until a stronger case for eastern-source glaciation can be built, we prefer the xenolith explanation. The clasts in this patch exhibit a wider range of forms than those around Lake Canopus (Figure 20a).

The patch of Peleus till that overlies OPWD and the patch that stretches from the Heimdall Glacier to the Conrow Glacier are considered a single unit. The Onyx ponds and Heimdall-Conrow Peleus do not exhibit as high a sand size mode as Peleus patches to the west (Figure 18b). Gravel percentages are relatively high at the low-elevation outcrops. The lithologic distributions, surface textures, and form of the gravel from these two patches are nearly identical (Figures 19 and 20). Both patches overlie striated and molded bedrock indicating

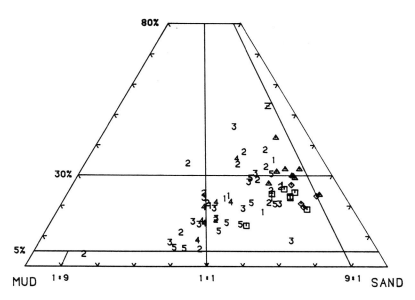

Fig. 17. Ternary diagram for gravel, sand, and mud percentages of central Wright Valley glacial drifts and colluvium. Numbers 1–5 identify groups of Peleus till samples with group number increasing from west to east. Groups are as follows: 1, excavations 791–794; 2, excavations 7927–7930; 3, excavations 7932–7941; 4, excavations 7949–79105; and 5, excavations 79121–79134, S8470, and 8447. Symbols represent colluvium and alpine glacier drift. Square, colluvium from excavations 8348–8351. Z, colluvium from excavation 841. Triangle, alpine III glacial drift from excavations S8422–S8449. Diamond, alpine III glacial drift from excavations 8446 and S8441–S8469.

ice flow from the west. The absence of metamorphic lithologies is consistent with a westerly ice source.

At Prospect Mesa, Peleus till is 6 m thick and directly overlies PMG. The till is poorly sorted but commonly exhibits a strong medium sand mode and no mud mode (Figure 18b). This trait is unlike Peleus at either the Onyx ponds or the Heimdall-Conrow patches. Similar strong sand modes are shown by Peleus till at Lake Canopus. A possible source of this sand population is the underlying PMG which exhibits this mode (Figure 15). The base of the Prospect Mesa Peleus till is moderately stratified as shown by the frequency distribution with the strong primary mode at small pebble size. Peleus till exhibits a platy structure and dispersed gravel fabric. This structure is more consistent with deposition from debris flow than from lodgement by basal ice. The lithologic distribution of Peleus gravel here is identical to that characterizing the Heimdall-Conrow patch. The degree of glacial marking of Peleus gravel is stronger at Prospect Mesa than elsewhere in the valley. Peleus clasts at Prospect Mesa are commonly compact bladed (Figure 20a).

Peleus till at Prospect Mesa contains a distinctive assemblage of marine diatoms and a few unidentifiable shell fragments (Tables 3 and 8). The critical species are *Nitzschia kerguelensis*, *Cosmiodiscus insignis*, *Thalassiosira vulnifica*, and *T. lentiginosa*. Because these same species occur within PMG, we infer that they are reworked into Peleus till from PMG or equivalent de-

posits in Wright Valley. The ages of these diatoms suggest that the Peleus till is younger than about 3 Ma.

A number of alternative interpretations for the deposition of Peleus till at Prospect Mesa have been proposed. One is that Peleus till here is ice-rafted debris. We do not favor this interpretation because of the absence of sedimentary structures in the Peleus diagnostic of subaqueous deposition. Another origin, as interglacial mudflow, was proposed by Calkin and cannot be ruled out. The strength of this hypothesis is that it best explains the gravel fabric and the outcrop pattern of the unit. By this hypothesis, the presence of glacially marked clasts in the Peleus indicates that the source material for the mudflows was previously deposited basal till. A flaw in this hypothesis is the presence of PMG diagnostic diatoms in the Peleus till meters above its base. Other weaknesses are the absence of surface-weathering features on the Peleus gravel and the lack of stratification through the 6-m thickness of the Peleus except at the base. A significant implication of the debris flow hypothesis for the Peleus till at Prospect Mesa is that the glaciation represented by the till could be older than PMG.

Peleus till between Conrow and Bartley glaciers as well as to the east exhibits fairly uniform grain size distributions similar to Peleus till to the west (Figure 18c). Peleus till from east of Bartley Glacier shows a primary mode in the silt sizes and represents the fine-grained extreme of the till. Gravel from the Conrow-

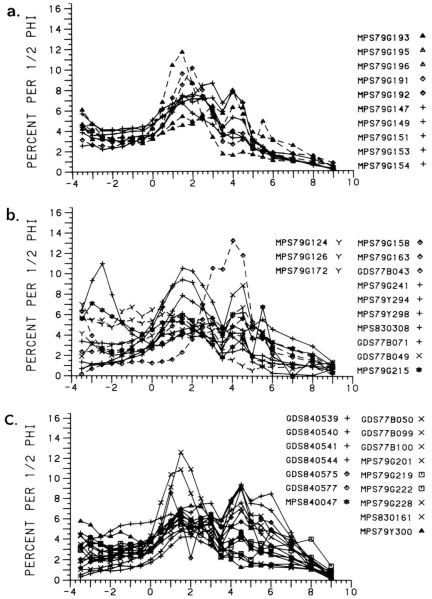

Fig. 18. Grain size frequency distributions for Peleus till from the (*a*) western field area, (*b*) central field area, and (*c*) eastern field area. The till crops out as separate patches. (*a*) MPS79G193 (excavation 791), MPS79G195 (excavation 791), MPS79G196 (excavation 794), MPS79G191 (excavation 793), and MPS79192 (excavation 793) are from the Lake Canopus patch. MPS79G147 (excavation 7929), MPS79G149 (excavation 7928), MPS79151 (excavation 7927), MPS79G153 (excavation 7927), and MPS79154 (excavation 7927) are from the Mount Jason patch. (*b*) MPS79G158 (excavation 7941), MPS79G163 (excavation 7944), and GDS77B043 (excavation 7940) are from the western Heimdall-Conrow patch. MPS79G241 (excavation 7979), MPS79Y294 (excavation 7995), MPS79Y298 (excavation 7972), MPS830308 (excavation 8356), and GDS77B071 (excavation 7972) are from Prospect Mesa. GDS77B049 (excavation 7963) is from the head of the Prospect fan. MPS79G215 (excavation 79109) is from the eastern Heimdall-Conrow patch. MPS79G124 (excavation 7933), MPS79G126 (excavation 7933), and MPS79G172 (excavation 7932) are from the Onyx ponds patch. (*c*) GDS77B050 (excavation 79129), GDS77B099 (excavation 79127), GDS77B100 (excavation 79127), MPS79G201 (excavation 79130), MPS79G219 (excavation 79121), MPS79G222 (excavation 79125), and MPS830161 (excavation S8338) are from the Conrow-Bartley patch. MPS79G228 (excavation 79133) and MPS79Y300 (excavation 79134) are from the Bartley and Meserve bend patches. GDS840539, GDS840540, and GDS840541 (all from excavation D8413) and GDS840544 (excavation S8449) are from the Bartley-Meserve patch. GDS840575 (excavation S8461) and GDS840577 (excavation S8461) are from the Meserve-Hart patch. MPS840047 (excavation 8444) is from the shoulder of the Olympus Range south of the eastern Clark Glacier snout and north of Denton Glacier (see Figure 2).

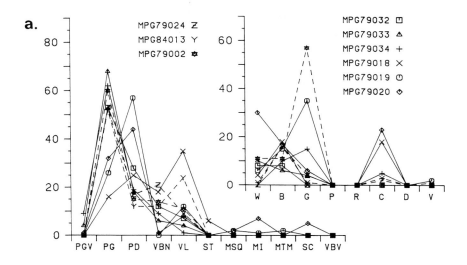

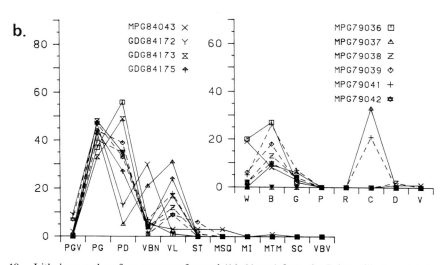

Fig. 19.   Lithology and surface texture of gravel (16–64 mm) from the Peleus till. Axes labels as in Figure 8. (*a*) MPG79032 (excavation 797, *n* = 50 clasts), MPG79033 (excavation 7910, *n* = 50 clasts), and MPG79034 (excavation 7916, *n* = 41 clasts) are from the Lake Canopus patch. MPG79018 (excavation 7929, *n* = 28 clasts), MPG79019 (excavation 7928, *n* = 52 clasts), and MPG79020 (excavation 7927, *n* = 53 clasts) are from the Mount Jason patch. MPG79024 (excavation 7933, *n* = 50 clasts) is from the Onyx ponds area. MPG84013 (excavation 8411, *n* = 74 clasts) is from Heimdall-Conrow patch across the Onyx River from Prospect Mesa. MPG79002 (excavation 7972, *n* = 54 clasts) is from Prospect Mesa. (*b*) MPG79036 (excavation 79118, *n* = 49 clasts), MPG79037 (excavation 79109, *n* = 33 clasts), MPG79038 (excavation 79121, *n* = 46 clasts), MPG79039 (excavation 79125, *n* = 51 clasts), MPG79041 (excavation 79130, *n* = 43 clasts), and MPG79042 (excavation 79131, *n* = 51 clasts) are from the Conrow-Bartley patch. GDG84172 (excavation D8413, *n* = 45 clasts), GDG84173 (excavation D8413, *n* = 46 clasts), and GDG84175 (excavation S8471, *n* = 41 clasts) are from east of Bartley Glacier. MPG84043 (excavation 8444, *n* = 197 clasts) is from the shoulder of the Olympus Range between the eastern snout of Clark Glacier and Wright Valley (see Figure 2).

Bartley patch is dominantly granite and dolerite with no metamorphic lithologies (Figure 19). With granite and dolerite cropping out only to the west and metamorphic rocks exposed only to the east of Bartley Glacier, the lithologic distributions support ice flow from the west. Interestingly, no volcanic fragments were found in the Peleus till which is overlain by alpine glacier drift rich in volcanics. Gravel from the Peleus in these patches

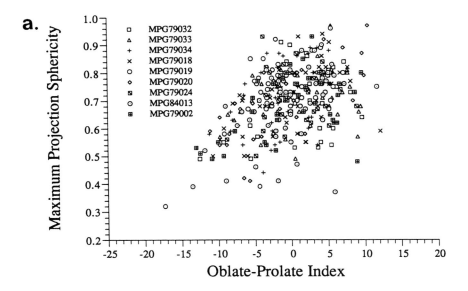

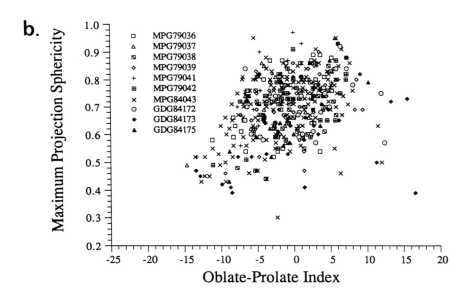

Fig. 20. Maximum projection sphericity and oblate-prolate index of granite (PG, PGV) and dolerite (PD) gravel (16–64 mm) from Peleus till samples described in Figure 19. (*a*) Western field area. (*b*) Eastern field area.

exhibits glacial markings but no evidence of exposure in a polar desert (Figure 19). The vast majority of clasts exhibit a compact-bladed form (Figure 20*b*).

Over the Conrow-Bartley patch, particularly below 500 masl, Peleus till is covered by a very thin coarser-grained deposit that is rich in dolerite (Figure 3). This deposit might reflect (1) an ablation till facies associated with the Peleus basal till, (2) a separate axial or alpine glaciation, or (3) ice-rafted detritus if a marine incursion accompanied the retreat of Peleus ice. We prefer the second interpretation because of the increased density

of this deposit below 500 m between two formerly more extensive alpine glaciers. Furthermore, there is an absence of such veils on Peleus till where there are no alpine glaciers.

Soils derived from Peleus till exhibit a well-developed desert pavement overlying a stained zone that averages 34 cm in thickness (Table 9). A weakly cemented salt pan averaging 13 cm in thickness occurs within this stained zone. Using criteria of *Campbell and Claridge* [1975, 1987], we assigned soils on Peleus till to weathering stage 5 (stage 1, least weathered; stage 6, most

TABLE 9. Morphological Properties of Central Wright Valley Soils

Deposit*	Depth of Stain, cm	Depth of Ghosts, cm	Depth of Visible Salts, cm	Depth of Coherence, cm	Salt Pan Thickness, cm	Salt Stage	Weathering Stage†	CDE§	N
RIP	$1 \pm 4\P$	4 ± 9	18 ± 8	18 ± 8	0	1	2	12	8
WLD	40 ± 1	26 ± 6	38 ± 53	70	0	1	4	13 ± 7	2
AIII	88 ± 17	16 ± 15	61 ± 24	91 ± 14	25 ± 14	5.2 ± 1	5.8 ± 0	13 ± 3	18
PMC	33 ± 10	19 ± 28	42 ± 35	37 ± 11	13 ± 8	4 ± 0	5 ± 1	9 ± 4	5
OC	23 ± 10	7 ± 6	16 ± 3	60 ± 35	16 ± 3	4 ± 1	5 ± 2	12 ± 0	3
PT	34 ± 16	16 ± 7	30 ± 29	92 ± 17	13 ± 9	4	5 ± 1	13 ± 8	10
Pt\PMC‡	20 ± 11	0	20 ± 11	54 ± 18	14 ± 2	5 ± 1	6 ± 1	8 ± 4	3
PT\AIIIb	44 ± 8	0	40 ± 3	44 ± 8	0	2 ± 1	6 ± 0	8 ± 0	2

*See Table 1.
†Weathering stages after *Campbell and Claridge* [1975].
‡Soil in PT (Peleus till) buried by PMC (Prospect Mesa colluvium).
§Color development equivalent [from *Buntley and Westlin*, 1965].
¶One standard deviation.

weathered). Peleus soils were also observed buried beneath younger deposits (Table 9). These buried Peleus soils exhibit less development than those subaerially exposed.

The character of Peleus till suggests that the depositing ice was substantially wet based up to the highest Peleus outcrop at 1150 masl. The striated bedrock, striated clasts, and abundant fine matrix indicate considerable basal melting and related sliding [*Boulton*, 1978]. Two explanations for this have been proposed. The first is glaciological and invokes significant ice thickness above 1150 masl and significant ice velocity to bring the ice base to the pressure melting point [*Denton et al.*, 1984]. The second is relatively thin ice and warmer-than-present climate [*Prentice et al.*, 1987]. This issue remains problematic but is critical to inferring the climate during the Peleus glaciation.

Prospect Mesa Upper Water-Laid Diamicton

The Prospect Mesa upper water-laid diamicton (Table 1) overlies Peleus till at Prospect Mesa (Figure 14). This deposit is fine grained (Table 2 and Figures 6 and 11) and slightly stratified, indicating subaqueous deposition. Isolated pockets of pebbly sand probably settled from floating ice. Gravel in this unit exhibits bladed to compact-bladed form (Figure 13). Granite and dolerite are common. Metamorphic rocks are absent (Figure 12). A search through three samples turned up no microfossils.

Prospect Mesa Colluvium

Prospect Mesa colluvium unconformably overlies both Peleus till and Prospect Mesa upper water-laid diamicton (Figure 14). Prospect colluvium exhibits a scarp that slopes down to the east and trends parallel to the mesa long axis. The coarse grain size (Figure 21), poor sorting, lack of stratification, and clast-supported

framework of Prospect colluvium all suggest a mass wastage origin. The location of this deposit within a large debris fan fed by flows from Bull Pass supports this interpretation. The lithologic distribution of the gravel is noteworthy because it contains very little granite, high percentages of both light- and dark-colored dike rocks, and no metamorphic rocks (Figure 16). The average form of granitic clasts from the colluvium is significantly less compact than that of clasts from Peleus till (Figure 22). Colluvium gravel lacks glacial markings but exhibits polar desert surface weathering features (Figure 16). This fact is evidence against a glacial origin for this deposit. Soils on Prospect Mesa colluvium were assigned to weathering stage 5 (Table 9).

Hart Ash

Volcanic ash crops out at three locations in eastern Wright Valley near the snout of Hart Glacier at elevations as low as 378 masl (Table 1 and Figure 2 [*Hall*, 1992; B. L. Hall et al., Late Tertiary Antarctic paleoclimate and ice sheet dynamics inferred from surficial deposits in Wright Valley, submitted to *Geografiska Annaler*]). The ash overlies colluvium and underlies drift deposited by Wright Lower Glacier [*Hall*, 1992]. The underlying colluvium contains gravel of only local lithologies. The Hart ash ranges up to 90 cm in thickness and consists of two distinct units. The lower unit is pure volcanic ash. Under SEM examination, the glass shards are angular and exhibit fragile spines (B. McIntosh, personal communication, 1986). Overlying the pure ash is a layer of ash mixed with some ventifacts and sand.

The pure composition and angularity of the lower ash unit as well as its sharp basal contact indicate that this ash is an in situ airfall deposit [*Hall*, 1992]. The upper unit is probably reworked. The poorly developed desert pavement beneath the ash suggests attainment of a polar

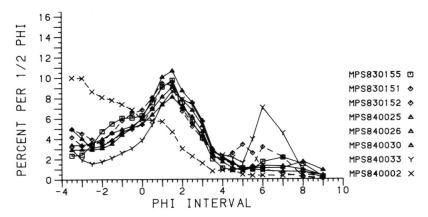

Fig. 21. Grain size frequency distributions for Prospect Mesa colluvium and Odin colluvium. MPS840002 (excavation 841) represents Prospect Mesa colluvium, the highest unit of the Prospect Mesa south as shown on map in Figure 14. MPS830151 (excavation 8348) and MPS830152 (excavation 8349) represent Odin colluvium above 500 masl, north of Siegfried Peak. MPS830155 (excavation 8351) represents a sandstone-rich debris flow shown in Figure 3. MPS840030 (excavation 8423) is from the westernmost sandstone-rich debris flow. MPS840025 (excavation 8418), MPS840026 (excavation 8419), and MPS840033 (excavation 8426) represent the easternmost sectors of Odin colluvium below 250 masl.

desert climate prior to the eruption. Two separate samples of glass from Hart ash yield a K/Ar mean age of 3.9 ± 0.3 Ma.

Because the Hart ash erupted and fell subaerially, no fjord reaching 378 masl could have existed here at 3.9 ± 0.3 Ma. Considering the 270-masl elevation of the valley mouth threshold, we cannot say that the Prospect Mesa

gravels fjord would have covered the location of the Hart ash. Hence this ash does not necessarily postdate PMG and the Prospect Fjord episode on the basis of position. We do suggest that Hart ash could not have survived the Peleus glaciation and therefore must postdate it. If correct, the age of Hart ash is minimum for the Peleus glaciation.

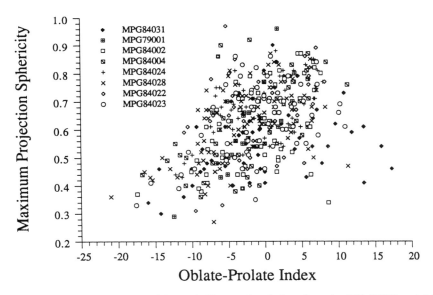

Fig. 22. Maximum projection sphericity and oblate-prolate index of granite (PG, PGV) and dolerite (PD) gravel (16–64 mm) from central Wright Valley colluvium. Colluvium units are Prospect Mesa (MPG79001, MPG84002, MPG84004) and Odin colluvium (MPG84022, MPG84024, MPG84028, MPG84031). Samples are described in Figure 16.

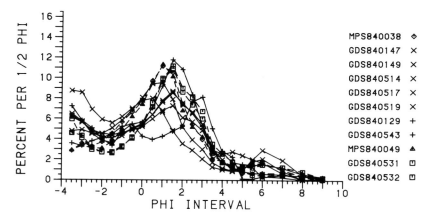

Fig. 23. Grain size frequency distributions for alpine IV and III glacier drift. MPS840038 (excavation 8433) represents left-lateral drift of Heimdall Glacier. GDS840129 (excavation S8422), GDS840147 (excavation S8426), and GDS840149 (excavation S8426) represent left-lateral drift of Bartley Glacier. GDS840514 (excavation S8436), GDS840517 (excavation S8437), GDS840519 (excavation S8438), and GDS840543 (excavation S8449) represent right-lateral drift of Bartley Glacier. GDS840531 and GDS840532, both from excavation D8417, represent left-lateral drift of Meserve Glacier. MPS840049 (excavation 8446) represents left-lateral drift of Conrow Glacier.

Alpine Glacier Drifts

Drifts deposited by alpine glaciers overlie Peleus till along the southern valley wall (Figure 3). These drifts have been subdivided into numerous drift sheets reflecting a long history of fluctuations on the basis of moraine morphology, soil stratigraphy, and radiometric dating [*Everett*, 1971; *Calkin and Bull*, 1972; *Hall*, 1992]. The outermost alpine drifts, reflecting alpine IV and III glaciations, are characterized by large and heavily dissected moraines [*Hall*, 1992]. Smaller, well-preserved moraines reflecting alpine II and I glaciations are set within the outermost limit of alpine IV and III drift.

Alpine IV and III drift is coarse grained and exhibits a dominant coarse-to-medium sand size mode (Figure 23 and Tables 1 and 2). Alpine III drift contains gravel size basalt, in some cases in high abundance (Figure 24). Outer alpine drift gravel also exhibits surface textures diagnostic of considerable polar desert surface weathering. Outer alpine drift plutonic clasts exhibit a wide variety of forms including the very platy and very bladed forms absent in Peleus till (Figure 25). We assigned soils on outermost alpine drift to weathering stage 6, the most strongly weathered category recognized by *Campbell and Claridge* [1975, 1987] (Table 9). At excavations S8341 and S8449, outer alpine drift overlies a soil horizon developed in Peleus till.

Basalt from Bartley and Meserve outer alpine drift yields K/Ar dates ranging from 3.8 Ma [*Armstrong*, 1978] to 3.4 Ma [*Fleck et al.*, 1972]. Basalt clasts from Meserve alpine III drift yielded ages based on the ^{40}Ar/^{39}Ar method of 3.5 ± 0.1 Ma [*Hall*, 1992]. These dates are all maximum for alpine IV drift.

Odin Colluvium

Odin colluvium (Table 1) is widespread and discontinuous on the southern wall of Wright Valley below Mount Odin (Figure 3). The edges of Odin colluvium

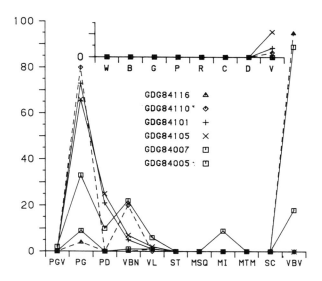

Fig. 24. Lithology and surface texture of gravel (16–64 mm) from alpine IV and III glacier drift. Axes labels and scaling as in Figure 8. GDG84116 (S8444, *n* = 51 clasts) is from the left-lateral drift of Meserve Glacier. GDG84110 (excavation S8440, *n* = 53 clasts), GDG84101 (excavation S8437, *n* = 48 clasts), and GDG84105 (excavation S8438, *n* = 54 clasts) represent right-lateral Bartley Glacier drift. GDG84005 (excavation S8422, *n* = 50 clasts) and GDG84007 (excavation S8427, *n* = 50 clasts) represent left-lateral drift of Bartley Glacier.

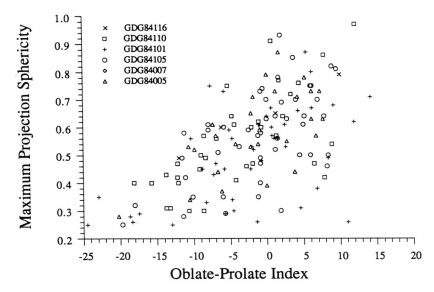

Fig. 25. Maximum projection sphericity and oblate-prolate index of granite (PG, PGV) and dolerite (PD) gravel (16–64 mm) from alpine glacier drift samples described in Figure 24.

patches are strikingly irregular. Odin colluvium directly overlies bedrock except at a small mesa where Odin colluvium overlies Peleus till (Figure 4, excavation 8421). Some Odin patches at higher elevation on the valley wall are clearly associated with debris fans emanating from various gullies on Mount Odin. The debris fan morphology of some Odin patches identifies them as colluvium.

By texture and gravel characteristics, Odin colluvium is clearly distinguishable from Peleus till but not alpine drift. Unlike gravel from Peleus till, Odin colluvium gravel contains a significant percentage of sandstone and quartzite (Figure 16). The quartzite was probably derived from sandstone on Mount Odin and became quartz cemented through long-term exposure in a polar desert climate [*Weed and Ackert*, 1986]. Hence quartzite is not an erratic here. Significant polar desert weathering is also suggested by the desert varnish, ventifaction, and pitting on Odin gravel (Figure 16). Unlike gravel in the Peleus till, Odin gravel exhibits many noncompact forms with low mps (Figure 22). The absence of erratics and glacial markings further suggest that Odin colluvium has not been glacially transported. Soils on Odin colluvium are assigned to weathering stage 5 and are similar to soils on Prospect Mesa colluvium (Table 9).

Wright Lower Glacier Drift

Thin discontinuous strips of diamicton that are gently concave toward the east mantle Peleus till and alpine drift in front of Bartley and Meserve glaciers (Figure 3 [*Hall*, 1992]). This drift is distinguished by its coarse texture [*Bockheim*, 1979], metamorphic gravel content

(Figure 26), and basalt clasts [*Hall*, 1992]. The form of the plutonic gravel is significantly less compact than that of Peleus till gravel (Figure 27). *Bockheim* [1979] referred to these drift patches as the "D" drift of the Wright Lower Glacier. *Hall* [1992] called this drift Wright drift and suggested that it is younger than Hart ash but older than alpine II drift. The metamorphic gravel in Wright drift indicates that the depositing ice

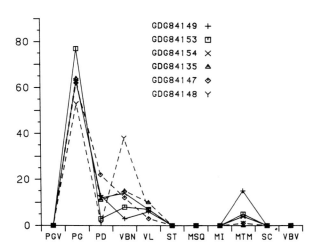

Fig. 26. Lithology of gravel (16–64 mm) from Wright Lower Glacier drift. Axes labels as in Figure 8. GDG84153 (*n* = 50 clasts) and GDG84154 (*n* = 46 clasts) represent drift across the Onyx River from Meserve Glacier. GDG84149 (excavation S8459, *n* = 49 clasts) and GDG84135 (excavation S8454, *n* = 52 clasts) represent drift in front of Hart Glacier. GDG84147 (*n* = 53 clasts) and GDG84148 (*n* = 43 clasts) are from excavation S8458 across the Onyx River from Meserve Glacier.

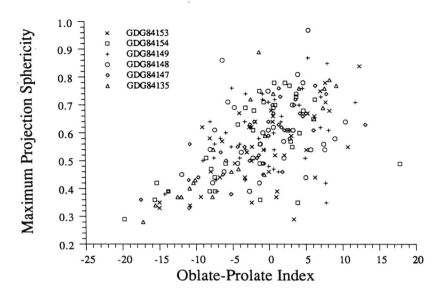

Fig. 27. Maximum projection sphericity and oblate-prolate index of granite (PG, PGV) and dolerite (PD) gravel (16–64 mm) from Wright Lower Glacier drift described in Figure 26.

advanced from the eastern end of Wright Valley where such rocks are extensively exposed.

Coarse Ripples

Coarse-grained isolated ripples overlie bedrock, Peleus till, central valley colluvium, and alpine IV–III drift (Figure 3 [*Prentice et al.*, 1985]). The ripples are up to 50 cm thick. Texturally, the ripples are bimodal and exhibit strong sand and gravel modes (Figure 28). Throughout the depth of the ripples, the gravel is highly ventifacted and varnished. This surface texture indicates that the gravel weathered in a polar desert climate before the ripples formed most likely by subaerial processes (Figure 29). Clast form is commonly bladed (Figure 30).

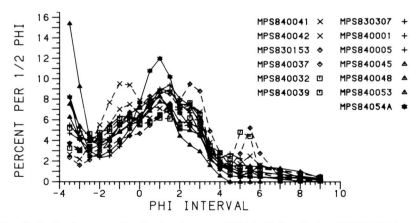

Fig. 28. Grain size frequency distributions for central Wright Valley ripples. MPS830307 (excavation 8355), MPS840001 (excavation 841), and MPS840005 (excavation 842) represent ripples on top of Prospect Mesa south. MPS840045 (excavation 8442), MPS840048 (excavation 8446), and MPS840053 (excavation 8447) represent ripples on alpine IV and III left-lateral moraines of Conrow Glacier. MPS840041 (excavation 8437) and MPS840042 (excavation 8438) are from the northeastern corner of Odin colluvium. MPS840032 (excavation 8426) and MPS840039 (excavation 8429) are from ripples just upslope from excavations 8437 and 8438. MPS840037 (excavation 8433) and MPS830153 (excavation 8349) represent ripples at 500 masl north of Siegfried Peak. For contrast, MPS84054A (excavation 8448) represents a flood plain ripple in front of Conrow IV drift.

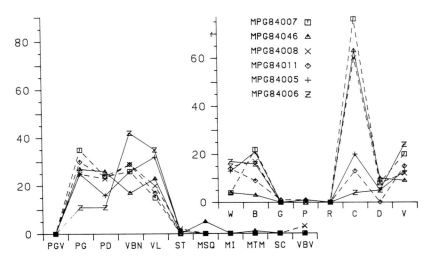

Fig. 29. Lithology and surface texture of gravel (16–64 mm) from ripples. Axes labels as in Figure 8. MPG84007 (*n* = 93 clasts) and MPG84008 (*n* = 78 clasts) are from excavation 844 south of Prospect Mesa. MPG84011 (excavation 848, *n* = 85 clasts) is from just above excavation 844. MPG84005 (excavation 842, *n* = 82 clasts) and MPG84006 (excavation 843, *n* = 92 clasts) are from the top of Prospect Mesa. MPG84046 (excavation 8448, *n* = 78 clasts) is from the flood plain below Conrow Glacier from Prospect Mesa.

Sediment and Bedrock Erosion Patterns

An important aspect of central Wright Valley geology is the array of discontinuous and complexly shaped patches of water-laid diamictons, Peleus till, colluvium, and alpine drifts (Figure 3). To what extent do these patterns represent subaerial erosion and deposition as opposed to differential subglacial erosion? Answering these questions is crucial to determining the origin of

some deposits as well as reconstructing the climates of subaerial reworking of the landscape.

Small patches of Peleus till are particularly difficult to explain. Examples are the small patch of Peleus till covering OPWD within an expanse of bare bedrock and the patch of Peleus till on PMG at Prospect Mesa. Was the Peleus till more widespread in these areas at the close of the Peleus glaciation with subsequent removal

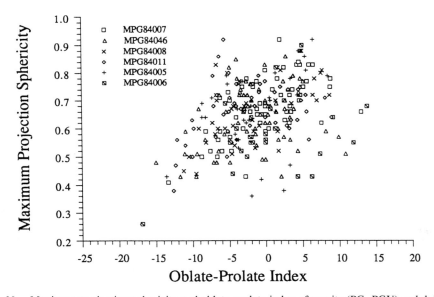

Fig. 30. Maximum projection sphericity and oblate-prolate index of granite (PG, PGV) and dolerite (PD) gravel (16–64 mm) from Wright Valley ripples. Samples are described in Figure 29.

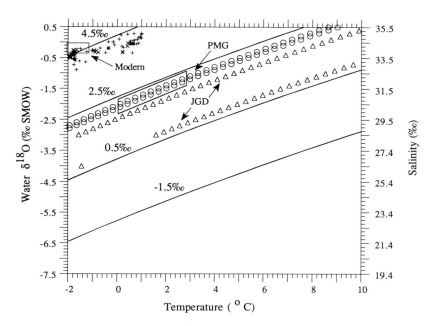

Fig. 31. Water temperature and water $\delta^{18}O$ (salinity) interpretations for Wright Valley carbonate $\delta^{18}O$. Sloping solid lines represent equilibrium calcite $\delta^{18}O$ calculated from the paleotemperature equation of *Epstein et al.* [1953]. The relation between water $\delta^{18}O$ and salinity was derived from *Jacobs et al.* [1985] and *Potter and Paren* [1985]. Modern *Adamussium colbecki* $\delta^{18}O$ average 4.5‰ and represent Ross Sea water conditions enclosed in the rectangle. Circles show the locus of water temperature and $\delta^{18}O$ in equilibrium with shell $\delta^{18}O$ from Prospect Mesa gravels. Triangles show the same for Jason glaciomarine diamicton. Pluses and crosses in upper left corner depict water temperature and $\delta^{18}O$ data from the northern edge of the George VI Ice Shelf at 70°S [*Potter and Paren*, 1985] and the Weddell Sea [*Weiss et al.*, 1979], respectively. Enclosed area around PMG $\delta^{18}O$ shows our best hypothesis for Prospect Fjord water temperature and salinity.

by subaerial processes, such as wind deflation, or even redeposition by debris flow? This is the traditional interpretation [e.g., *Webb*, 1972] of central Wright Valley erosional patterns. This view supports interpretation of Peleus till at Prospect Mesa as reworked and subaerial processes are very important. In this view, Peleus till patch geometry would have much less to do with the Peleus glaciation.

The alternative interpretation is that the discontinuity of the Peleus till patches reflects selective glacial erosion during the final stages of the Peleus glaciation. Where the basal Peleus ice thermal regime became wet-freezing, the underlying sediment may have been frozen into the overriding ice and removed from the region. We favor this mechanism primarily because of the evidence of erratic-free bedrock landscapes adjacent to Peleus till. Given the significant abundance of resistant boulders in Peleus till, it is difficult to imagine that the till was originally more extensive around these patches and then completely stripped back, including the large resistant boulders, by deflation. More likely, the complexly shaped patches primarily reflect glacial erosional and depositional processes.

DISCUSSION

Late Miocene Jason Fjord Episode

We infer glacier recession from Wright Valley during the Jason Fjord episode at about 9 ± 1.5 Ma. We interpret the negative $\delta^{18}O$ of Jason marine carbonates relative to modern carbonates from McMurdo Sound as reflecting Jason Fjord waters at least slightly warmer and less saline than local seawater at present. Assuming equilibrium $\delta^{18}O$ precipitation, a range of water temperatures and salinities are consistent with the data (Figure 31). We cannot estimate the water depth of shell precipitation within the fjord to constrain the ambiguity in interpretation because the shell fragments are not in situ. Even if JGD carbonates precipitated near the fjord surface where salinity is expected to be at a minimum, the low-shell $\delta^{18}O$ is not consistent with a climate similar to the present. Salinities in the region today [e.g., *Jacobs et al.*, 1985] are never less than 30‰. Salinity and water $\delta^{18}O$ lowering sufficient to explain the low JGD carbonate $\delta^{18}O$ values apparently require some warming from the modern climate to increase melting.

From Jason Fjord data, we estimate the elevation of Wright Valley in the late Miocene and from this the tectonic uplift rate of this region since this time. The fragmentation of marine diatoms in Jason diamicton and the abundance of benthic species suggest that the Jason Fjord was shallow. Benthic diatoms are especially important as they are restricted to the photic zone, which we estimate on the basis of modern fjords and the character of JGD as less than 20 m. However, benthic diatoms are commonly reworked within fjords and concentrated in fossil assemblages because of their increased resistance relative to planktonic taxa. Hence water depth cannot be inferred from basin center deposits of JGD such as that at the bottom of Lake Vanda in DVDP Hole 4A. Given JGD cropping out as high as 250 masl and adding 50 m for a photic zone and minimal connection with the Ross Sea over the valley mouth threshold, we estimate the present elevation of the Jason Fjord shoreline at about 300 m. Assuming sea level highstands in the late Miocene were not significantly different from today [*Schroeder and Greenlee*, 1993], Jason Fjord data suggest a maximum uplift of the area of 400 m since 9 ± 1.5 Ma. Suggested Cenozoic uplift rates of the Transantarctic Mountains range from 100 m/m.y. [*Gleadow and Fitzgerald*, 1987] to about 600 m/m.y. [*Webb et al.*, 1986].

The interpretation that the local mountains were only slightly lower in elevation at 9 ± 1.5 Ma than they are today has important implications for the size of the Antarctic ice sheet prior to this time. Numerous workers [e.g., *Brady*, 1982] have inferred that the valley was carved by an outlet glacier of the East Antarctic Ice Sheet. *Denton et al.* [1984], on the other hand, argued that Wright Valley was carved beneath a local ice cap. Although the valley itself does not provide unequivocal evidence for either glaciation style, we are inclined toward ice sheet erosion to produce the overdeepened longitudinal profile. If the mountains were not much lower at 9 Ma than at present, the East Antarctic Ice Sheet had to have attained dimensions at least as large as at present prior to 9 ± 1.5 Ma to accomplish the overdeepening. This requirement conflicts with an alternative hypothesis that the mountains were considerably lower in elevation in the late Miocene [*Fitzgerald*, 1992] and that a smaller-than-present "dwarf" ice sheet overrode low mountains to erode the Wright Valley trough [*Webb et al.*, 1986].

Latest Miocene Prospect Fjord Episode

Climate. We infer an interval of glacier recession from Wright Valley during the Prospect Fjord episode in the early Pliocene. On the basis of the flora, fauna, and faunal isotopic evidence, we infer that the waters of the Prospect Fjord were warmer than present-day local waters. *Webb* [1972] previously inferred that Prospect Fjord bottom waters might have been as warm as 10°C

but that surface salinities were low enough to exclude planktonic foraminifers.

The marine diatoms in PMG and the absence of coccoliths suggest that Prospect Fjord water temperatures were 0°C to <3°C. The critical marine diatoms present are *N. kerguelensis*, *T. lentiginosa*, and *T. nitzschioides* (Table 3). The first two species dominate surface water and sediment between the Subantarctic front and the Antarctic slope front where annual water temperatures range between 0°C and 8°C [*Burckle*, 1984; *Burckle et al.*, 1987]. The third species occurs today in Subantarctic waters but may enter the Ross Sea during warm summers. If temperature primarily controls the distribution of these species and if their temperature preferences have not changed, these marine diatoms imply water temperatures from 0°C to 8°C. However, if water temperatures were higher than 3°C, coccoliths should be present, and they are not. Coccoliths will live at 0°C but will only secrete skeletons at 3°C or above (McIntyre, quoted by *Burckle and Pokras* [1991]).

Species of the genus *Chlamys* were common in the Southern Ocean from the early Cenozoic through the Pliocene but were restricted to southern temperate latitudes in the Pleistocene [*Speden*, 1962; *Beu*, 1985]. One explanation is that they became intolerant of polar temperatures south of the Antarctic Convergence. With the increase in calcium carbonate solubility as water temperatures decreased, *Chlamys* may have had difficulty precipitating its thick shell. Alternatively, other nearshore environmental changes, such as more extensive and persistent sea ice, may have contributed to the extinction of *Chlamys tuftsensis*. As both of these changes suggest conditions warmer in the early Pliocene than at present, we think it likely that Prospect Fjord waters were warmer than McMurdo Sound waters today.

Overall, we consider *C. tuftsensis* $\delta^{18}O$ to support these inferences for a Prospect Fjord warmer than local waters today. The average of the acceptable $\delta^{18}O$ is 2.2‰, which is about 2.3‰ more negative than modern scallop $\delta^{18}O$. Figure 31 shows the locus of water temperatures and $\delta^{18}O$ that, assuming equilibrium precipitation, are consistent with the observed $\delta^{18}O$. This average calcite $\delta^{18}O$ would indicate a temperature of 5°–6°C for Prospect Fjord waters, assuming no water $\delta^{18}O$ or salinity change. Alternatively, if temperature did not change but remained near freezing, the salinity of Prospect Fjord waters would have been as low as ~29.5‰. To resolve the ambiguity, we assume that the paleo–water depth at Prospect Mesa was at least 100 m [*Webb*, 1972, 1974] and that an estuarine circulation prevailed. Hence PMG should have been bathed by inflowing relatively dense water from McMurdo Sound. We consider it unlikely for intermediate-depth waters in McMurdo Sound to achieve salinities roughly 4‰ more negative than they are in the present climate without some climate warming. Even water near the melting bases of the Ross Ice Shelf or the George VI Ice Shelf today is no less saline than 32.5‰ [*Jacobs et al.*, 1985; *Potter and Paren*, 1985]. More likely, water salinities significantly lower than today required warmer-than-present seawater. We consider the minimum water temperature suggested on the basis of the marine diatoms, 0°C, to be the minimum consistent with *C. tuftsensis* $\delta^{18}O$.

Age. As was stated above, the average $^{87}Sr/^{86}Sr$ ratio of the two unaltered shells, 0.708975, suggests an age of 5.5 ± 0.4 Ma using the Site 758 $^{87}Sr/^{86}Sr$ record. Given the maximum reported stratigraphic range for the critical marine diatoms, we feel that the diatom ranges can accommodate the $^{87}Sr/^{86}Sr$-based age estimate. Both *Nitzschia kerguelensis* and *Thalassiosira insigna* are reported from early Pliocene sediments in the Pacific sector of the Antarctic Ocean [*Abrams*, 1986], while the latter species is also reported from upper Miocene sediments in the Indian sector [*Baldauf and Barron*, 1991]. Hence our best estimate for the age of PMG is 5.5 ± 0.4 Ma.

This age estimate for PMG contrasts with previous estimates. For instance, *Webb and Andreasen* [1986] proposed a middle-to-late Pliocene age for the PMG primarily because of the similarity of its foraminiferal fauna with that of the Scallop Hill Formation from the nearby coastal region. The Scallop Hill Formation was K/Ar dated as younger than 2.6 Ma [*Eggers*, 1979; *Webb and Andreasen*, 1986]. Because we suspect that the Scallop Hill Formation occurs only as glacial erratics and that its components were gathered from a variety of source beds, we cannot accept this age estimate for the Prospect gravels. *Brady* [1979] proposed an early Pliocene age for PMG on the basis of close similarities between the PMG foraminiferal fauna and that from "Zone II" [*Webb and Wrenn*, 1982] in DVDP Hole 10 from Taylor Valley (Figure 1). Zone II is referred to as the *Ammoelphidiella antarctica* Zone and is bounded between unconformities H2 and H3 [*Ishman and Rieck*, 1992]. Zone II is barren of diatoms but is sandwiched, below and above the unconformities, by sediments with early Pliocene diatoms. More recent work [e.g., *Baldauf and Barron*, 1991] indicates that the critical diatom species upon which the early Pliocene age assignment rests, *Thalassiosira oestrupi*, first appeared in the Southern Ocean at 5.1 ± 0.4 Ma. *Ishman and Rieck* [1992] suggested a middle Pliocene age, 3.7–3.8 Ma to 3.4 Ma, for the *A. antarctica* Zone in Taylor Valley, but this was based on their age interpretation of PMG.

Uplift. Similarities between PMG foraminiferal fauna and that living in the estuaries of the Pacific Northwest prompted *Webb* [1972, 1974] to suggest that the depth of the Prospect Fjord above Prospect Mesa did not exceed 100 m. This water depth estimate suggests that the modern elevation of the Prospect Fjord shoreline is about 300 masl or just above the elevation of the valley mouth threshold at 270 masl. We agree with this estimate based on the diatom flora. *Mayewski and Goldthwait* [1985], on the other hand, suggested that the Prospect Fjord shoreline presently occurs at 600 masl because that is the elevation of a deposit on the southern side of the valley that they correlated to PMG. We think that this other deposit of Mayewski and Goldthwait is Peleus till. On this basis, we suggest a maximum uplift of 400 m since 5–3.9 Ma, consistent with the minimal uplift estimated from JGD.

Neogene Peleus Glaciation Episode

Glaciation style and climate. Presently available evidence favors a thick valley glacier alternative for the Peleus ice mass at peak glaciation (M. L. Prentice et al., Neogene extreme glaciation: An ice sheet and climate model study, submitted to *Journal of Geophysical Research*, 1993) as opposed to a thick ice sheet alternative [*Denton et al.*, 1984]. We prefer the conservative interpretation of minimum dimensions for the Peleus ice mass. However, even the minimum configuration implies that the inland surface of the East Antarctic Ice Sheet was higher than it is now assuming negligible uplift since the Peleus glaciation. Modeling studies [e.g., *Prentice et al.*, 1993] indicate that warmer-than-present temperatures and higher accumulation rates are a possible mechanism for causing such ice sheet expansion. We think that a warmer-than-present climate at least partially explains the wet base of the Peleus valley glacier at 1150 masl, very near the upper ice surface. *Wilch et al.* [1992] inferred a warmer-than-present climate to explain Pliocene till with striated clasts up to similar elevations directly to the south in Taylor Valley.

An alternative explanation, assuming a valley glacier model, is that increased ice velocity producing strain heating and basal friction warmed the base of Peleus ice to the pressure melting point even at its upper edge. In favor of the glaciological explanation, some polar glaciers in the region today are partially wet based. *Robinson* [1984] suggested that Taylor Glacier, a polar valley glacier in adjacent Taylor Valley, is wet based over a substantial portion of its ablation zone (Figure 1). *MacPherson* [1987] suggested the same for the base of the Mackay Glacier to the north. In these examples, the thawed beds are restricted to the valley floor under a thick ice overburden and resulting high ice hydrostatic pressure.

The difference between these cases and the case of the Peleus till is that, for the Peleus till, the thawed ice bed ranged up to high elevation on the valley wall, very near the upper ice surface. Given the high inland threshold of Wright Valley over which Peleus ice from East Antarctica passed, Peleus ice velocities were not likely to be sufficient for strain heating to generate a wet bed. This unlikelihood implies that, during the Peleus glaciation, the climate was at least a little warmer than at present.

We know virtually nothing of the advance and retreat conditions for the Peleus glacier. We do not know whether the glacier advanced into the Prospect Fjord or a subaerially exposed valley. We do not know whether the climate was cooling or warming. Likewise, there is no discernible record of Peleus ice retreat. The retreat may have occurred subaerially or in conjunction with marine incursion. The climate may have warmed or cooled during retreat.

Age. We think that the Peleus till is older than 3.9 ± 0.3 Ma, the age of Hart ash. We are less certain about the maximum age of the Peleus till. As was stated previously, Peleus till might be in place at Prospect Mesa and so may be younger than PMG. Alternatively, Peleus till might be reworked at Prospect Mesa and so might be older than 5.5 ± 0.4 Ma.

The interpretation that Peleus till is younger than PMG has weaknesses. Those weaknesses at Prospect Mesa pertaining to the dispersed fabric of the gravel and platy structure have been mentioned. Another is that an age between 5 ± 1 and 3.9 Ma for Peleus till conflicts with evidence from the western Asgard Range (Figure 2, C-6) inferred to suggest that the entire region including Wright Valley has not been glaciated by a large ice mass since 13.6 Ma [*Ackert*, 1990; D. R. Marchant et al., Miocene glacial stratigraphy and landscape evolution of the western Asgard Range, Antarctica, submitted to *Geografiska Annaler*]. A strength of the early Pliocene age for Peleus till is that Taylor Glacier thickened to 1080 masl in the adjacent part of Taylor Valley during the late and possibly early Pliocene [*Wilch*, 1991]. Like the Peleus glaciation, these Taylor glaciations featured ice that was wet based up to its upper margin.

On the other hand, interpretation of Peleus till as older than 13.6 Ma also has weaknesses. One is the lack of glaciomarine or other sediments on top of Peleus till. Considering that Wright Valley was occupied by a fjord during much of the late Miocene and early Pliocene, it is odd that no sediment covers Peleus till if it is mid-Miocene in age. Because each interpretation for maximum age is problematic, we cannot presently specify the age of Peleus till more closely than Neogene, older than 3.9 Ma.

Post–Prospect Fjord Episode

We believe that outer alpine drift, Prospect Mesa colluvium, and Odin colluvium are middle-to-late Pliocene in age. As was explained above, alpine III drift age is constrained by radiometrically dated volcanics in Wright and Taylor valleys [*Hall*, 1992]. This age range is extended to Prospect Mesa colluvium and Odin colluvium on the basis of similar soil development, classified as weathering stages 5 and 6. Any of these deposits could be as old as 5.5 ± 0.4 Ma. Alternatively, they might only date to about 2.5 Ma.

The colluvium and alpine drift and the rest of the exposed central valley evidence suggest that the climate was commonly polar and that any warmings were at most modest. Gravel and soil weathering features associated with the post-Peleus deposits show significant evidence [*Bockheim and Ugolini*, 1990] of polar desert climate. For the gravel the evidence is ventifaction, desert varnish, and thick weathering rinds. The salt morphology of the Peleus soils also suggests long-duration exposure to aridity (Table 9).

The formation of the colluvium and advance of alpine glaciers to the outer alpine drift limits are evidence for a slightly warmer-than-present climate. Colluvium such as caps Prospect Mesa is not forming today and probably requires more running water which implies warmer-than-present temperatures. Advance to the outer alpine drift limit probably requires an increase in precipitation over present levels and so a warmer-than-present climate. The significant dissection of some of the central valley deposits such as the outer alpine moraines probably requires more running water than exists today.

The lack of evidence for glaciation of Wright Valley from East Antarctica in post-Peleus time may have as much to do with increased downcutting of the major through-glaciers to the north and south of the dry valleys (e.g., David Glacier) as with mountain uplift or climatically driven lowering in the maximum size of the East Antarctic Ice Sheet. The western threshold to Wright Valley is high in relation to thresholds of the major glacial valleys into interior East Antarctica. Late Pliocene and Pleistocene climatic fluctuations likely caused many East Antarctic Ice Sheet glaciations of other valleys, such as Taylor Valley, but not Wright Valley. In this view, the Wright Valley record only reflects the major glaciations of the East Antarctic Ice Sheet.

Implications for Late Neogene Antarctic Climate

The warmer-than-present climate that we infer for the Prospect Fjord episode and also the Peleus glacial episode is not consistent with the hypothesis for continuous polar desert climate like the present climate since the middle Miocene. Reconciliation may be possible if lapse rates were steeper in the Neogene than at present because the evidence for Miocene polar desert climate is from above 1200 masl [e.g., *Marchant et al.*, 1993]. Our evidence is also inconsistent with the extreme climatic warmth inferred by some [e.g., *Webb and Harwood*, 1991; *Harwood*, 1991] for the early Pliocene from marine diatoms and also *Nothofagus* twigs in Sirius Group deposits.

Central Wright Valley evidence is inconsistent with major East Antarctic Ice Sheet glaciation of the valley in the late Pliocene, as has been proposed by *Webb and Harwood* [1991], among others, from Sirius Group data. There is no evidence in the valley for major glaciation from the west since 3.9 Ma. East Antarctic Ice Sheet expansions since 3.9 Ma were not sufficient to surmount the high-elevation western threshold to Wright Valley.

CONCLUSIONS

We recognize four climate episodes in the Neogene history of central Wright Valley, Antarctica. At 9 ± 1.5 Ma, a shallow fjord was established in which was deposited the Jason glaciomarine diamicton. This age estimate is based on the $^{87}Sr/^{86}Sr$ ratio (0.708875) of Jason shells and is consistent with Jason marine diatoms. We infer that the local Transantarctic Mountains were only slightly lower than at present during the Jason Fjord episode. Hence the East Antarctic Ice Sheet, involved with carving the valley, could have exceeded its present dimensions prior to 9 ± 1.5 Ma. At 5.5 ± 0.4 Ma, another fjord was established in which the Prospect Mesa gravels were deposited. This age estimate derives from the $^{87}Sr/^{86}Sr$ ratio of pecten shells (0.708975) and marine diatoms. The presence of marine diatoms from the Antarctic Convergence and thick-shelled pectens, restricted today to temperate latitudes, with $\delta^{18}O$ values 2‰ more negative than characterize modern local pectens, coupled with the absence of coccolithophores suggest water temperatures in the Prospect Fjord were 0°C to <3°C. This level of warmth is inconsistent both with hypotheses for extreme early Pliocene warmth inferred from Sirius Group evidence and also with hypotheses for continuous polar desert climate since the middle Miocene.

In the Neogene before 3.9 Ma, the age of an in situ airfall ash deposit, a largely wet-based glacier draining the East Antarctic Ice Sheet filled Wright Valley depositing Peleus till. We infer that the climate during the Peleus glacial episode was warmer than at present. As the local mountains were near their present elevations, the Peleus glaciation suggests a larger-than-present East Antarctic Ice Sheet. During the post–Prospect Fjord episode, extensive colluvium and alpine glacier drift, possibly as old as 5.5 ± 0.4 Ma, were deposited. These deposits as well as the morphology of the Peleus till surface suggest a dominance of polar desert conditions with only modest climate warmings.

Acknowledgments. We thank R. Ackert, H. Conway, E. Downes, J. Leide, D. Marchant, N. Potter, B. Weed, and the New Zealand Antarctic Research Program for assistance in the field. M. Alford and C. Fink did the surveying. R. Laskowski, S. Wharton, R. Mendez, M. Boxwell, L. Caron, and R. Pershken helped with the lab work. D. Kelly drafted many figures. T. Kellogg and P. Webb identified Prospect gravel foraminifers. We thank D. Marchant and B. Hall for discussions. This manuscript has benefited from reviews by P. Barrett, P. Calkin, G. Denton, J. Kennett, P. Mayewski, and A. Stroeven. We thank G. Denton for inspiring the project and providing Wright Valley radiometric dates. M.L.P.'s research was supported by the Office of Polar Programs (OPP-9020975), National Science Foundation. C.S. was supported by the Swiss National Science Foundation (21-28971.90). This research benefitted greatly from OPP support to Denton.

REFERENCES

Abrams, N. J., Paleoceanographic and paleoclimate study of the Pacific sector of the Southern Ocean using Pliocene marine diatoms, M.S. thesis, 215 pp., Rutgers Univ., Newark, N. J., 1986.

Ackert, R. P., Jr., Surficial geology and geomorphology of Njord Valley and adjacent areas of the western Asgard Range, Antarctica: Implications for late Tertiary glacial history, M.S. thesis, Univ. of Maine, Orono, 1990.

Allan, J. R., and R. K. Matthews, Isotopic signatures associated with early meteoric diagenesis, *Sedimentology*, 29, 797–817, 1982.

Anderson, J. B., S. S. Shipp, L. R. Bartek, and D. E. Reid, Evidence for a grounded ice sheet on the Ross Sea continental shelf during the late Pleistocene and preliminary paleodrainage reconstruction, in *Contributions to Antarctic Research III*, *Antarct. Res. Ser.*, vol. 57, edited by D. H. Elliot, pp. 39–62, AGU, Washington, D. C., 1992.

Armstrong, R. L., K-Ar dating: Late Cenozoic McMurdo volcanic group and dry valley glacial history, Victoria Land, Antarctica, *N. Z. J. Geol. Geophys.*, 21, 685–698, 1978.

Baldauf, J. G., and J. A. Barron, Diatom biostratigraphy: Kerguelen Plateau and Prydz Bay regions of the Southern Ocean, *Proc. Ocean Drill. Program Sci. Results*, 119, 547–598, 1991.

Banerjee, I., and B. C. Mcdonald, Glaciofluvial and glaciolacustrine sedimentation, in *Nature of Esker Sedimentation*, *Spec. Publ. 23*, edited by A. V. Jopling and B. C. Mcdonald, pp. 132–154, Society of Economic Paleontologists and Mineralogists, Tulsa, Okla., 1975.

Barrera, E., M. J. S. Tevesz, and J. G. Carter, Variations in oxygen and carbon isotopic compositions and microstructure of the shell of *Adamussium colbecki* (Bivalvia), *Palaios*, 5, 149–159, 1990.

Barrett, P. J., and M. J. Hambrey, Plio-Pleistocene sedimentation in Ferrar Fjord, Antarctica, *Sedimentology*, 39, 109–123, 1992.

Barrett, P. J., D. P. Elston, D. M. Harwood, B. C. McKelvey, and P.-N. Webb, Cenozoic glaciation, tectonism, and sea-level change from MSSTS-1 on the margin of the Victoria Land Basin, Antarctica, *Geology*, 15, 634–637, 1987.

Barrett, P. J., C. J. Adams, W. C. McIntosh, C. C. Swisher III, and G. S. Wilson, Geochronological evidence supporting Antarctic deglaciation three million years ago, *Nature*, 359, 816–818, 1992.

Beu, A. G., Pleistocene *Chlamys patagonica delicutula* (Bivalvia: Pectinidae) off southeastern Tasmania, and history of its species group in the Southern Ocean, *Spec. Publ. South Aust. Dep. Mines Energy*, 5, 1–11, 1985.

Bockheim, J. G., Relative age and origin of soils in eastern Wright Valley, *Soil Sci.*, 128, 142–152, 1979.

Bockheim, J. G., Use of soils in studying the behaviour of the McMurdo ice dome, in *Antarctic Earth Science*, edited by R. L. Oliver, P. R. James, and J. B. Jago, pp. 457–460, Australian Academy of Science, Canberra, 1983.

Bockheim, J. G., and F. C. Ugolini, A review of pedogenic zonation in well-drained soils of the southern circumpolar region, *Quat. Res.*, 34, 47–66, 1990.

Boulton, G. S., Boulder shapes and grain-size distributions of debris as indicators of transport paths through a glacier and till genesis, *Sedimentology*, 25, 773–799, 1978.

Brady, H. T., Late Neogene diatom biostratigraphy and paleoecology of the dry valleys and McMurdo Sound, Antarctica, M.S. thesis, Northern Ill. Univ., De Kalb, 1977.

Brady, H. T., A diatom report on DVDP cores 3, 4A, 12, 14, 15, and other related surface sections, in *Proceedings of Seminar III on the Dry Valley Drilling Project, 1978, Memoirs of the National Institute of Polar Research*, edited by T. Nagata, pp. 165–175, Tokyo, 1979.

Brady, H. T., Late Cenozoic history of Taylor and Wright valleys and McMurdo Sound inferred from diatoms in Dry Valley Drilling Project cores, in *Antarctic Geoscience*, edited by C. Craddock, pp. 1123–1131, University of Wisconsin Press, Madison, 1982.

Brady, H. T., and B. McKelvey, Some aspects of the Cenozoic glaciation of southern Victoria Land, Antarctica, *J. Glaciol.*, *29*, 343–349, 1983.

Brenchley, P. J., and G. Newall, Flume experiments on the orientation and transport of models and shell valves, *Palaeogeogr. Palaeoclimatol. Palaeoecol.*, *7*, 185–220, 1970.

Brooks, H. K., A fjord deposit in Wright Valley, Antarctica, *Antarct. J. U. S.*, *7*, 241–243, 1972.

Brotherhood, G. R., and J. C. Griffiths, Mathematical derivation of the unique frequency curve, *J. Sediment. Petrol.*, *17*, 77–82, 1947.

Bullivant, J. S., Photographs of the bottom fauna in the Ross Sea, *N. Z. J. Sci.*, *2*, 485–497, 1959.

Bullivant, J. S., Photographs of Antarctic bottom fauna, *Polar Rec.*, *10*, 505–508, 1961.

Buntley, G. J., and F. C. Westin, A comparative study of developmental color in a chestnut-chernozem-brunizem soil climosequence, *Soil Sci. Soc. Am. Proc.*, *29*, 579–582, 1965.

Burckle, L. H., Diatom distribution and paleoceanographic reconstruction in the Southern Ocean—Present and last glacial maximum, *Mar. Micropaleontol.*, *9*, 241–261, 1984.

Burckle, L. H., and B. M. Pokras, Implications of a Pliocene stand of Nothofagus (southern beech) with 500 kilometres of the South Pole, *Antarct. Sci.*, *3*, 389–403, 1991.

Burckle, L. H., M. L. Prentice, and G. H. Denton, Neogene Antarctic glacial history: New evidence from marine diatoms in continental deposits (abstract), *Eos Trans. AGU*, *67*, 295, 1986.

Burckle, L. H., S. S. Jacobs, and R. B. McLaughlin, Late austral spring diatom distribution between New Zealand and the Ross Ice Shelf, Antarctica: Hydrographic and sediment correlations, *Micropaleontology*, *33*, 74–81, 1987.

Calkin, P. E., Subglacial geomorphology surrounding the ice-free valleys of southern Victoria Land, Antarctica, *J. Glaciol.*, *13*, 415–429, 1974.

Calkin, P. E., and C. Bull, Interaction of the East Antarctic Ice Sheet, alpine glaciations and sea-level in the Wright Valley area, southern Victoria Land, in *Antarctic Geology and Geophysics*, edited by R. J. Adie, pp. 435–440, Universitetsforlaget, Oslo, 1972.

Calkin, P. E., R. E. Behling, and C. Bull, Glacial history of Wright Valley, southern Victoria Land, Antarctica, *Antarct. J. U. S.*, *5*, 22–27, 1970.

Campbell, I. B., and G. G. C. Claridge, Morphology and age relationships of Antarctic soils, in *Quaternary Studies*, edited by R. P. Suggate and M. M. Cresswell, pp. 83–88, New Zealand Royal Society, Wellington, 1975.

Campbell, I. B., and G. G. C. Claridge, *Antarctica: Soils, Weathering Processes and Environment*, 368 pp., Elsevier, New York, 1987.

Ciesielski, P. F., and F. M. Weaver, Early Pliocene temperature changes in the Antarctic seas, *Geology*, *2*, 511–515, 1974.

Ciesielski, P. F., M. T. Ledbetter, and B. E. Brooks, The development of Antarctic glaciation and the Neogene paleoenvironment of the Maurice Ewing Bank, *Mar. Geol.*, *46*, 1–51, 1982.

Clemens, S., J. Farrell, and L. P. Gromet, Milankovitch—To subtectonic-scale covariation between seawater $^{87}Sr/^{86}Sr$ and planktonic $\delta^{18}O$ (abstract), *Eos Trans. AGU*, in press, 1993.

Dayton, P. K., G. A. Robilliard, and R. T. Paine, Benthic faunal zonation as a result of anchor ice at McMurdo Sound, Antarctica, *Antarct. Ecol.*, *1*, 244–258, 1970.

Denton, G. H., and R. L. Armstrong, Glacial geology and chronology of the McMurdo Sound region, *Antarct. J. U. S.*, *3*, 99–101, 1968.

Denton, G. H., M. L. Prentice, D. E. Kellogg, and T. B. Kellogg, Late Tertiary history of the Antarctic ice sheet: Evidence from the dry valleys, *Geology*, *12*, 263–267, 1984.

Denton, G. H., M. L. Prentice, and L. H. Burckle, Cainozoic history of the Antarctic ice sheet, in *The Geology of Antarctica*, edited by R. J. Tingey, pp. 365–433, Clarendon, Oxford, 1991.

Dobkins, J. E., and R. L. Folk, Shape development on Tahiti-nui, *J. Sediment. Petrol.*, *40*, 1167–1203, 1970.

Dreimanis, A., Tills: Their genetic terminology and classification, in *Genetic Classification of Glacigenic Deposits*, edited by R. P. Goldthwait and C. L. Matsch, pp. 17–84, A. A. Balkema, Rotterdam, Netherlands, 1988.

Dreimanis, A., and C. Schluchter, Field criteria for the recognition of till or tillite, *Palaeogeogr. Palaeoclimatol. Palaeoecol.*, *51*, 7–14, 1985.

Eggers, A. J., Scallop Hill Formation, Brown Peninsula, McMurdo Sound, Antarctica, *N. Z. J. Geol. Geophys.*, *22*, 353–361, 1979.

Elverhøi, A., Glacigenic and associated marine sediments in the Weddell Sea, fjords of Spitsbergen and the Barents Sea, a review, *Mar. Geol.*, *57*, 53–88, 1984.

Elverhøi, A., O. Lonne, and R. Seland, Glaciomarine sedimentation in a modern fjord environment, Spitsbergen, *Polar Res.*, *1*, 127–149, 1983.

Epstein, S., R. Buchsbaum, H. Lowenstam, and H. C. Urey, Revised carbonate water isotopic temperature scale, *Geol. Soc. Am. Bull.*, *64*, 1315–1326, 1953.

Everett, K. R., Soils of the Meserve Glacier area, Wright Valley, south Victoria Land, Antarctica, *Soil Sci.*, *112*, 425–438, 1971.

Fitzgerald, P. G., The Transantarctic Mountains of southern Victoria Land: The application of apatite fission track analysis to a rift shoulder uplift, *Tectonics*, *11*, 634–662, 1992.

Fleck, R. J., L. M. Jones, and R. E. Behling, K-Ar dates of the McMurdo volcanics and their relation to the glacial history of Wright Valley, *Antarct. J. U. S.*, *7*, 245–246, 1972.

Folk, R. L., *Petrology of Sedimentary Rocks*, 182 pp., Hemphill, Austin, Tex., 1974.

Galehouse, J. S., Sedimentation analysis, in *Procedures in Sedimentary Petrology*, edited by R. E. Carver, pp. 69–94, John Wiley, New York, 1971.

Gersonde, R., and L. H. Burckle, Neogene diatom biostratigraphy of ODP Leg 113, Weddell Sea (Antarctic Ocean), *Proc. Ocean Drill. Program Sci. Results*, *113*, 761–789, 1990.

Gleadow, A. J. W., and P. G. Fitzgerald, Uplift history of the Transantarctic Mountains: New evidence from fission track dating of basement apatites in the dry valleys area, southern Victoria Land, *Earth Planet. Sci. Lett.*, *82*, 1–14, 1987.

Hall, B., Surficial Geology and Geomorphology of eastern Wright Valley, Antarctica: Implications for Plio-Pleistocene ice-sheet dynamics, M.S. thesis, Univ. of Maine, Orono, 1992.

Haq, B. U., J. Hardenbol, and P. R. Vail, Mesozoic and Cenozoic chronostratigraphy and eustatic cycles, in *Sea-Level Changes: An Integrated Approach*, edited by C. S. Wilgus, B. S. Hastings, C. G. St. C. Kendall, H. Posamentier, J. V. Wagoner, and C. A. Ross, pp. 71–108, Society of Economic Paleontologists and Mineralogists, Tulsa, Okla., 1988.

Harwood, D. M., Diatom biostratigraphy and paleocology with a Cenozoic history of Antarctic ice sheets, Ph.D. dissertation, Ohio State Univ., Columbus, 1986.

Harwood, D. M., Cenozoic diatom biogeography in the southern high latitudes: Inferred biogeographic barriers and progressive endemism, in *Geological Evolution of Antarctica*, edited by M. R. A. Thomson, J. A. Crame, and J. W. Thomson, pp. 667–673, Cambridge University Press, New York, 1991.

Harwood, D. M., and P.-N. Webb, Recycled marine microfossils from basal debris-ice in ice-free valleys of southern Victoria Land, *Antarct. J. U. S.*, *11*, 87–88, 1986.

Hodell, D. A., P. A. Mueller, and J. R. Garrido, Variations in the strontium isotopic composition of seawater during the Neogene, *Geology*, *19*, 24–27, 1991.

Ishman, S. E., and H. J. Rieck, A late Neogene Antarctic glacio-eustatic record, Victoria Land Basin margin, Antarctica, in *The Antarctic Paleoenvironment: A Perspective on Global Change, Part One, Antarct. Res. Ser.*, vol. 56, edited by J. P. Kennett and D. A. Warnke, pp. 327–347, AGU, Washington, D. C., 1992.

Jackson, M. L., L. D. Whitting, and R. P. Pennington, Segregation procedures for mineralogical analysis of soils, *Soil Sci. Soc. Am. Proc.*, *14*, 77–81, 1949.

Jacobs, S. S., R. F. Fairbanks, and Y. Horibe, Origin and evolution of water masses near the Antarctic continental margin: Evidence from $H_2^{18}O/H_2^{16}O$ ratios in seawater, in *Oceanology of the Antarctic Continental Shelf, Antarct. Res. Ser.*, vol. 43, edited by S. S. Jacobs, pp. 59–86, AGU, Washington, D. C., 1985.

Kelling, G., and J. Holroyd, Clast size, shape, and composition in some ancient and modern fan gravels, in *Sedimentation in Submarine Canyons, Fans, and Trenches*, edited by D. J. Stanley and G. Kelling, pp. 138–159, Dowden, Hutchinson, and Ross, Stroudsburg, Pa., 1978.

Kellogg, T. B., and R. S. Truesdale, Late Quaternary paleoecology and paleoclimatology of the Ross Sea: The diatom record, *Mar. Micropaleontol.*, *4*, 137–158, 1979.

Kennett, J. P., The development of planktonic biogeography in the Southern Ocean during the Cenozoic, *Mar. Micropaleontol.*, *3*, 301–345, 1978.

Krumbein, W. C., Measurement and geological significance of shape and roundness of sedimentary particles, *J. Sediment. Petrol.*, *11*, 64–72, 1941.

Landim, P. M. B., and L. A. Frakes, Distinction between tills and other diamictons based on textural characteristics, *J. Sediment. Petrol.*, *38*, 1213–1223, 1968.

Lawson, D. E., Sedimentological analysis of the western terminus region of the Matanuska Glacier, Alaska, *CRREL Rep. 79-9*, Cold Regions Res. and Eng. Lab., U.S. Army Corps of Eng., Hanover, N. H., 1979.

Leroy, S. D., Grain-size and moment measures: A new look at Karl Pearson's ideas on distributions, *J. Sediment. Petrol.*, *51*, 625–630, 1981.

Macpherson, A. J., Glaciological, oceanographic and sedimentological data from Mackay Glacier and Granite Harbour Antarctica, Antarct. Res. Centre, Victoria Univ., Wellington, New Zealand, 1987.

Marchant, D. R., C. C. Swisher III, D. R. Lux, D. P. West, Jr., and G. H. Denton, Pliocene paleoclimate and East Antarctic Ice-Sheet history from surficial ash deposits, *Science*, *260*, 667–670, 1993.

Mayewski, P. A., and R. P. Goldthwait, Glacial events in the Transantarctic Mountains: A record of the East Antarctic Ice Sheet, in *Geology of the Central Transantarctic Mountains, Antarct. Res. Ser.*, vol. 36, edited by M. D. Turner and J. F. Splettstoesser, pp. 275–324, AGU, Washington, D. C., 1986.

McBride, E. F., Mathematical treatment of size distribution data, in *Procedures in Sedimentary Petrology*, edited by R. E. Carver, pp. 109–127, John Wiley, New York, 1971.

McKelvey, B. C., and P.-N. Webb, Geological investigations in southern Victoria Land, Antarctica, Part 3, Geology of Wright Valley, *N. Z. J. Geol. Geophys.*, *5*, 143–162, 1962.

McKelvey, B. C., P.-N. Webb, D. M. Harwood, and M. C. G.

Mabin, The Dominion Range Sirius Group: A record of the late Pliocene–early Pleistocene Beardmore Glacier, in *Geological Evolution of Antarctica*, edited by M. R. A. Thomson, J. A. Crame, and J. W. Thomson, pp. 675–682, Cambridge University Press, New York, 1991.

McSaveney, M. J., and E. R. McSaveney, A reappraisal of the Pecten glacial episode, Wright Valley, Antarctica, *Antarct. J. U. S.*, *7*, 235–240, 1972.

Menard, H. W., and A. J. Boucot, Experiments on the movement of shells by water, *Am. J. Sci.*, *249*, 131–151, 1951.

Mercer, J. H., Glacial development and temperature trends in the Antarctic and in South America, in *Antarctic Glacial History and World Paleoenvironments*, edited by E. M. van Zinderen Bakker, pp. 73–93, A. A. Balkema, Rotterdam, Netherlands, 1978.

Middleton, G. V., and M. A. Hampton, Subaqueous sediment transport and deposition by sediment gravity flows, in *Marine Sediment Transport and Environmental Management*, edited by D. J. Stanley and D. J. P. Swift, pp. 197–218, John Wiley, New York, 1976.

Mills, H. H., Textural characteristics of drift from some representative Cordilleran glaciers, *Geol. Soc. Am. Bull.*, *88*, 1135–1143, 1977.

Nichols, R. L., Multiple glaciation in the Wright Valley, McMurdo Sound, Antarctica, in *Abstracts of Symposium Papers, 10th Pacific Science Congress of the Pacific Science Association*, p. 317, Pacific Science Association, Honolulu, Ha., 1961.

Nichols, R. L., Geology of Lake Vanda, Wright Valley, south Victoria Land, Antarctica, in *Antarctic Research, Geophys. Monogr. Ser.*, vol. 7, edited by H. Wexler et al., pp. 47–52, AGU, Washington, D. C., 1962.

Nichols, R. L., Antarctic interglacial features, *J. Glaciol.*, *5*, 433–449, 1965.

Nichols, R. L., Glacial geology of the Wright Valley, McMurdo Sound, in *Research in the Antarctic*, edited by L. O. Quam, pp. 293–340, American Association for the Advancement of Science, Washington, D. C., 1971.

Pickard, J., et al., Early Pliocene marine sediments, coastline, and climate of East Antarctica, *Geology*, *165*, 158–161, 1988.

Potter, J. R., and J. G. Paren, Interaction between ice shelf and ocean in George VI Sound, Antarctica, in *Oceanology of the Antarctic Continental Shelf, Antarct. Res. Ser.*, vol. 43, edited by S. S. Jacobs, pp. 35–58, AGU, Washington, D. C., 1985.

Powell, R. D., Glacimarine processes and inductive lithofacies modelling of ice sheets and tidewater glacier sediments based on Quaternary examples, *Mar. Geol.*, *57*, 1–52, 1984.

Powell, R. D., Glacimarine processes at grounding-line fans and their growth to ice-contact deltas, in *Glacimarine Environments: Processes and Sediments*, edited by J. A. Dowdeswell and J. D. Scourse, pp. 53–73, Geological Society of London, London, 1990.

Prentice, M. L., Surficial geology and stratigraphy in central Wright Valley, Antarctica: Implications for Antarctic Tertiary glacial history, M.S. thesis, Univ. of Maine, Orono, 1982.

Prentice, M. L., Peleus glaciation of Wright Valley, Antarctica, *S. Afr. J. Sci.*, *81*, 241–243, 1985.

Prentice, M. L., and G. H. Denton, The deep-sea oxygen isotope record, the global ice sheet system, and hominid evolution, in *The Evolutionary History of the Robust Australopithecines*, edited by F. Grine, pp. 383–403, Aldine de Gruyter, New York, 1988.

Prentice, M. L., and R. K. Matthews, Cenozoic ice-volume history: Development of a composite oxygen isotope record, *Geology*, *16*, 963–966, 1988.

Prentice, M. L., and R. K. Matthews, Tertiary ice sheet

dynamics: The snow gun hypothesis, *J. Geophys. Res.*, *96*, 6811–6827, 1991.

Prentice, M. L., S. C. Wilson, J. G. Bockheim, and G. H. Denton, Geological evidence for pre–late Quaternary East Antarctic glaciation of central and eastern Wright Valley, *Antarct. J. U. S.*, *19*, 61–62, 1985.

Prentice, M. L., G. H. Denton, T. V. Lowell, H. Conway, and L. E. Heusser, Pre–late Quaternary glaciation of the Beardmore Glacier region, Antarctica, *Antarct. J. U. S.*, *21*, 95–98, 1986.

Prentice, M. L., G. H. Denton, L. H. Burckle, and D. Hodell, Evidence from Wright Valley for the response of the Antarctic ice sheet to climate warming, *Antarct. J. U. S.*, *22*, 56–59, 1987.

Prentice, M. L., J. L. Fastook, and R. Oglesby, Early Pliocene Antarctic interglaciations: Climate and ice-sheet modeling results, *Antarct. J. U. S.*, in press, 1993.

Quilty, P. G., The geology of Marine Plain, Vestfold Hills, East Antarctica, in *Geological Evolution of Antarctica*, edited by M. R. A. Thomson, J. A. Crame, and J. W. Thomson, pp. 683–686, Cambridge University Press, New York, 1991.

Robinson, P. L., Ice dynamics and thermal regime of Taylor Glacier, south Victoria Land, Antarctica, *J. Glaciol.*, *30*, 133–160, 1984.

Rust, B. R., and R. Romanelli, Late Quaternary subaqueous outwash deposits near Ottawa, Canada, in *Glaciofluvial and Glaciolacustrine Sedimentation*, edited by A. V. Jopling and B. C. McDonald, pp. 177–192, Society of Economic Paleontologists and Mineralogists, Tulsa, Okla., 1975.

Savin, S. M., R. G. Douglas, and F. G. Stehli, Tertiary marine paleotemperatures, *Geol. Soc. Am. Bull.*, *86*, 1490–1510, 1975.

Schrader, H. J., Proposal for a standardized method of cleaning diatom-bearing deep sea and land exposed marine sediments, in *Third Symposium on Recent and Fossil Diatoms*, edited by R. Simonson, pp. 403–409, Nova Hedwigia, Beiheft, 1974.

Schroeder, F. W., and S. M. Greenlee, Testing eustatic curves based on Baltimore Canyon Neogene stratigraphy: An example application of basin-fill simulation, *AAPG Bull.*, *77*, 638–656, 1993.

Shackleton, N. J., and J. P. Kennett, Paleotemperature history of the Cenozoic and the initiation of Antarctic glaciation: Oxygen and carbon isotopic analyses in DSDP sites 277, 279, and 281, *Initial Rep. Deep Sea Drill. Proj.*, *29*, 743–756, 1975.

Sneed, E. D., and R. L. Folk, Pebbles in the lower Colorado River, Texas, a study in particle morphogenesis, *J. Geol.*, *66*, 114–150, 1958.

Speden, I. G., Fossiliferous Quaternary marine deposits in the McMurdo Sound region, Antarctica, *N. Z. J. Geol. Geophys.*, *5*, 746–777, 1962.

Stanley, D. J., and R. Unrug, Submarine channel deposits, fluxo-turbidites, and other indicators of slope and base-of-slope environments in modern and ancient marine basin, in *Recognition of Ancient Sedimentary Environments*, edited by J. K. Rigby and W. K. Hamblin, pp. 287–340, Society of Economic Paleontologists and Mineralogists, Tulsa, Okla., 1972.

Stanley, D. J., H. D. Palme, and R. F. Dill, Coarse sediment transport by mass flow and turbidity current processes, and downslope transformations in Annot and Sandstone canyon-fan valley systems, in *Sedimentation in Submarine Canyons, Fans, and Trenches*, edited by D. J. Stanley and G. Kelling, pp. 85–115, Dowden, Hutchinson, and Ross, Stroudsburg, Pa., 1978.

Sugden, D. E., G. H. Denton, and D. R. Marchant, Subglacial

meltwater channel systems and ice sheet overriding, Asgard Range, Antarctica, *Geogr. Ann.*, *73A*, 109–121, 1991.

Turner, R. D., A new species of fossil Chlamys from Wright Valley, McMurdo Sound, Antarctica, *N. Z. J. Geol. Geophys.*, *10*, 446–455, 1967.

Vorren, T. O., M. Hald, M. Edvardsen, and L. Odd-Willy, Glacigenic sediments and sedimentary environments on continental shelves: General principles with a case study from the Norwegian shelf, in *Glacial Deposits in North West Europe*, edited by J. Ehlers, pp. 61–73, Geological Survey, Hamburg, Germany, 1983.

Vucetich, C. G., and W. W. Topping, A fjord origin for the pecten deposits, Wright Valley, Antarctica, *N. Z. J. Geol. Geophys.*, *15*, 660–673, 1972.

Walker, R. G., Generalized facies model for resedimented conglomerates of turbidite association, *Geol. Soc. Am. Bull.*, *86*, 737–748, 1975.

Webb, P.-N., Wright Fjord, Pliocene marine invasion of an Antarctic dry valley, *Antarct. J. U. S.*, *7*, 227–234, 1972.

Webb, P.-N., Micropaleontology, paleoecology, and correlation of the Pecten gravels, Wright Valley, Antarctica, and description of *Trochoelphidiella onyxi* n. gen., n. sp., *J. Foraminiferal Res.*, *4*, 185–189, 1974.

Webb, P.-N., and J. E. Andreasen, Potassium/argon dating of volcanic material associated with the Pliocene Pecten Conglomerate (Cockburn Island) and Scallop Hill Formation (McMurdo Sound), *Antarct. J. U. S.*, *21*, 59, 1986.

Webb, P.-N., and D. M. Harwood, Late Cenozoic glacial history of the Ross Embayment, Antarctica, *Quat. Sci. Rev.*, *10*, 215–224, 1991.

Webb, P.-N., and J. H. Wrenn, Upper Cenozoic micropaleontology and biostratigraphy of eastern Taylor Valley, Antarctica, in *Antarctic Geoscience*, edited by C. Craddock, pp. 1117–1122, University of Wisconsin Press, Madison, 1982.

Webb, P.-N., D. M. Harwood, B. C. McKelvey, J. H. Mercer, and L. D. Stott, Cenozoic marine sedimentation and ice-volume variation on the East Antarctic craton, *Geology*, *12*, 287–291, 1984.

Webb, P.-N., D. M. Harwood, B. C. McKelvey, M. C. G. Mabin, and J. H. Mercer, Late Cenozoic tectonic and glacial history of the Transantarctic Mountains, *Antarct. J. U. S.*, *21*, 99–100, 1986.

Weed, R., and R. P. Ackert, Jr., Chemical weathering of Beacon Supergroup sandstones and implications for Antarctic glacial chronology, *S. Afr. J. Sci.*, *82*, 513–516, 1986.

Weiss, R. F., H. G. Ostlund, and H. Craig, Geochemical studies of the Weddell Sea, *Deep Sea Res.*, *26*, 1093–1120, 1979.

Wilch, T. I., The surficial geology and geochronology of middle Taylor Valley, Antarctica: Implications for Plio-Pleistocene Antarctic glacial history, M.S. thesis, Univ. of Maine, Orono, 1991.

Wilch, T. I., D. R. Lux, W. C. McIntosh, and G. H. Denton, Plio-Pleistocene uplift of the McMurdo dry valley sector of the Transantarctic Mountains, *Antarct. J. U. S.*, *24*, 30–33, 1989.

Wilch, T. I., G. H. Denton, D. R. Lux, and D. P. West, Jr., The surficial geology of middle Taylor Valley, Antarctica: Evidence for limited climatic warming in early Pliocene (abstract), *Eos Trans. AGU*, *73*, 169, 1992.

Zachos, J. C., J. R. Breza, and W. W. Sherwood, Early Oligocene ice-sheet expansion on Antarctica: Stable isotope and sedimentological evidence from Kerguelen Plateau, southern Indian Ocean, *Geology*, *20*, 569–573, 1992.

(Received July 2, 1992;
accepted April 28, 1993.)

THE ANTARCTIC PALEOENVIRONMENT: A PERSPECTIVE ON GLOBAL CHANGE
ANTARCTIC RESEARCH SERIES, VOLUME 60, PAGES 251–264

COASTAL EAST ANTARCTIC NEOGENE SECTIONS AND THEIR CONTRIBUTION TO THE ICE SHEET EVOLUTION DEBATE

PATRICK G. QUILTY

Australian Antarctic Division, Kingston, Tasmania, Australia 7050
Antarctic Cooperative Research Centre, University of Tasmania, Hobart, Tasmania, Australia 7000

Several localities in coastal East Antarctica have yielded thin sediment sequences of Pliocene and younger ages. Marine Plain in the Vestfold Hills contains a 9-m-thick section of diatomite and sandstone from the interval 4.2–3.5 Ma. The locality is notable in containing abundant cetacean remains including a new genus and species of dolphin and at least three other species of whale. No glacial debris has been found in these sediments, and some parameters suggest a water temperature of about 5°C. The East Antarctic ice sheet margin was inland of its present position, and the ice sheet was smaller. At Stornes Peninsula in the Larsemann Hills a small area contains a 40-cm-thick unit containing well-preserved Pliocene benthic foraminifera. Diatoms indicate that this deposit is 3 to 2 m.y. old. Sediments near Casey Station in the Windmill Islands, have yielded diatom floras, mostly marine, but one lacustrine. These deposits formed 2–1 m.y. ago and lack diatom taxa typical of today's very cold coastal conditions. Reworked sediment in the base of an ice cliff north of Casey is much younger. In the Heidemann Valley in the Vestfold Hills, marine sediments contain diverse faunas and floras, including both planktonic and benthic foraminifera. Various dating techniques provide an age estimated at between 1 Ma and 300 ka. Oxygen isotope analysis suggests a water temperature slightly higher than that of the present day. Late Pleistocene and Holocene sediments are widespread elsewhere in the region, mostly deposited in marine environments. These coastal sequences probably are biased toward deposition during intervals of relatively warm water conditions, which may also imply high sea level. They do, however, provide a source of data for hypotheses that are apparently in conflict with some of the ideas generated from deep-sea sections.

INTRODUCTION

This paper summarizes what is known of Pliocene and younger sediment sections and paleoenvironments of coastal East Antarctica. It is based on evolving study of a series of thin (<9 m) sediment sections from a variety of locations (Figure 1), mostly marginal to Prydz Bay, and within a few kilometers of the present coast. The thickest section (Marine Plain (68°37.7'S, 78°07.8'E), Vestfold Hills) is only about 9 m thick, but most are much thinner, often only a few tens of centimeters thick.

Recent reviews of the late Neogene history have been given by *Quilty* [1990, 1991a, 1992]. This review does not cover the area as far west as Queen Maud Land and thus does not attempt to incorporate the Fukushima and Yamato glaciations [*Yoshida*, 1983; *Moriwaki et al.*, 1992], which have been identified and named, but their age and significance remain undefined.

The results of analyses of these sections must be integrated regionally with studies of more continuous sequences offshore recorded by either local coring or by major programs such as the Ocean Drilling Program (ODP) which was active in the region during Leg 119 [*Barron et al.*, 1991] and Leg 120 [*Wise et al.*, 1992].

To contribute to our understanding of the total history of Antarctica over this interval, the marginal East Antarctic history must be combined with that of the few other areas where similar studies are possible including the Ross Sea region [*Webb and Harwood*, 1991], Transantarctic Mountains [*Webb et al.*, 1986], northern Antarctic Peninsula [*Webb and Andreasen*, 1986], inland Antarctica [*Bardin*, 1982; *McKelvey and Stephenson*, 1990], and offshore by ODP [*Barker et al.*, 1990; *Barron et al.*, 1991; *Wise et al.*, 1992] and the Deep Sea Drilling Project (DSDP) [*Hayes et al.*, 1975]. *Webb* [1990] has recently attempted this synthesis, but the discipline is very dynamic, and much has been achieved since that review.

The sections studied so far are no older than 4.2 m.y., the oldest age estimate for the Marine Plain section, but previously unrecorded sections are being discovered regularly and the potential is high for identifying further records.

Although the portion of the last 4.2 m.y. represented by sediment sections may be small, the amount of information obtained is significant because of the continental margin location, the presence of macrofauna, and the providing of information on lateral sedimentary variation.

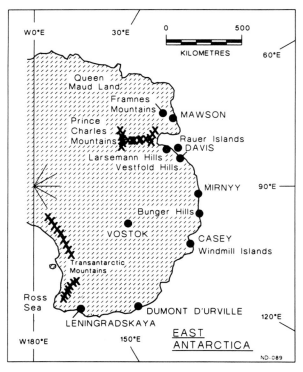

Fig. 1. Locations mentioned in the text.

Significance

The value of knowledge of the Pliocene and younger interval has increased in recent years for several reasons. First, it is now important to know the path of evolution of the global environment to provide a measure of the rate and amplitude of natural environmental variation as a baseline, so that anthropogenically induced change can be placed in context. This is particularly true in the polar regions because models suggest that many elements of change are greatest there [e.g., *Rind*, 1984]. Second, any attempt to develop an understanding of global evolution needs data from all parts of the world, and the Antarctic is a large, poorly known region. Third, there is now interest [*Intergovernmental Panel on Climate Change*, 1990, hereinafter IPCC] in identifying possible paleoanalogs of the environment into which the Earth may be evolving under anthropogenically induced ''Greenhouse Effect'' changes. Fourth, from the viewpoint purely of Antarctica, the Pliocene and younger is the interval during which, according to one hypothesis, the continent seems to have evolved from one with some significant vegetation [*Carlquist*, 1987; *Webb and Harwood*, 1987] and probably relatively small ice volume, to one that is ice covered and devoid of vegetation other than mosses and lichens. This hypothesis is in conflict with existing hypotheses [e.g., *Kennett*, 1977] and has been disputed strongly by *Burckle and Pockras* [1991] and *Prentice*

and Matthews [1991]. Better understanding of the evolution of continental glaciation of the Antarctic during the late Cenozoic is necessary within the context of global environmental development and for better understanding of the origin and evolution of earlier glaciations, for example, those of the late Precambrian and the Permian. What was an area of academic concern in the past has now become of mainstream interest.

Chronologic Framework

Dating of these sequences depends on biostratigraphic zonations developed elsewhere, particularly in the Ross Sea region, and is based on diatoms [*Harwood*, 1986] and to a lesser extent on foraminifera [*Ishman and Webb*, 1988]. Questions regarding the correlation of Antarctic diatom-based biostratigraphic schemes with zonal schemes elsewhere, and their chronology, continue to exist [e.g., *Burckle and Pokras*, 1991], but the author believes that the problems remaining relate to minor adjustment rather than major miscorrelation [e.g., *Barrett et al.*, 1991, 1992]. The East Antarctic sections described here add little to this debate because they come from a tectonically very stable region and thus do not include volcanic glass shards or lavas that can be dated radiometrically.

The Marine Plain section does, however, have the potential to contribute significantly to the debate over the existence of a relatively warm interval in the Antarctic during the Pliocene, prior to the major expansion of the Antarctic ice sheets after approximately 2.5 Ma [*Burckle et al.*, 1990; *Hodell et al.*, 1991; *Hodell and Ciesielski*, 1991; *Hodell and Venz*, 1992].

SECTIONS IN THE VESTFOLD HILLS

Sequences of several ages occur in the Vestfold Hills. Only one (Marine Plain) has been studied in any detail, but even there, most work remains to be done. Other localities are Heidemann Valley (68°34.3'S, 78°01.9'E) (Airport Road), numerous widely scattered in situ Holocene outcrops, and scattered redistributed sediments. The most comprehensive reviews of the distribution of the Neogene of the Vestfold Hills are by *Pickard* [1986] and *Quilty* [1992].

Marine Plain: Pliocene

Marine Plain lies about 10 km south southeast of Davis Station (Figures 2, 3, and 4). The highest elevation of sediments lies less than 15 m above sea level and is overlain by a thin Holocene glacial veneer (see below). The lithostratigraphy was discussed by *Adamson and Pickard* [1986], *Zhang and Peterson* [1984], and *Zhang* [1989], and the general geology was discussed by *Quilty* [1991b]. Summary stratigraphic sections were included in the work of *Adamson and Pickard* [1986] and *Zhang and Peterson* [1984].

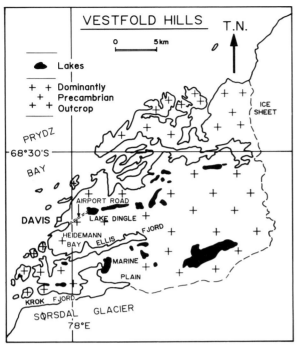

Fig. 2. Location of sections in the Vestfold Hills.

Pliocene [*Harwood*, 1986, personal communication, 1993; *Pickard et al.*, 1986, 1988] sediments occur in sections up to about 9 m thick. Two adjacent localities, Marine Plain and Poseidon Basin, contain the sequences. They are separated by a north-south trending ridge of Precambrian metamorphics [*Collerson and Sheraton*, 1986]. The sediments are isolated at the northern end of both areas of occurrence, but they unite south of the ridge and thus were part of one depositional unit in the Pliocene. They probably continue to the south to the present sea, in the vicinity of Iceberg Graveyard, via a gap between other Precambrian rocks. They probably also continue to the west into the area of Burton Lake. Rocks of this age have not been found elsewhere in the Vestfold Hills. Maximum thickness has been measured only near the margins of Marine Plain and Poseidon Basin, and greater thickness may exist in the center of the areas. *Zhang* [1985, 1989] and *Zhang and Peterson* [1984] erroneously referred to this Pliocene sequence as of Quaternary age.

The sediments are characterized by generally horizontal bedding and are undisturbed by tectonic features. They are, particularly in the northern part of Marine Plain, cut by almost vertical sandstone dykes. The origin of these structures is unknown. They also contain

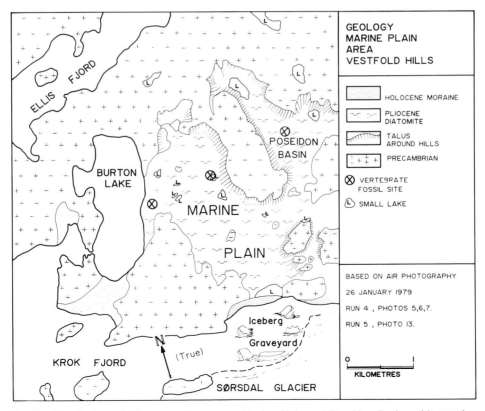

Fig. 3. Distribution of Pliocene sediments on Marine Plain and Poseidon Basin, with vertebrate localities highlighted.

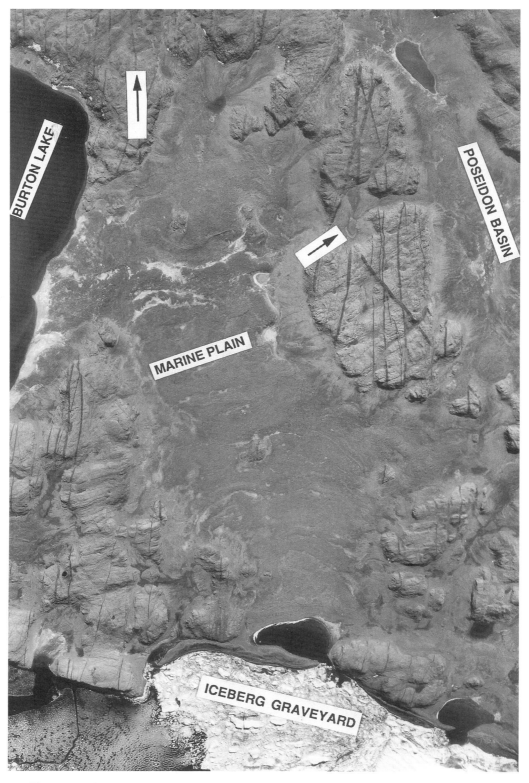

Fig. 4. Vertical air photo of Marine Plain. Precambrian outcrop elevated and marked by black dolerite dikes. Marine Plain surface marked by concave-to-south ridges in Holocene glacials. Pliocene sequence underlies the Holocene. Lake of Figure 5f is indicated by arrow. Distance across image is 2.8 km.

uncommon wedges of Holocene marine coarse-grained glacial debris probably filling frost wedge structures.

The rocks are poorly indurated, buff colored diatomite, siltstone, and fine sandstone with a high fossil content, dominantly diatoms and sponge spicules. Microfossils are abundant in some samples. In outcrop and in excavations to over 2 m deep, carbonate fossils generally have been removed. Diagenesis has been active, as weathering now is. Sulfur gases are released from the surface during disturbance by walking, and it thus appears that pyrite was abundant in the fresh sediment. Abundant gypsum in dried out ponds, on the surface, and as crystals in the section suggests that oxidation of the pyrite generates sulfuric acid that reacts with the carbonate to produce the gypsum at the expense of any carbonate fossils.

In the upper few meters of the section, there are limestone concretions and lenses, and in places these are almost continuous enough to constitute a limestone bed up to 10 cm thick. In this limestone, carbonate fossils are preserved, sporadically very well. The limestone is dark and contains mollusks, especially bivalves [*Adamson and Pickard*, 1986] including *Chlamys tuftsensis* Turner which was the first clue to the Pliocene age of the sequence (Dell in the work of *Pickard et al.* [1986, 1988]). Thin sections contain echinoid spines and benthic foraminifera, including miliolids, but no planktonic foraminifera have been identified in thin sections yet, although a single overgrown specimen of *Neogloboquadrina pachyderma* (Ehrenberg) was seen in residues of surface samples.

The age of the sediments is taken to be early Pliocene [*Pickard et al.*, 1988] on the basis of the presence of the diatom *Nitzschia praeinterfrigidaria* McCollum and the absence of *Actinocyclus actinochilus* (Ehrenberg), *Thalassiosira kolbei* (Jousé), *T. vulnificus* (Gombos), and *Cosmiodiscus insignis* Jousé.

A small area at the northern end of Marine Plain has yielded a diverse benthic fauna of ophiuroids, asteroids, echinoids, bivalves, gastropods, and bryozoa. This fauna has not been studied yet. A very different fauna occurs adjacent to a small island of Precambrian metamorphics in the southern end of Marine Plain, but the details of the differences have not been documented. The bivalve fauna is different, and nonbivalve elements are less diverse.

The most spectacular discoveries in the area are of vertebrate fossils, especially cetaceans, which are the principal reason for the nomination of the area as Site of Special Scientific Interest number 25 under the Antarctic Treaty. The localities from which these remains have been recovered are shown in Figure 3. Although study is not yet complete, the cetacean fauna includes a new genus and species of dolphin, a right whale, and at least two other species. The dolphin is estimated to have been about 4.5 m long, whereas other cetaceans known so far are about 8–9 m long (Figures 5a and 5b). The

dolphin skull was illustrated by Fordyce in the work of *Harrison and Bryden* [1989], and its ear bones were illustrated by *Fordyce* [1989]. R. E. Fordyce (personal communication, 1992) states that there is nothing to indicate that they are functionally adapted to cold water, nor that they are obviously related to extant cold water forms, and that the entire cetacean fauna appears noncryophilic (non-ice-dwelling) and lived in conditions with a water temperature similar to that now found in the vicinity of the Antarctic Convergence (about 5°C), considerably above the present. A preliminary oxygen isotopic temperature (Chivas, in the work of *Quilty* [1991b]) is much higher (10.5°C) than that of the present day and consistent with the general indications of warmer-than-now conditions documented by *Pickard et al.* [1986, 1988]. The oxygen isotope analysis was conducted on a single surface specimen of *Chlamys tuftsensis* Turner. (The 10.5°C calculated temperature is derived from a $\delta^{18}O$ value of 1.55‰. This compares with calculated temperatures of 3.7°C for modern mollusks (*Latemula elliptica*) and 5.3°C for a Holocene specimen of the same species. The gas extraction of the Pliocene specimen gave a poor yield and cannot be considered to be precise. The material was not X rayed to determine whether it is calcite and/or aragonite. The calculated temperatures are based on the assumption that $\delta^{18}O$ of the seawater during the Pliocene interval was 0‰ (standard mean ocean water). This is undoubtedly incorrect because of the addition of freshwater from ice. Also, no adjustment was made for any possible ocean difference in $\delta^{18}O$ during the Pliocene. The calculated temperatures are maxima.) Careful analysis is needed of additional, well-preserved material, as well as consideration of salinity effects on oxygen isotopic values.

The benthic diatom flora indicates that deposition occurred in fully marine conditions shallower than 75 m, and lithological changes and fauna, including those in the limestone, suggest that water depth shallowed with time and was only a few meters deep during deposition of the uppermost part of the section [*Harwood*, 1986]. This change is indicated by an upsection increase in the content of ice-related species *Eucampia antarctica* (Castracane), *Nitzschia curta* (Van Heurck), and *Odontella weissflogii* (Janisch) that may be related to developing glaciation [*Pickard et al.*, 1988], and the sequence may represent the sediment filling of a complex series of embayments that eventually became dry land.

Airport Road, Heidemann Valley

The sediments recorded here occupy a valley (probably glacially formed) that trends east northeast–west southwest from Heidemann Bay immediately south of Davis, to Dingle Lake (Figure 2). The valley is straight, and its surface is within 5 m of modern sea level.

The sediments that fill the valley have been studied by *Hirvas and Nenonen* [1989a] and *Hirvas et al.* [1992]

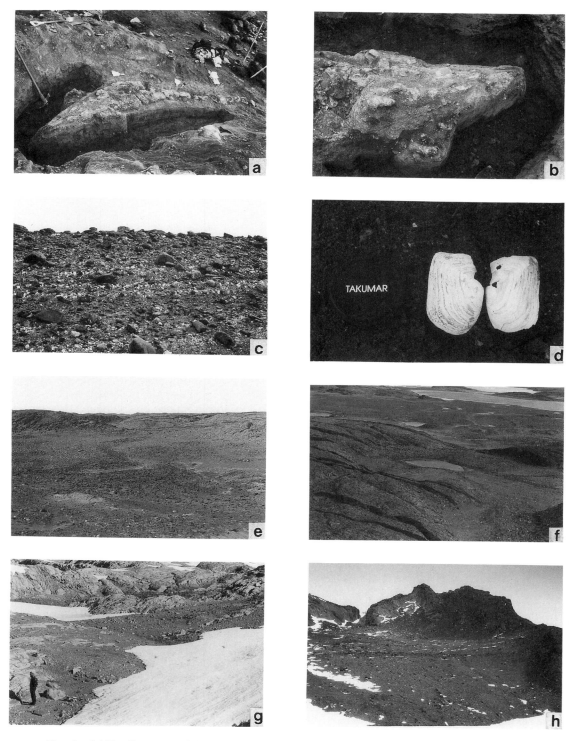

Fig. 5. (*a*) Fossil cetacean skeleton, about 8 m long, during excavation, eastern side of Marine Plain. Curvature exaggerated by wide angle lens. Skull to left. (*b*) Enlargement of skull of cetacean in Figure 5*a*. (*c*) *Laternula elliptica* King and Broderip in living position in two clearly defined beds. Location is on western edge of Marine Plain, near Burton Lake. (*d*) *Laternula elliptica* King and Broderip from upper of beds in Figure 5*c*. (*e*) Stream valley leading from hill of Precambrian rock to eastern edge of Marine Plain. Path of water flow indicated by darker edges of gully. This view lies to west of lake indicated in Figure 4. (*f*) Note small lake set in top of hill of Precambrian. This lake is source of water in the past for the stream in Figure 5*e*. (*g*) Pliocene sediment occurrence in the Larsemann Hills. Section lies in upper right of image. Person in lower left for scale. (*h*) Sediment infill of a cirque immediately south of Rumdoodle Peak in the Framnes Mountains south of Mawson. Sediment as yet undated.

using field methods described by *Hirvas and Nenonen* [1990]. The sediments have been documented in detail only in excavations on the northern side of the valley where the section is up to 4.5 m thick. It may be thicker in the valley center.

The sequence consists of three units, a lower and upper till separated by poorly sorted dark gray-green marine sand and silt which in places is gravelly with shells of *Laternula elliptica* and *Hiatella arctica* [*Hirvas and Nenonen*, 1989*b*]. The upper unit seems to be part of a Holocene glacial sequence (see below) as *Adamson and Pickard* [1986] recorded radiocarbon dates (7–5 ka) from what seems to be this unit from the nearby Death Valley system. A stratigraphic section for this sequence is shown in the work of *Hirvas et al.* [1992].

The older sequence is marine and contains abundant microfossils (diatoms, sponge spicules, radiolaria, benthic and planktonic foraminifera, including *Neogloboquadrina pachyderma*). Bivalves from the middle unit include *Hiatella arctica* [*Hirvas et al.*, 1992] that yield oxygen isotope evidence consistent with a water temperature of 2°C, warmer than at present, and hinting at interglacial deposition. An alternative explanation of the isotopic results is that seawater was diluted by some 3% glacial meltwater [*Hirvas and Nenonen*, 1989*b*]. Bivalves are in their living position and articulated, indicating quiet conditions at the time and no significant disruption since. The top unit forms the surface of Heidemann Valley and consists of coarse glacial debris which *Hirvas et al.* [1992] interpreted as basal till. In places, it contains abundant indices of deposition in a marine environment.

The age of the sediments is only poorly known. The upper unit may be Holocene, but the lower units are too old for radiocarbon techniques. Other techniques applied include thermoluminescence, amino acid racemization, and diatom biostratigraphy. No one technique provides a reliable age estimate, but results are consistent with the rocks being deposited during the interval 1 Ma to 300 ka, i.e., during an interglacial interval in the Pleistocene. This interglacial period has been named the Davis Interglacial by *Hirvas and Nenonen* [1989*b*]. The sediments were deposited under shallow, fully marine conditions when the region must have been a deeply embayed coast following a glacial phase. Water was shallow, but deep enough to allow planktonic foraminifera and a few radiolaria to be incorporated in the sediment. A water depth of 50–25 m is consistent with the microfossil assemblages.

It is probable that other straight valleys, covered with glacial debris, near the coast of the Vestfold Hills are filled with similar sediments.

On the southern side of Heidemann Bay lies the Heidemann Moraine, a ridge that contains shell fragments with an age older than radiocarbon techniques can define [*Adamson and Pickard*, 1986] and these may

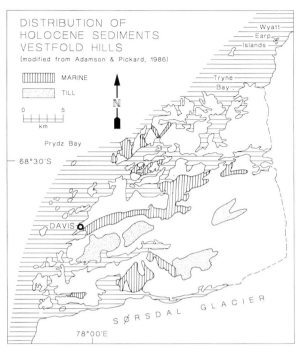

Fig. 6. Distribution of in situ Holocene deposits in the Vestfold Hills [after *Adamson and Pickard*, 1986].

be reworked material of the sort excavated on Airport Road.

Subglacial Aragonite

Aharon [1988] has documented an episode of aragonite deposition in a closed system at 35 ka. He suggested that deposition was subglacial and related to the last glacial maximum. Since Aharon documented the distribution of this material, other occurrences have been identified in the southeastern corner of the Vestfold Hills. Wherever it occurs, it is as a thin patchy veneer, firmly adhering to the Precambrian basement or as flakes of the same material that have broken free.

Holocene Sediments

Holocene sediments are widespread and diverse throughout the Vestfold Hills [*Adamson and Pickard*, 1986]. The study of these sediments has advanced a little by further radiocarbon dating and analysis of lake and fjord sediments [e.g., *Bird et al.*, 1991]. In Figure 6 the distribution of in situ Holocene sediments is shown. All are below 10 m above modern sea level and thus comparable with or of lower elevation than the Pliocene of Marine Plain. Other scattered deposits, including those discussed in the next section, may be Holocene but have not yet been shown to be so. Foraminifera from various Holocene sediments were recorded by

Crespin [1960], *Setty et al.* [1980], and *Quilty* [1988]. *Adamson and Pickard* [1986] reported an uncorrected radiocarbon age range of approximately 8–2.5 ka on a variety of fossils including mollusks and algae.

In addition to these clearly marine sediments, there are widespread moraines and marine till covering valley floors (for example, Marine Plain, Poseidon Basin, and Heidemann Valley). Some, such as at Marine Plain, have marine molluscs, especially *Laternula elliptica* King and Broderip (Figures 5c and 5d) in living position and in great abundance. Dating of these molluscs gives an uncorrected radiocarbon age of 6490 ± 130 years B.P. (Adamson, in the work of *Quilty* [1991b]), suggesting that these glacial deposits represent the latest phase of glaciation and withdrawal of the marine environment, perhaps due in part to isostatic rebound.

All such deposits are within a few meters of present sea level. The area is rising isostatically in relation to modern sea level at 1–1.5 mm/yr [*Adamson and Pickard*, 1986] and thus may be an absolute 2–2.5 mm/yr, if estimates of general sea level rise of 1–2 mm/yr over the last century (Warrick and Oerlemans, in the work of *IPCC* [1990]) are correct.

Other Sediments

In recent years it has become obvious that there is a discontinuous thin veneer of reworked sediment widely distributed throughout the Vestfold Hills. It is under study at present (D. Gore, personal communication, 1993). It occurs at altitudes to almost 160 m, indicating that it was deposited from the base of the retreating ice sheet when it was larger than at present and covering the Vestfold Hills. No age information is yet available, but it must predate the in situ Holocene sediment so widespread in the Vestfold Hills and probably was emplaced immediately following a glacial maximum. The sediments contain marine indices such as sponge spicules, foraminifera (including fragmentary *Cibicides*), and mollusk fragments. The sediments are clayey tills and not a windblown deposit. The source for reworking is not clear but could be from many localities in the region.

Since deposition of some Holocene sediments, there has been an interval of minor fluvial activity. A ridge of Precambrian rock, separating Marine Plain and Poseidon Basin, contains a small lake in its center (Figure 5f) and well above the elevation of the Holocene and Pliocene sediment sections (for example, Figure 5c). It is situated in a large basin. This lake was once much larger, and water flowed from it to the west (Figure 5e), over a lip and down into what is now a series of ponds on the eastern side of Marine Plain.

The deepest point in the area is in northeastern Marine Plain and lies immediately below the lake illustrated in Figure 5f and referred to above. It now has no water in it, nor pale, fine-grained sediments as several active ponds in the area now do. Nevertheless, this position marks the downslope end of a small "river" valley. The sediments cut by the valley are those that constitute the "talus skirt" of *Quilty* [1991b] and are thought to be the same age as the 6490-years B.P. glacial sediments covering Marine Plain. The age of this fluvial activity is unknown but is about or younger than 6500 years.

Since retreat of the marine environment there has been deposition of gray-green, fresh, windblown, well-sorted quartz/feldspar sand on the lee side of many hills, in the lakes, and adjacent to embayments. The sediment is apparently lithologically identical regardless of the age and environment of deposition, and a variety of ages may be represented.

LARSEMANN HILLS

On the eastern side of Stornes (Peninsula), in the vicinity of Jennings Bluff (informal name used by *Quilty et al.* [1990]), Pliocene sediments occur over about one hectare (Figure 5g) and have a maximum thickness of only 40 cm, although they are present over an elevation range of perhaps 1.5 m. They occur on a narrow bench at about 55 m above sea level, the highest occurrence of Neogene sediment in the region of the margin of Prydz Bay. This bench is quite widespread throughout the Larsemann Hills.

The sediment is essentially unbedded, poorly sorted, gray, fine to coarse sand, coarser near the base of the section. There is some internal rough stratification [*Quilty et al.*, 1990]. Other than coarse terrigenous material, the most prominent feature of the sediment is abundant broken bivalve fragments, probably representing two species. The shell material is unidentifiable. Fragments are usually 5–10 mm in diameter and angular. There is no evidence of any diagenetic alteration to the shell fragments, nor have they been subject to any obvious rounding by turbulence.

The original environment of deposition of the sediment was high-energy shallow water, perhaps in a depth of <50 m and probably 10–20 m.

The sediment is Pliocene in age based on the well-preserved abundant benthic foraminifera fauna [*Quilty et al.*, 1990] which includes common *Ammoelphidiella antarctica* Conato and Segre (previously employed by *Ishman and Webb* [1988] and other authors as a Pliocene zone fossil, under the name *Trochoelphidiella onyxi* Webb) and other calcareous forms. There are no agglutinated benthics, nor planktonics. There are also small, fragmented diatoms and silicoflagellates that D. Harwood (personal communication, 1990) believes indicate an age of 3–2 Ma. Some samples are barren, but others contain a rich flora. The consistency of the ages by the different fossil groups indicates that the material is in situ or, at least, is not mixed with sediment of any other age. It is possible that the sediment has been redeposited during glacial advance or retreat and that the original bivalves were fragmented during this process and not rounded by this action. The foraminifera are not known well enough through the Antarctic for their use in paleotemperature reconstruction. The diatoms are not well preserved, and study has not advanced to the point where comments about paleoenvironment can be made.

The unique occurrence and small area are the basis for steps toward nomination of the area for protection of the site as an Antarctic Specially Protected Area under the Protocol on Environmental Protection to the Antarctic Treaty.

WINDMILL ISLANDS

Late Neogene sediments have been known from the Windmill Islands since the International Geophysical Year [*Cameron et al.*, 1959] and are widespread, but the range of ages and significance were not realized until very recently. Details of their origin (in glacial terms) have been discussed by *Goodwin* [1992]. Study of the Neogene of the region is in its infancy. No map of their distribution is available, as there has been no systematic study.

Recent interest in the sediments was fostered following the statement by *Lewis-Smith* [1986] that samples from the vicinity of Casey Station contained spicules which he interpreted as fibrous anthropogenic pollutants. The spicules are sponge spicules and led to the search for more sediments and examination of their fossil content.

Fine sand and silt occurs entrained in moving ice in ice cliffs near Jacks Donga north of Casey Station, and sand that may have been reworked and redeposited during glaciation is found as thin deposits close to Casey Station. Whether or not only one age is represented by these sediments has yet to be determined. The sediments occur as thin deposits up to about 40 m above sea level in the station limits within Site of Special Scientific Interest number 16, where they act as a medium for growth for extensive moss beds, described as the most extensive on continental Antarctica outside Palmer Land [*Lewis-Smith*, 1985, 1988], and the sediments may be one of the main reasons for the spectacular moss development.

The few samples examined to date are devoid of calcareous fossils, either shell or microfossil, but contain a diverse shallow water siliceous biota. Sponge spicules are most prominent, but there are also diatoms that have been examined briefly (D. Harwood, personal communication, 1990). These samples indicate an age of 2–1 Ma (late Pliocene–early Pleistocene) for the higher-elevation sediments at Casey Station and include indications of both marine and lacustrine environments of deposition. D. Harwood (personal communication, 1990) states that the diatom assemblage indicates water

temperatures between 2° and 6°C, with an absence of species that live at lower temperatures.

MAWSON REGION

The main interest in sediments from the Mawson region lies in an occurrence in the Framnes Mountains south of Mawson. Here there is at least one cirque (Figure 5*h*), now well above modern ice level that is filled with sediment with a marked change of slope at its distal end. This change of slope probably represents the old ice level. If these sediments can be dated and elevation above modern ice sheet level identified, the results will be of value in documenting the past configuration of the ice sheet in this region.

OTHER LOCALITIES

This paper has concentrated on the coastal sections from East Antarctica, but there are others farther inland in the region that are the subject of study at present. The most critical are those of the Prince Charles Mountains [*Bardin*, 1982; *McKelvey and Stephenson*, 1990; *McKelvey et al.*, 1991] which may provide important information for elucidation of the history of the East Antarctic Ice Sheet. The Pagodroma Tillite, discussed briefly and very succinctly by *McKelvey and Stephenson* [1990], has yielded reworked late Miocene and middle Pliocene marine diatoms (in addition to possible lacustrine diatoms and other marine microfossils). The youngest reworked material is 3.5 Ma and thus coeval with or younger than the youngest Marine Plain sediment. It indicates that emplacement of this sediment section was latest Pliocene or Pleistocene, and the authors suggested it is an equivalent of the Sirius Group of the Transantarctic Mountains and formed by the same expansion of the East Antarctic Ice Sheet, an expansion greater in scope than any since.

Holocene sediments are widespread in outcrops in the East Antarctic region, but studies to date have been rather sporadic, and much remains to be done in documenting distribution and in subsequent detailed study. There has been virtually no study of occurrences in the Rauer Islands. They contain bivalve debris, particularly of *Laternula elliptica*.

Those in the Bunger Hills have been documented in some detail and discussed by *Colhoun and Adamson* [1992], who recorded that they are up to 7.5 ± 1 m above sea level and yield radiocarbon ages in the range 9–5 ka (uncorrected). *Colhoun and Adamson* [1992] compared the Bunger Hills results with those from several other Antarctic localities. They then suggested that ice was 150–400 m thick over this area at the last glacial maximum, and 450 m thick over the Windmill Islands region.

Bolshiyanov et al. [1991] also have reported on sediments from lakes in the Bunger Hills and showed a range of ages similar to those of *Colhoun and Adamson* [1992] but have documented a marine phase in sediments at the base of some of the many lakes in the area. This feature is taken to indicate higher mid-Holocene sea level. They also note marine indices (foraminifera, mollusks, and sponge spicules) on terraces to 40 m above sea level.

DISCUSSION

The analysis of the coastal East Antarctic sections discussed above is still very preliminary, and some sections have yielded information only on age so far. Some sections do provide data relevant to hypotheses on the evolution of the Antarctic glaciation over the last 5 m.y. or so. It is to be expected that further sections will be found and contribute to the debate.

Sections studied are all <160 m above sea level, and most are within 10–15 m. In situ Pliocene of Marine Plain is less than 15 m above sea level. The Pliocene from Larsemann Hills is at 55 m above sea level but may be ex situ. Little information is available on the influence of isostasy other than that contained in the work of *Adamson and Pickard* [1986], which suggests land elevation in the range 1–1.5 mm/yr for the Vestfold Hills over the last few thousand years, and in the work of *Colhoun and Adamson* [1992], which refers to uplift rates of 1.4 mm/year in the Bunger Hills. Considerable research is being conducted into this issue at present.

The Marine Plain section is the most studied to date. The assessment of higher water temperature is based on preliminary analysis of oxygen isotopes from mollusks, features of the cetacean fauna, and the general lack of glacial indices in the sediments. This evidence of a warmer climate for some interval in the early Pliocene in the Antarctic supports the hypothesis, proposed earlier and reviewed by *Webb and Harwood* [1991], of higher water and air temperatures at times within the early Pliocene but is independent of estimates based on oxygen isotope data from deep-sea benthic foraminifera. This scenario of generally higher temperatures has serious consequences for estimates of early Pliocene ice volume, and thus global sea level. *Webb and Harwood* [1991], restating the estimate by *Harwood and Webb* [1990], suggested that at this time, Antarctic ice volume was one-third its present value and reviewed some oxygen isotope evidence [*McKenzie et al.*, 1984] for a sea level that may have been 70 m higher than now, commencing at about 4.3 Ma, also roughly consistent with other independent evidence [*Vail et al.*, 1977].

Prentice and Matthews [1991] provided an alternative hypothesis that accepts the seawater temperature figures but suggests that this can be made consistent with an ice sheet of roughly present volume but with a form that differs from that of the modern by being thicker in central Antarctica and thinner on the margins. Their hypothesis (the "snow gun" hypothesis) envisages warmer oceanic conditions, higher snow precipitation, and perhaps more mobile ice. A key difference between the hypotheses lies in the question of sea level history in the Miocene-Pliocene and the existence or not of oceanic conditions on the East Antarctic Craton during the early Pliocene, to provide a source for the reworked material found in the Sirius Group [*Webb and Harwood*, 1991]. The Prentice and Matthews hypothesis has some similarities with some scenarios of future anthropogenically induced global change which suggest that an early consequence of increased oceanic and atmospheric temperatures in the Antarctic will be an increase in snow precipitation on the continent, thus reducing the rate of rise in sea level otherwise estimated [*IPCC*, 1990].

Denton et al. [1991] discussed in detail the history of the Antarctic ice sheet but, for the Pliocene, concentrated on the evidence provided by the sections in the Ross Sea region, especially from the Sirius Formation and equivalents. The temperature implications from the Marine Plain section are consistent with the reconstruction of a reduced ice sheet at the time [*Denton et al.*, 1991, Figure 10.28] but may indicate a somewhat higher seawater temperature than suggested there.

The coastal East Antarctic sections do not resolve the two opposing hypotheses on minimum ice sheet configuration, and the resolution depends on gaining central Antarctic information, including from under the ice overlying the Wilkes and Pensacola basins, and on the validity or otherwise of the independent sea level estimates of *Vail et al.* [1977] and *Haq et al.* [1987].

The author believes results so far obtained from the Marine Plain section reinforce the evidence of vegetation from the Sirius Group [*Carlquist*, 1987; *Webb and Harwood*, 1987; R. Hill and E. M. Truswell, personal communication, 1991] that there was a time during the early Pliocene when Antarctic air and seawater temperatures were significantly warmer than now, that there was a much smaller ice sheet, and if so, that sea level should have been higher globally. There is evidence from marine seismic studies that sea level was higher [*Vail et al.*, 1977; *Haq et al.*, 1987] by about 60 m, itself consistent with there being a much reduced ice sheet during the early Pliocene. However, it was suggested by *Harwood et al.* [1992], *Harwood and Webb* [1992], and *Webb* [1992] that the Sirius Formation is late Pliocene–early Pleistocene age and thus considerably younger than the Marine Plain sediments.

This warm interval is also recorded elsewhere, for example, widely in Australia. *Quilty* [1974] recorded a warm interval off northwestern Australia where coiling ratio changes in planktonic foraminifera indicate that the early Pliocene was warmer than the late Miocene and late Pliocene or younger. In the Murray Basin of southeastern Australia, a *Nothofagus*-dominated forest regime returned in the early Pliocene for a "brief" interval consistent with high, nonseasonal precipitation [*Martin*, 1989]. This may have been equivalent to what has been described in the Lake George region of eastern Australia as the "warm, wet phase" [*Truswell*, 1990]. The Murray Basin vegetation changed in the middle and late Pliocene to wet sclerophyll forest as it evolved to its current dry phase.

The information available to date from the Pagodroma Tillite in the northern Prince Charles Mountains is consistent with an expansion of the East Antarctic Ice Sheet late in the Pliocene or early in the Pleistocene, younger than deposition of the Marine Plain section which then can be taken to represent the interval during which the Pliocene ice sheet was at a minimum, sea level was highest, and marine conditions were most widespread. The Pagodroma Tillite reworked microfossils occur in "diatomaceous sediment clumps and as isolated specimens" and may have been derived from a local Lambert Graben source [*McKelvey and Stephenson*, 1990].

It is not yet clear how the Larsemann Hills Pliocene sediments relate to the evolving story, but their age (3–2 Ma) would indicate that they may have been deposited prior to the expansion of the ice sheet toward the end of the Pliocene. Although *Quilty* [1990, 1992] has concluded that these sediments are in situ because there is no evidence of reworking, a reworking hypothesis cannot be ruled out.

The sediments from the Windmill Islands are very poorly known, but the 2–1 Ma age and lack of cold water indices make it difficult to reconcile them with present scenarios of Antarctic glacial history.

Several sections such as those from Airport Road, Vestfold Hills, and the episode of aragonite deposition cannot yet be placed into an East Antarctic evolutionary history. Placement in context will require better age definition for the Airport Road material.

The Holocene sequences known all postdate the last glacial maximum, but they represent a variety of ages and environments of deposition and are concentrated in the interval 7.5–4.5 ky B.P. To date the principal studies have related to documenting age and occurrence.

SUMMARY

East Antarctica has provided sections representing a variety of ages (Figure 7) and environments over the last 4.2 Ma, and most contain microfossils or other fossils that yield data for interpretation of age and paleoenvironment. The key results to date are as follows:

1. In the interval 4.2–3.5 Ma the Vestfold Hills provides some evidence of marine water temperatures of about 5°C. On the basis of this evidence and data from outside the Antarctic, I infer that there was a significantly smaller ice sheet and higher sea level at the time.

2. The same sequence contains excellent cetacean fossils that are under study at present.

3. The Larsemann Hills has a thin section with very

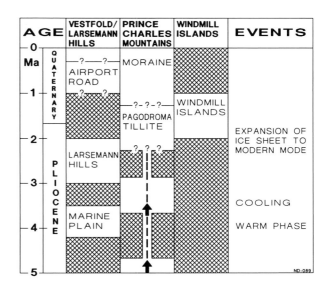

Fig. 7. Correlation diagram for sections discussed in text. Blank boxes in the Prince Charles Mountains column refer to the record of fossils of this age but of, as yet unknown, significance.

well preserved foraminiferal fauna and fragmented mollusks dated on the basis of foraminifera and diatoms at 3–2 Ma.

4. The Pagodroma Tillite of the Prince Charles Mountains contains evidence of reworking of Neogene marine microfossils during an ice sheet expansion in the latest Pliocene or earliest Pleistocene.

5. The Windmill Islands region has yielded marine sediments of 2–1 Ma which contain no evidence of modern style cold water diatom flora. Other ages may be represented, and environments represented include lacustrine conditions.

6. The Vestfold Hills also has sections of a variety of late Neogene ages including 1 Ma to 300 ka, 35 ka, widespread undated late Pleistocene, and a variety of Holocene ages.

7. Other widely spaced sequences exist, but their full potential is not known. Most appear to be Holocene in age.

8. Study of all these sections is in its infancy.

FUTURE RESEARCH

All sediment sections identified to date need a great deal of work to document their distribution, age, facies variation, stable isotope record, and paleoenvironmental significance and eventually to have the information integrated into the evolving story of the Antarctic glaciation. There is also a need to search for additional sources of information.

In the long run, there will be a need for further ocean

drilling to provide a continuous sequence against which the onshore record can be compared.

Perhaps the most important area requiring detailed study is that in the Windmill Islands, where the sediments known to date indicate great potential for data on a time for which little other information is available.

For the late Pleistocene and Holocene there is an exciting and unique opportunity to integrate the results of study of sections representing approximately the last 500 ka through material found in outcrops, deep-sea drilling, and the Antarctic ice sheet [e.g., *Jouzel et al.*, 1987]. Each provides its own type of information and holds hope for documenting a history in a detail not possible for any other part of the time scale.

Perhaps the most challenging area for the future is in integrating the documented change in environmental parameters with changes in the Antarctic biota over the same time. With environmental change great, and perhaps fast in the polar regions, there are many lessons to be learned about evolutionary processes during intervals of significant environmental change.

Acknowledgments. Many people have contributed to this paper, especially in finding localities where Neogene sediments occur. Here I must mention J. Burgess and collaborators, F. Baciu and J. Pickard. D. Harwood has provided a continuous supply of diatom-based dates when asked. J. Cox and B. Hansen produced the figures. D. Hodell, M. Prentice, and an anonymous reviewer made remarks that have improved the paper.

REFERENCES

Adamson, D. A., and J. Pickard, Cainozoic history of the Vestfold Hills, in *Antarctic Oasis*, edited by J. Pickard, pp. 63–97, Academic, San Diego, Calif., 1986.

Aharon, P., Oxygen, carbon and U-series isotopes of aragonites from Vestfold Hills, Antarctica: Clues to geochemical processes in subglacial environments, *Geochim. Cosmochim. Acta*, 52, 2321–2331, 1988.

Bardin, V. I., Composition of some East Antarctic moraines and some problems of Cenozoic history, in *Antarctic Geoscience*, edited by C. Craddock, pp. 1069–1076, University of Wisconsin Press, Madison, 1982.

Barker, P. F., et al., Leg 113, *Proc. Ocean Drill. Program Sci. Results*, 113, 1033 pp., 1990.

Barrett, P. G., G. S. Wilson, C. J. Adams, G. J. Gossan, D. M. Harwood, and A. R. Pyne, Radiometric dating of volcanic ash in Ferrar Fjord and its bearing on Pliocene deglaciation of Antarctica (abstract), in *Abstracts, Sixth International Symposium on Antarctic Earth Sciences*, p. 38, National Institute of Polar Research, Tokyo, 1991.

Barrett, P. G., C. J. Adams, W. C. McIntosh, C. C. Swisher III, and G. S. Wilson, Geochronological evidence supporting Antarctic deglaciation three million years ago, *Nature*, 359, 816–818, 1992.

Barron, J., et al., Leg 119, *Proc. Ocean Drill. Program Sci. Results*, 119, 1003 pp., 1991.

Bird, M. I., A. R. Chivas, C. J. Radnell, and H. R. Burton, Sedimentological and stable-isotope evolution of lakes in the Vestfold Hills, Antarctica, *Palaeogeogr. Palaeoclimatol. Palaeoecol.*, 84, 109–130, 1991.

Bolshiyanov, D., S. Verkulich, Z. Pushina, and E. Kirienko,

Some features of the late Pleistocene and Holocene history of the Bunger Hills (East Antarctica), in *Abstracts, Sixth International Symposium on Antarctic Earth Sciences*, pp. 66–71, National Institute of Polar Research, Tokyo, 1991.

Burckle, L. H., and E. M. Pokras, Implications of a Pliocene stand of *Nothofagus* (southern beech) within 500 kilometres of the south pole, *Antarct. Sci.*, 3, 389–403, 1991.

Burckle, L. H., R. Gersonde, and N. Abrams, Late Pliocene–Pleistocene paleoclimate in the Jane Basin region: ODP Site 697, *Proc. Ocean Drill. Program Sci. Results*, 113, 803–809, 1990.

Cameron, R. L., O. Loken, and J. Molholm, Wilkes Station glaciological data 1957–58, *Rep. 825-1*, part 3, Ohio State Univ. Res. Found., Columbus, 1959.

Carlquist, S., Pliocene *Nothofagus* wood from the Transantarctic Mountains, *Aliso*, 11, 571–583, 1987.

Colhoun, E. A., and D. A. Adamson, Raised beaches of the Bunger Hills, *ANARE Sci. Rep.*, 136, 1–47, 1992.

Collerson, K. D., and J. W. Sheraton, Bedrock geology and crustal evolution of the Vestfold Hills, in *Antarctic Oasis*, edited by J. Pickard, pp. 21–62, Academic, San Diego, Calif., 1986.

Crespin, I., Some recent foraminifera from Vestfold Hills, Antarctica, *Sci. Rep. Tohoku Univ. Ser. 2*, 4, 19–31, 1960.

Denton, G. H., M. L. Prentice, and L. H. Burckle, Cainozoic history of the Antarctic ice sheet, in *The Geology of Antarctica*, edited by R. J. Tingey, pp. 365–433, Clarendon, Oxford, 1991.

Fordyce, R. E., Origins and evolution of Antarctic marine mammals, Origins and Evolution of the Antarctic Biota, *Spec. Publ. Geol. Soc. London*, 47, 269–281, 1989.

Goodwin, I., Holocene deglaciation, sea level change and the emergence of the Windmill Islands, Budd Coast, Antarctica, *Quat. Res.*, 40, 70–80, 1992.

Haq, B. U., J. Hardenbol, and P. R. Vail, The chronology of fluctuating sea level since the Triassic, *Science*, 235, 1156–1167, 1987.

Harrison, R., and M. M. Bryden, *Whales, Dolphins and Porpoises*, Weldon Owen, Sydney, Australia, 1989.

Harwood, D. M., Cretaceous and Cenozoic siliceous microfossil biostratigraphy and Antarctic glacial and marine geologic history, Ph.D. dissertation, 590 pp., Ohio State Univ., Columbus, 1986.

Harwood, D. M., and P.-N. Webb, Early Pliocene deglaciation of the Antarctic ice sheet and late Pliocene onset of bipolar glaciation (abstract), *Eos Trans. AGU*, 71, 538–539, 1990.

Harwood, D. M., and P.-N. Webb, Pliocene Antarctic deglaciation and high-latitude warming: SIRIUS Group evidence (abstract), in *Cenozoic Glaciations and Deglaciations*, Geological Society of London, London, 1992.

Harwood, D. M., P.-N. Webb, and P. J. Barrett, The search for consistency between several indices of Antarctic Cenozoic glaciation (abstract), in *Cenozoic Glaciations and Deglaciations*, Geological Society of London, London, 1992.

Hayes, D. E., et al., Leg 28, *Initial Rep. Deep Sea Drill. Proj.*, 28, 1017 pp., 1975.

Hirvas, H., and K. Nenonen, Moreenin kerrosjärjestyksestä ja jäätiköitymishistoriasta Vestfold Hillsin alueella itäAntarktisella (On till stratigraphy and glacial history of the Vestfold Hills, East Antarctica), *Geologi*, 8, 151–156, 1989a.

Hirvas, H., and K. Nenonen, Glacial history and paleoclimates of the Vestfold Hills area, East Antarctica, in *Antarctic Reports of Finland, FINNARP-89 Report 1*, pp. 31–38, Ministry of Trade and Industry, Helsinki, 1989b.

Hirvas, H., and K. Nenonen, Field methods for glacial indicator tracing, in *Glacial Indicator Tracing*, edited by R. Kujansuu and M. Saarnisto, pp. 217–248, A. A. Balkema, Rotterdam, Netherlands, 1990.

Hirvas, H., K. Nenonen, and P. G. Quilty, Till stratigraphy

and glacial history of the Vestfold Hills area, East Antarctica, in *INQUA Quaternary International*, Pergamon, New York, 1992.

Hodell, D. A., and P. F. Ciesielski, Stable isotope and carbonate stratigraphy of the late Pliocene and Pleistocene of Hole 704A: Eastern Subantarctic South Atlantic, *Proc. Ocean Drill. Program Sci. Results*, 114, 409–435, 1991.

Hodell, D. A., and K. Venz, Toward a high-resolution stable isotopic record of the Southern Ocean during the Pliocene-Pleistocene (4.8 to 0.8 Ma), in *The Antarctic Paleoenvironment: A Perspective on Global Change, Part 1, Antarct. Res. Ser.*, vol. 56, edited by J. P. Kennett and D. A. Warnke, pp. 265–310, AGU, Washington, D. C., 1992.

Hodell, D. A., D. W. Müller, P. F. Ciesielski, and G. A. Mead, Synthesis of oxygen and carbon isotopic results from Site 704: Implications for major climatic-geochemical transitions during the late Neogene, *Proc. Ocean Drill. Program Sci. Results*, 114, 475–480, 1991.

Intergovernmental Panel on Climate Change, *Climate Change: The IPCC Scientific Assessment*, edited by J. T. Houghton, G. J. Jenkins, and J. J. Ephraums, Cambridge University Press, New York, 1990.

Ishman, S. E., and P.-N. Webb, Late Neogene benthic foraminifera from the Victoria Land Basin margin, Antarctica: Application to glacio-eustatic and tectonic events, *Rev. Paleobiol.*, Spec. Vol., 2, 523–551, 1988.

Jouzel, J., C. Lorius, J. R. Petit, C. Genthon, N. Barkov, V. M. Kotlyakov, and V. N. Petrov, Vostok ice core: A continuous isotope temperature record over the last climatic cycle (160,000 years), *Nature*, 327, 403–409, 1987.

Kennett, J. P., Cenozoic evolution of Antarctic glaciation, the circum-Antarctic Ocean, and their impact on global paleoceanography, *J. Geophys. Res.*, 82, 3843–3860, 1977.

Lewis-Smith, R. I., Nutrient cycling in relation to biological productivity in Antarctic and Sub-Antarctic terrestrial and freshwater ecosystems, in *Antarctic Nutrient Cycles and Food Webs*, edited by W. R. Siegfried, P. R. Condy, and R. M. Laws, pp. 138–155, Springer-Verlag, New York, 1985.

Lewis-Smith, R. I., Report of biological programme at Casey Station, October 1985–January 1986, Australian Antarctic Division, Tasmania, 1986.

Lewis-Smith, R. I., Classification and ordination of cryptogamic communities in Wilkes Land, *Vegetatio*, 76, 155–166, 1988.

Martin, H. A., Vegetation and climate of the late Cainozoic in the Murray Basin and their bearing on the salinity problem, *BMR J. Aust. Geol. Geophys.*, 11, 291–299, 1989.

McKelvey, B. C., and N. C. N. Stephenson, A geological reconnaissance of the Radok Lake area, Amery Oasis, Prince Charles Mountains, *Antarct. Sci.*, 2, 53–66, 1990.

McKelvey, B. C., M. C. G. Mabin, D. M. Harwood, and P.-N. Webb, The Pagodroma Event—A late Pliocene major expansion of the ancestral Lambert Glacier system (abstract), in *Abstracts, Sixth International Symposium on Antarctic Earth Sciences*, p. 403, National Institute of Polar Research, Tokyo, 1991.

McKenzie, J. A., H. Weissert, R. Z. Poore, R. C. Wright, F. S. J. Percival, H. Oberhansli, and M. Casey, Paleoceanographic implications of stable-isotope data from upper Miocene–lower Pliocene sediments from the southeast Atlantic Deep Sea Drilling Project Site 519, *Initial Rep. Deep Sea Drill. Proj.*, 73, 717–724, 1984.

Moriwaki, K., Y. Yoshida, and D. M. Harwood, Cenozoic glacial history of Antarctica—A correlative synthesis, in *Proceedings of the Sixth SCAR Symposium on Antarctic Earth Science*, edited by Y. Yoshida and K. Kaminuma, pp. 773–780, National Institute of Polar Research, Tokyo, 1992.

Pickard, J. (Ed.), *Antarctic Oasis*, Academic, San Diego, Calif., 1986.

Pickard, J., D. A. Adamson, D. M. Harwood, G. H. Miller, P. G. Quilty, and R. K. Dell, Early Pliocene marine sediments in the Vestfold Hills, East Antarctica: Implications for coastline, ice sheet and climate, *S. Afr. J. Sci.*, 82, 520–521, 1986.

Pickard, J., D. A. Adamson, D. M. Harwood, G. H. Miller, P. G. Quilty, and R. K. Dell, Early Pliocene marine sediments, coastline, and climate of East Antarctica, *Geology*, 16, 158–161, 1988.

Prentice, M. L., and R. K. Matthews, Tertiary ice sheet dynamics: The snow gun hypothesis, *J. Geophys. Res.*, 96, 6811–6827, 1991.

Quilty, P. G., Tertiary stratigraphy of Western Australia, *J. Geol. Soc. Aust.*, 21, 301–318, 1974.

Quilty, P. G., Foraminiferida from Neogene sediments Vestfold Hills, Antarctica, in *Biology of the Vestfold Hills, Antarctica*, edited by J. M. Ferris, H. R. Burton, G. W. Johnstone, and I. E. Bayly, pp. 213–220, Kluwer Academic, Hingham, Mass., Dordrecht, 1988.

Quilty, P. G., New evidence for changes in the Antarctic environment over the last five million years, in *Antarctic Ecosystems*, edited by K. R. Kerry and G. Hempel, pp. 3–8, Springer-Verlag, New York, 1990.

Quilty, P. G., Implications of late Neogene sediments, East Antarctica, in *Proceedings of the First Latinamerican Conference on Antarctic Geophysics, Geodesy and Space Sciences*, edited by O. Schneider, pp. 19–24, Centro Latino-Americano de Física, Buenos Aires, 1991a.

Quilty, P. G., The geology of Marine Plain, Vestfold Hills, East Antarctica, in *The Geological Evolution of Antarctica*, edited by M. R. A. Thomson, J. A. Crame, and J. W. Thomson, pp. 683–686, Cambridge University Press, New York, 1991b.

Quilty, P. G., Sources of information on the late Neogene of coastal East Antarctica, in *Proceedings of the Sixth SCAR Symposium on Antarctic Earth Science*, edited by Y. Yoshida and K. Kaminuma, pp. 699–705, National Institute of Polar Research, Tokyo, 1992.

Quilty, P. G., D. Gillieson, J. Burgess, G. Gardiner, A. Spate, and R. Pidgeon, *Ammoelphidiella* from the Pliocene of Larsemann Hills, East Antarctica, *J. Foraminiferal Res.*, 20, 1–7, 1990.

Rind, D., Global climate in the 21st century, *Ambio*, 13, 148–151, 1984.

Setty, M. G. A. P., R. Williams, and K. R. Kerry, Foraminifera from Deep Lake terraces, Vestfold Hills, Antarctica, *J. Foraminiferal Res.*, 10, 303–312, 1980.

Truswell, E. M., Australian rainforests: The 100 million year record, in *Australian Tropical Rainforests: Science-value-meaning*, edited by L. J. Webb and J. Kikkawa, pp. 7–22, Commonwealth Scientific and Industrial Research Organization, Melbourne, Australia, 1990.

Vail, P. R., R. M. Mitchum, Jr., R. G. Todd, J. M. Widmier, S. Thompson III, J. B. Sangree, J. N. Bubb, and W. G. Hatlelid, Seismic stratigraphy and global changes of sea level, in *Seismic Stratigraphy—Applications to Hydrocarbon Exploration*, edited by C. E. Payton, pp. 49–212, American Association of Petroleum Geologists, Tulsa, Okla., 1977.

Webb, P.-N., The Cenozoic history of Antarctica and its global impact, *Antarct. Sci.*, 2, 3–21, 1990.

Webb, P.-N., The terrestrial and marine record of Cenozoic glaciation in the southern hemisphere (abstract), in *Fourth International Conference on Paleoceanography*, Kiel, Germany, 1992.

Webb, P.-N., and J. E. Andreasen, Potassium/argon dating of volcanic material associated with the Pliocene Pecten Conglomerate (Cockburn Island) and Scallop Hill Formation (McMurdo Sound), *Antarct. J. U. S.*, 22, 59, 1986.

Webb, P.-N., and D. M. Harwood, The terrestrial flora of the Sirius Formation: Its significance in interpreting Late Cenozoic glacial history, *Antarct. J. U. S.*, *22*, 7–11, 1987.

Webb, P.-N., and D. M. Harwood, Late Cenozoic glacial history of the Ross Embayment, Antarctica, *Quat. Sci. Rev.*, *10*, 215–223, 1991.

Webb, P.-N., D. M. Harwood, B. C. McKelvey, M. C. G. Mabin, and J. H. Mercer, Late Cenozoic tectonic and glacial history of the Transantarctic Mountains, *Antarct. J. U. S.*, *21*, 99–100, 1986.

Wise, S. W., Jr., et al., Leg 120, *Proc. Ocean Drill. Program Sci. Results*, *120*, 1155 pp., 1992.

Yoshida, Y., Physiography of the Prince Olav and Prince Harald coasts, East Antarctica, *Mem. Natl. Inst. Polar Res. Ser. C.*, *13*, 76–83, 1983.

Zhang, Q., Quaternary stratigraphy of Vestfold Hills, Antarctica, in *Selected Papers of Antarctic Research*, pp. 27–38, Chinese Academy of Sciences, Beijing, 1985.

Zhang, Q., Evolution of the Antarctic ice sheet since late Pleistocene, in *Proceedings of the International Symposium on Antarctic Research*, pp. 67–73, China Ocean Press, Tianjin, 1989.

Zhang, W., and J. A. Peterson, A geomorphology and late Quaternary geology of the Vestfold Hills, Antarctica, *ANARE Sci. Rep.*, *133*, 1–84, 1984.

(Received July 15, 1992;
accepted January 26, 1993.)

THE ANTARCTIC PALEOENVIRONMENT: A PERSPECTIVE ON GLOBAL CHANGE
ANTARCTIC RESEARCH SERIES, VOLUME 60, PAGES 265–272

300-YEAR CYCLICITY IN ORGANIC MATTER PRESERVATION IN ANTARCTIC FJORD SEDIMENTS

Eugene W. Domack, Tracy A. Mashiotta,[1] and Lewis A. Burkley

Department of Geology, Hamilton College, Clinton, New York 13323

Scott E. Ishman

U.S. Geological Survey, Reston, Virginia 22092

Total organic carbon and biogenic silica analyses were conducted on samples from a 9-m-long piston core collected in Andvord Bay, Antarctica. Accelerator mass spectrometer radiocarbon analyses were conducted on organic matter fractions and foraminifers and resulted in an excellent chronology for the past 3000 years. Fluctuations in the preserved total organic carbon and biogenic silica record approximately 12 cycles over the same time period. The cycles are the result of either temporal variations in the vertical carbon flux, as controlled by variable productivity, or changes in the terrigenous sediment supply, as controlled by variable meltwater input or variations in bottom sediment resuspension. Hence the complete 3000-year record of fluctuating total organic carbon and biogenic silica may record a unique signal of paleoclimatic and/or oceanographic changes for this portion of the Antarctic Peninsula.

INTRODUCTION

There is an urgent need for high-resolution paleoclimatic records from Antarctica that span the last 10,000 years. Some ice cores do exist [*Mosley-Thompson et al.*, 1990], but the duration and resolution are not sufficient to ideally resolve variations within a significant portion of the Holocene record. At the same time there has been a great expansion of our knowledge of the primary productivity signal in modern circum-Antarctic waters [*Dunbar et al.*, 1989; *Bodungen et al.*, 1986; *Leventer*, 1991; *Smith et al.*, 1988; *Wefer et al.*, 1990], so that the spatial and temporal variation in particle flux has been resolved for several offshore areas around the continent. What is needed are records that illuminate the fluctuation of primary productivity for the most recent past and that can serve as paleoclimatic proxies over time scales appropriate to understanding rapid change.

Marine sediment cores from the Antarctic continental shelf contain a useful high-resolution record of climatic fluctuations along the periphery of the continent [*Domack et al.*, 1991]. To date the record in such sediment has remained largely unexamined because of the complexity of glacial-marine sedimentation and because of the limited availability of suitable core mate-

rial. The complexity of sedimentation is illustrated by the fact that at least four major sediment contributors can be recognized, including ice rafting and biogenic, meltwater, and eolian processes. The ability to isolate the paleoclimatic signal depends upon thorough analyses of all components of the sediment and consequently an understanding of the relative contributions of each to the sediment record. It is important that such attempts be made, since Antarctica has relatively few high-resolution climatic time series that extend back more than a few hundred years.

The purpose of this paper is to demonstrate the utility of glacial-marine sediment cores, recently recovered from fjords of the Antarctic Peninsula, in the reconstruction of past productivity changes. Our time resolution is of the order of decades and includes a continuous record of the past 3000 years. Hence the record is ideally suited to resolving the nature of very high resolution changes on a time scale similar to that of the Little Ice Age.

METHODS

Core 22 was collected in 1988 from the central basin (water depth of 440 m) of Andvord Bay along the southern Danco Coast, western Antarctic Peninsula (Figures 1 and 2). The core is curated at the Antarctic Research Facility of Florida State University. The core was X rayed from a split half and then sampled every 10

---

[1] Now at Department of Geological Sciences, University of California, Santa Barbara, Santa Barbara, California 93106.

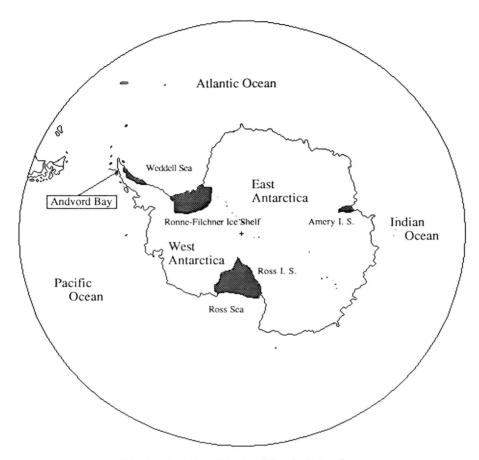

Fig. 1.   Location of Andvord Bay in Antarctica.

cm for total organic carbon analyses. A second set of samples was collected every 20 cm for bulk density and biogenic silica analyses. Total organic carbon (TOC) was determined by combustion in a LECO induction furnace after first removing the very coarse sand and gravel fraction from the sediment. Biogenic silica was determined following the procedures of *DeMaster* [1979]. Five samples were selected for ^{14}C analyses including three organic matter dates and two foraminifer samples. The samples for total organic dating were taken from the core on the ship and were frozen while being transported to the University of Arizona accelerator laboratory. Benthic and planktonic foraminifers were removed from sieved and dried samples of the sand fraction. We assume no difference in ^{14}C ages for benthic and planktonic species.

RESULTS

X Radiographs

Results of the X radiographs as shown in Figure 3 demonstrate a uniform distribution of ice-rafted mate-

rial. In places, concentrations of ice-rafted debris occur, and these concentrations are indicated in Figure 3. The uniform distribution of coarse material and lack of current bedding structures suggest continuous deposition without significant hiatuses.

Chronology

Five radiocarbon accelerator ages were obtained from three stratigraphic levels within the core (Table 1). The ages range from 2025 ± 60 years B.P., at 66-cm depth, to 4480 ± 75 years B.P., at 778- to 798-cm depth. Reservoir corrections of ~1200 years are based upon an age of 1240 ± 85 years for a living mollusk recovered in a grab sample taken near the core site (Table 1 and Figure 2). Hence the sediment record in core 22 represents approximately the last 3000 years as the ^{14}C dates result in a linear depth/age relationship of some 0.31 cm/yr (Figure 4). *Harden et al.* [1992] report rates of 0.01 and 0.08 cm/yr for glacial-marine sediments in the offshore regions of the Gerlache Strait. Interval sedimentation rates for core 22, as determined between

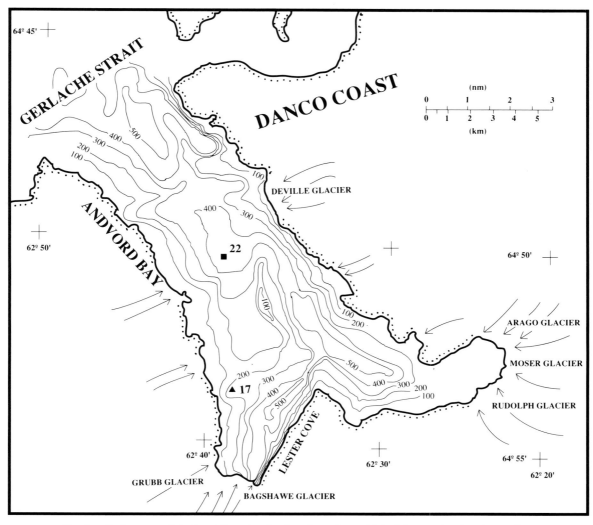

Fig. 2.   Bathymetry of Andvord Bay showing locations of core 22 (solid square) and grab sample 17 (solid triangle).

adjacent sample depths, are 0.23 and 0.52 cm/yr for the upper and lower portions of the core, respectively. If a constant rate of sedimentation is assumed, the 10-cm sampling for TOC has a temporal resolution of some 30 years, while that for biogenic silica is approximately 60 years. The ^{14}C ages are considered to be especially accurate, since both foraminifer and organic matter dates produced near-identical results (Table 1). For example, the age of 3825 ± 65 years on organic matter at 487-cm depth corresponds to an age of 3750 ± 65 years on foraminifers between 472- and 510-cm depth. This means that there has been little reworking of significantly "old" organic matter within these intervals of core 22.

Upward extrapolation of the linear sedimentation rate results in an assumed core top age of ~2000 years (Figure 4). Such an age is significantly older than the living mollusk (reservoir) age of 1240 ± 85 years. From this information alone it appears that a significant reduction in the sedimentation rate occurs in the uppermost 50 cm of the core. However, surface (0–5 cm) organic matter taken from a box core collected in the same basin has a ^{14}C age of 2320 ± 120 years [*Harden et al.*, 1992]. Taken together, these data suggest that there may be a difference between calcite and organic matter ages in the upper portion of the core. This may reflect a recent increase in reworking of "old" organic matter within the basin. Large foraminifers were not abundant enough in the upper portion of the core so that comparative calcite and organic matter ages were unobtainable.

Core USAP-87 22

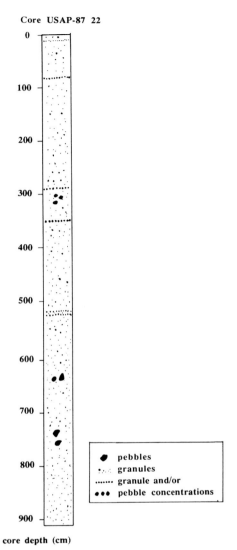

core depth (cm)

Fig. 3. Log of X ray radiograph of core 22 showing internal textural variations.

Bulk Density and Particle Flux

Dry bulk densities ranged from 0.48 to 0.87 g/cm^3 with an average of 0.71 g/cm^3 for the entire core. This allows for calculations of particle flux which ranged

TABLE 1. Carbon 14 Results for Core 22, Andvord Bay

Sample Number	Depth, cm	^{14}C Age, Years B.P.	Carbon Source
AA-5215	0 (grab 17)	1240 ± 85	mollusk
AA-4751	66	2025 ± 60	organic matter
AA-5210	472–510	3750 ± 65	foraminifers
AA-4752	487	3825 ± 65	organic matter
AA-5209	778–798	4480 ± 75	foraminifers
AA-4753	792	4415 ± 60	organic matter

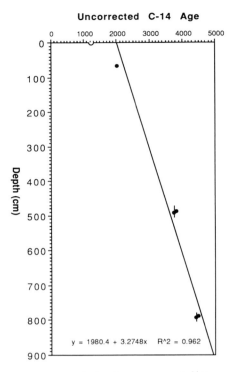

Fig. 4. Downcore variation in uncorrected ^{14}C age. Depth ranges of foraminifer samples are shown by short vertical lines through the sample location. Reservoir age is shown by open dot at ~1200 years B.P. Rate of sediment accumulation is shown as the inverse to the slope of the best fit line, at 0.31 cm/yr.

from 1500 to 2600 g m^{-2} yr^{-1} with an average of 2200 g m^{-2} yr^{-1}. These results are 2 orders of magnitude higher than particle fluxes determined for open ocean Antarctic sites [*Wefer et al.*, 1990] but several orders of magnitude less than fjord sedimentation in temperate oceanic areas such as in Alaska [*Cowan and Powell*, 1991]. The high particle flux is a reflection of the nearshore fjord environment where productivity is enhanced [*Mashiotta*, 1992] and terrigenous sources are in close proximity [*Domack and Williams*, 1990].

Modern Sedimentation and Preservation of Organic Matter

Modern surface sediments from Andvord Bay have TOC contents that range from 0.1% to 1.4% [*Domack and Ishman*, 1993]. Those samples with TOC contents greater than 1.0% are found within the central basin of Andvord Bay near the site of core 22 (Figure 2). The TOC content of the modern sediments is controlled by the vertical flux of organic carbon and terrigenous sedimentation (dilution) associated with ice-proximal meltwater plumes and reworking of bottom sediments.

The vertical flux of organic carbon in the central basin of Andvord Bay is relatively high (up to 290 mg C m^{-2}

d⁻¹ [*Mammone*, 1992]) as determined by sediment trap experiments. Such high fluxes are related to elevated productivity within the central basin where an eddy-type circulation favors physical stability and warming of the surface layer [*Mammone*, 1992; *Domack and Ishman*, 1993]. Organic matter from surface production is diluted by terrigenous sedimentation in the head regions of Andvord Bay such as in Lester Cove (Figure 2). Terrigenous sediment is transported and deposited from midwater cold tongues derived from basal meltwater of the fjord-head glaciers [*Domack and Williams*, 1990]. At present, transport of this sediment is restricted by bottom topography (Figure 2) and the dominance of midwater versus surface layer transport [*Domack and Williams*, 1990]. Organic matter is also diluted by near-bottom transport of resuspended particulate matter which is enriched in terrigenous material [*Mammone*, 1992]. This process is presently occurring along the bottom of the central basin [*Mammone*, 1992].

The total organic carbon (TOC) and biogenic silica content of the core is illustrated in Figure 5. The core is marked by subtle yet cyclical fluctuations in TOC with generally higher values in the lower third of the core. Biogenic silica contents range from 12% to 21%, which falls within the range of a diatomaceous mud. The downcore pattern in biogenic silica parallels that of the TOC, with a few minor exceptions. Of note is the general first-order trend of increasing silica and TOC with depth and the second-order variations which define cycles of silica and TOC abundance that are well outside of the analytical error of the methods. In Figure 5 it is clear to see that there are 12 peaks and 11 troughs in the TOC content of the entire 9-m core. With a constant rate of sedimentation the average duration of each cycle is ~268 years, with the dominant period being close to 300 years.

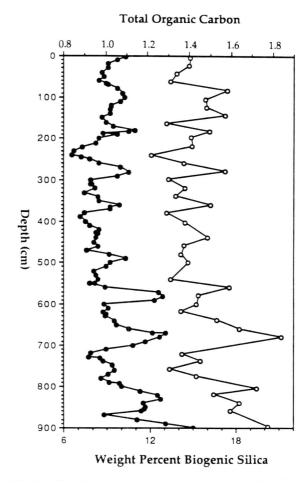

Fig. 5. Plot of downcore variation in total organic carbon (solid dots) and biogenic silica (open dots). A smoothing function has been applied to the organic carbon data. The smoothing function is a running average of every three samples.

Benthic Foraminifer Data

Benthic foraminifer distribution within the bays and fjords of the Antarctic Peninsula coast is strongly controlled by the organic carbon concentration within the surface sediments [*Ishman*, 1990]. Thirty-two taxonomic groups representing 35 species were identified from eight sediment samples from core 22 (Table 2). Species diversity (number of species present) ranged from 12 to 4 species. The number of foraminifers per gram of sediment ranged from 23.90 to 1.67 foraminifers per gram (Figure 6).

The upper part of core 22 (<300 cm) is dominated by agglutinated benthic taxa (Table 2). Significant taxa include *Spiroplectammina biformis*, *Verneuilina minuta*, *Portatrochammina* spp., and *Miliammina* spp. These species predominate in modern sediments from Marguerite Bay and Bransfield Strait, regions with low biogenic-dominated sedimentation rates. In contrast, the lower portion of core 22 (>300 cm) is dominated by the calcareous benthic taxa *Globocassidulina biora* and *Fursenkoina* spp. These species are more typical of fjord depositional systems in the northern region of the Antarctic Peninsula, where high biogenic productivity results in the deposition of surface sediments rich in organic carbon.

The number of benthic foraminifers per unit of sediment (grams) is indicative of relative sedimentation rates. Increased sedimentation rates result in the dilution of the benthic foraminifer number per unit of sediment. The foraminifers-per-gram values for core 22 show a decrease with depth (Figure 6). In particular, the values for the upper part of core 22 (<300 cm) indicate sedimentation rates reduced from those indicated from the foraminifers-per-gram values at depths greater than 300 cm in the core.

TABLE 2.   Benthic Foraminifer Distribution in Core 22, Andvord Bay

SAMPLE	Spiroplectammina biformis	Bolivina pseudopunctata	Trochammina intermedia	Adercotryma glomeratum	Verneulina minuta	Portatrochammina elatninae	Textularia antarctica	Textularia tenuissima	Textularia wiesneri	Reophax ovicula	Bulimina aculeata	Neogloboquadrina pachyderma	Cassidulinoides parkerianus	Globocassidulina subglobosa	Astrononion echolsi	Portatrochammina wiesneri	Portatrochammina antarctica	Trochammina bullata	Globocassidulina biora	Conotrochammina alternans	Uvigerina sp.	Cassidulinoides porrectus	Trifarina earlandi	Pullenia sallisburyi	Cribrostomoides jefferysi	Fursenkoina spp.	Nonionella spp.	Nodosariids	Miliammina spp.	TOTAL	SED. WT.	Forams/gram
40-90											1								2				1					1	52	57		
101-103	153	6		45	10	2	1		6							12	14			1					2				3	255	10.67	23.90
201-203	69	12	1	11	9	2			2							19	10	6		3									60	204	10.67	19.12
302-304	16		20	3	6	1										2				1									13	62	10.73	5.78
398-400	5		11	5	2			1	5							1													16	46	10.92	4.21
472-510		1									1	2	1		4				413			4	1	2		8	3		140	580		
549-551											1						2	1											23	27	12.18	2.22
618-620		1			1						1							1											12	16	9.57	1.67
698-700	3		3	2	1	1				1						3	1	1											18	34	10.64	3.20
778-798											14	1		1	6				290		1		2			25	4		22	366		
891-892	4	1		2													1												15	23	11.33	2.03

DISCUSSION

The cycles of TOC and silica might be an indication of dilution due to terrigenous sedimentation as from meltwater. In this case the TOC minima would indicate climatic warm periods, and the maxima would indicate colder periods. Important in this regard is the relationship of sedimentation rate and TOC contents in modern Antarctic fjord sediments. Our data indicate that as the sedimentation rate increases, the TOC content increases, indicating a primary link between sedimentation rate and productivity. This interpretation is consistent with the benthic foraminifer data that suggest higher sedimentation rates associated with higher TOC content. In melt-dominated systems such as those in Baffin Island, Arctic Canada, the situation is reversed where the sedimentation rate increases with decreasing TOC contents [*Syvitski et al.*, 1990]. Since Andvord Bay is clearly not dominated by meltwater processes today [*Griffith and Anderson*, 1989], it appears that the TOC minima do in fact represent cold periods when primary productivity was limited. This conclusion is further supported by the presence of a benthic foraminifer assemblage similar to modern assemblages from Marguerite Bay and Bransfield Strait, where primary productivity is reduced in comparison with the Gerlache Strait and fjord systems of the Antarctic Peninsula. However, under slightly warmer conditions, meltwater processes such as surface overflow plumes may become established. This process would result in more effective down-fjord transport of terrigenous sediment and increased sedimentation within the central basin. Resolu-

tion of this interpretation depends upon a finer-scale chronology than is currently available for core 22.

Temporal variations in bottom sediment reworking might also be responsible for the variations in preserved

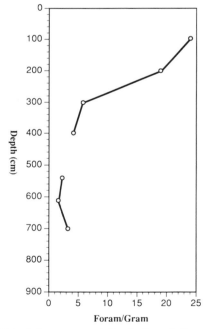

Fig. 6.   Plot of downcore variation in foraminifers per gram. Note that three samples were omitted from this plot (40–90, 472–510, and 778–798), owing to their extended sampling interval.

organic matter. Similar to the meltwater model, increased rates of sedimentation would be associated with TOC minima. Changes in bottom current velocity would be one obvious factor that could lead to variations in near-bottom suspended sediment transport and deposition.

It is possible that core 22 records past fluctuations in primary productivity as recorded by the biogenic silica and TOC contents. If one assumes a constant sedimentation rate, then the first trough on the TOC curve has a minimum at ~197 years and terminates with a peak at ~328 years. If we also add 133 years of missing time due to piston core blowout of the uppermost part of the section [*Mashiotta*, 1992], we get a minimum in TOC and biogenic silica at about 330 years B.P. This corresponds to the peak of the Little Ice Age, which dates from A.D. 1600 to 1860. Subsequent maxima and minima cannot be associated with any known global climatic events. However, the persistent pattern to the data would suggest a repetition of the kind of climatic conditions experienced during the last Little Ice Age.

Of all the variables influencing primary productivity in Antarctic surface waters, sea ice extent is the most important. Intense blooms of diatoms occur along the receding edge of the pack ice during the austral spring. Such blooms can be even more intense where surface mixing of the low-salinity surface layer is inhibited. These conditions are favored in nearshore and fjord regions where the absence of open ocean waves and winds prevents rapid disintegration of surface layer characteristics. Although nutrient limitations to productivity are generally not found in Antarctic waters, some nutrient depletion can be observed where productivity is high [*Holm-Hansen and Mitchell*, 1991]. Nutrient depletion is one of the major stresses placed upon planktonic populations of diatoms which forces them into spore formation. Over 90% of the diatoms in core 22 are *Chaetoceros* spores [*Mashiotta*, 1992], which are believed to be related to excesses in primary productivity in nearshore waters of the Gerlache Strait [*Leventer*, 1991].

We propose that fjords along the Danco Coast like Andvord Bay are ideal settings in which to preserve changes in primary productivity as influenced by climatic variations. Today the waters of Andvord Bay undergo a seasonal fluctuation in sea ice cover which leads to a strongly pulsed primary productivity signal to the bottom. It is not unreasonable to imagine a slightly warmer climate whereby the sea ice is not as extensive and may not even form. It is also not unreasonable to imagine a more severe regime in which the sea ice persists longer into the summer season than it presently does. Such changes in sea ice cover could significantly influence the flux of primary production to the bottom as noted by *Smith et al.* [1988]. We believe therefore that the downcore trends in TOC and biogenic silica are proxy records of paleoclimate variations that occurred

over several centuries. In fact, it is more unreasonable to assume that no changes in primary productivity have taken place over the past 3000 years. It is also possible that temporal variations in terrigenous sediment supply have taken place. At present we have no explanation for the forcing mechanism responsible for the pronounced cyclicity, as comparative paleoclimatic records over this same time span are rare for the southern hemisphere. However, we would suggest that additional marine sediment records for Antarctica should be examined for such fluctuations.

Acknowledgments. This work was supported by a grant from the National Science Foundation's Division of Polar Programs under the Research in Undergraduate Institutions program (DPP 86-13565 and DPP 89-15977) and from National Science Foundation Division of Polar Programs grant DPP 89-17200. We would like to thank Jill Singer (State University of New York at Buffalo) for the use of her laboratory in order to conduct the biogenic silica analyses. We thank the crew of the R/V *Polar Duke* during cruise 87-III for their support. We also thank Dave DeMaster and Amy Leventer for their helpful review of this paper.

REFERENCES

Bodungen, D. V., V. S. Smetacek, M. M. Tilzer, and B. Zeitzschel, Primary production and sedimentation during spring in the Antarctic Peninsula region, *Deep Sea Res.*, *33*, 177–194, 1986.

Cowan, E. A., and R. D. Powell, Ice-proximal sediment accumulation rates in a temperate glacial fjord, southeastern Alaska, Glacial Marine Sedimentation: Paleoclimatic Significance, *Spec. Pap. Geol. Soc. Am.*, *261*, 61–73, 1991.

DeMaster, D. J., The marine budgets of silica and ^{32}Si, Ph.D. thesis, Yale Univ., New Haven, Conn., 1979.

Domack, E. W., and S. Ishman, Oceanographic and physiographic controls on modern sedimentation within Antarctic fjords, *Geol. Soc. Am. Bull.*, in press, 1993.

Domack, E. W., and C. R. Williams, Fine structure and suspended sediment transport in three Antarctic fjords, in *Contributions to Antarctic Research I*, Antarct. Res. Ser., vol. 50, pp. 71–89, AGU, Washington, D. C., 1990.

Domack, E. W., A. J. T. Jull, and S. Nakao, Advance of East Antarctic outlet glaciers during the hypsithermal: Implications for the volume state of the Antarctic ice sheet under global warming, *Geology*, *19*, 1059–1062, 1991.

Dunbar, R. B., A. R. Leventer, and W. L. Stockton, Biogenic sedimentation in McMurdo Sound, Antarctica, *Mar. Geol.*, *85*, 155–179, 1989.

Griffith, T. W., and J. B. Anderson, Climatic control of sedimentation in bays and fjords of the northern Antarctic Peninsula, *Mar. Geol.*, *85*, 181–204, 1989.

Harden, S. L., D. J. DeMaster, and C. A. Nittrouer, Developing sediment geochronologies for high-latitude continental shelf deposits: A radiochemical approach, *Mar. Geol.*, *103*, 69–97, 1992.

Holm-Hansen, O., and B. G. Mitchell, Spatial and temporal distribution of phytoplankton and primary production in the western Bransfield Strait region, *Deep Sea Res.*, *38*, 961–980, 1991.

Ishman, S. E., Quantitative analysis of Antarctic benthic foraminifera: Application to paleoenvironmental interpretations, Ph.D. thesis, Ohio State Univ., 266 pp., Columbus, Ohio, 1990.

Leventer, A., Sediment trap diatom assemblages from the northern Antarctic Peninsula region, *Deep Sea Res.*, *38*, 1127–1143, 1991.

Mammone, K. A., Modern particle flux and productivity in Andvord Bay, Antarctica, B.A. thesis, Hamilton Coll., Clinton, N. Y., 1992.

Mashiotta, T. A., Biogenic sedimentation in Andvord Bay, Antarctica: A 3,000 year record of paleoproductivity, B.A. thesis, Hamilton Coll., Clinton, N. Y., 1992.

Mosley-Thompson, E., L. G. Thompson, P. M. Grootes, and N. Gundestrup, Little Ice Age (neoglacial) paleoenvironmental conditions at Siple Station, Antarctica, *Ann. Glaciol.*, *14*, 199–204, 1990.

Smith, W. O., N. K. Keene, and J. C. Comiso, Interannual variability in estimated primary productivity of the Antarctic marginal ice zone, in *Antarctic Ocean and Resources Variability*, edited by D. Sahrhage, pp. 131–139, Springer-Verlag, New York, 1988.

Syvitski, J. P. M., K. LeBlanc, G. William, and R. E. Cranston, The flux and preservation of organic carbon in Baffin Island fjords, Glaciomarine Environments: Processes and Sediments, *Geol. Soc. Spec. Publ. London*, *53*, 177–199, 1990.

Wefer, G., G. Fischer, D. K. Futterer, R. Gersonde, S. Honjo, and D. Ostermann, Particle sedimentation and productivity in Antarctic waters of the Atlantic Sector, in *Geological History of the Polar Oceans: Arctic Versus Antarctic*, edited by U. Bleil and J. Thiede, pp. 363–379, Kluwer Academic, Hingham, Mass., 1990.

(Received April 20, 1992;
accepted October 26, 1992.)

DEEP SEA DRILLING PROJECT AND OCEAN DRILLING PROGRAM CRUISES SHOWING THE TWO OR THREE CO-CHIEF SCIENTISTS ON EACH LEG

Initial Reports of the Deep Sea Drilling Project are available from the U.S. Government Printing Office, Washington, D. C.

Initial Reports and Scientific Results of the Ocean Drilling Program are available from The Ocean Drilling Program, College Station, Texas.

Barker, P. F., R. L. Carlson, D. A. Johnson, et al., *Initial Rep. Deep Sea Drill. Proj.*, 72, 1024 pp., 1983

Barker, P. F., I. W. D. Dalziel, et al., *Initial Rep. Deep Sea Drill. Proj.*, 36, 1079 pp., 1977.

Barker, P. F., J. P. Kennett, et al., *Proc. Ocean Drill. Program Initial Rep.*, 113, 785 pp., 1988.

Barker, P. F., J. P. Kennett, et al., *Proc. Ocean Drill. Program Sci. Results*, 113, 1033 pp., 1990.

Barron, J. A., B. Larsen, et al., *Proc. Ocean Drill. Program Initial Rep.*, 119, 42 pp., 1989.

Barron, J. A., B. Larsen, et al., *Proc. Ocean Drill. Program Sci. Results*, 119, 1003 pp., 1991.

Bolli, H. M., W. B. F. Ryan, et al., *Initial Rep. Deep Sea Drill. Proj.*, 40, 1079 pp., 1978.

Bougault, H., S. C. Cande, et al., *Initial Rep. Deep Sea Drill. Proj.*, 82, 667 pp., 1985.

Burns, R. E., J. E. Andrews, et al., *Initial Rep. Deep Sea Drill. Proj.*, 21, 931 pp., 1973.

Ciesielski, P. F., Y. Kristoffersen, et al., *Proc. Ocean Drill. Program Initial Rep.*, 114, 815 pp., 1988.

Ciesielski, P. F., Y. Kristoffersen, *Proc. Ocean Drill. Program Sci. Results*, 114, 826 pp., 1991.

De Graciansky, P. C., C. W. Poag, et al., *Initial Rep. Deep Sea Drill. Proj.*, 80, 1258 pp., 1985.

Duncan, R. A., J. Backman, L. C. Peterson, et al., *Proc. Ocean Drill. Program Sci. Results*, 115, 887 pp., 1990.

Hayes, D. E., L. A. Frakes, et al., *Initial Rep. Deep Sea Drill. Proj.*, 28, 1017 pp., 1975.

Hays, J. D., H. E. Cook, et al., *Initial Rep. Deep Sea Drill. Proj.*, 9, 1205 pp., 1972.

Hollister, C. D., C. Craddock, et al., *Initial Rep. Deep Sea Drill. Proj.*, 35, 129 pp., 1976.

Hsü, K. J., J. L. La Breque et al., *Initial Rep. Deep Sea Drill. Proj.*, 73, 798 pp., 1984.

Kennett, J. P., R. E. Houtz, et al., *Initial Rep. Deep Sea Drill. Proj.*, 29, 1197 pp., 1975.

Kennett, J. P., C. C. von der Borch, et al., *Initial Rep. Deep Sea Drill. Proj.*, 90, 744 pp., 1986.

Larson, R. L., R. Moberly, et al., *Initial Rep. Deep Sea Drill. Proj.*, 32, 980 pp., 1975.

Lewis, B. T. R., P. Robinson, et al., *Initial Rep. Deep Sea Drill. Proj.*, 65, 752 pp., 1983.

Ludwig, W. J., V. A. Krasheninnikov, et al., *Initial Rep. Deep Sea Drill. Proj.*, 71, 1187 pp. 1983.

Mayer, L. A., F. Theyer, et al., *Initial Rep. Deep Sea Drill. Proj.*, 85, 1021 pp., 1985.

Moore, T. C., Jr., P. D. Rabinowitz, et al., *Initial Rep. Deep Sea Drill. Proj.*, 74, 894 pp., 1984.

Pierce, J., J. Weissel, et al., *Proc. Ocean Drill. Program Initial Rep.*, 121, 1000 pp., 1989.

Ruddiman, W. F., R. B. Kidd, E. Thomas, et al., *Initial Rep. Deep Sea Drill. Proj.*, 94, 1261 pp., 1987.

Schlich, R., S. W. Wise, Jr., et al., *Proc. Ocean Drill. Program Initial Rep.*, 120, 648 pp., 1989.

Wise, S. W., Jr., R. Schlich, et al., *Proc. Ocean Drill. Program Sci. Results*, 120, 1155 pp., 1992.